Residuated Structures in Algebra and Logic

Mathematical
Surveys
and
Monographs

Volume 277

Residuated Structures in Algebra and Logic

George Metcalfe
Francesco Paoli
Constantine Tsinakis

AMERICAN
MATHEMATICAL
SOCIETY
Providence, Rhode Island

2020 *Mathematics Subject Classification.* Primary 03B47, 03C05, 03F52, 03G10, 06D35, 06F05, 06F15.

For additional information and updates on this book, visit
www.ams.org/bookpages/surv-277

Library of Congress Cataloging-in-Publication Data

Names: Metcalfe, George, author. | Paoli, Francesco, author. | Tsinakis, Constantine, author.
Title: Residuated structures in algebra and logic / George Metcalfe, Francesco Paoli, Constantine Tsinakis.
Description: Providence, Rhode Island : American Mathematical Society, [2023] | Series: Mathematical surveys and monographs, 0076-5376 ; Volume 277 | Includes bibliographical references and index.
Identifiers: LCCN 2023030386 | ISBN 9781470469856 (paperback) | ISBN 9781470475512 (ebook)
Subjects: LCSH: Algebraic logic. | Algebra, Universal. | AMS: Mathematical logic and foundations – General logic – Substructural logics (including relevance, entailment, linear logic, Lambek calculus, BCK and BCI logics). | Mathematical logic and foundations – Model theory – Equational classes, universal algebra. | Mathematical logic and foundations – Proof theory and constructive mathematics – Linear logic and other substructural logics. | Mathematical logic and foundations – Algebraic logic – Lattices and related structures. | Order, lattices, ordered algebraic structures – Distributive lattices – MV-algebras. | Order, lattices, ordered algebraic structures – Ordered structures – Ordered semigroups and monoids. | Order, lattices, ordered algebraic structures – Ordered structures – Ordered groups.
Classification: LCC QA10 .M48 2023 | DDC 511.3/24–dc23/eng/20231004
LC record available at https://lccn.loc.gov/2023030386

∞ The paper used in this book is acid-free and falls within the guidelines
established to ensure permanence and durability.
Visit the AMS home page at `https://www.ams.org/`

10 9 8 7 6 5 4 3 2 1 28 27 26 25 24 23

George: *to my wife Nia*
Francesco: *to my wife Alessia*
Costas: *to my wife Annell and the memory of my parents Andreas and Dorothea*

Contents

Introduction

This book is about fascinating and deep connections that exist between *algebra* — understood, crudely, as the study of structure — and *logic*, understood as the study of consequence. Our key contention is that a common thread running through and binding together these two disparate fields is the notion of a *residuated structure*.

Rather than give a formal definition at this stage, let us first consider some notable examples of residuated structures arising in algebra and logic.

Algebras of logic. The following relationship between subsethood $\subseteq$, intersection $\cap$, union $\cup$, and complementation ^{c} holds for all subsets X, Y, Z of a set A:

$$X \cap Y \subseteq Z \iff X \subseteq Y^c \cup Z.$$

For any non-empty set B of subsets of A closed under $\cap$, $\cup$, and ^{c}, the structure $\langle B, \cap, \cup, ^c, \emptyset, A\rangle$ is a *Boolean algebra*, one of the "algebras of logic" originating in George Boole's pioneering work on the laws of thought. Indeed, by a theorem of Marshall Stone, every Boolean algebra admits such a representation and can be viewed as a residuated structure, where the defined operation $Y \to Z := Y^c \cup Z$ is said to be the *residual* of $\cap$. More generally, algebras corresponding to a broad family of non-classical logics admit a lattice-order $\leq$ and binary operations $\cdot$ and $\to$ on a set L that satisfy, for all elements a, b, c of L,

$$a \cdot b \leq c \iff a \leq b \to c.$$

For example, a bounded distributive lattice $\langle L, \wedge, \vee, 0, 1\rangle$ with a binary operation $\to$ satisfying this condition for $a \cdot b := a \wedge b$ is called a *Heyting algebra*, while the real unit interval $[0,1]$ with binary operations $a \cdot b := \max(0, a + b - 1)$ and $a \to b := \min(1, 1 - a + b)$, and constants 0 and 1, also satisfies the condition and forms an important instance of an *MV-algebra*.

Logical consequence. Logical consequence $\vdash$, conjunction $\wedge$, and implication $\to$ are related in *classical propositional logic* by a "deduction theorem" asserting that, for any formulas α, β, γ,

$$\alpha \wedge \beta \vdash \gamma \iff \alpha \vdash \beta \to \gamma.$$

The similarity to the residuation condition satisfied by Boolean algebras is of course no coincidence, but consequence in classical propositional logic can be defined in many different ways: *semantically*, via truth tables, sets, or Boolean algebras, or *syntactically*, via axiom systems, sequent calculi, or tableaux. Moreover, analogous relationships hold between consequence and connectives in other non-classical logics defined using a wide range of different classes of algebras and proof systems.

The Lambek calculus. *Categorial grammar* aims to analyze natural language by assigning grammatical categories or *types* to linguistic expressions that describe their syntactic role — e.g., as nouns, verbs, noun phrases, verb phrases, or sentences. New types can then be constructed using operations $\cdot$, $\backslash$, and $/$, where w has type $\alpha \cdot \beta$ whenever $w = uv$, with u of type α and v of type β, and type $\alpha\backslash\beta$ (respectively, type β/α) if whenever the expression u has type α, then the expression uw (respectively, wu) has type β. Combinations of types and their results are expressed using the symbol $\Rightarrow$ to obtain, for example, $\alpha \cdot \alpha\backslash\beta \Rightarrow \beta$ or $\alpha/\beta \cdot \beta \Rightarrow \alpha$. Allowing hypothetical reasoning then yields, for all types α, β, γ,

$$\beta \Rightarrow \alpha\backslash\gamma \iff \alpha \cdot \beta \Rightarrow \gamma \iff \alpha \Rightarrow \gamma/\beta.$$

These observations led Joachim Lambek to develop a calculus for deciding *sequents* of the form $\Gamma \Rightarrow \alpha$, where Γ is any finite sequence of types and α is any type. Later, this calculus was extended with rules for further constant symbols and lattice connectives, giving what is now known as the *full Lambek calculus*, a fundamental example of a *substructural logic*.

Ideals of rings. The set $I(\mathbf{R})$ of two-sided ideals of any unital ring $\mathbf{R}$ is lattice-ordered by set-inclusion and admits an associative binary operation

$$I \cdot J := \{a_1b_1 + \cdots + a_nb_n \mid a_1, \ldots, a_n \in I \text{ and } b_1, \ldots, b_n \in J\}.$$

This operation has neutral element R and is residuated in the sense that there exist binary operations $\backslash$ and $/$ satisfying, for all $I, J, K \in I(\mathbf{R})$,

$$J \subseteq I\backslash K \iff I \cdot J \subseteq K \iff I \subseteq K/J.$$

More concretely, for any $I, J \in I(\mathbf{R})$,

$$I\backslash J = \{a \in R \mid ba \in J, \text{ for all } b \in I\} \quad \text{and} \quad J/I = \{a \in R \mid ab \in J, \text{ for all } b \in I\}.$$

Abstracting from this example, Morgan Ward and R.P. Dilworth arrived at the notion of a *residuated lattice*, which — modulo a few more recent changes — consists of a lattice equipped with a monoidal multiplication $\cdot$ and neutral element e, and binary operations $\backslash$ and $/$ that satisfy, for all elements a, b, c,

$$b \leq a\backslash c \iff a \cdot b \leq c \iff a \leq c/b.$$

Lattice-ordered groups. The notion of a group equipped with a lattice-order preserved by the group multiplication plays a fundamental role in several areas of mathematics, not least in the study of algebras of logic. These *lattice-ordered groups*, or *ℓ-groups* for short, include number systems such as the real numbers or integers with the natural order, and groups of automorphisms of a totally ordered set with a pointwise defined order. They can be viewed as residuated lattices by observing that, for any elements a, b, c of an ℓ-group,

$$b \leq a^{-1} \cdot c \iff a \cdot b \leq c \iff a \leq c \cdot b^{-1}.$$

Remarkably, as will become clear in the course of this book, many properties and methods developed for the study of ℓ-groups extend to broad classes of residuated lattices. In some cases, there exist even closer relationships in the form of categorical equivalences between categories of residuated lattices and ℓ-groups equipped with an additional unary operation.

The concept of a residuated structure provides the glue that binds together these disparate examples. Such structures may then in turn be studied in a uniform logical framework using the concept of a *consequence relation*: a relation between (sets of) elements of some set of formulas, equations, or other objects, that tells us "what follows from what". Any class of algebras of the same signature can be associated in a natural way with a family of consequence relations on equations. Also, any proof system gives rise to a consequence relation on formulas, sequents, or more complex objects, via derivability in the system.

Consequence relations for classes of algebras and proof systems can be related via a natural notion of *equivalence*. Famously, consequence relations on formulas defined via an axiom system for classical propositional logic are equivalent to both the equational consequence relation for Boolean algebras and the consequence relation on sequents obtained from the propositional part of Gentzen's sequent calculus LK. Similarly, equivalences are established in this book between equational consequence relations for classes of residuated structures and consequence relations on sequents obtained from extensions of the full Lambek calculus.

The bridge between algebra and logic provided by the concept of equivalence offers benefits to both sides. Proof-theoretic methods can be used to tackle a wide range of problems in algebra, while new insights into logical systems are obtained by studying associated classes of algebras. Indeed, results of one domain can often be imported directly into the other using so-called "bridge theorems". In particular, algebraic properties for an equational class of algebras such as amalgamation or generation by a subclass can be related to properties of an equational consequence relation such as interpolation or admissibility of a certain rule. A suitable proof system — roughly, one where derivations are constructed from elements already occurring in the structure to be proved — can then be used, not only to transfer algorithmic properties like decidability, but also to establish algebraic properties by proving a corresponding syntactic property for the system.

In the case of residuated structures, the interplay of algebra and logic has benefits that extend well beyond the transfer of methods for solving particular problems: they provide an umbrella concept for understanding results originally obtained for specific classes of algebras or logical systems. Notably, certain fundamental structure theorems for ℓ-groups depend only on purely order-theoretical properties shared by other classes of residuated lattices, while familiar properties of ℓ-groups, such as divisibility and cancellativity, play a crucial role in the structure theory of algebras for many non-classical logics. In particular, bounded divisible residuated lattices encompass algebras of intuitionistic propositional logic (Heyting algebras), Łukasiewicz logic (MV-algebras), and Hájek's basic fuzzy logic (BL-algebras), and admit powerful decomposition theorems. Similarly, cancellativity is encountered at every turn when studying irreducible components of these decompositions, and certain classes of cancellative residuated lattices may be viewed — via categorical equivalence — as algebraic semantics for modal logics based on ℓ-groups.

Overview of the book

This book is an introduction to residuated structures that appear in and connect algebra and logic. For the reader's convenience, a concise primer on basic notions from universal algebra is given in Appendix B, referring to [**24, 53, 65**] for more detailed treatments. In the main text, we assume no prior knowledge of order and residuation, proof systems, or consequence relations. Chapters 1, 2, and 3 therefore provide introductions to these topics that are intended to be accessible to the general reader. We also conclude each of these first three chapters with an excursus that puts the preceding development into some historical context, exploring the origins of residuated structures in algebra, sources of substructural logics, and development of abstract algebraic logic.

In subsequent chapters, we pick up the pace a little. In Chapter 4, we develop the structure theory of residuated lattices, focusing on the fundamental role of (normal) convex subuniverses. In Chapters 5, 6, and 7, we investigate in detail the structure of residuated lattices satisfying semilinearity and distributivity, cancellativity, and divisibility, respectively. In Chapter 8, we direct our attention to bridge theorems between algebra and logic, focusing on relationships between amalgamation and interpolation, and in Chapter 9, we study finite embeddability and finite model properties for varieties of residuated lattices and related decision problems. Finally, in Appendix A, we briefly describe some open problems in the area.

A central theme of this book is the interaction between algebra and logic, and the mutual benefits that accrue from this interaction. Nevertheless, more narrow readings focused on one of these fields, are also possible:

- *Algebraic track.* A reader wishing to focus only on algebraic aspects of residuated structures may do so fairly easily by skipping Chapters 2 and 3, and accepting without proof certain theorems in Chapters 5, 8, and 9.
- *Algorithmic track.* A proof-theoretic-oriented reading of the book, avoiding heavy algebraic machinery, is also possible: proceed directly to Chapters 2 and 3, referring only to Section 1.4 for the definition of a residuated lattice, and then extract applications of proof theory from Chapters 5, 8, and 9.

Finally, let us say what we leave out. Since the beginning of the 21st century, the field of residuated structures — encompassing the study of substructural logics and varieties of residuated lattices — has exploded, leading to many fascinating new directions and developments. It would be impossible to cover all of these in one book and burdensome on the reader to attempt even a modest coverage. Our aim is therefore to focus on core ideas and methods rather than attempt an encyclopaedic account of the area. We often give proofs only for significant cases instead of striving for the most general presentation. From an algebraic perspective, we give preference to those properties that have stood the test of time for their importance and mathematical depth — semilinearity, divisibility, and cancellativity — and classes of algebras that span well-studied families such as Heyting algebras, MV-algebras, and ℓ-groups. From a proof-theoretic perspective, we focus on classes that can be captured with single-conclusion sequent or hypersequent calculi, omitting multiple-conclusion calculi and other formalisms such as display calculi, nested sequents, and proof nets. We also omit recent work on residuated frames that aims to define relational semantics for substructural logics. Such structures provide a uniform framework for presenting algebraic proofs of results first proved proof-theoretically, but do not yet cover significant properties such as cancellativity and divisibility, or important classes such as ℓ-groups and MV-algebras.

To avoid diverting the reader's attention from the flow of topics in a chapter, we omit all references to the literature, and keep the use of footnotes to a minimum. All results and concepts covered, unless appearing in this book for the first time or belonging to the folklore of the discipline, are duly credited to their authors in the bibliographical remarks at the end of each chapter, which also contain suggestions for further reading. An exception to this policy is made only for the historical excursuses at the end of the first three chapters, where all bibliographical references are included unless referring to famous texts from the history of mathematics.

Acknowledgments. We would like to express our gratitude to the University of Bern, University of Cagliari, and Vanderbilt University for their support in the form of sabbaticals and funds for hosting visitors. For useful comments on the text, we would like to thank Almudena Colacito, Wesley Fussner, José Gil-Férez, Adam Přenosil, Simon Santschi, Naomi Tokuda, Olim Tuyt, and Line van den Berg. Finally we would also like to express our appreciation to Charles Holland for his enthusiastic support during the initial stages of planning and writing this book.

CHAPTER 1

Order and residuation

The key ideas of residuation emerge from several intertwined — conceptually and historically — strands of research in the fields of order, algebra, and logic. The aim of this chapter is to recall some of the most relevant results from these fields, and at the same time, give perhaps even an experienced reader a fresh perspective on their connections.

1.1. Partially ordered sets and lattices

Ordered sets and their well-behaved maps, order-homomorphisms, are central to our endeavors in this book. Let us begin by establishing some basic notation and terminology. A *partially ordered set* (or *poset*) is an ordered pair $\mathbf{P} = \langle P, \leq^{\mathbf{P}} \rangle$, where P is a set and $\leq^{\mathbf{P}}$ is a partial order on P, that is, $\leq^{\mathbf{P}}$ is a binary relation on P that is reflexive, antisymmetric, and transitive. In this case, $x <^{\mathbf{P}} y$ denotes $x \leq^{\mathbf{P}} y$ and $x \neq y$, and we define also $x \geq^{\mathbf{P}} y :\iff y \leq^{\mathbf{P}} x$ and $x >^{\mathbf{P}} y :\iff y <^{\mathbf{P}} x$.

A *subposet* of $\mathbf{P}$ is a poset $\mathbf{Q} = \langle Q, \leq^{\mathbf{Q}} \rangle$, where $Q \subseteq P$ and $\leq^{\mathbf{Q}} := \leq^{\mathbf{P}} \cap Q^2$. In particular, for any $a, b \in P$, the *intervals* $[a,b] := \{x \in P \mid a \leq^{\mathbf{P}} x \leq^{\mathbf{P}} b\}$, $(a,b] := \{x \in P \mid a <^{\mathbf{P}} x \leq^{\mathbf{P}} b\}$, $[a,b) := \{x \in P \mid a \leq^{\mathbf{P}} x <^{\mathbf{P}} b\}$, and $(a,b) := \{x \in P \mid a <^{\mathbf{P}} x <^{\mathbf{P}} b\}$ all form subposets of $\mathbf{P}$.

The *dual* of a poset $\mathbf{P}$, denoted by $\mathbf{P}^\partial = \langle P, \leq^{\mathbf{P}^\partial} \rangle$, is the poset obtained by reversing the partial order on $\mathbf{P}$, i.e., $x \leq^{\mathbf{P}^\partial} y$ if and only if $y \leq^{\mathbf{P}} x$. For any statement S about posets, the dual statement S^∂ is obtained by replacing each occurrence of $\leq^{\mathbf{P}}$ in S by $\leq^{\mathbf{P}^\partial}$. It is easy to see that if a statement S is true for all posets, then its dual S^∂ is also true for all posets. This fact, referred to as the *Duality Principle*, is very helpful in formulating definitions and stating theorems for ordered sets.

Let $\mathbf{P}$ be any poset and let $X \subseteq P$. An element $a \in P$ is an *upper bound* of X in $\mathbf{P}$ if $x \leq^{\mathbf{P}} a$, for all $x \in X$; it is the *least upper bound* of X in $\mathbf{P}$ if a is an upper bound of X in $\mathbf{P}$ and $a \leq^{\mathbf{P}} b$, for any upper bound b of X in $\mathbf{P}$. The notions of lower bound and greatest lower bound are defined dually. The least upper bound and greatest lower bound of X in $\mathbf{P}$, when they exist, are called the *join* and *meet* of X in $\mathbf{P}$ and denoted by $\bigvee^{\mathbf{P}} X$ and $\bigwedge^{\mathbf{P}} X$, respectively. If $X = \{a, b\}$, then we switch to infix notation and write $a \vee^{\mathbf{P}} b$ and $a \wedge^{\mathbf{P}} b$ for the join and meet of X.

A poset $\mathbf{P}$ is called *upper-bounded* if it contains a *greatest element* $\top^{\mathbf{P}}$ satisfying $a \leq \top^{\mathbf{P}}$, for all $a \in P$, *lower-bounded* if it contains a *least element* $\bot^{\mathbf{P}}$ satisfying $\bot^{\mathbf{P}} \leq a$, for all $a \in P$, and *bounded* if it is both upper-bounded and lower-bounded. For $X \subseteq P$, we denote the greatest element of the subposet of $\mathbf{P}$ formed by X by $\max X$ and its least element by $\min X$, whenever they exist.

Two elements a and b of a poset $\mathbf{P}$ are *comparable* if $a \leq^{\mathbf{P}} b$ or $b \leq^{\mathbf{P}} a$; otherwise, they are *incomparable*. A poset $\mathbf{P}$ in which any two elements are comparable is called a *chain* or a *totally ordered set*, and $\leq^{\mathbf{P}}$ is said to be a *total order* on P; a poset in which any two distinct elements are incomparable is called an *antichain*. A chain $\mathbf{P}$ with the property that every non-empty subposet of $\mathbf{P}$ has a least element is called a *well-ordered set*, and $\leq^{\mathbf{P}}$ is said to be a *well-order* on P. *Dually well-ordered sets* and *dual well-orders* are defined analogously.

From now on, we will omit the subscript "$\mathbf{P}$" when no confusion can occur.

EXAMPLE 1.1.1. Given posets $\mathbf{P}$ and $\mathbf{Q}$, define for $x_1, x_2 \in P$ and $y_1, y_2 \in Q$,

$$\langle x_1, y_1 \rangle \leq \langle x_2, y_2 \rangle \;:\Longleftrightarrow\; x_1 <^{\mathbf{P}} x_2 \text{ or } (x_1 = x_2 \text{ and } y_1 \leq^{\mathbf{Q}} y_2).$$

Then $\langle P \times Q, \leq \rangle$ is a poset, called the *lexicographic product* of $\mathbf{P}$ and $\mathbf{Q}$. Moreover, if $\mathbf{P}$ and $\mathbf{Q}$ are both chains or well-ordered sets, then $\langle P \times Q, \leq \rangle$ is also a chain or well-ordered set, respectively.

An element a of a poset $\mathbf{P}$ is said to *cover* an element $b \in P$, written $b \prec a$, if $b < a$ and, whenever $b \leq c$ and $c \leq a$, either $b = c$ or $a = c$. If $\mathbf{P}$ has a least element $\bot$, then $a \in P$ is called an *atom* of $\mathbf{P}$ if $\bot \prec a$. An element $a \in P$ is said to be *maximal* in $\mathbf{P}$ if $a \leq b$ implies $a = b$, for all $b \in P$. The sets of atoms and maximal elements of $\mathbf{P}$ form antichains. *Coatoms* and *minimal* elements are defined dually.

The following principle is equivalent to the Axiom of Choice and Well-Ordering theorem in Zermelo-Fraenkel Set Theory and is assumed throughout this book:

Zorn's lemma: Any non-empty poset, in which every chain has an upper bound, has a maximal element.

A poset $\mathbf{P}$ in which each $\{a, b\} \subseteq P$ has an upper bound is called an *up-directed set*, and a non-empty poset $\mathbf{P}$ in which each $\{a, b\} \subseteq P$ has a join is called a *join-semilattice*. By induction, every finite non-empty subset of a join-semilattice has a join. A non-empty subposet $\mathbf{Q}$ of a join-semilattice $\mathbf{P}$ that is closed under taking non-empty finite joins in $\mathbf{P}$ is called a *join-subsemilattice* of $\mathbf{P}$. *Down-directed sets*, *meet-semilattices*, and *meet-subsemilattices* are defined dually.

A non-empty poset $\mathbf{P}$ in which each $\{a, b\} \subseteq P$ has a join and a meet is called a *lattice*. A non-empty subposet $\mathbf{Q}$ of a lattice $\mathbf{P}$ that is closed under taking non-empty finite meets and joins in $\mathbf{P}$ is called a *sublattice* of $\mathbf{P}$.

A poset in which every subset has a join and a meet is called a *complete lattice*. In particular, every complete lattice $\mathbf{P}$ is bounded: it possesses a greatest element $\top := \bigwedge \emptyset = \bigvee P$ and a least element $\bot := \bigvee \emptyset = \bigwedge P$. The following well-known result provides a useful criterion for checking that a poset is a complete lattice.

PROPOSITION 1.1.2. *Let $\mathbf{P}$ be any poset. The following statements are equivalent:*

(1) *$\mathbf{P}$ is a complete lattice.*
(2) *The meet of every subset of P exists.*
(3) *The join of every subset of P exists.*

A lattice $\mathbf{P}$ is said to be *distributive* if, for all $a, b, c \in P$,

$$a \wedge (b \vee c) = (a \wedge b) \vee (a \wedge c) \quad \text{and} \quad a \vee (b \wedge c) = (a \vee b) \wedge (a \vee c).$$

It is called a *frame* if it is complete and finite meets distribute over arbitrary joins, i.e., for any $a \in P$ and $X \subseteq P$,

$$a \wedge \bigvee X = \bigvee \{a \wedge x \mid x \in X\}.$$

Similarly, a complete lattice in which finite joins distribute over arbitrary meets is called a *co-frame.*

The following well-known result, proved in any textbook on ordered structures, shows that lattices may also be presented as an equational class of algebras.

PROPOSITION 1.1.3. *There exists a bijective correspondence between lattices and algebras $\langle L, \wedge, \vee \rangle$ of signature $\langle 2, 2 \rangle$ satisfying the equations*

(L1.1)	$x \wedge (y \wedge z) \approx (x \wedge y) \wedge z$	(L2.1)	$x \vee (y \vee z) \approx (x \vee y) \vee z$
(L1.2)	$x \wedge y \approx y \wedge x$	(L2.2)	$x \vee y \approx y \vee x$
(L1.3)	$x \wedge x \approx x$	(L2.3)	$x \vee x \approx x$
(L1.4)	$x \wedge (x \vee y) \approx x$	(L2.4)	$x \vee (x \wedge y) \approx x.$

The correspondence is implemented by the map associating to any lattice $\langle L, \leq \rangle$, the algebra $\langle L, \wedge, \vee \rangle$, where $\wedge$ and $\vee$ are the binary meet and join operations of $\langle L, \leq \rangle$, respectively, and the inverse map associating to any such algebra $\langle L, \wedge, \vee \rangle$, the poset $\langle L, \leq \rangle$, where $a \leq b :\Longleftrightarrow a \wedge b = a$.

EXAMPLE 1.1.4. The set $\mathrm{Con}(\mathbf{A})$ of *congruences* of any algebra $\mathbf{A}$ forms a complete lattice ordered by set-inclusion with greatest element A^2 and least element $\Delta_A := \{\langle a, a \rangle \mid a \in A\}$, and yields the algebra $\langle \mathrm{Con}(\mathbf{A}), \cap, \vee \rangle$, with

$$\Theta_1 \vee \Theta_2 = \bigcup \{\Theta_{i_0} \circ \cdots \circ \Theta_{i_n} \mid n \in \mathbb{N} \text{ and } i_0, \ldots, i_n \in \{1, 2\}\}.$$

Lattices may therefore be viewed as algebras or relational structures, depending on the preferred vantage point. Throughout the rest of the book, we use the term "lattice" interchangeably, according to the context, to denote one of these meanings. A similar policy, justified by the next proposition, will be adopted for join-semilattices and meet-semilattices.

PROPOSITION 1.1.5. *There are bijective correspondences between join-semilattices, meet-semilattices, and algebras $\langle L, * \rangle$ of signature $\langle 2 \rangle$ that satisfy the equations $x * (y * z) \approx (x * y) * z$, $x * y \approx y * x$, and $x * x \approx x$.*

Since join-semilattices do not differ algebraically from meet-semilattices — only the maps that implement the correspondences with their relational counterparts are different — we will use the term *semilattice* to refer to both concepts.

Let $\mathbf{P}$ be a poset. A subset $I \subseteq P$ is said to be an *order-ideal* of $\mathbf{P}$ if whenever $x \in I$, $y \in P$, and $y \leq x$, then $y \in I$. Note that the empty set is an order-ideal of $\mathbf{P}$ and that the set of all order-ideals of $\mathbf{P}$ forms a complete lattice $\mathcal{L}(\mathbf{P})$ ordered by set-inclusion, with meet and join operations $\cap$ and $\cup$, respectively.

A *principal* order-ideal of $\mathbf{P}$ is an order-ideal of the form ${\downarrow}a := \{x \in P \mid x \leq a\}$, for some $a \in P$. More generally, the smallest order-ideal containing some $A \subseteq P$ is ${\downarrow}A := \{x \in P \mid x \leq a, \text{ for some } a \in A\}$. The *order-filters* of $\mathbf{P}$ are defined dually. We also write $\mathcal{U}(\mathbf{P})$ to denote the complete lattice of order-filters of $\mathbf{P}$ ordered by set-inclusion, and define ${\uparrow}a := \{x \in P \mid a \leq x\}$, for any $a \in P$, and ${\uparrow}A := \{x \in P \mid a \leq x, \text{ for some } a \in A\}$, for any $A \subseteq P$.

EXAMPLE 1.1.6. A *(finite) tree* is a (finite) poset $\mathbf{T} = \langle T, \leq \rangle$ with a least element, called the *root* of $\mathbf{T}$, such that $\langle {\downarrow}t, \leq \cap ({\downarrow}t)^2 \rangle$ is a chain, for each $t \in T$. The members of T are called *nodes* and each maximal node t is called a *leaf*. For each leaf t, the set ${\downarrow}t$ is called a *branch*, and a node t_1 covered by a node t_2 is called a *parent* of the *child* t_2. When $\mathbf{T}$ is finite, its *height* is defined as $\max\{|{\downarrow}t| \mid t \in T\}$.

A non-empty order-ideal I of a lattice $\mathbf{L}$ is said to be a *lattice-ideal* of $\mathbf{L}$ if $a, b \in I$ implies $a \vee b \in I$; it is called *prime* if $I \neq L$ and whenever $a, b \in L$ and $a \wedge b \in I$, then $a \in I$ or $b \in I$. If $\mathbf{L}$ is distributive, then, for any $a, b \in L$ such that $a \neq b$, there exists a prime lattice-ideal I of $\mathbf{L}$ such that $a \in I$ and $b \notin I$. This result is proved using Zorn's lemma and known as the *Prime ideal separation theorem* for distributive lattices. *Lattice-filters* and *prime lattice-filters* are defined dually, and satisfy a dual *Prime filter separation theorem.*

Let $\mathbf{P}$ and $\mathbf{Q}$ be posets. A map $\varphi \colon P \to Q$ is said to be *order-preserving* and called an *order-homomorphism* from $\mathbf{P}$ to $\mathbf{Q}$, denoted by $\varphi \colon \mathbf{P} \to \mathbf{Q}$, if $x \leq y$ implies $\varphi(x) \leq \varphi(y)$, for all $x, y \in P$. It is called an *order-embedding* of $\mathbf{P}$ into $\mathbf{Q}$ if it satisfies $x \leq y$ if and only if $\varphi(x) \leq \varphi(y)$, for all $x, y \in P$. It is clear that any order-embedding of $\mathbf{P}$ into $\mathbf{Q}$ is an injective order-homomorphism from $\mathbf{P}$ to $\mathbf{Q}$, but the converse need not hold.

An order-homomorphism φ from $\mathbf{P}$ to $\mathbf{Q}$ is called an *order-isomorphism* from $\mathbf{P}$ to $\mathbf{Q}$ if it is bijective and $\varphi^{-1} \colon Q \to P$ is an order-homomorphism from $\mathbf{Q}$ to $\mathbf{P}$. In this case, we write $\varphi \colon \mathbf{P} \cong \mathbf{Q}$ or simply $\mathbf{P} \cong \mathbf{Q}$, and say that $\mathbf{P}$ and $\mathbf{Q}$ are *order-isomorphic*. An order-homomorphism (order-isomorphism) from $\mathbf{P}$ to $\mathbf{P}$ is called an *order-endomorphism* (*order-automorphism*) of $\mathbf{P}$.

A map $\varphi \colon P \to Q$ is said to be *order-reversing* and called a *dual order-homomorphism* from $\mathbf{P}$ to $\mathbf{Q}$ if it is an order-homomorphism from $\mathbf{P}$ to $\mathbf{Q}^\partial$, i.e., $x \leq y$ implies $\varphi(x) \geq \varphi(y)$, for all $x, y \in P$. The dual notions to order-embeddings, order-isomorphisms, etc. are defined analogously.

For maps φ and ψ of appropriate domain and range, we will often use the shorthand $\varphi\psi$ to denote their composition $\varphi \circ \psi$.

Order-homomorphisms abound in mathematics. For now, let us just mention a couple of examples that will play a key role in what follows.

EXAMPLE 1.1.7. Every map $\varphi \colon X \to Y$ between sets X and Y induces the following order-homomorphisms between $\langle \wp(X), \subseteq \rangle$ and $\langle \wp(Y), \subseteq \rangle$:

(i) the *direct image* map $\varphi[\,] \colon \wp(X) \to \wp(Y)$; $A \mapsto \{\varphi(a) \in Y \mid a \in A\}$;
(ii) the *inverse image* map $\varphi^{-1}[\,] \colon \wp(Y) \to \wp(X)$; $B \mapsto \{x \in X \mid \varphi(x) \in B\}$.

The verification of the following result is straightforward.

LEMMA 1.1.8. *Let $\varphi \colon P \to Q$ be any map between posets $\mathbf{P}$ and $\mathbf{Q}$. The following statements are equivalent:*

(1) *φ is an order-homomorphism.*
(2) *For any order-ideal I of $\mathbf{Q}$, its inverse image $\varphi^{-1}[I]$ is an order-ideal of $\mathbf{P}$.*
(3) *For any order-filter F of $\mathbf{Q}$, its inverse image $\varphi^{-1}[F]$ is an order-filter of $\mathbf{P}$.*

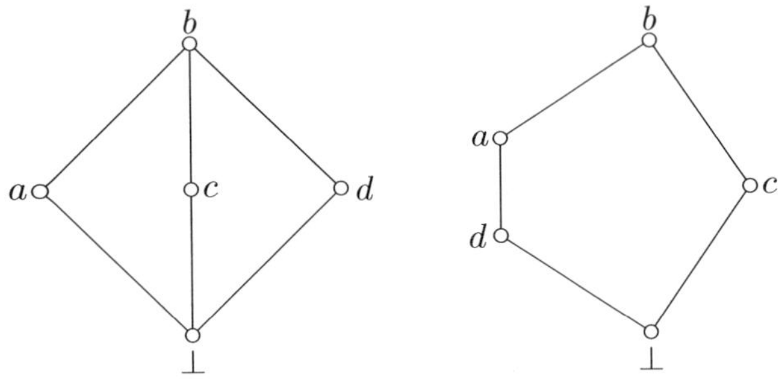

FIGURE 1.1. The lattices $\mathbf{M}_5$ and $\mathbf{N}_5$

EXAMPLE 1.1.9. Lemma 1.1.8 has a nice topological interpretation. Consider the topological space on P whose open sets are the order-filters of $\mathbf{P}$. This is an example of an *Alexandrov space*, characterized by the property that the intersection of any family of open sets is an open set. Of course, the closed sets of this space are the order-ideals of $\mathbf{P}$. Conditions (2) and (3) of Lemma 1.1.8 then state that the map φ between the topological spaces on P and Q defined in this way is continuous.

For lattices, we often consider more specialized classes of order-homomorphisms. A map $\varphi\colon L \to M$ between lattices $\mathbf{L}$ and $\mathbf{M}$ is called a *join-homomorphism* if $\varphi(x \vee y) = \varphi(x) \vee \varphi(y)$, for all $x, y \in L$, *meet-homomorphism* if $\varphi(x \wedge y) = \varphi(x) \wedge \varphi(y)$, for all $x, y \in L$, and *lattice homomorphism* if it is both a join-homomorphism and a meet-homomorphism. Any join-homomorphism or meet-homomorphism is an order-homomorphism, but the converse is not true.

The following proposition makes use of these notions and a familiar result from lattice theory to provide useful necessary and sufficient conditions for a lattice to be distributive.

PROPOSITION 1.1.10. *Let* $\mathbf{L}$ *be any lattice. The following statements are equivalent:*

(1) $\mathbf{L}$ *is distributive.*
(2) $\mathbf{L}$ *does not contain a sublattice that is order-isomorphic to one of the lattices* $\mathbf{M}_5$ *or* $\mathbf{N}_5$ *depicted in Figure* 1.1.
(3) *For all* $a, b \in L$ *with* $a \leq b$, *there exists a join-homomorphism* $\varphi\colon \mathbf{L} \to \mathbf{L}$ *such that* $\varphi(b) = a$ *and* $\varphi(x) \leq x$, *for all* $x \in L$.
(4) *For all* $a, b \in L$ *with* $a \leq b$, *there exists a meet-homomorphism* $\varphi\colon \mathbf{L} \to \mathbf{L}$ *such that* $\varphi(a) = b$ *and* $x \leq \varphi(x)$, *for all* $x \in L$.

PROOF.

(1) $\Leftrightarrow$ (2) Proved in any lattice theory or universal algebra textbook.

(1) $\Rightarrow$ (3) Suppose that $\mathbf{L}$ is distributive and consider any $a, b \in L$ with $a \leq b$. It is easily checked that the map $\varphi\colon L \to L$; $x \mapsto a \wedge x$ is a join-homomorphism of $\mathbf{L}$ such that $\varphi(b) = a$ and $\varphi(x) \leq x$, for all $x \in L$.

(3) $\Rightarrow$ (2) Assume (3) and suppose that $\mathbf{L}$ contains a sublattice with elements $a, b, c, d, \bot$ that is order-isomorphic to one of the lattices $\mathbf{M}_5$ or $\mathbf{N}_5$ depicted in Figure 1.1. By assumption, there exists a join-homomorphism $\varphi\colon \mathbf{L} \to \mathbf{L}$ such that

$\varphi(b) = a$ and $\varphi(x) \leq x$, for all $x \in L$. Then $a = \varphi(b) = \varphi(c \vee d) = \varphi(c) \vee \varphi(d)$ and $\varphi(c) \leq a$. But $\varphi(c) \leq c$, so $\varphi(c) \leq a \wedge c = \bot$ and $a = \varphi(c) \vee \varphi(d) = \varphi(d) \leq d$, a contradiction in both cases.

(1) $\Leftrightarrow$ (4) Follows by the duality principle. $\square$

Finally, for the sake of completeness, we mention a closely related result from lattice theory: a lattice $\mathbf{L}$ does not contain a sublattice that is order-isomorphic to the lattice $\mathbf{N}_5$ if and only if $\mathbf{L}$ is *modular*, i.e., satisfies the quasi-equation

$$x \leq y \implies x \vee (y \wedge z) \approx y \wedge (x \vee z).$$

1.2. Residuated maps

In this section, we define the concept of a residuated map, one of the central notions investigated in this book, and present a first installment of its basic properties.

Let $\mathbf{P}$ and $\mathbf{Q}$ be posets. A map $\varphi\colon P \to Q$ is said to be *residuated* if there exists a map $\psi\colon Q \to P$, called a *residual* of φ, such that, for all $x \in P$ and $y \in Q$,

$$\varphi(x) \leq y \iff x \leq \psi(y).$$

We refer to $\langle \varphi, \psi \rangle$ as a *residuated pair* between $\mathbf{P}$ and $\mathbf{Q}$.

As the next two examples indicate, residuated maps abound in mathematics.

EXAMPLE 1.2.1. For any sets X and Y and map $\varphi\colon X \to Y$, the direct image and inverse image maps $\varphi[\,]\colon \wp(X) \to \wp(Y)$ and $\varphi^{-1}[\,]\colon \wp(Y) \to \wp(X)$, defined in Example 1.1.7, form a residuated pair between $\langle \wp(X), \subseteq \rangle$ and $\langle \wp(Y), \subseteq \rangle$.

EXAMPLE 1.2.2. The preceding example generalizes to an arbitrary non-empty binary relation $R \subseteq X \times Y$. Let $R^{\smile} := \{\langle y, x \rangle \mid \langle x, y \rangle \in R\}$ be the converse relation of R, and let ^{c} denote complementation in $\wp(X)$ and $\wp(Y)$. We claim that the map between the posets $\langle \wp(X), \subseteq \rangle$ and $\langle \wp(Y), \subseteq \rangle$ defined by

$$\varphi_R\colon \wp(X) \to \wp(Y); \quad A \mapsto \{y \in Y \mid \langle x, y \rangle \in R, \text{ for some } x \in A\}$$

is residuated and that its residual is the map

$$\psi_R\colon \wp(Y) \to \wp(X); \quad B \mapsto (\varphi_{R^{\smile}}(B^c))^c.$$

We show that $\varphi_R(A) \subseteq B$ if and only if $A \subseteq \psi_R(B)$, for all $A \subseteq X$ and $B \subseteq Y$, or, which amounts to the same thing, $\varphi_R(A) \cap B^c = \emptyset$ if and only if $A \cap \varphi_{R^{\smile}}(B^c) = \emptyset$. For the right-to-left direction, suppose contrapositively that $y \in \varphi_R(A) \cap B^c$. Then $y \in B^c$ and there exists $x \in A$ such that $\langle x, y \rangle \in R$, or, equivalently, $\langle y, x \rangle \in R^{\smile}$. So $x \in A \cap \varphi_{R^{\smile}}(B^c)$. The left-to-right direction follows symmetrically.

It is worth noting that *every* residuated map $\varphi\colon \wp(X) \to \wp(Y)$ arises as described in Example 1.2.2; that is, $R := \{\langle x, y \rangle \mid x \in X \text{ and } y \in \varphi(\{x\})\}$ yields $\varphi = \varphi_R$.

We continue by proving three simple, but useful, lemmas.

LEMMA 1.2.3. *$\langle \varphi, \psi \rangle$ is a residuated pair between posets $\mathbf{P}$ and $\mathbf{Q}$ if and only if $\langle \psi, \varphi \rangle$ is a residuated pair between $\mathbf{Q}^\partial$ and $\mathbf{P}^\partial$.*

PROOF. Follows immediately from the definition of a residuated pair. $\square$

LEMMA 1.2.4. *Let $\langle \varphi, \psi \rangle$ be a residuated pair between posets* $\mathbf{P}$ *and* $\mathbf{Q}$.

(a) $x \leq \psi\varphi(x)$, *for any* $x \in P$, *and* $\varphi\psi(y) \leq y$, *for any* $y \in Q$.
(b) φ *and* ψ *are order-homomorphisms.*
(c) $\varphi\psi\varphi = \varphi$ *and* $\psi\varphi\psi = \psi$.
(d) φ *is injective if and only if* ψ *is surjective, and* φ *is surjective if and only if* ψ *is injective.*

PROOF.

(a) Let $x \in P$ and $y \in Q$. Then $\varphi(x) \leq \varphi(x)$ and $\psi(y) \leq \psi(y)$, so, by residuation, $x \leq \psi\varphi(x)$ and $\varphi\psi(y) \leq y$.

(b) Let $x_1, x_2 \in P$ with $x_1 \leq x_2$. Then $x_1 \leq x_2 \leq \psi(\varphi(x_2))$, by (a), and, by residuation, $\varphi(x_1) \leq \varphi(x_2)$. So φ is an order-homomorphism and, by Lemma 1.2.3, also ψ is an order-homomorphism.

(c) We verify $\varphi\psi\varphi = \varphi$, observing that the second equality then follows using Lemma 1.2.3. Let $x \in P$. Then $x \leq \psi\varphi(x)$, by (a), and $\varphi(x) \leq \varphi\psi\varphi(x)$, by (b). But also $\varphi\psi\varphi(x) \leq \varphi(x)$, by (a), so $\varphi\psi\varphi(x) = \varphi(x)$.

(d) Suppose first that φ is injective. It follows using (c) that $\varphi\psi\varphi(x) = \varphi(x)$ and hence $\psi\varphi(x) = x$, for any $x \in P$. So ψ is surjective. For the converse, suppose that ψ is surjective and consider any $x_1, x_2 \in P$ such that $\varphi(x_1) = \varphi(x_2)$. By the surjectivity of ψ, there exist $y_1, y_2 \in Q$ such that $x_1 = \psi(y_1)$ and $x_2 = \psi(y_2)$, and $x_1 = \psi(y_1) = \psi\varphi\psi(y_1) = \psi\varphi(x_1) = \psi\varphi(x_2) = \psi\varphi\psi(y_2) = \psi(y_2) = x_2$, using (c). So φ is injective. The second claim follows similarly. □

LEMMA 1.2.5. *Let* $\varphi\colon P \to Q$ *and* $\psi\colon Q \to P$ *be maps between posets* $\mathbf{P}$ *and* $\mathbf{Q}$. *The following statements are equivalent:*

(1) $\langle \varphi, \psi \rangle$ *is a residuated pair between* $\mathbf{P}$ *and* $\mathbf{Q}$.
(2) ${\downarrow}\psi(y) = \varphi^{-1}[{\downarrow}y]$, *for all* $y \in Q$.
(3) ${\uparrow}\varphi(x) = \psi^{-1}[{\uparrow}x]$, *for all* $x \in P$.

PROOF.

(1) $\Leftrightarrow$ (2) Observe that

$$
\begin{aligned}
\langle \varphi, \psi \rangle \text{ is a residuated pair} &\iff (\varphi(x) \leq y \Leftrightarrow x \leq \psi(y)), \text{ for all } x \in P,\ y \in Q \\
&\iff (\varphi(x) \in {\downarrow}y \Leftrightarrow x \in {\downarrow}\psi(y)), \text{ for all } x \in P,\ y \in Q \\
&\iff (x \in \varphi^{-1}[{\downarrow}y] \Leftrightarrow x \in {\downarrow}\psi(y)), \text{ for all } x \in P,\ y \in Q \\
&\iff {\downarrow}\psi(y) = \varphi^{-1}[{\downarrow}y], \text{ for all } y \in Q.
\end{aligned}
$$

(1) $\Leftrightarrow$ (3) Follows from the equivalence of (1) and (2) and Lemma 1.2.3. □

The following analogue of Lemma 1.1.8 is an immediate consequence of Lemma 1.2.5.

COROLLARY 1.2.6. *A map* $\varphi\colon P \to Q$ *between posets* $\mathbf{P}$ *and* $\mathbf{Q}$ *is residuated if and only if* $\varphi^{-1}[{\downarrow}y]$ *is a principal order-ideal of* $\mathbf{P}$, *for all* $y \in Q$.

We also obtain an element-wise description of the members of a residuated pair.

COROLLARY 1.2.7. *Let $\langle\varphi, \psi\rangle$ be a residuated pair between posets* **P** *and* **Q**.

(a) $\varphi(x) = \min \psi^{-1}[\uparrow x]$, *for all* $x \in P$.
(b) $\psi(y) = \max \varphi^{-1}[\downarrow y]$, *for all* $y \in Q$.

Hence each member of a residuated pair uniquely determines the other.

A residuated map and its residual are not only order-homomorphisms, as shown in Lemma 1.2.4, they also satisfy stronger preservation properties. A map $\varphi\colon P \to Q$ between posets **P** and **Q** is said to *preserve arbitrary joins* if for any (possibly empty) subset X of P such that the join of X in **P** exists, the join of $\varphi[X]$ in **Q** exists, and $\varphi(\bigvee^{\mathbf{P}} X) = \bigvee^{\mathbf{Q}} \varphi[X]$. *Preservation of arbitrary meets* is defined dually.

LEMMA 1.2.8. *Let $\langle\varphi, \psi\rangle$ be a residuated pair between posets* **P** *and* **Q**.

(a) *If* **P** *has a least element, then* **Q** *has a least element* $\bot^{\mathbf{Q}} = \varphi(\bot^{\mathbf{P}})$. *Similarly, if* **Q** *has a greatest element, then* **P** *has a greatest element* $\top^{\mathbf{P}} = \psi(\top^{\mathbf{Q}})$.
(b) φ *preserves arbitrary joins, and* ψ *preserves arbitrary meets.*

PROOF.

(a) Suppose that **P** has a least element $\bot^{\mathbf{P}} = \bigvee^{\mathbf{P}} \emptyset$. Then $\bot^{\mathbf{P}} \le \psi(y)$ and hence $\varphi(\bot^{\mathbf{P}}) \le y$, for any $y \in Q$. So $\bot^{\mathbf{Q}} = \varphi(\bot^{\mathbf{P}})$ is the least element of **Q**. The case where **Q** has a greatest element is symmetrical.

(b) We first prove that φ preserves arbitrary joins. The case for the empty set follows directly from (a), so let $\emptyset \ne X \subseteq P$ and assume that $x_0 := \bigvee^{\mathbf{P}} X$ exists. The fact that φ is an order-homomorphism ensures that $\varphi(x_0)$ is an upper bound for the set $\varphi[X]$. Let y be any upper bound for $\varphi[X]$. Then $\varphi(x) \le y$ and hence $x \le \psi(y)$, for all $x \in X$. But then $\psi(y)$ is an upper bound for X and so $x_0 \le \psi(y)$ and $\varphi(x_0) \le y$. Hence $\varphi(x_0) = \bigvee^{\mathbf{Q}} \varphi[X]$.

That ψ preserves arbitrary meets follows from Lemma 1.2.3 and the fact that φ preserves arbitrary joins. □

The following result is a partial converse to the preceding lemma.

LEMMA 1.2.9. *If* **P** *is a complete lattice,* **Q** *is a poset, and* $\varphi\colon P \to Q$ *is a map that preserves arbitrary joins, then* φ *is residuated.*

PROOF. For each $y \in Q$, let $\psi(y) := \bigvee^{\mathbf{P}} \varphi^{-1}[\downarrow y]$. We claim that ψ is the residual of φ. Let $x \in P$ and $y \in Q$. If $\varphi(x) \le y$, then $x \in \varphi^{-1}[\downarrow y]$, so $x \le \psi(y)$. Conversely, if $x \le \psi(y)$, then, since φ is order-preserving and preserves arbitrary joins, $\varphi(x) \le \varphi(\psi(y)) = \varphi(\bigvee^{\mathbf{P}} \varphi^{-1}[\downarrow y]) = \bigvee^{\mathbf{Q}} \varphi[\varphi^{-1}[\downarrow y]] \le y$. □

EXAMPLE 1.2.10. Let φ be any order-endomorphism of a poset **P**. Given any order-ideal I of **P**, represented as $I = \bigcup_{x \in A} \downarrow x$, for some $A \subseteq P$, let

$$\widehat{\varphi}(I) := \bigcup_{x \in A} \downarrow \varphi(x).$$

Let us check that $\widehat{\varphi}(I)$ is well-defined. Suppose that $\bigcup_{x \in A} \downarrow x = \bigcup_{y \in B} \downarrow y$, for some $A, B \subseteq P$. Let $x \in A$. Since $x \in \bigcup_{y \in B} \downarrow y$, it follows that $\downarrow x \subseteq \downarrow y_0$, for some

$y_0 \in B$. But φ is an order-endomorphism, so ${\downarrow}\varphi(x) \subseteq {\downarrow}\varphi(y_0) \subseteq \bigcup_{y\in B} {\downarrow}\varphi(y)$. Hence $\bigcup_{x\in A} {\downarrow}\varphi(x) \subseteq \bigcup_{y\in B} {\downarrow}\varphi(y)$ and the reverse inclusion is symmetrical.

Since the join of order-ideals of $\mathbf{P}$ is their union, $\widehat{\varphi}$ is an order-endomorphism of the complete lattice $\mathcal{L}(\mathbf{P})$ that preserves arbitrary joins. Hence, by Lemma 1.2.9, it is residuated. Note also that if $\widehat{\varphi}_1 = \widehat{\varphi}_2$ for order-endomorphisms φ_1, φ_2 of $\mathbf{P}$, then ${\downarrow}\varphi_1(x) = \widehat{\varphi}_1({\downarrow}x) = \widehat{\varphi}_2({\downarrow}x) = {\downarrow}\varphi_2(x)$ and hence $\varphi_1(x) = \varphi_2(x)$, for any $x \in P$. That is, the correspondence $\varphi \mapsto \widehat{\varphi}$ is one-to-one.

1.3. Closure and co-closure operators

Certain order-homomorphisms on posets play a central role in algebra and logic — notably, in the theory of residuated structures — and therefore merit a closer look. We introduce these maps first for power sets ordered by set-inclusion, then lift the definition to a more abstract setting.

Let A be any set. A *closure operator* on $\langle \wp(A), \subseteq \rangle$ is an order-endomorphism γ of $\langle \wp(A), \subseteq \rangle$ satisfying, for all $X \subseteq A$,

(i) $X \subseteq \gamma(X)$: γ is enlarging;
(ii) $\gamma(\gamma(X)) = \gamma(X)$: γ is idempotent.

A subposet $\langle \mathcal{C}, \subseteq \rangle$ of $\langle \wp(A), \subseteq \rangle$ is a *closure system* of $\langle \wp(A), \subseteq \rangle$ if it is closed under arbitrary intersections, that is, $\bigcap \mathcal{D} \in \mathcal{C}$ whenever $\mathcal{D} \subseteq \mathcal{C}$. By Proposition 1.1.2, a closure system $\langle \mathcal{C}, \subseteq \rangle$ is a complete lattice. However, it need not be a sublattice of $\langle \wp(A), \subseteq \rangle$; in general, the join of $\mathcal{D} \subseteq \mathcal{C}$ is $\bigvee \mathcal{D} = \bigcap \{X \in \mathcal{C} \mid \bigcup \mathcal{D} \subseteq X\}$.

EXAMPLE 1.3.1. For any poset $\mathbf{P}$, the map sending $X \subseteq P$ to the order-ideal ${\downarrow}X$ of $\mathbf{P}$ is a closure operator on $\langle \wp(P), \subseteq \rangle$, and the complete lattice of order-ideals $\mathcal{L}(\mathbf{P})$ is a closure system of $\langle \wp(P), \subseteq \rangle$.

EXAMPLE 1.3.2. For any algebra $\mathbf{A}$, the map sending $X \subseteq A$ to the subuniverse $\mathrm{Sg}^{\mathbf{A}}(X)$ of $\mathbf{A}$ generated by X is a closure operator on $\langle \wp(A), \subseteq \rangle$, and the complete lattice of subuniverses of $\mathbf{A}$ is a closure system of $\langle \wp(A), \subseteq \rangle$.

These examples prompt the question as to whether a closure system can always be matched with a corresponding closure operator, and vice versa. Before showing that this is indeed the case, let us first extend these concepts to arbitrary posets.

A *closure operator* on a poset $\mathbf{P}$ is an order-endomorphism γ of $\mathbf{P}$ satisfying, for all $x \in P$,

(i) $x \leq \gamma(x)$: γ is enlarging;
(ii) $\gamma(\gamma(x)) = \gamma(x)$: γ is idempotent.

Such a map is determined by its image $P_\gamma := \gamma[P]$; that is, for each $x \in P$,

$$\gamma(x) = \min\{y \in P_\gamma \mid x \leq y\}.$$

A *closure system* of $\mathbf{P}$ is a subposet $\mathbf{C}$ of $\mathbf{P}$ such that $\min\{y \in C \mid x \leq y\}$ exists, for all $x \in P$. It is easily seen that when $\mathbf{P} = \langle \wp(A), \subseteq \rangle$, for some set A, this definition coincides with the notion of a closure system of $\mathbf{P}$ given above.

We leave to the reader the proof of the following result, which throws some light on Examples 1.3.1 and 1.3.2.

PROPOSITION 1.3.3. *There exists a bijective correspondence between the closure operators γ on a poset $\mathbf{P}$ and the closure systems $\mathbf{C}$ of $\mathbf{P}$. More precisely, if $\mathbf{C}$ is a closure system of $\mathbf{P}$, then the map $\gamma_{\mathbf{C}}\colon P \to P;\ x \mapsto \min\{y \in C \mid x \leq y\}$ is a closure operator on $\mathbf{P}$. Conversely, if $\gamma\colon \mathbf{P} \to \mathbf{P}$ is a closure operator on $\mathbf{P}$, then $\mathbf{P}_\gamma = \langle P_\gamma, \leq\rangle$ is a closure system of $\mathbf{P}$. Moreover, $\mathbf{P}_{\gamma_{\mathbf{C}}} = \mathbf{C}$ and $\gamma_{\mathbf{P}_\gamma} = \gamma$.*

Residuated pairs are a significant source of closure operators.

PROPOSITION 1.3.4. *If $\langle \varphi, \psi\rangle$ is a residuated pair between posets $\mathbf{P}$ and $\mathbf{Q}$, then $\psi\varphi$ is a closure operator on $\mathbf{P}$.*

PROOF. It follows immediately from Lemma 1.2.4 that $\psi\varphi$ is both enlarging and idempotent. That $\psi\varphi$ is an order-endomorphism of $\mathbf{P}$ follows from the fact that it is the composition of two order-homomorphisms. $\square$

On the other hand, given any closure operator γ on a poset $\mathbf{P}$, it is easily checked that, for all $x \in P$ and $y \in P_\gamma$,

$$\gamma(x) \leq y \iff x \leq y.$$

Hence every closure operator on a poset can be viewed as a residuated map.

COROLLARY 1.3.5. *If γ is a closure operator on a poset $\mathbf{P}$, then the map $\gamma\colon P \to P_\gamma$ from $\mathbf{P}$ to $\mathbf{P}_\gamma$ is residuated, and its residual is the inclusion map from P_γ to P.*

Using Corollary 1.3.5 and Lemma 1.2.8, we obtain the following result for the preservation of joins and meets.

PROPOSITION 1.3.6. *Let γ be a closure operator on a poset $\mathbf{P}$, and let $\mathbf{C}$ be a closure system of $\mathbf{P}$ (e.g., $\mathbf{P}_\gamma$).*

(a) *If $\bigvee^{\mathbf{P}} X$ exists for $X \subseteq P$, then $\bigvee^{\mathbf{P}_\gamma} \gamma[X]$ exists and $\gamma(\bigvee^{\mathbf{P}} X) = \bigvee^{\mathbf{P}_\gamma} \gamma[X]$.*
(b) *If $\bigwedge^{\mathbf{C}} Y$ exists for $Y \subseteq C$, then $\bigwedge^{\mathbf{P}} Y$ exists and $\bigwedge^{\mathbf{P}} Y = \bigwedge^{\mathbf{C}} Y$.*

An element a of a complete lattice $\mathbf{L}$ is said to be *compact* if for any subset $S \subseteq L$ satisfying $a \leq \bigvee S$, there exists a finite subset $S' \subseteq S$ such that $a \leq \bigvee S'$. The subposet $\mathcal{K}(\mathbf{L})$ of compact elements of $\mathbf{L}$ is always a join-subsemilattice of $\mathbf{L}$, but may not be a sublattice.

Let A be a set and let $\mathbf{C}$ be a closure system of $\langle \wp(A), \subseteq\rangle$. Then each compact element Y of $\mathbf{C}$ is *finitely generated*, that is, there exists some finite $X \subseteq A$ such that $\gamma_{\mathbf{C}}(X) = Y$. The converse need not be true in general, but holds for an important special case. Let us call $\mathbf{C}$ *algebraic* if $\bigcup \mathcal{D} \in C$ for any up-directed subposet $\langle \mathcal{D}, \subseteq\rangle$ of $\mathbf{C}$. We then call a closure operator γ on $\langle \wp(A), \subseteq\rangle$ *algebraic* if $\langle \wp(A), \subseteq\rangle_\gamma$ is an algebraic closure system. The following proposition, whose proof is left to the reader, characterizes such operators.

PROPOSITION 1.3.7. *Let A be a set and let γ be a closure operator on $\langle \wp(A), \subseteq\rangle$. The following statements are equivalent:*

(1) *γ is algebraic.*
(2) *$\gamma(X) = \bigcup\{\gamma(Y) \mid Y \subseteq X$ and Y is finite$\}$, for each $X \subseteq A$.*

(3) *γ preserves up-directed unions; that is, if $\mathcal{D} \subseteq \wp(A)$ and $\langle \mathcal{D}, \subseteq \rangle$ is up-directed, then $\gamma(\bigcup \mathcal{D}) = \bigcup\{\gamma(X) \mid X \in \mathcal{D}\}$.*
(4) *Every finitely generated member of $\langle \wp(A), \subseteq \rangle_\gamma$ is compact.*

The concepts of an algebraic closure system and algebraic closure operator extend, following the correspondence established in Proposition 1.3.7, to arbitrary complete lattices. We say that a closure operator γ on a complete lattice $\mathbf{L}$ is *algebraic* if it preserves up-directed *joins*, and that a closure system $\mathbf{C}$ of $\mathbf{L}$ is *algebraic* if $\gamma_{\mathbf{C}}$ is an algebraic closure operator on $\mathbf{L}$.

Every algebraic closure system is an *algebraic lattice*: a complete lattice such that each of its elements is a join of compact elements. Algebraic lattices are ubiquitous in mathematics and will be encountered at every turn throughout this book. Before describing their most important properties, we first recall some further key notions.

An element a of a lattice $\mathbf{L}$ is said to be *meet-irreducible* if $b \wedge c = a$ implies $b = a$ or $c = a$, for any $b, c \in L$, and *meet-prime* if $b \wedge c \leq a$ implies $b \leq a$ or $c \leq a$, for any $b, c \in L$. It is easily proved that every meet-prime element is meet-irreducible, and that if $\mathbf{L}$ is distributive, the converse also holds.

An element a of a lattice $\mathbf{L}$ is said to be *completely meet-irreducible* if $\bigwedge X = a$ implies $a \in X$, for any $X \subseteq L$, and *completely meet-prime* if $\bigwedge X \leq a$ implies $x \leq a$, for some $x \in X$, for any $X \subseteq L$. Again, it is easily proved that every completely meet-prime element is completely meet-irreducible, and that if $\mathbf{L}$ is a co-frame, the converse also holds.

(Completely) join-irreducible and *join-prime* elements are defined dually and satisfy analogous properties.

PROPOSITION 1.3.8. *Let $\mathbf{L}$ be an algebraic lattice.*

(a) *Every element of $\mathbf{L}$ is a meet of completely meet-irreducible elements.*
(b) *If $\mathbf{L}$ is distributive, then it is a frame.*

PROOF. We confine ourselves to a proof of (a), leaving the proof of (b) as an exercise. Assume that $\mathbf{L}$ is an algebraic lattice and consider any $b \in L$. It suffices to prove the following:

Claim. For any $a \in L$ satisfying $a \not\leq b$, there is a completely meet-irreducible element $q \in {\uparrow}b$ such that $a \not\leq q$.

We then consider $a := \bigwedge\{q \in {\uparrow}b \mid q \text{ is completely meet-irreducible}\}$. Clearly, $b \leq a$ and applying the claim to $a \not\leq b$ leads to a contradiction, so $b = a$ is a meet of completely meet-irreducible elements.

To prove the claim, let $a \in L$ satisfy $a \not\leq b$. Since $\mathbf{L}$ is algebraic, a is a join of compact elements, and, since $a \not\leq b$, not all these elements can be less than or equal to b. So there exists a compact element $c \in {\downarrow}a$ such that $c \not\leq b$. Now let

$$X := \{x \in {\uparrow}b \mid c \not\leq x\}.$$

Clearly, $b \in X$ and $X \cap {\uparrow}a = \emptyset$. An application of Zorn's lemma to the poset $\langle X, \leq \rangle$ yields a maximal element $q \in X$. So $q \in {\uparrow}b$ and $a \not\leq q$, and it remains to show that q is completely meet-irreducible. Suppose that $q = \bigwedge S$, for some $S \subseteq L$. Since

$c \not\leq q$, there exists $y \in S$ such that $y \in {\uparrow}b$ and $c \not\leq y$. Hence $y \in X$ and, since q is maximal, $y = q$. □

EXAMPLE 1.3.9. Let $\mathbf{A}$ be any algebra. The map $\mathrm{Cg}^{\mathbf{A}}$ sending $R \subseteq A \times A$ to the congruence $\mathrm{Cg}^{\mathbf{A}}(R) := \bigcap\{\Theta \in \mathrm{Con}(\mathbf{A}) \mid R \subseteq \Theta\}$ of $\mathbf{A}$ generated by R is an algebraic closure operator on $\langle A \times A, \subseteq\rangle$ and hence its associated closure system, the congruence lattice $\langle \mathrm{Con}(\mathbf{A}), \subseteq\rangle$, is algebraic.

Dual to the notion of a closure operator is the notion of a *co-closure operator*: an order-endomorphism δ of a poset $\mathbf{P}$ satisfying, for all $x \in P$,

(i) $\delta(x) \leq x$: δ is contracting;

(ii) $\delta(x) = \delta(\delta(x))$: δ is idempotent.

Again, δ is determined by its image $\mathbf{P}_\delta := \langle \delta[P], \leq\rangle$; that is, for each $x \in P$,

$$\delta(x) = \max\{y \in P_\delta \mid y \leq x\}.$$

A *co-closure system* of a poset $\mathbf{P}$ is a subposet $\mathbf{O}$ of $\mathbf{P}$ such that $\max\{y \in O \mid y \leq x\}$ exists, for all $x \in P$. The following results are dual to Propositions 1.3.3, 1.3.4, and 1.3.6, respectively.

PROPOSITION 1.3.10. *There exists a bijective correspondence between the co-closure operators δ on a poset $\mathbf{P}$ and the co-closure systems $\mathbf{O}$ of $\mathbf{P}$. More precisely, if $\mathbf{O}$ is a co-closure system of $\mathbf{P}$, then the map $\delta_{\mathbf{O}} \colon P \to P$; $x \mapsto \max\{y \in O \mid y \leq x\}$ is a co-closure operator on $\mathbf{P}$. Conversely, if $\delta \colon \mathbf{P} \to \mathbf{P}$ is a co-closure operator on $\mathbf{P}$, then $\mathbf{P}_\delta$ is a co-closure system of $\mathbf{P}$. Moreover, $\mathbf{P}_{\delta_{\mathbf{O}}} = \mathbf{O}$ and $\delta_{\mathbf{P}_\delta} = \delta$.*

PROPOSITION 1.3.11. *If $\langle \varphi, \psi\rangle$ is a residuated pair between posets $\mathbf{P}$ and $\mathbf{Q}$, then $\varphi\psi$ is a co-closure operator on $\mathbf{Q}$.*

PROPOSITION 1.3.12. *Let δ be a co-closure operator on a poset $\mathbf{P}$, and let $\mathbf{O}$ be a co-closure system of $\mathbf{P}$ (e.g., $\mathbf{P}_\delta$).*

(a) *If $\bigwedge^{\mathbf{P}} X$ exists for $X \subseteq P$, then $\bigwedge^{\mathbf{P}_\delta} \delta[X]$ exists and $\delta(\bigwedge^{\mathbf{P}} X) = \bigwedge^{\mathbf{P}_\delta} \delta[X]$.*

(b) *If $\bigvee^{\mathbf{O}} Y$ exists for $Y \subseteq O$, then $\bigvee^{\mathbf{P}} Y$ exists and $\bigvee^{\mathbf{P}} Y = \bigvee^{\mathbf{O}} Y$.*

EXAMPLE 1.3.13. For any topological space X, the map sending $Y \subseteq X$ to its interior is a co-closure operator on $\langle \wp(X), \subseteq\rangle$, and the complete lattice of open sets of X is the associated co-closure system of $\wp(X)$.

1.4. Residuated lattices: the algebras of logic

In this section, we introduce the central objects of study of this book: namely, *residuated lattices*. The defining properties for the class $\mathcal{RL}$ of residuated lattices are few and easy to grasp, and concrete examples are readily constructed that illustrate their key features. However, the theory is also sufficiently robust for $\mathcal{RL}$ to encompass a large portion of the ordered algebras arising in logic. Notably, the rich algebraic theory of residuated lattices has produced powerful tools for the comparative study of substructural logics (see Chapters 2 and 3). On the other hand, bridges provided by algebraic logic have provided new and far-reaching methods for addressing purely algebraic problems.

A binary operation $\cdot$ on a poset $\mathbf{P} = \langle P, \leq \rangle$ is said to be *order-preserving* if $x \leq y$ implies $z \cdot x \leq z \cdot y$ and $x \cdot z \leq y \cdot z$, for all $x, y, z \in P$. Equivalently, $\cdot$ is order-preserving if and only if, for each $a \in P$, the maps $x \mapsto a \cdot x$ and $x \mapsto x \cdot a$ are order-endomorphisms of $\mathbf{P}$, in the sense of Section 1.1.

An algebra $\langle S, \cdot \rangle$ of signature $\langle 2 \rangle$ is a *semigroup* if $\cdot$ is associative, and an algebra $\langle M, \cdot, \mathrm{e} \rangle$ of signature $\langle 2, 0 \rangle$ is a *monoid* if $\langle M, \cdot \rangle$ is a semigroup and e is a neutral element of $\cdot$, i.e., $x \cdot \mathrm{e} = \mathrm{e} \cdot x = x$, for all $x \in M$.

A *partially ordered monoid* (or *pomonoid*) is a structure $\mathbf{P} = \langle P, \cdot, \mathrm{e}, \leq \rangle$ such that $\langle P, \cdot, \mathrm{e} \rangle$ is a monoid, $\langle P, \leq \rangle$ is a poset, and $\cdot$ is order-preserving on $\langle P, \leq \rangle$. We also say in this case that $\leq$ is a *partial order on* $\langle P, \cdot, \mathrm{e} \rangle$, and call $P^- := \{x \in P \mid x \leq \mathrm{e}\}$ and $P^+ := \{x \in P \mid \mathrm{e} \leq x\}$ the *negative cone* and *positive cone*, respectively, of $\leq$ and, by extension, the pomonoid $\mathbf{P}$.

EXAMPLE 1.4.1. Any partial order $\leq$ on a group $\langle G, \cdot, \mathrm{e} \rangle$ is uniquely determined by its negative (or positive) cone; that is, $x \leq y$ if and only if $y^{-1} \cdot x \in G^-$. Moreover, the negative cone S of any partial order $\leq$ on $\langle G, \cdot, \mathrm{e} \rangle$ satisfies

(i) $\{x \in S \mid x^{-1} \in S\} = \{\mathrm{e}\}$;

(ii) $x \cdot y \in S$, for all $x, y \in S$;

(iii) $y \cdot x \cdot y^{-1} \in S$, for all $x \in S$ and $y \in G$.

Conversely, any subset $S \subseteq G$ satisfying (i)–(iii) is the negative cone of the partial order on $\langle G, \cdot, \mathrm{e} \rangle$ defined by $x \leq y :\iff y^{-1}x \in S$.

A binary operation $\cdot$ on a poset $\mathbf{P} = \langle P, \leq \rangle$ is said to be *residuated* if there exist binary operations $\backslash$ and $/$ on P such that, for all $x, y, z \in P$,

$$y \leq x \backslash z \iff x \cdot y \leq z \iff x \leq z / y,$$

referred to as the *left residual* and the *right residual* of $\cdot$, respectively.

Equivalently, $\cdot$ is residuated if and only if, for each $y \in P$, the maps $x \mapsto y \cdot x$ and $x \mapsto x \cdot y$ on P are residuated in the sense of Section 1.2, where the residuals are the maps $x \mapsto y \backslash x$ and $x \mapsto x / y$ on P, respectively. Hence $\cdot$ is residuated if and only if it is order-preserving and, for any $x, y \in P$, the sets $\{z \in P \mid x \cdot z \leq y\}$ and $\{z \in P \mid z \cdot x \leq y\}$ contain greatest elements, $x \backslash y$ and y / x, respectively. In particular, $\cdot$ and $\leq$ uniquely determine $\backslash$ and $/$.

It is suggestive to think of the residuals as generalized division operations. The expression y/x is read as "y over x" and $x \backslash y$ is read as "x under y." In both cases, y is considered the *numerator* and x the *denominator*. Note also that any statement about residuated structures has a "mirror image" obtained by replacing $x \cdot y$ by $y \cdot x$ and interchanging x/y with $y \backslash x$.

As usual, we write xy for $x \cdot y$ and adopt the convention that, in the absence of parentheses, $\cdot$ is performed first, followed by $\backslash$ and $/$, and finally $\vee$ and $\wedge$. We also define $x^1 := x$ and $x^{n+1} := x^n \cdot x$ for $n \in \mathbb{N}^{>0}$, assuming $x^0 := \mathrm{e}$ when considering monoids. The notation $\prod_{j=1}^{n} x_j$ will sometimes be used to denote $x_1 \cdots x_n$.

The following result is a direct consequence of Lemma 1.2.8.

PROPOSITION 1.4.2. *Let $\cdot$ be a residuated binary operation on a poset $\mathbf{P} = \langle P, \leq \rangle$ with residuals $\backslash$ and $/$.*

(a) *The operation $\cdot$ preserves all existing joins in each argument. That is, if $\bigvee X$ and $\bigvee Y$ exist for $X, Y \subseteq P$, then $\bigvee\{xy \mid x \in X,\ y \in Y\}$ exists and satisfies*
$$(\bigvee X)(\bigvee Y) = \bigvee\{xy \mid x \in X,\ y \in Y\}.$$
(b) *The residuals preserve all existing meets in the numerator, and convert all existing joins in the denominator to meets. That is, if $\bigvee X$ and $\bigwedge Y$ exist for $X, Y \subseteq P$, then $\bigwedge\{y/x \mid x \in X,\ y \in Y\}$ exists and satisfies*
$$(\bigwedge Y)/(\bigvee X) = \bigwedge\{y/x \mid x \in X,\ y \in Y\},$$
and the mirror statement holds for $\backslash$.

A *residuated partially ordered monoid* (or *residuated pomonoid*) is a structure
$$\mathbf{P} = \langle P, \cdot, \backslash, /, \mathrm{e}, \leq \rangle$$
such that $\langle P, \cdot, \mathrm{e}, \leq\rangle$ is a pomonoid and $\cdot$ is a residuated binary operation on $\langle P, \leq\rangle$ with residuals $\backslash$ and $/$.

If the underlying poset of $\mathbf{P}$ is a lattice, this residuated pomonoid may be viewed (as in the case of lattices) as a purely algebraic structure. That is, a *residuated lattice* is an algebra of signature $\langle 2, 2, 2, 2, 2, 0\rangle$
$$\mathbf{L} = \langle L, \wedge, \vee, \cdot, \backslash, /, \mathrm{e}\rangle$$
such that $\langle L, \wedge, \vee\rangle$ is a lattice, $\langle L, \cdot, \mathrm{e}\rangle$ is a monoid, and $\cdot$ is a residuated binary operation on $\langle L, \leq\rangle$ with residuals $\backslash$ and $/$.

The description of certain residuated structures — typically those arising as algebras of logic — requires an extra constant symbol. We call an algebra $\langle L, \wedge, \vee, \cdot, \backslash, /, \mathrm{e}, \mathrm{f}\rangle$ of signature $\langle 2, 2, 2, 2, 2, 0, 0\rangle$ a *pointed residuated lattice* if $\langle L, \wedge, \vee, \cdot, \backslash, /, \mathrm{e}\rangle$ is a residuated lattice, noting that f is any element of L.

A (pointed) residuated lattice is said to be *distributive* or *complete* if its lattice reduct is, respectively, distributive or complete.

EXAMPLE 1.4.3. Useful examples of residuated lattices arise by considering subsets of an arbitrary monoid. In particular, given any monoid $\mathbf{M} = \langle M, \cdot, \mathrm{e}\rangle$, there is a natural way to define a (complete and distributive) residuated lattice on the power set of M. That is, let
$$\wp(\mathbf{M}) := \langle \wp(M), \cap, \cup, \cdot, \backslash, /, \{\mathrm{e}\}\rangle,$$
where $X \cdot Y := \{xy \mid x \in X,\ y \in Y\}$ and the resulting residuals are
$$X\backslash Y := \{z \in M \mid xz \in Y, \text{ for all } x \in X\}$$
$$Y/X := \{z \in M \mid zx \in Y, \text{ for all } x \in X\}.$$

We use the symbols $\mathcal{RL}$ and $\mathcal{PRL}$ to denote the classes of all residuated lattices and pointed residuated lattices, respectively. Note that a statement is true in all (pointed) residuated lattices if and only if its mirror image is true, and we therefore often state results in only one form. Note also that for classes of algebras with a lattice reduct, such as $\mathcal{RL}$ and $\mathcal{PRL}$, we write $s \leq t$ for formulas s, t to denote the equation $s \wedge t \approx s$.

The following lemma collects basic properties of residuated lattices, most of which can by now be ascribed to the "folklore" of the subject.

PROPOSITION 1.4.4. *The following equations (and their mirror images) are satisfied by every (pointed) residuated lattice:*

(i) $x(y/z) \leq xy/z$;

(ii) $x/y \leq xz/yz$;

(iii) $(x/y)(y/z) \leq x/z$;

(iv) $x/yz \approx (x/z)/y$;

(v) $x\backslash(y/z) \approx (x\backslash y)/z$;

(vi) $(\mathrm{e}/x)(\mathrm{e}/y) \leq \mathrm{e}/yx$;

(vii) $(x/x)x \approx x$;

(viii) $(x/x)^2 \approx x/x$.

If a residuated lattice $\mathbf{L}$ has a least element $\bot$, then it also has a greatest element $\top = \bot\backslash\bot$, and it follows that $x\bot = \bot = \bot x$ and $\bot\backslash x = \top = x\backslash\top$, for each $x \in L$.

For the purposes of this book, we define a *bounded residuated lattice* to be a pointed residuated lattice with a least element $\bot = \mathrm{f}$, and hence also a greatest element $\top = \mathrm{f}\backslash\mathrm{f}$. We denote the class of bounded residuated lattices by $\mathcal{BRL}$, noting that this class is defined relative to $\mathcal{PRL}$ by the equation $\mathrm{f} \leq x$. We emphasize that "bounded" implies that the constant f representing the least element is in the signature, but that the neutral element e may not be the greatest element.

The following lemma and proposition, whose proofs are left to the reader, show that $\mathcal{RL}$ and $\mathcal{PRL}$ are finitely based equational classes.

LEMMA 1.4.5. *An algebra $\langle L, \wedge, \vee, \cdot, \backslash, /, \mathrm{e}\rangle$ is a residuated lattice if and only if $\langle L, \wedge, \vee\rangle$ is a lattice, $\langle L, \cdot, \mathrm{e}\rangle$ is a monoid, and, for all $a, b \in L$,*

(i) *the maps $x \mapsto ax$ and $x \mapsto xa$ preserve binary joins;*

(ii) *the maps $x \mapsto a\backslash x$ and $x \mapsto x/a$ are order-homomorphisms;*

(iii) $a(a\backslash b) \leq b \leq a\backslash ab$ *and* $(b/a)a \leq b \leq ba/a$.

PROPOSITION 1.4.6. *$\mathcal{RL}$ and $\mathcal{PRL}$ are finitely based equational classes. A finite set of defining equations for these classes consists of the defining equations for lattices and monoids together with*

(RL1)	$x(y \vee z) \approx xy \vee xz$	(RL4)	$y/x \leq (y \vee z)/x$
(RL2)	$(y \vee z)x \approx yx \vee zx$	(RL5)	$x(x\backslash y) \leq y \leq x\backslash xy$
(RL3)	$x\backslash y \leq x\backslash(y \vee z)$	(RL6)	$(y/x)x \leq y \leq yx/x$.

Sometimes it can also be convenient to think of $\mathcal{RL}$ as the (equational) class of pointed residuated lattices satisfying the equation $\mathrm{e} \approx \mathrm{f}$.

By the HSP theorem (Theorem B.12), a class of algebras of the same signature $\mathcal{K}$ is an equational class if and only if it is a *variety*: a class of algebras of the same signature that is closed under taking homomorphic images, subalgebras, and products. Proposition 1.4.6 therefore tells us that $\mathcal{RL}$ and $\mathcal{PRL}$ are both varieties.

We can also — in advance of the structure theory to come in Chapter 4 — identify some key properties of (pointed) residuated lattices.

PROPOSITION 1.4.7. *$\mathcal{RL}$ and $\mathcal{PRL}$ are congruence-distributive and congruence-permutable varieties.*

PROOF. Every (pointed) residuated lattice has a lattice reduct and is therefore congruence-distributive (Proposition B.5). To establish congruence-permutability, it suffices to observe that the formula $\alpha(x, y, z) := (((y/z) \wedge \mathrm{e})\backslash x) \wedge (((y/x) \wedge \mathrm{e})\backslash z)$ yields $\mathcal{RL} \models \alpha(x, x, z) \approx z$ and $\mathcal{RL} \models \alpha(x, z, z) \approx x$ (Proposition B.13). □

The concept of a residuated lattice is rather broad, covering a vast landscape of well-known and less well-known algebraic structures. To aid the reader in navigating this landscape, we introduce below some of the most notable examples and subclasses that will be encountered in this book.

EXAMPLE 1.4.8. As pointed out in the introduction, ring theory is a historically important source for residuated structures (see Subsection 1.6.3 for further details). Let $\mathbf{R}$ be any unital ring and let $I(\mathbf{R})$ be the set of two-sided ideals of $\mathbf{R}$. Then $\mathbf{I}(\mathbf{R}) = \langle I(\mathbf{R}), \cap, \vee, \cdot, \backslash, /, R\rangle$ is a residuated lattice, where, for any $I, J \in I(\mathbf{R})$,

$$\begin{aligned}
I \vee J &:= \{r_1 a_1 s_1 + \cdots + r_n a_n s_n \mid r_1, s_1, \ldots, r_n, s_n \in R \text{ and } a_1, \ldots, a_n \in I \cup J\}\\
I \cdot J &:= \{a_1 b_1 + \cdots + a_n b_n \mid a_1, \ldots, a_n \in I \text{ and } b_1, \ldots, b_n \in J\}\\
I\backslash J &:= \{a \in R \mid ba \in J, \text{ for all } b \in I\}\\
J/I &:= \{a \in R \mid ab \in J, \text{ for all } b \in I\}.
\end{aligned}$$

For a related interesting example, consider any integral domain $\mathbf{R}$ and its field of quotients $\mathbf{K}$, and let $L(\mathbf{K})$ denote the set of $\mathbf{R}$-submodules of $\mathbf{K}$. Then $\mathbf{L}(\mathbf{K}) = \langle L(\mathbf{K}), \cap, \vee, \cdot, \backslash, /, R\rangle$ is a residuated lattice, with $\vee$, $\cdot$, $\backslash$, and $/$ defined as above.

EXAMPLE 1.4.9. Let $\mathbf{L} = \langle L, \wedge, \vee, \cdot, \backslash, /, \mathrm{e}\rangle$ be any residuated lattice. We define the *negative cone* of $\mathbf{L}$ to be the residuated lattice $\mathbf{L}^-$ with universe L^-, monoid and lattice operations obtained as the restrictions to L^- of the corresponding operations in $\mathbf{L}$, and residuals given by $x\backslash_- y := (x\backslash y) \wedge \mathrm{e}$ and $y/_- x := (y/x) \wedge \mathrm{e}$.

A (pointed) residuated lattice $\mathbf{L}$ is said to be *commutative* if it satisfies the equation $xy \approx yx$. In this case, $x\backslash y = y/x$, for all $x, y \in L$, and we therefore often write $x \to y$ to denote the common result. The classes $\mathcal{CRL}$ of commutative residuated lattices and $\mathcal{CPRL}$ of commutative pointed residuated lattices are subvarieties of $\mathcal{RL}$ and $\mathcal{PRL}$, respectively, but we abuse notation slightly by listing only one occurrence of $\to$ in describing their members.

The subvarieties of $\mathcal{CPRL}$ described in the next three examples are of particular importance in the study of propositional logics.

EXAMPLE 1.4.10. Recall that a *Boolean algebra* is traditionally defined as an algebra $\mathbf{B} = \langle B, \wedge, \vee, \neg, 0, 1\rangle$ of signature $\langle 2, 2, 1, 0, 0\rangle$ such that $\langle B, \wedge, \vee, 0, 1\rangle$ is a bounded distributive lattice and $\mathbf{B}$ satisfies the equations $x \wedge \neg x \approx 0$ and $x \vee \neg x \approx 1$. The variety of Boolean algebras may also be identified, however, with the variety, which we may harmlessly call $\mathcal{BA}$, of bounded residuated lattices satisfying the equations $x \wedge y \approx xy$ and $x \vee y \approx (x \to y) \to y$. More formally, the subvariety $\mathcal{BA}$ of $\mathcal{CPRL}$ is said to be *term-equivalent* to the variety of Boolean algebras. Every Boolean algebra $\mathbf{B} = \langle B, \wedge, \vee, \neg, 0, 1\rangle$ satisfies the defining equations of $\mathcal{BA}$ with defined operations $x \cdot y := x \wedge y$, $x \to y := \neg x \vee y$, $\mathrm{e} := 1$, and $\mathrm{f} := 0$. Conversely, if a pointed residuated lattice $\mathbf{L}$ satisfies these equations and we define $\neg x := x \to \mathrm{f}$, $0 := \mathrm{f}$, and $1 := \mathrm{e}$, then $\langle L, \wedge, \vee, \neg, 0, 1\rangle$ is a Boolean algebra.

EXAMPLE 1.4.11. *Brouwerian algebras* and *Heyting algebras* may be defined in our setting as residuated lattices and bounded residuated lattices, respectively, that satisfy the equation $xy \approx x \wedge y$. Since the meet operation of a Brouwerian algebra or Heyting algebra $\mathbf{H}$ is residuated, the map $x \mapsto y \wedge x$ preserves arbitrary existing joins, for each $y \in H$. In particular, $\mathbf{H}$ is distributive, and if it is also complete, the underlying lattice is a frame. On the other hand, any frame $\mathbf{L}$ can be endowed with a Heyting implication; that is, its meet operation has left and right residuals with a common result given by $x \to y = \bigvee\{z \in L \mid x \wedge z \leq y\}$. The varieties $\mathcal{HA}$ and $\mathcal{B}rw$ of Heyting algebras and Brouwerian algebras form algebraic semantics for intuitionistic propositional logic and its positive fragment, respectively.

EXAMPLE 1.4.12. MV-algebras are the algebraic counterparts of the infinite-valued Łukasiewicz propositional logic. Traditionally, an MV-algebra is defined to be an algebra $\langle M, \oplus, \neg, 0\rangle$ of signature $\langle 2, 1, 0\rangle$ satisfying the equations

(MV1)	$x \oplus (y \oplus z) \approx (x \oplus y) \oplus z$	(MV4)	$\neg\neg x \approx x$
(MV2)	$x \oplus y \approx y \oplus x$	(MV5)	$x \oplus \neg 0 \approx \neg 0$
(MV3)	$x \oplus 0 \approx x$	(MV6)	$\neg(\neg x \oplus y) \oplus y \approx \neg(\neg y \oplus x) \oplus x.$

However, the variety of MV-algebras is also term-equivalent to the subvariety $\mathcal{MV}$ of commutative bounded residuated lattices satisfying $x \vee y \approx (x \to y) \to y$. More precisely, if $\langle M, \oplus, \neg, 0\rangle$ is an MV-algebra and we define $x \cdot y := \neg(\neg x \oplus \neg y)$, $x \to y := \neg x \oplus y$, $x \wedge y := x \cdot (x \to y)$, $x \vee y := (x \to y) \to y$, $\mathrm{e} := \neg 0$, and $\mathrm{f} := 0$, then $\langle M, \wedge, \vee, \cdot, \to, \mathrm{e}, \mathrm{f}\rangle \in \mathcal{MV}$. Conversely, if $\mathbf{L} \in \mathcal{MV}$ and we define $\neg x := x \to \mathrm{f}$, $x \oplus y := \neg(\neg x \cdot \neg y)$, and $0 := \mathrm{f}$, then $\langle L, \oplus, \neg, 0\rangle$ is an MV-algebra.

Examples 1.4.10, 1.4.11, and 1.4.12 imply that Boolean algebras may be viewed both as MV-algebras for which the meet $x \wedge y$ coincides with the product xy, and as Heyting algebras for which the join $x \vee y$ is definable as $(x \to y) \to y$.

In addition to being commutative, $\mathcal{BA}$ and $\mathcal{MV}$ (but not $\mathcal{HA}$) possess a further key feature. Let $\mathbf{L}$ be any (pointed) residuated lattice. An element $d \in L$ is said to be *cyclic* if $d/x = x\backslash d$, for all $x \in L$; we then denote this residual by $x \to d$, and say that $\mathbf{L}$ is *d-cyclic*. We also call an element $d \in L$ *dualizing* if $d/(x\backslash d) = (d/x)\backslash d = x$, for all $x \in L$. Clearly, the least element f of any MV-algebra — in particular, any Boolean algebra — is both cyclic and dualizing.

More generally, we call a pointed residuated lattice $\mathbf{L}$ an *involutive residuated lattice* if f is both cyclic and dualizing.[1] The expression "involutive" reflects the fact that the defined map $'\colon L \to L;\ x \mapsto x \to \mathrm{f}$ is then an *involution* on the lattice $\langle L, \wedge, \vee\rangle$; that is, for all $x, y \in L$,

$$x \leq y \iff y' \leq x' \quad \text{and} \quad x'' = x.$$

Indeed, the variety of involutive residuated lattices is term-equivalent to the variety of algebras $\langle L, \wedge, \vee, \cdot, ', \mathrm{e}\rangle$ of signature $\langle 2, 2, 2, 1, 0\rangle$ such that $\langle L, \wedge, \vee\rangle$ is a lattice, $\langle L, \cdot, \mathrm{e}\rangle$ is a monoid, $'$ is an involution on $\langle L, \wedge, \vee\rangle$, and, for all $x, y, z \in L$,

$$y \leq (z'x)' \iff x \cdot y \leq z \iff x \leq (yz')'.$$

[1] Note that in **[145]** (and some other works in the literature) the definition of an involutive residuated lattice does not assume that f is cyclic; in this case, neither of the maps $x \mapsto x\backslash\mathrm{f}$, $x \mapsto \mathrm{f}/x$ is an involution in general.

Just observe that the last condition means that $\cdot$ is residuated with residuals $x\backslash z := (z'x)'$ and $z/y := (yz')'$.

EXAMPLE 1.4.13. Lattice-ordered groups play a fundamental role in the study of algebras of logic, as will become clear in Chapters 6 and 7. (A brief account of their role in mathematics is provided in Subsection 1.6.5 of the historical excursus.) A *lattice-ordered group* (or *ℓ-group*) is traditionally defined as an algebra $\mathbf{G} = \langle G, \wedge, \vee, \cdot, ^{-1}, \mathrm{e}\rangle$ such that (i) $\langle G, \wedge, \vee\rangle$ is a lattice; (ii) $\langle G, \cdot, ^{-1}, \mathrm{e}\rangle$ is a group; and (iii) $\cdot$ is order-preserving in each argument (equivalently, $\mathbf{G}$ satisfies the equation $x(y \vee z)w \approx xyw \vee xzw$). However, the variety of ℓ-groups is also term-equivalent to the variety $\mathcal{LG}$ of residuated lattices defined relative to $\mathcal{RL}$ by the equation $(\mathrm{e}/x)x \approx \mathrm{e}$. More precisely, if $\mathbf{G}$ is an ℓ-group and we define $x/y := xy^{-1}$ and $y\backslash x := y^{-1}x$, then $\langle G, \wedge, \vee, \cdot, \backslash, /, \mathrm{e}\rangle \in \mathcal{LG}$. Conversely, if $\mathbf{L} \in \mathcal{LG}$ and we define $x^{-1} := \mathrm{e}/x$, then $\langle L, \wedge, \vee, \cdot, ^{-1}, \mathrm{e}\rangle$ is an ℓ-group. Similarly, the class $\mathcal{LG}^-$ of negative cones of members of $\mathcal{LG}$ (shown in Chapter 7 to be a variety) can be identified with the class of negative cones of ℓ-groups.

Lattice-ordered groups and their negative cones are cancellative as monoids. A monoid $\mathbf{M}$ is said to be *left (right) cancellative* if, for all $a, b, c \in M$, whenever $ab = ac$ ($ba = ca$), then $b = c$, and *cancellative* if it is left and right cancellative. *Cancellative residuated lattices* — namely, residuated lattices whose monoid reducts are cancellative — have received considerable attention in the literature. Since the cancellativity condition is expressed by quasi-equations, it is natural to ask whether the class $\mathcal{CanRL}$ of cancellative residuated lattices is a variety. That this is indeed the case will be proved in Chapter 6: a residuated lattice $\mathbf{L}$ is cancellative if and only if it satisfies the equations $xy/y \approx x$ and $x \approx y\backslash yx$. Let us just note here that some of the most interesting members of this variety lie outside the families of ℓ-groups and their negative cones, as illustrated by the following example.

EXAMPLE 1.4.14. Let $\mathbf{F}_2 = \langle F_2, \cdot, \mathrm{e}\rangle$ be the free commutative monoid over the set $\{a, b\}$ consisting of sequences $a^i b^j$ with $i, j \in \mathbb{N}$, where e denotes the empty sequence. For $i_1, i_2, j_1, j_2 \in \mathbb{N}$, define

$$a^{i_1}b^{j_1} \leq a^{i_2}b^{j_2} :\Longleftrightarrow i_1 + j_1 > i_2 + j_2 \text{ or } (i_1 + j_1 = i_2 + j_2 \text{ and } i_1 \leq i_2).$$

Hence, in particular,

$$\mathrm{e} > a > b > a^2 > ab > b^2 > a^3 > a^2b > ab^2 > b^3 > \cdots .$$

We leave it to the reader to verify that $\cdot$ is a residuated operation on $\langle F_2, \leq\rangle$ with residuals $\backslash$ and $/$ such that $\langle F_2, \wedge, \vee, \cdot, \backslash, /, \mathrm{e}\rangle$ is a totally ordered, commutative, cancellative residuated lattice with greatest element e. However, it cannot be the negative cone of any ℓ-group: such an algebra satisfies the equation $(x/y)y \approx x \wedge y$, whereas $(b/a)a = a^2 \neq b = a \wedge b$.

A further important property of ℓ-groups, possessed also by all Heyting algebras and MV-algebras, is divisibility. A pomonoid $\langle M, \cdot, \mathrm{e}, \leq\rangle$ is said to be *divisible* if, for all $a, b \in M$ such that $a \leq b$, there exist $c, d \in M$ such that $bc = db = a$. We will show in Chapter 7 that the pomonoid reduct of a residuated lattice $\mathbf{L}$ is divisible if and only if $\mathbf{L}$ satisfies the equations $y(y\backslash(x \wedge y)) \approx x \wedge y$ and $((x \wedge y)/y)y \approx x \wedge y$. Such residuated lattices are known as *GBL-algebras* (short for *generalized basic logic algebras*) and form a variety $\mathcal{GBL}$.

EXAMPLE 1.4.15. The operation $*$ of a commutative pomonoid $\langle [0,1], *, 1, \leq \rangle$, known in the fuzzy logic literature as a *t-norm*, has a residual $\to_*$ if and only if it is left-continuous. In this case, the residuated lattice $\langle [0,1], \min, \max, *, \to_*, 1 \rangle$, is a GBL-algebra if and only if $*$ is continuous. Notable examples of continuous t-norm algebras, expanded with the constant 0, include the Heyting algebra obtained when $x*y := \min(x,y)$ and the MV-algebra obtained when $x*y := \max(0, x+y-1)$. The former generates the variety $\mathcal{GA}$ of *Gödel algebras*, defined relative to $\mathcal{HA}$ by the equation $(x \to y) \vee (y \to x) \approx \mathrm{e}$, while the latter generates the variety $\mathcal{MV}$ of *MV-algebras* considered in Example 1.4.12. The class of *all* continuous t-norm algebras expanded with the constant 0 generates the variety $\mathcal{BL}$ of *BL-algebras*: commutative bounded residuated lattices with a GBL-reduct satisfying $(x \to y) \vee (y \to x) \approx \mathrm{e}$.

We conclude this section by mentioning two further significant varieties of (pointed) residuated lattices. The first encompasses all Boolean algebras, Heyting algebras, and MV-algebras, but excludes all non-trivial ℓ-groups. A (pointed) residuated lattice $\mathbf{L}$ is called *integral* if it satisfies the equations $x \backslash \mathrm{e} \approx \mathrm{e}$ and $\mathrm{e}/x \approx \mathrm{e}$, or, equivalently, e is the greatest element of $\mathbf{L}$.

Secondly, many (pointed) residuated lattices — all Boolean algebras, Abelian ℓ-groups, and MV-algebras, but not all Heyting algebras or ℓ-groups — are *semilinear*: isomorphic to a subdirect product of totally ordered algebras. Crucially, every variety of semilinear (pointed) residuated lattices is generated by its totally ordered members. An equational basis for the variety of semilinear residuated lattices will be provided in Chapter 5, but let us just mention here that a simpler equational basis for the variety of commutative, semilinear residuated lattices relative to $\mathcal{CRL}$ is given by the *commutative prelinearity law* $((x \to y) \wedge \mathrm{e}) \vee ((y \to x) \wedge \mathrm{e}) \approx \mathrm{e}$. The reason for this name is that the equation specializes to the commutative case a number of properties commonly known as *prelinearity laws* that will be introduced and thoroughly discussed in Chapter 4.

Given any variety $\mathcal{V}$ of (pointed) residuated lattices, let us use the prefixes $\mathcal{C}$, $\mathcal{C}an$, $\mathcal{S}em$, and $\mathcal{I}$ to denote the commutative, cancellative, semilinear, and integral members, respectively, of $\mathcal{V}$. For example, $\mathcal{CS}em\mathcal{IBRL}$ denotes the variety of commutative, semilinear, integral bounded residuated lattices, also known in the literature as MTL-algebras. We also write $\mathcal{TRIV}$ to denote the varieties of trivial (one-element) algebras defined by the equation $x \approx y$. Figures 1.2 and 1.3 illustrate inclusions between some of the most important varieties of residuated lattices and pointed residuated lattices, respectively, studied in this book.

1.5. Nuclei and co-nuclei

Any residuated pomonoid — in particular, any residuated lattice — possesses an underlying partial order that can support closure or co-closure operators. However, these maps need not, in general, bear any special relationship to the product, residuals, or neutral element of the structure. In this section, we focus on closure and co-closure operators that are well-behaved with respect to this richer setting.

A *nucleus* on a pomonoid $\mathbf{P} = \langle P, \cdot, \mathrm{e}, \leq \rangle$ is a closure operator γ on $\langle P, \leq \rangle$ such that $\gamma(x)\gamma(y) \leq \gamma(xy)$, for all $x, y \in P$. It is clear that a closure operator γ on $\langle P, \leq \rangle$ is a nucleus if and only if $\gamma(\gamma(x)\gamma(y)) = \gamma(xy)$, for all $x, y \in P$. If $\mathbf{A}$ is

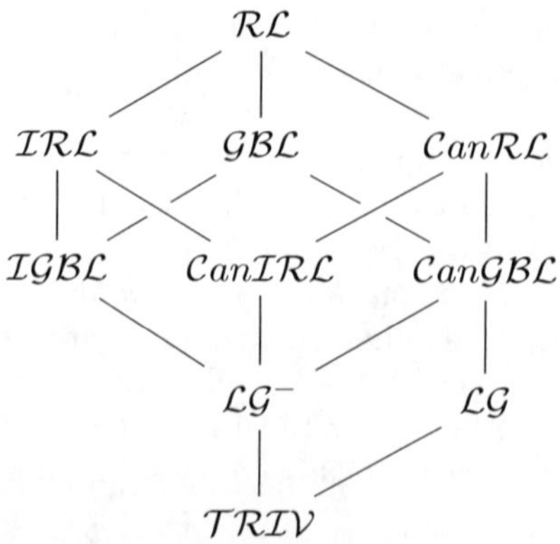

FIGURE 1.2. Varieties of residuated lattices

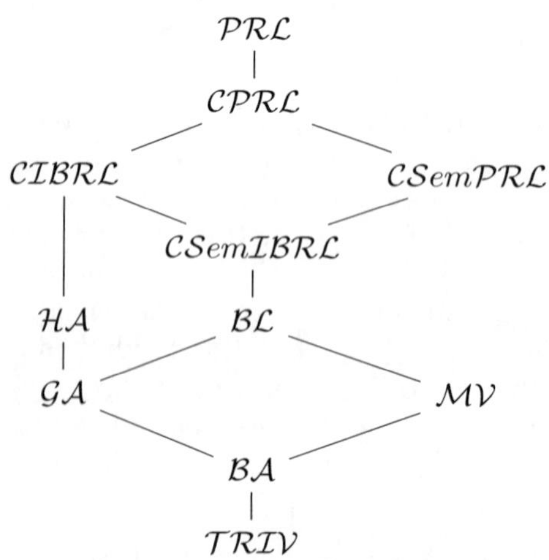

FIGURE 1.3. Varieties of pointed residuated lattices

any structure with an underlying pomonoid $\mathbf{P}$, a nucleus on $\mathbf{P}$ will be called, by extension, a nucleus on $\mathbf{A}$.

EXAMPLE 1.5.1. If u is any element of an integral residuated lattice $\mathbf{L}$, then

$$\alpha_u \colon L \to L; \quad x \mapsto x \vee u$$

is a nucleus on $\mathbf{L}$ and L_{α_u} is the principal order-filter $\uparrow u = \{x \in L \mid u \leq x\}$.

Let $\mathbf{P} = \langle P, \cdot, \backslash, /, \mathrm{e}, \leq \rangle$ be a residuated pomonoid. A closure system $\mathbf{C}$ of $\langle P, \leq \rangle$ is called a *nucleus-system* of $\mathbf{P}$ if $x/y, y\backslash x \in C$, for all $x \in C$ and $y \in P$.

The next result describes the relationship between nuclei and nucleus-systems.

LEMMA 1.5.2. *Let $\mathbf{P} = \langle P, \cdot, \backslash, /, \mathrm{e}, \leq \rangle$ be a residuated pomonoid and let γ be a closure operator on $\langle P, \leq \rangle$ with an associated closure system $\mathbf{P}_\gamma$. The following statements are equivalent:*

(1) *γ is a nucleus on $\mathbf{P}$.*

(2) $\gamma(x)/y$, $y\backslash\gamma(x) \in P_\gamma$, *for all* $x, y \in P$.

(3) $\mathbf{P}_\gamma$ *is a nucleus-system of* $\mathbf{P}$.

(4) $\gamma(x)/y = \gamma(x)/\gamma(y)$ *and* $y\backslash\gamma(x) = \gamma(y)\backslash\gamma(x)$, *for all* $x, y \in P$.

PROOF. Let $x, y \in P$.

(1) $\Rightarrow$ (2) Assuming that γ is a nucleus on $\mathbf{P}$,

$$\gamma(\gamma(x)/y)y \leq \gamma(\gamma(x)/y)\gamma(y) \leq \gamma((\gamma(x)/y)y) \leq \gamma(\gamma(x)) = \gamma(x).$$

So, by residuation, $\gamma(\gamma(x)/y) \leq \gamma(x)/y$. Since the reverse inequality follows from the fact that γ is enlarging, we obtain $\gamma(x)/y = \gamma(\gamma(x)/y) \in P_\gamma$. The result for the other residual is obtained similarly.

(2) $\Rightarrow$ (3) This follows from the definition of a nucleus-system.

(3) $\Rightarrow$ (4) Since $y(y\backslash\gamma(x)) \leq \gamma(x)$, by residuation, $y \leq \gamma(x)/(y\backslash\gamma(x))$, and hence, assuming that $\mathbf{P}_\gamma$ is a nucleus-system of $\mathbf{P}$,

$$\gamma(y) \leq \gamma(\gamma(x)/(y\backslash\gamma(x))) = \gamma(x)/(y\backslash\gamma(x)).$$

So, by residuation, $y\backslash\gamma(x) \leq \gamma(y)\backslash\gamma(x)$. The reverse inequality follows from the fact that γ is enlarging, and the result for the other residual is obtained similarly.

(4) $\Rightarrow$ (1) Applying residuation and (4) twice, we get successively, $xy \leq \gamma(xy), y \leq x\backslash\gamma(xy)$, $y \leq \gamma(x)\backslash\gamma(xy)$, $\gamma(x)y \leq \gamma(xy)$, $\gamma(x) \leq \gamma(xy)/y$, $\gamma(x) \leq \gamma(xy)/\gamma(y)$, and, finally, $\gamma(x)\gamma(y) \leq \gamma(xy)$. Hence γ is a nucleus on $\mathbf{P}$. □

Recall from Proposition 1.3.3 that the closure operators on a pomonoid $\mathbf{P}$ are in bijective correspondence with the closure systems of $\mathbf{P}$. The next result specializes this correspondence to residuated pomonoids.

PROPOSITION 1.5.3. *There exists a bijective correspondence between the nuclei* γ *on a residuated pomonoid* $\mathbf{P}$ *and the nucleus-systems* $\mathbf{C}$ *of* $\mathbf{P}$. *More precisely, if* $\mathbf{C}$ *is a nucleus-system of* $\mathbf{P}$, *then the map* $\gamma_{\mathbf{C}}\colon P \to P$; $x \mapsto \min\{y \in C \mid x \leq y\}$ *is a nucleus on* $\mathbf{P}$. *Conversely, if* $\gamma\colon \mathbf{P} \to \mathbf{P}$ *is a nucleus on* $\mathbf{P}$, *then* $\mathbf{P}_\gamma$ *is a nucleus-system of* $\mathbf{P}$. *Moreover,* $\mathbf{P}_{\gamma_{\mathbf{C}}} = \mathbf{C}$ *and* $\gamma_{\mathbf{P}_\gamma} = \gamma$.

PROOF. Building on Proposition 1.3.3 and Lemma 1.5.2, it suffices to observe that $P_{\gamma_{\mathbf{C}}} = C$ and that $\mathbf{C}$ is a nucleus-system of $\mathbf{P}$. □

We show now that any nucleus-system of a residuated pomonoid (residuated lattice) forms a residuated pomonoid (residuated lattice) in its own right.

LEMMA 1.5.4. *Let* $\mathbf{P} = \langle P, \cdot, \backslash, /, \mathrm{e}, \leq\rangle$ *be a residuated pomonoid and let* γ *be a nucleus on* $\mathbf{P}$. *The structure (with a slight notational abuse)*

$$\mathbf{P}_\gamma = \langle P_\gamma, \circ_\gamma, \backslash, /, \gamma(\mathrm{e}), \leq\rangle,$$

with $x \circ_\gamma y := \gamma(x \cdot y)$ *for* $x, y \in P_\gamma$, *is a residuated pomonoid. Moreover, if* $\mathbf{L} = \langle L, \wedge, \vee, \cdot, \backslash, /, \mathrm{e}\rangle$ *is a residuated lattice, then* $\mathbf{L}_\gamma = \langle L_\gamma, \wedge, \vee_\gamma, \circ_\gamma, \backslash, /, \gamma(\mathrm{e})\rangle$, *with* $x \vee_\gamma y := \gamma(x \vee y)$ *for* $x, y \in L_\gamma$, *is a residuated lattice.*

PROOF. It is easily checked that $\circ_\gamma$ is associative and has neutral element $\gamma(\mathrm{e})$. To confirm that $\circ_\gamma$ is residuated, recall that P_γ is closed under the residual operations by Lemma 1.5.2, and hence, for any $x, y, z \in P_\gamma$,

$$x \circ_\gamma y \le z \iff \gamma(xy) \le z \iff xy \le z \iff y \le x\backslash z \iff x \le z/y.$$

Also, if $\mathbf{L}$ is a residuated lattice, then $\langle L_\gamma, \le\rangle$, as the image of a closure operator on $\mathbf{L}$, is a lattice with meets and joins given by, respectively, $\wedge$ and $\vee_\gamma$. □

In light of Lemma 1.5.4, we may henceforth also refer to the residuated lattice $\mathbf{L}_\gamma$ as a nucleus-system of the residuated lattice $\mathbf{L}$.

We obtain the following analogue of Proposition 1.3.4 for residuated pomonoids.

PROPOSITION 1.5.5. *Let $\langle \varphi, \psi\rangle$ be a residuated pair between residuated pomonoids $\mathbf{P}$ and $\mathbf{Q}$. If φ is a monoid homomorphism, then $\psi\varphi$ is a nucleus on $\mathbf{P}$.*

PROOF. If φ is a monoid homomorphism, then, for any $x, y \in P$, by Lemma 1.2.4(c),

$$\varphi(\psi\varphi(x)\psi\varphi(y)) = \varphi\psi\varphi(x)\varphi\psi\varphi(y) = \varphi(x)\varphi(y) = \varphi(xy),$$

and hence, since $\psi\varphi$ is a closure operator and ψ is an order-homomorphism,

$$\psi\varphi(x)\psi\varphi(y) \le \psi\varphi(\psi\varphi(x)\psi\varphi(y)) \le \psi\varphi(xy).$$ □

We also obtain the following analogue of Corollary 1.3.5.

COROLLARY 1.5.6. *If γ is a nucleus on a residuated pomonoid $\mathbf{P}$, then the map $\gamma\colon P \to P_\gamma$ is a residuated monoid homomorphism, and its residual is the inclusion map from P_γ to P.*

PROOF. It remains to show that any nucleus γ on a residuated pomonoid $\mathbf{P}$ is a monoid homomorphism, to which end, we simply observe that, for any $x, y \in P$,

$$\gamma(xy) = \gamma(\gamma(x)\gamma(y)) = \gamma(x) \circ_\gamma \gamma(y).$$ □

For convenience, we will abuse notation somewhat when referring to a (residuated) pomonoid or residuated lattice $\mathbf{P}$ and denote also by $\mathbf{P}$ either its underlying poset or monoid reduct.

The following example will be of some importance in later chapters of this book.

EXAMPLE 1.5.7. Let $\mathbf{M} = \langle M, \cdot, \mathrm{e}\rangle$ be any monoid and consider the residuated lattice $\wp(\mathbf{M}) = \langle \wp(M), \cap, \cup, \cdot, \backslash, /, \{\mathrm{e}\}\rangle$ defined in Example 1.4.3. Given any nucleus γ on $\wp(\mathbf{M})$, it follows from Lemma 1.5.4 that $\wp(\mathbf{M})_\gamma = \langle \wp(M)_\gamma, \cap, \cup_\gamma, \circ_\gamma, \backslash, /, \mathrm{e}_\gamma\rangle$, with $X \circ_\gamma Y := \gamma(X \cdot Y)$, $X \cup_\gamma Y := \gamma(X \cup Y)$, and $\mathrm{e}_\gamma := \gamma(\{\mathrm{e}\})$ is also a residuated lattice. In particular, for any pomonoid $\mathbf{P} = \langle P, \cdot, \mathrm{e}, \le\rangle$, the map

$$\gamma_\downarrow\colon \wp(P) \to \mathcal{L}(\mathbf{P}); \quad X \mapsto {\downarrow}X$$

is a nucleus on $\wp(\mathbf{P})$ with image $\mathcal{L}(\mathbf{P})$ and yields a residuated lattice, where the product of two order-ideals $X, Y \in \mathcal{L}(\mathbf{P})$ is given by $X \circ_{\gamma_\downarrow} Y := {\downarrow}(X \cdot Y)$. We also note, for future reference, that the product of two principal order-ideals ${\downarrow}x, {\downarrow}y \in \mathcal{L}(\mathbf{P})$ is the principal order-ideal ${\downarrow}(xy)$.

The next lemma supplies a particularly interesting family of nuclei.

LEMMA 1.5.8. *Let* $\mathbf{P}$ *be a residuated pomonoid and let* $d \in P$.

(a) $\gamma_d \colon P \to P;\ x \mapsto d/(x\backslash d)$ *is a closure operator on* $\mathbf{P}$.

(b) $\gamma_d(x)y \leq \gamma_d(xy)$*, for all* $x, y \in P$.

Suppose also that $d/(x\backslash d) = (d/x)\backslash d$*, for all* $x \in P$.

(c) γ_d *is a nucleus on* $\mathbf{P}$ *and* $P_{\gamma_d} = \{d/x \mid x \in P\} = \{x\backslash d \mid x \in P\}$.

(d) γ_d *is the largest nucleus on* $\mathbf{P}$ *that fixes* d*; that is, if* γ *is a nucleus on* $\mathbf{P}$ *such that* $\gamma(d) = d$*, then* $\gamma(x) \leq \gamma_d(x)$*, for all* $x \in P$.

PROOF. Let $x, y \in P$.

(a) If $x \leq y$, then $y\backslash d \leq x\backslash d$ and $\gamma_d(x) = d/(x\backslash d) \leq d/(y\backslash d) = \gamma_d(y)$. So γ_d is an order-endomorphism of $\mathbf{P}$. Also, $x(x\backslash d) \leq d$, so $x \leq d/(x\backslash d) = \gamma_d(x)$; that is, γ_d is enlarging. To confirm that γ_d is idempotent, observe first that $x\backslash d = (d/(x\backslash d))\backslash d$, since $(d/(x\backslash d))(x\backslash d) \leq d$ yields $x\backslash d \leq (d/(x\backslash d))\backslash d$, and $x \leq d/(x\backslash d)$ yields $(d/(x\backslash d))\backslash d \leq x\backslash d$. Hence $\gamma_d(x) = d/(x\backslash d) = d/((d/(x\backslash d))\backslash d) = \gamma_d(\gamma_d(x))$.

(b) Observe that $\gamma_d(x)y(xy\backslash d) = \gamma_d(x)y(y\backslash(x\backslash d)) \leq (d/(x\backslash d))(x\backslash d) \leq d$, and hence $\gamma_d(x)y \leq d/(xy\backslash d) = \gamma_d(xy)$.

(c) By dual arguments to those given for (a) and (b), $\gamma'_d \colon P \to P;\ z \mapsto (d/z)\backslash d$ is a closure operator on $\mathbf{P}$ such that $x\gamma'_d(y) \leq \gamma'_d(xy)$. By assumption, $\gamma_d = \gamma'_d$, so $\gamma_d(x)\gamma_d(y) \leq \gamma_d(\gamma_d(x)y) \leq \gamma_d(xy)$. That $P_{\gamma_d} = \{d/z \mid z \in P\} = \{z\backslash d \mid z \in P\}$ follows from the fact that $z\backslash d = (d/(z\backslash d))\backslash d$ and $d/z = d/((d/z)\backslash d)$, for any $z \in P$.

(d) Observe first that $d = \mathrm{e}\backslash d \in P_{\gamma_d}$, by (c), so $\gamma_d(d) = d$. Now let γ be any nucleus on $\mathbf{P}$ such that $\gamma(d) = d$. Then $x\backslash d = x\backslash\gamma(d) = \gamma(x)\backslash\gamma(d) = \gamma(x)\backslash d$, by Lemma 1.5.2. So $\gamma_d(x) = d/(x\backslash d) = d/(\gamma(x)\backslash d) = \gamma_d(\gamma(x))$ and $\gamma(x) \leq \gamma_d(x)$. □

Note that if $\mathbf{L}$ is an integral residuated lattice satisfying $y/(x\backslash y) \approx x \vee y$, then for any $d \in L$, the closure operator γ_d defined in the previous lemma coincides with the map α_d defined in Example 1.5.1.

Using Lemma 1.5.8, we obtain a characterization of nucleus-systems on a residuated lattice that can be expanded to an involutive residuated lattice.

PROPOSITION 1.5.9. *Let* $\mathbf{L}$ *be a residuated lattice and let* γ *be any nucleus on* $\mathbf{L}$. *The nucleus-system* $\mathbf{L}_\gamma$ *is the residuated lattice reduct of an involutive residuated lattice if and only if there exists a cyclic element* d *of* $\mathbf{L}$ *such that* $\gamma = \gamma_d$.

PROOF. Suppose first that d is a cyclic element of $\mathbf{L}$ such that $\gamma = \gamma_d$. It follows from Lemma 1.5.8.(c) that γ_d is a nucleus on $\mathbf{L}$ and d is a cyclic dualizing element of $\mathbf{L}_{\gamma_d}$. Hence $\mathbf{L}_{\gamma_d}$ expanded with the constant d is an involutive residuated lattice.

For the converse, suppose that the nucleus-system $\mathbf{L}_\gamma$ is the residuated lattice reduct of an involutive residuated lattice, and let $d \in L_\gamma$ be its cyclic dualizing element. Then $x\backslash d = \gamma(x)\backslash d = d/\gamma(x) = d/x$, for any $x \in L$, by Lemma 1.5.2. Hence d is a cyclic element of $\mathbf{L}$ and, for any $x \in L$,

$$\gamma_d(x) = (x \to d) \to d = (\gamma(x) \to d) \to d = \gamma(x). \qquad \square$$

Just as nuclei are the well-behaved closure operators on residuated pomonoids, so co-nuclei are the well-behaved co-closure operators. A co-closure operator δ on a pomonoid $\mathbf{P}$ is called a *co-nucleus* on $\mathbf{P}$ if $\delta(\mathrm{e}) = \mathrm{e}$ and $\delta(x)\delta(y) \leq \delta(xy)$, for all $x, y \in P$, where the latter condition is clearly equivalent to $\delta(\delta(x)\delta(y)) = \delta(x)\delta(y)$. A co-closure system of a residuated pomonoid $\mathbf{P}$ that contains e and is closed under taking products is called a *co-nucleus-system* of $\mathbf{P}$.

We obtain the following analogues of Proposition 1.5.3 and Lemma 1.5.4.

PROPOSITION 1.5.10. *There exists a bijective correspondence between the co-nuclei δ on a residuated pomonoid $\mathbf{P}$ and the co-nucleus-systems $\mathbf{O}$ of $\mathbf{P}$. More precisely, if $\mathbf{O}$ is a co-nucleus-system of $\mathbf{P}$, then the map $\delta_{\mathbf{O}} \colon P \to P$; $x \mapsto \max\{y \in O \mid y \leq x\}$ is a co-nucleus on $\mathbf{P}$. Conversely, if $\delta \colon \mathbf{P} \to \mathbf{P}$ is a co-nucleus on $\mathbf{P}$, then $\mathbf{P}_\delta$ is a co-nucleus-system of $\mathbf{P}$. Moreover, $\mathbf{P}_{\delta_{\mathbf{O}}} = \mathbf{O}$ and $\delta_{\mathbf{P}_\delta} = \delta$.*

LEMMA 1.5.11. *Let $\mathbf{P} = \langle P, \cdot, \backslash, /, \mathrm{e}, \leq \rangle$ be a residuated pomonoid and let δ be a co-nucleus on $\mathbf{P}$. The structure (with a slight notational abuse)*

$$\mathbf{P}_\delta = \langle P_\delta, \cdot, \backslash_\delta, /_\delta, \mathrm{e}, \leq \rangle,$$

with $x \backslash_\delta y := \delta(x \backslash y)$ and $x /_\delta y := \delta(x/y)$, for all $x, y \in P_\delta$, is a residuated pomonoid. Moreover, if $\mathbf{L} = \langle L, \wedge, \vee, \cdot, \backslash, /, \mathrm{e} \rangle$ is a residuated lattice, then $\mathbf{L}_\delta = \langle L_\delta, \wedge_\delta, \vee, \cdot, \backslash_\delta, /_\delta, \mathrm{e} \rangle$, with $x \wedge_\delta y := \delta(x \wedge y)$, is also a residuated lattice.

PROOF. $\mathbf{P}_\delta$ is obviously closed under $\cdot$, $\backslash_\delta$, and $/_\delta$. Moreover, $\cdot$ is residuated with residuals $\backslash_\delta$ and $/_\delta$: for all $x, y, z \in P_\delta$, using for the first and last equivalences, the fact that δ is a contracting order-endomorphism of $\mathbf{P}$ and $y = \delta(y)$, $x = \delta(x)$,

$$y \leq x \backslash_\delta z \iff y \leq x \backslash z \iff xy \leq z \iff x \leq z/y \iff x \leq z /_\delta y.$$

Finally, if $\mathbf{L}$ is a residuated lattice, then $\wedge_\delta$ and $\vee$ are, respectively, the meet and join operations of the lattice-order of $\mathbf{L}_\delta$, by Proposition 1.3.12. □

Following Lemma 1.5.11, we also refer, similarly to the case of nucleus systems, to the residuated lattice $\mathbf{L}_\delta$ as a co-nucleus-system of the residuated lattice $\mathbf{L}$.

1.6. Historical excursus

In this section, we provide a brief overview of the development of the concept of a residuated structure in algebra: from its origins in Galois theory and ring theory to more recent appearances in universal algebra and category theory. Instead of a strictly chronological account, we follow, for the sake of greater readability, six separate thematic threads, each of which has made a significant contribution to moulding the concept of a residuated structure as it is understood today.

A word of caution is in order here, however. Although residuated maps and Galois correspondences abound in mathematics, we circumscribe our survey to examples of residuation related to the main themes of this book. We therefore focus on the development of residuated lattices, disregarding, e.g., plentiful and important examples of residuated pairs between two different posets. The next two subsections form the only exception to this policy, motivated by the historical relevance of Galois theory as the first significant appearance of residuation outside elementary arithmetic, and the importance of adjoint functors in the more general framework of category theory.

1.6.1. Galois theory. Following proofs given by mathematicians Scipione del Ferro, Niccolò Tartaglia, Girolamo Cardano, and Lodovico Ferrari of the Italian Renaissance that cubic and quartic equations are solvable by radicals using a general formula, algebraists in subsequent centuries strived to obtain similar results for polynomial equations of degree 5 and higher. These efforts came to an abrupt end in 1824, when Niels Abel (repairing an earlier incorrect proof by Paolo Ruffini) showed that such equations have no general solution by radicals. However, since it was well-known that certain equations of degree 5 and higher could indeed be solved, the question remained open as to whether there exists a general characterization of the polynomial equations for which this is the case.

In 1832, Évariste Galois (just before meeting his untimely end in a duel) found the right approach to settle this issue. He associated to each polynomial equation ε a permutation group (now called a *Galois group*) $\mathbf{G}_\varepsilon$ consisting of those permutations of the solutions of ε having the property that every algebraic equation satisfied by the solutions is still satisfied after the solutions have been permuted. Galois established that ε is solvable by radicals if and only if $\mathbf{G}_\varepsilon$ is *solvable*, that is, when there exist subgroups $\mathbf{G}_0, \ldots, \mathbf{G}_n$ of $\mathbf{G}_\varepsilon$ satisfying

$$\mathbf{G}_0 \subseteq \mathbf{G}_1 \subseteq \cdots \subseteq \mathbf{G}_{n-1} \subseteq \mathbf{G}_n,$$

such that $\mathbf{G}_0$ is the trivial subgroup, $\mathbf{G}_n = \mathbf{G}_\varepsilon$, $\mathbf{G}_{i-1}$ is normal in $\mathbf{G}_i$ and $\mathbf{G}_i/\mathbf{G}_{i-1}$ is Abelian, for each $i \in \{1, \ldots, n\}$.

Permutations that respect algebraic equations satisfied by the solutions of ε can be viewed as automorphisms of the extension field obtained by adjoining the solutions of ε to the base field. This approach generalizes to other field extensions, and it is this abstract perspective that underlies present-day Galois theory (see e.g. [**214**]), where Galois groups are viewed as field automorphisms of an extension field $\mathbf{L}$ of a given base field $\mathbf{F}$. Let us now stipulate that

(i) $S(\mathbf{L}, \mathbf{F})$ is the set of all subfields of $\mathbf{L}$ that contain $\mathbf{F}$;
(ii) $\mathbf{Gal}_\mathbf{M}(\mathbf{L})$, where $\mathbf{M} \in S(\mathbf{L}, \mathbf{F})$, is the group of all field automorphisms φ of $\mathbf{L}$ such that $\varphi{\restriction}M$ is the identity map on M;
(iii) $S(\mathbf{Gal}_\mathbf{F}(\mathbf{L}))$ is the set of all subgroups of $\mathbf{Gal}_\mathbf{F}(\mathbf{L})$.

Then the following maps between $\mathbf{P} := \langle S(\mathbf{L}, \mathbf{F}), \subseteq\rangle$ and the order dual $\mathbf{Q}^\partial$ of $\mathbf{Q} := \langle S(\mathbf{Gal}_\mathbf{F}(\mathbf{L})), \subseteq\rangle$ form a residuated pair:

$$\varphi\colon \mathbf{P} \to \mathbf{Q}^\partial; \quad \mathbf{M} \mapsto \mathbf{Gal}_\mathbf{M}(\mathbf{L})$$
$$\psi\colon \mathbf{Q}^\partial \to \mathbf{P}; \quad \mathbf{H} \mapsto \{a \in L \mid \varphi(a) = a, \text{ for all } \varphi \in H\}.$$

These examples of residuated pairs are indeed so important that any residuated pair $\langle\varphi, \psi\rangle$ between posets $\mathbf{P}$ and $\mathbf{Q}^\partial$ is commonly referred to as a *Galois correspondence* between $\mathbf{P}$ and $\mathbf{Q}$.

1.6.2. From residuated pairs to adjoint functors. As we have just seen, an important source of residuated pairs was discovered already in the first half of the nineteenth century. An abstract treatment of this concept, however, did not emerge for another hundred years. The transition to the fully-fledged notion of a residuated pair presented in Section 1.2 was completed in three stages [**120**]:

1. In the 1940 edition of his *Lattice Theory* **[32]**, Garrett Birkhoff studied Galois correspondences between families of sets partially ordered by set-inclusion, under the name of *polarities*. Given sets X and Y and a binary relation $R \subseteq X \times Y$, he considered the maps (compare with Example 1.2.2)

$$\varphi\colon \wp(X) \to \wp(Y); \quad A \mapsto \{y \in Y \mid \langle x, y\rangle \in R, \text{ for all } x \in A\}$$
$$\psi\colon \wp(Y) \to \wp(X); \quad B \mapsto \{x \in X \mid \langle x, y\rangle \in R, \text{ for all } y \in B\}.$$

Birkhoff observed that φ and ψ are order-reversing and that the maps $\varphi\psi$ and $\psi\varphi$ are closure operators. He also proved that φ and ψ induce a dual order-isomorphism between the closure systems $\langle \wp(X), \subseteq\rangle_{\psi\varphi}$ and $\langle \wp(Y), \subseteq\rangle_{\varphi\psi}$. As examples of this construction, he cited the Dedekind-MacNeille completion of a poset and the Galois correspondence mentioned in the previous subsection, as well as other examples from ring theory and topology. The name "polarity" stems from a further example to be found in linear algebra: if $X = Y$ is the universe of a finite-dimensional vector space with scalar product and $R := \{\langle x, y\rangle \in X^2 \mid (x|y) = 0\}$, then $\varphi(A) = \psi(A)$ is the *polar* $A^\perp$ of A.

2. In a 1944 paper **[302]**, Øystein Ore lifted Galois correspondences from the concrete set-theoretic setting to a more abstract framework allowing *any* pair of posets. Ore considered pairs of order-reversing maps $\varphi\colon \mathbf{P} \to \mathbf{Q}$ and $\psi\colon \mathbf{Q} \to \mathbf{P}$ between posets $\mathbf{P}$ and $\mathbf{Q}$ satisfying $x \le \psi\varphi(x)$, for all $x \in P$, and $y \le \varphi\psi(y)$, for all $y \in Q$. He did not consider the equivalent characterization

$$x \le \psi(y) \iff y \le \varphi(x),$$

showing that $\langle \varphi, \psi\rangle$ is a residuated pair between $\mathbf{P}$ and $\mathbf{Q}^\partial$, but established the bijective correspondence between closure operators on a poset and closure systems of the same poset (Proposition 1.3.3), and a representation theorem for Galois correspondences in terms of Birkhoff's polarities.

3. Constructions of residuated pairs appeared frequently between the end of the 1940s and beginning of the 1950s in topology and analysis papers by Georg Nöbeling, Mihail Benado, and Günter Pickert. However, the first in-depth study of this concept can be credited to Jürgen Schmidt, who introduced residuated pairs in his 1953 article on "Galois correspondence of mixed type" (*Galois-Korrespondenz vom gemischten Typus*) **[335]**. Schmidt provided a proof of Lemma 1.2.8 and generalized Ore's result to show that each member of a residuated pair completely determines the other (see Lemma 1.2.5). He also discussed and investigated in detail several applications to topology.

In the 1950s, the concept of a residuated pair crossed over to the even more abstract setting of *category theory*, which had recently emerged from algebraic topology, thanks largely to work by Samuel Eilenberg and Saunders Mac Lane, with the goal of investigating properties of related mathematical structures by considering structure-preserving maps between them. We will not recall here all the basic definitions of category theory (the reader may consult, e.g., **[242]** or **[13]**), just the fundamental concept of a *functor*: a map from a category $\mathcal{C}$ to a category $\mathcal{D}$ that associates to every object $\mathbf{C}$ of $\mathcal{C}$, an object $F(\mathbf{C})$ of $\mathcal{D}$, and to every morphism φ of $\mathcal{C}$, a morphism $F(\varphi)$ of $\mathcal{D}$, such that, for all objects $\mathbf{C}, \mathbf{C}', \mathbf{C}''$ of $\mathcal{C}$ and morphisms $\varphi \in \hom_{\mathcal{C}}(\mathbf{C}, \mathbf{C}'), \psi \in \hom_{\mathcal{C}}(\mathbf{C}'', \mathbf{C})$,

$$F(id_{\mathbf{C}}) = id_{F(\mathbf{C})} \quad \text{and} \quad F(\varphi \circ \psi) = F(\psi) \circ F(\varphi).$$

Residuated pairs entered this picture when, in a 1958 paper **[213]**, the algebraic topologist Daniel Kan first introduced the concept of an *adjoint functor*.[2] Given categories $\mathcal{C}$ and $\mathcal{D}$ and functors $F\colon \mathcal{C} \to \mathcal{D}$ and $G\colon \mathcal{D} \to \mathcal{C}$, the functor G is said to be a *left adjoint* to F, and F a *right adjoint* to G, if there exists a family of bijections that induces a natural isomorphism between $\hom_{\mathcal{C}}(G(\mathbf{D}), \mathbf{C})$ and $\hom_{\mathcal{D}}(\mathbf{D}, F(\mathbf{C}))$, for all $\mathbf{C}$ in $\mathcal{C}$ and $\mathbf{D}$ in $\mathcal{D}$. When G is a left adjoint to F, we also say that F and G form an *adjoint pair*.

Observe that every poset $\mathbf{P}$ can be considered as a special example of a category whose objects are the members of P, and such that, for all $a, b \in P$, there is a single morphism $\varphi\colon a \to b$ if and only if $a \leq b$. If $\langle \varphi, \psi \rangle$ is a residuated pair between posets $\mathbf{P}$ and $\mathbf{Q}$, then φ and ψ — viewed as functors from the category $\mathbf{P}$ to the category $\mathbf{Q}$ and back again — form an adjoint pair according to the previous definition. Therefore, Kan's notion of adjoint functors generalizes the concept of a residuated pair from the setting of posets to the framework of an arbitrary pair of categories. Important examples of adjoint pairs of functors abound in algebra, topology, and algebraic geometry.

1.6.3. Ideal theory of rings. Polynomial equations are a historically significant source for residuated structures in one further respect. Around the middle of the nineteenth century, it was observed by Ernst Kummer that unique factorization into primes, which holds for integers by virtue of the fundamental theorem of arithmetic, fails in general for algebraic integers, i.e., roots of monic polynomials with integer coefficients. Let $\mathbb{Z}[\sqrt{-n}]$ denote the quadratic integer ring of complex numbers of the form $a + b\sqrt{-n}$, with $a, b \in \mathbb{Z}$ and $n \in \mathbb{N}$. Unique factorization fails in $\mathbb{Z}[\sqrt{-n}]$ for certain instances of n, although it holds in some special cases, e.g., for *Gauss integers* where $n = 1$. Kummer tried to recover a weakened form of this fundamental property with his theory of *ideal numbers*, but the modern abstract viewpoint — leading to the birth of contemporary ring theory and the ideal theory of rings — first appeared in Richard Dedekind's 1871 X Supplement to the second edition of Gustav Lejeune Dirichlet's *Zahlentheorie*. Dedekind introduced the concepts of a ring and a ring ideal in what essentially is their modern usage (the term *ring*, however, was first coined by Hilbert much later) and proved that every ideal of the ring of algebraic integers is uniquely representable (up to permutation of factors) as a product of prime ideals. Unique factorization is therefore recovered at the level of ideals (see, e.g., **[119]**).

The ideal theory of commutative rings was intensively investigated early in the twentieth century by Emanuel Lasker and Francis Macaulay, who generalized the results by Dedekind to polynomial rings, and in the 1920s by Emmy Noether and Wolfgang Krull. In particular, Noether proved the celebrated theorem establishing that in any commutative ring whose lattice of ideals satisfies the ascending chain condition, ideals decompose into intersections of finitely many primary ideals. This theorem is typical of Noether's general approach. The system $I(\mathbf{R})$ of ideals of a commutative ring $\mathbf{R}$ is viewed, first and foremost, as an example of a lattice endowed with an additional binary operation: namely, the *product* of ideals I and

[2] The definition of adjoint functors is implicit in a famous 1957 paper by Grothendieck **[173]**, where he discusses at length, among many other things, natural equivalences. However, according to **[222]**, Kan should be credited with the first explicit introduction of the concept.

J of $\mathbf{R}$ is the ideal

$$I \cdot J := \{a_1b_1 + \cdots + a_nb_n \mid a_1, \ldots, a_n \in I \text{ and } b_1, \ldots, b_n \in J\}.$$

The same viewpoint was taken and extended by Morgan Ward and his student R.P. Dilworth in a series of papers during the late 1930s [**101–103, 365, 367, 368**], focusing on a further binary operation on ideals: namely, given ideals I and J of $\mathbf{R}$, the *residual* of J with respect to I is the ideal

$$I \to J := \{a \in R \mid ab \in J, \text{ for all } b \in I\}.$$

This residual has the key property that for any ideal K of the ring, $J \cdot K \subseteq I$ if and only if $K \subseteq J \to I$. Abstracting from the properties of products of ideals, Ward and Dilworth introduced and investigated in some detail a certain class of lattice-ordered structures with a product and residual, that they called *residuated lattices*. They were then able to extend some of the results obtained by Noether and Krull for lattices of ideals of commutative rings, including Noether's decomposition theorem, to a purely lattice-theoretic setting.

Ward's and Dilworth's papers did not have much of an immediate impact, but began a line of research that would crop up repeatedly in the following decades. In particular, the notion of a residuated lattice re-emerged, occasionally under the name of *residuated lattice-ordered monoid*, in the study of semantics for fuzzy logics (see, e.g., [**174, 185**]) and substructural logics (see, e.g. [**294, 300**]), and papers with a more pronounced universal-algebraic flavor (see, e.g., [**44, 206**]), these two streams converging eventually in the monograph [**145**].

In fact, none of the definitions of a residuated lattice given in these works (or in this chapter) coincides exactly with the original definition given by Ward and Dilworth. Differences concern both the signature and, less superficially, properties characterizing the respective algebras. The residuated lattices studied in [**174, 185, 294, 300**] are invariably both bounded as lattices and integral as pomonoids. Moreover, the product operation is commutative. Residuated lattices as defined in this book and [**44, 145, 206**] are not necessarily bounded and, even when this is the case, need not be integral; also, the product operation is not required to be commutative. The original definition lies somewhere in between; Ward and Dilworth do not assume the existence of a greatest or least element in the underlying lattice, but if there is a greatest element, then it must be the neutral element of the product operation, which is assumed to be commutative. It is only fair to note that Dilworth also introduced in [**102**] a non-commutative variant of his notion of a residuated lattice, inspired by the residuated lattice of two-sided ideals of a non-commutative ring (see Example 1.4.8), but to the best of our knowledge this generalization was not taken up again until the work reported in [**44**].

1.6.4. Boolean and Heyting algebras. One of the first classes of residuated lattices to receive considerable attention in their own right was of course the class of Boolean algebras. Although their name might suggest that their inventor was George Boole, author of *The Mathematical Analysis of Logic* (1847) and of the *Laws of Thought* (1854) — who sought no less than to "investigate the fundamental laws of those operations of the mind by which reasoning is performed, to give expression to them in the symbolical language of a calculus, and upon this foundation to establish the science of Logic and construct its method" — it is now commonly accepted by the scholarly community that this is not the case. Not that it is always easy to clearly understand what Boole had in mind; as Theodore Hailperin puts it ([**177**, p. 61]):

> If we look carefully at what Boole actually did [...], we find him carrying out operations, procedures, and processes of an algebraic character, often inadequately justified by present-day standards and, at times, making no sense to a modern mathematician. [...] Boole considered this acceptable so long as the end result could be given a meaning.

The reason why Boole's calculus cannot be considered as a first incarnation of a Boolean algebra of sets is the fact that the two operations of *combination* of two classes x and y (written xy in Boole's notation) and of *aggregation* of x and y (noted $x + y$) do correspond to intersection and union, respectively, but the latter only makes sense when x and y are disjoint classes. The algebras to be found in Boole's work bear some resemblance to partial algebras in the modern sense, except for the fact that Boole often disregarded his disjointness condition in calculations, seemingly finding such a procedure unobjectionable provided the final result did not itself violate the condition.

A closer approximation to the modern understanding of Boolean algebras can be found in the writings of Ernst Schröder. In particular, in his *Operationskreis des Logikkalkuls* (1877), he introduces something resembling an equational basis for Boolean algebras, replacing Boole's partial aggregation operation by the lattice join, and comes very close to realizing that Boolean algebras are residuated structures. He observes, in fact, that the equation $x \wedge b = a$ is solvable if and only if $a \wedge \neg b = 0$, and possesses in this case a smallest solution $x = a$ and largest solution $x = a \vee \neg b$. What he apparently misses, however, is the fact that $x = a \vee \neg b$ is also a largest solution for the inequation $x \wedge b \leq a$, or, in other words, that such an operation is the residual of the meet in Boolean algebras (see [**120**] for further details).

A proper treatment of Boolean algebras as *algebras* was given at the turn of the twentieth century, when Edward V. Huntington provided a first equational basis,[3] followed in the next thirty years by more and more economical bases with different choices of primitive operations, due, e.g., to Benjamin Bernstein and Henry Sheffer. Most of the fundamental results of the structure theory for Boolean algebras were, however, established in the 1930s by Marshall Stone [**348**]: in particular,

(i) the term-equivalence between Boolean rings (i.e., idempotent unital rings) and Boolean algebras;
(ii) the representation of Boolean algebras as algebras of sets, which can be seen as a corollary of a more comprehensive duality between Boolean algebras and a class of topological spaces (Boolean spaces, i.e., totally disconnected compact Hausdorff spaces).

Both results are extremely powerful. Regarding the former, it is striking that a class of algebras introduced for the purpose of systematizing logical reasoning turns out

[3]In his seminal 1904 paper on the subject [**198**], Huntington includes an axiom to the effect that his algebras cannot be trivial; hence, the class he defined fails to be closed under quotients. Strictly speaking, therefore, he does not provide a basis for the *variety* of Boolean algebras.

to be a very natural subclass of a class of structures central to classical algebra and other parts of "standard" mathematics.

The second result, on the other hand, fleshes out a natural intuition about Boolean algebras as being essentially algebras of sets. Let $\mathbf{B} = \langle B, \wedge, \vee, \neg, 0, 1\rangle$ be any Boolean algebra and let $U(\mathbf{B})$ denote the set of ultrafilters on $\mathbf{B}$, that is, the set of lattice-filters F of $\langle B, \leq\rangle$ such that $0 \notin F$ and either $b \in F$ or $\neg b \in F$ for any $b \in B$. Stone showed that $\mathbf{B}$ embeds into the Boolean algebra (of sets) $\langle \wp(U(\mathbf{B})), \cap, \cup, ^c, \emptyset, U(\mathbf{B})\rangle$ via the map

$$\varphi\colon B \to \wp(U(\mathbf{B})); \quad x \mapsto \{F \in U(\mathbf{B}) \mid x \in F\}.$$

Around the same time, due to the independent work of Valery Glivenko [**166**], Arend Heyting [**183**]), and others, *Heyting algebras* were introduced as the algebraic counterparts of intuitionistic propositional logic (see Example 1.4.11). As observed earlier in this chapter, the varieties of Heyting algebras and Boolean algebras are both term-equivalent to varieties of pointed residuated lattices.

1.6.5. Lattice-ordered groups. The name of Richard Dedekind occurs quite often in these historical notes; indeed, he invented, anticipated, or investigated first at the abstract level, some of the most fundamental notions in contemporary abstract algebra. We have already mentioned his contributions to ring theory in his X Supplement to the second edition of Dirichlet's *Zahlentheorie* (1871). In his Göttingen classnotes on algebra, written between 1858 and 1868 but only published more than a century later [**334**], Dedekind studied the abstract concept of a group beyond their concrete representations as groups of permutations. In a 1897 paper *Über Zerlegung von Zahlen durch ihre größten gemeinsamen Teiler*, he essentially introduced and investigated the notion of a lattice, as well as combining the two concepts into what is now known as a *lattice-ordered group* (or ℓ-group) [**120**]. Dedekind's approach to lattices is more algebraic than order-theoretic: he views lattices as sets equipped with two binary operations each satisfying associativity, commutativity, and idempotency, and linked by the absorption law, rather than as posets where every pair of elements has a meet and a join. Moreover, since his study of lattices is motivated mainly by number-theoretic interests, the privileged examples of meet and join he has in mind are not the set-theoretic operations of intersection and union but the arithmetical operations of greatest common divisor and least common multiple [**215**].

At least for the commutative case, the definition of ℓ-groups given in Example 1.4.13 is implicit in Dedekind's 1897 paper. He considered an Abelian group $\langle A, \cdot, ^{-1}, \mathrm{e}\rangle$ endowed with an additional semilattice operation $\vee$ such that $\cdot$ distributes over $\vee$, that is, such that the algebra $\langle A, \vee, \cdot, ^{-1}, \mathrm{e}\rangle$ satisfies the equation $x(y \vee z) \approx xy \vee xz$. He then observed that the defined operation $a \wedge b = ab(a \vee b)^{-1}$ turns $\langle A, \wedge, \vee\rangle$ into a distributive lattice and $\langle A, \wedge, \vee, \cdot, ^{-1}, \mathrm{e}\rangle$ into an Abelian ℓ-group. The same proof carries over to the non-commutative case if the meet is defined as $a \wedge b = a(a \vee b)^{-1}b$, but Dedekind himself did not observe this fact.

In the late 1920s and early 1930s, functional analysts such as Frigyes Riesz, Hans Freudenthal, and Leonid Kantorovich developed significant parts of ℓ-group theory in the context of their investigations into vector lattices. These different streams converged in Birkhoff's 1940 work *Lattice Theory* [**32**], which contains a chapter on

ℓ-groups where the concept is defined precisely and Birkhoff's own contributions to the subject, as well as the main results proved in various diverse research areas, are collected and systematized.

The relevance of this account to residuated lattices lies in the fact that the group operation of any ℓ-group $\mathbf{L}$ is residuated; that is, for any $a, b, c \in L$,

$$b \leq a^{-1}c \iff ab \leq c \iff a \leq cb^{-1}.$$

To the best of our knowledge, the first author to explicitly notice this property was Jeremiah Certaine in his PhD thesis [**63**]. It was only much later, however, that an explicit proof was given that the variety of ℓ-groups is term-equivalent to the variety of residuated lattices satisfying $(\mathrm{e}/x)x \approx \mathrm{e}$ via the correspondences provided in Example 1.4.13 [**43, 44**].

1.6.6. Birkhoff's problem. Boolean algebras and ℓ-groups contributed towards the historical development of the theory of residuated structures also in other ways. In the 1940 edition of his *Lattice Theory* [**32**, Problem 108], Birkhoff suggested the following project:

> Develop a common abstraction which includes Boolean algebras (rings) and lattice-ordered groups as special cases.

Over subsequent decades, several mathematicians tried their hands at Birkhoff's intriguing problem. A minimal requirement for a suitable class of structures is that it should include all Boolean algebras and ℓ-groups as instances, and it is clear that such a desideratum can be satisfied by very different concepts. Indeed, the suggestions advanced in response to Birkhoff's challenge have included such disparate items as classes of partial algebras [**90, 340, 341, 374**] and classes of structures with multi-valued operations [**284**].

K.L.N. Swamy was the first author to observe, in this context, that both Boolean algebras and ℓ-groups are dually residuated structures. In the 1960s [**350–352**] he tackled a problem that is in some aspects more general and in others more specific than Birkhoff's problem: he aimed to develop a common generalization of Brouwerian algebras and Abelian ℓ-groups. To this end, he introduced the notion of a *dually residuated lattice-ordered semigroup* (or *DRℓ-semigroup*): an algebra $\mathbf{A} = \langle A, \wedge, \vee, +, -, 0\rangle$ of signature $\langle 2, 2, 2, 2, 0\rangle$ such that

(i) $\langle A, +, 0\rangle$ is a commutative monoid;

(ii) $\langle A, \wedge, \vee\rangle$ is a lattice, and $+$ distributes over both lattice operations;

(iii) $a - b = \bigwedge\{x \in A \mid a \leq b + x\}$, for all $a, b \in A$: in other words, subtraction is the residual of addition with respect to the order dual to $\leq$;

(iv) $((a - b) \vee 0) + b \leq a \vee b$ and $0 \leq a - a$, for all $a, b \in A$.

Swamy observed that Abelian ℓ-groups are DRℓ-semigroups and Boolean algebras are DRℓ-semigroups when the lattice join is identified with addition and difference $(a \wedge \neg b)$ with subtraction; he also showed that DRℓ-semigroups form a variety and provided equational bases for the subvarieties of Boolean algebras and Abelian ℓ-groups. Finally, he listed necessary and sufficient conditions for a DRℓ-semigroup to be decomposed as a direct product of a Boolean algebra and an Abelian ℓ-group.

In his third paper of this series, Swamy proposed a non-commutative generalization of DRℓ-semigroups of greater interest from the viewpoint of this book. Indeed, if a binary operation symbol for the "right coresidual" is added to the signature, the resulting variety is term-equivalent to the variety of *GBL-algebras*, introduced in [**206**] (see Chapter 7). Swamy's sadly neglected work on these non-commutative structures has been taken up again only in relatively recent times [**219, 223**].

While Swamy's motivation was rooted in pure mathematics, Ettore Casari [**61**] was drawn to Birkhoff's problem by linguistical and philosophical considerations. He was interested in giving an account of such natural language comparative sentences as "c is at most as P as d is Q", where c, d are names and P, Q are predicates. If we accept that sentences may admit different "degrees of truth" (a widely held view in the area of many-valued logics), the aforementioned sentence can be considered true when "c is P" is at most as true as "d is Q". To formalize this idea, Casari developed a *comparative logic* with an implication connective that is true exactly when its antecedent is at most as true as its consequent, whose equivalent algebraic semantics is the variety of *Abelian ℓ-pregroups*, term-equivalent to a certain subvariety of commutative involutive residuated lattices.

Finally, developing an initial suggestion in [**304**], two of the authors of this book [**306**] argued that the best possible candidate for the requested common abstraction is the *join* of Boolean algebras and ℓ-groups in the lattice of subvarieties of $\mathcal{PRL}$. They provided an equational basis for this variety and investigated some of its properties. The same paper also contains a more detailed historical account of this problem.

1.7. Bibliographical remarks

Good primers on order and residuation, covering the same ground as the first three sections of this chapter but in greater detail, are [**157**, Chapter 0] and [**45**]. For the material on residuated maps, we refer the reader also to [**46**]. Proposition 1.1.10, a generalization of a result of Berman [**26**], is taken from [**16**], while Lemmas 1.2.5 and 1.2.8 were first proved by Schmidt [**335**].

The concept of a closure operator on a power set lattice has its antecedents in the work of Moore [**276**]; the abstract version presented here was introduced by Ward [**366**]. Proposition 1.3.3 was proved by Ore [**302**] and Proposition 1.3.8 is a celebrated result of Birkhoff [**31**].

The presentation of residuated lattices in this chapter draws substantially from the papers [**44, 180, 206, 260**]; in particular, see [**206**] for the equational basis provided by Proposition 1.4.6. References for specific subvarieties of $\mathcal{RL}$ will be given in later chapters, but for involutive, idempotent, conic, and integrally closed residuated lattices, which will not be studied in great detail, we refer the reader already to [**149, 303, 371**], [**150, 161, 321**], [**68, 197**], and [**160, 313**], respectively.

For the details of the term-equivalence between the variety of MV-algebras and the subvariety $\mathcal{MV}$ of $\mathcal{PRL}$ given in Example 1.4.12, we refer to [**73**, Section 4.5] or [**371**, Subsection 3.4.5]. A first explicit proof of the term-equivalence between the variety of ℓ-groups and the subvariety $\mathcal{LG}$ of $\mathcal{RL}$ presented in Example 1.4.13 was given in [**43, 44**]. Example 1.4.14 is taken from [**16**].

Lemma 1.5.2 can be found in [**329**, p. 31] or [**152**, Lemma 3.1]; we refer also to [**339**, Corollary 3.7] for an earlier related result in the setting of Brouwerian meet-semilattices. Example 1.5.7 has been presented in several papers; an early reference is, e.g., [**298**]. Proposition 1.5.9 belongs to the folklore of the subject and generalizes the classical Glivenko-Frink theorem for Brouwerian lattices ([**166**] and [**135**]; see also [**339**, Theorem 4.1] and [**329**, p. 142]).

Let us remark finally that a significant portion of the theory of residuated pomonoids transfers to the non-associative setting of *residuated* ℓ*-groupoids* (see [**76**, **147**]). Lattice-ordered groupoids with only one residual include term-equivalent versions of *orthomodular lattices* [**25**], studied in connection with mathematical foundations of quantum mechanics; see, e.g., [**64**] for a proof of this term-equivalence.

CHAPTER 2

Proof systems

The goal of this chapter is to develop the theory of residuated structures from a *proof-theoretic* perspective. A variety can be defined as a class of structures satisfying certain equations, but may also be presented via a proof system that generates the equations satisfied by all of its members. In certain cases, derivations in the system contain only elements that occur in their conclusions. Such systems not only provide an elegant and strikingly different perspective on a variety, they can also be useful, as will become evident in subsequent chapters, for establishing properties such as decidability, generation by subclasses, and amalgamation.

We focus on the close relationship between proof theory and algebra, motivating our design choices for proof systems by relevant algebraic properties. Historically, however, these systems developed largely independently of algebraic considerations, motivated rather by issues arising in linguistics, many-valued logics, set theory, philosophy, and computer science. We describe this development and the related literature in the historical excursus at the end of the chapter.

2.1. Rules and derivations

Let us begin with some general definitions. Since the objects of a proof system need not be equations — in particular, they can be formulas or more complex entities such as sequents or hypersequents — we assume only that they form a non-empty set A of *structures*. An *inference* for A is then an ordered pair $\langle\{a_1, \dots, a_n\}, a\rangle$ consisting of a finite (possibly empty) set of *premises* $\{a_1, \dots, a_n\} \subseteq A$ and a *conclusion* $a \in A$. Such an inference may also be written $a_1, \dots, a_n \,/\, a$ or

$$\frac{a_1 \ \dots \ a_n}{a.}$$

An inference $\emptyset \,/\, a$ with no premises, typically denoted just by a, is an *axiom* for A. A *rule* r for A is a set of inferences for A, referred to as *instances* of r.

A *proof system* C is an ordered pair $\langle A, \mathcal{R}\rangle$ consisting of a non-empty set A of structures and a set $\mathcal{R}$ of rules for A. We sometimes write $\mathsf{C} + \mathcal{R}'$ to denote the proof system $\langle A, \mathcal{R} \cup \mathcal{R}'\rangle$ obtained by adding a set of rules $\mathcal{R}'$ for A to C.

Intuitively, $a \in A$ "follows from" $X \subseteq A$ in C if, starting with members of X and axioms of C, instances of rules from $\mathcal{R}$ can be used to obtain a. More formally, a *finite tree labeled by a set A* is an ordered pair consisting of a finite tree $\langle T, \leq\rangle$ (see Example 1.1.6) and a function $\varphi\colon T \to A$, where each node $t \in T$ is said to be *labeled by* $\varphi(t) \in A$. A C-*derivation* d of $a \in A$ from $X \subseteq A$ is a finite tree $\langle T, \leq\rangle$ labeled by A such that

(A1)	$\alpha \to (\beta \to \alpha)$	(A9)	$(\alpha \wedge \beta) \to \alpha$
(A2)	$(\alpha \to (\beta \to \gamma)) \to ((\alpha \to \beta) \to (\alpha \to \gamma))$	(A10)	$(\alpha \wedge \beta) \to \beta$
(A3)	$\neg\neg\alpha \to \alpha$	(A11)	$\alpha \to (\alpha \vee \beta)$
(A4)	$(\alpha \to \beta) \to ((\alpha \to \neg\beta) \to \neg\alpha)$	(A12)	$\beta \to (\alpha \vee \beta)$
(A5)	$\alpha \to (\neg\alpha \to \beta)$	(A13)	$\neg 1 \to 0$
(A6)	$1 \to (\alpha \to \alpha)$	(A14)	$0 \to \neg 1$
(A7)	$(\alpha \to \beta) \to ((\alpha \to \gamma) \to (\alpha \to (\beta \wedge \gamma)))$	(A15)	$(\alpha \to \alpha) \to 1$
(A8)	$(\alpha \to \gamma) \to ((\beta \to \gamma) \to ((\alpha \vee \beta) \to \gamma))$		

$$\frac{\alpha \quad \alpha \to \beta}{\beta}\ (\text{MP})$$

FIGURE 2.1. The Hilbert system HCL

(i) the root of T is labeled by a;

(ii) each node of T labeled by b is either a leaf node with $b \in X$, or has exactly $n \in \mathbb{N}$ child nodes labeled by $b_1, \ldots, b_n$ such that $b_1, \ldots, b_n \,/\, b$ is an instance of a rule of C.

The *height* of d is the height of $\langle T, \leq \rangle$, denoted by $\mathrm{h}(d)$.

Note that the leaves of d are labeled either by members of X or by axioms of C. In particular, a C-derivation of $a \in A$ from $\emptyset$ is a finite tree where all the leaves are labeled by axioms of C and all other nodes are obtained from these axioms by applying rules of C; in this case, a is called a *theorem* of C.

If there exists a C-derivation d of $a \in A$ from $X \subseteq A$, then a is said to be C-*derivable from* X, denoted by $d; X \vdash_{\mathsf{C}} a$ or $X \vdash_{\mathsf{C}} a$. For $X = \emptyset$, we write just $d \vdash_{\mathsf{C}} a$ or $\vdash_{\mathsf{C}} a$. A rule r is called C-*derivable* if $\{a_1, \ldots, a_n\} \vdash_{\mathsf{C}} a$ for each instance $a_1, \ldots, a_n \,/\, a$ of r.

These definitions are very general. In practice, most proof systems, and indeed all those encountered in this book, deal with sets of structures built in some way from a set of *formulas* $\mathrm{Fm}_{\mathcal{L}}(X)$: the universe of a formula algebra $\mathbf{Fm}_{\mathcal{L}}(X)$ of signature $\mathcal{L}$ over a set of variables X. (For details of the construction of these algebras, see Appendix B.) If the signature is clear from the context, we omit the subscript and write $\mathbf{Fm}(X)$, or just $\mathbf{Fm}$ when X is a fixed countably infinite set. A *substitution* on $\mathbf{Fm}(X)$ is an endomorphism $\sigma \colon \mathbf{Fm}(X) \to \mathbf{Fm}(X)$.

We use (subscripted) symbols x, y, z and α, β, γ as metavariables to denote arbitrary elements of a set X and formulas in $\mathrm{Fm}(X)$, respectively. We also define the *complexity* of a formula $\alpha \in \mathrm{Fm}(X)$ as the total number of occurrences of operation symbols in α, denoted by $\mathrm{cp}(\alpha)$.

Proof systems where the set of structures is Fm for some signature are often referred to as *Hilbert systems*. Typically, such systems are presented schematically: each rule is defined via an inference with variables replaced by metavariables that denotes the set of inferences obtained by replacing the metavariables uniformly with arbitrary formulas. In practice, Hilbert systems tend to have many axiom schemas and few other rule schemas — often just one: the modus ponens rule defined schematically as $\alpha, \alpha \to \beta \,/\, \beta$. For such systems, it is common to display derivations as lists of formulas, indicating for each item the previous items from which it was obtained.

EXAMPLE 2.1.1. Figure 2.1 displays a Hilbert system HCL for classical propositional logic in a signature $\langle 2,2,2,1,0,0\rangle$ with operation symbols $\wedge$, $\vee$, $\to$, $\neg$, 0, and 1. Given any formulas α, β of this signature, the following HCL-derivation establishes $\vdash_{\mathsf{HCL}} (\alpha \wedge \beta) \to (\beta \wedge \alpha)$:

1.	$(\alpha \wedge \beta) \to \alpha$	(A9)
2.	$(\alpha \wedge \beta) \to \beta$	(A10)
3.	$((\alpha \wedge \beta) \to \beta) \to (((\alpha \wedge \beta) \to \alpha) \to ((\alpha \wedge \beta) \to (\beta \wedge \alpha)))$	(A7)
4.	$((\alpha \wedge \beta) \to \alpha) \to ((\alpha \wedge \beta) \to (\beta \wedge \alpha))$	(MP) 2, 3
5.	$(\alpha \wedge \beta) \to (\beta \wedge \alpha)$	(MP) 1, 4.

Hilbert systems are useful in many respects. They are usually easy to relate to a corresponding class of algebras, and provide a framework for establishing properties by induction on the height of a derivation. However, in cases where the structure of derivations plays a critical role, Hilbert systems may be found wanting. Critically, a derivation of a formula β in a proof system with the modus ponens rule may involve deriving an unrelated — i.e., containing different operation symbols or variables — formula α and the implication $\alpha \to \beta$. It therefore makes sense to seek proof systems that are *analytic* in the sense that a derivation of a structure is built in some way from the elements of that structure.

A first step beyond Hilbert systems may be taken by considering ordered pairs of formulas — denoting, e.g., equations or inequations — and then, more generally, ordered pairs of finite sequences of formulas called sequents, and finite multisets of sequents called hypersequents. We define here analytic systems based on these structures first for the variety of lattices and then for a broad range of varieties of residuated lattices.

First, however, we remind the reader of a very general but distinctly non-analytic approach to deriving equations from sets of equations. Let $\mathrm{Eq}_{\mathcal{L}}(X)$ denote the set of $\mathcal{L}$-equations (ordered pairs of $\mathcal{L}$-formulas) for a signature $\mathcal{L}$ over a set of variables X, writing $\mathrm{Eq}(X)$ when the signature is clear from the context, or just Eq when X is a fixed countably infinite set of variables. Members of $\mathrm{Eq}(X)$ are written either as ordered pairs $\langle \alpha, \beta\rangle$ or using the more customary notation $\alpha \approx \beta$.

Equational logic for a signature $\mathcal{L}$ can be described as a proof system EL consisting of the rules displayed in Figure 2.2, where σ stands for an arbitrary substitution on $\mathbf{Fm}$ and $\gamma[x/\alpha]$ denotes the formula obtained by replacing a specified occurrence of a variable x in γ with α. It can be proved that, for any $\Sigma \cup \{\alpha \approx \beta\} \subseteq \mathrm{Eq}$,

$$\Sigma \vdash_{\mathsf{EL}} \alpha \approx \beta \iff \Sigma \models \alpha \approx \beta,$$

where $\models$ denotes logical consequence in first-order logic.

2.2. A proof system for lattices

We consider now a simple but relevant case study. Recall from Proposition 1.1.3 that lattices can be defined in two ways: order-theoretically as non-empty posets $\langle L, \leq\rangle$ containing all meets and joins of pairs of elements from L, and algebraically as algebras $\langle L, \wedge, \vee\rangle$ with binary operation symbols $\wedge$ and $\vee$ satisfying a given set of equations. Let us denote the variety of lattices by $\mathcal{L}at$ and write $\vdash_{\mathcal{L}at} \alpha \approx \beta$ if the equation $\alpha \approx \beta$ is satisfied by all lattices.

$$\frac{}{\alpha \approx \alpha}\ (\text{reflexivity}) \qquad \frac{\beta \approx \alpha}{\alpha \approx \beta}\ (\text{symmetry}) \qquad \frac{\alpha \approx \beta \quad \beta \approx \gamma}{\alpha \approx \gamma}\ (\text{transitivity})$$

$$\frac{\alpha \approx \beta}{\sigma(\alpha) \approx \sigma(\beta)}\ (\text{substitution}) \qquad \frac{\alpha \approx \beta}{\gamma[x/\alpha] \approx \gamma[x/\beta]}\ (\text{replacement})$$

FIGURE 2.2. Equational logic EL

Our goal is to provide an *algorithmic* presentation of $\mathcal{L}at$: a proof system for determining when an equation is satisfied by all lattices. Rather than deal with equations directly, we consider inequations, writing $\alpha \leq \beta$ as an abbreviation for $\alpha \wedge \beta \approx \alpha$ and noting that $\vdash_{\mathcal{L}at} \alpha \approx \beta$ if and only if $\vdash_{\mathcal{L}at} \alpha \leq \beta$ and $\vdash_{\mathcal{L}at} \beta \leq \alpha$. To emphasize that the meaning of such expressions in proof systems arises solely from their role in derivations and changes as the rules change, we also replace $\leq$ with the more generic "sequent arrow" symbol $\Rightarrow$. A *unary $\mathcal{L}$-sequent* for a signature $\mathcal{L}$ is therefore an ordered pair $\langle \alpha, \beta \rangle$ where $\alpha, \beta \in \mathrm{Fm}$, written $\alpha \Rightarrow \beta$.

Assuming (without further mention in this section) that $\mathcal{L}$ is the signature $\langle 2, 2 \rangle$ of lattices with operation symbols $\wedge$ and $\vee$, our set of *identity axioms* consists of all unary sequents of the form $\alpha \Rightarrow \alpha$. We then choose rules for each operation symbol that introduce the symbol on either side of the sequent arrow, e.g., for $\wedge$,

$$\frac{\alpha_1 \Rightarrow \beta}{\alpha_1 \wedge \alpha_2 \Rightarrow \beta}\ (\wedge\Rightarrow)_1 \qquad \frac{\alpha_2 \Rightarrow \beta}{\alpha_1 \wedge \alpha_2 \Rightarrow \beta}\ (\wedge\Rightarrow)_2 \qquad \frac{\alpha \Rightarrow \beta_1 \quad \alpha \Rightarrow \beta_2}{\alpha \Rightarrow \beta_1 \wedge \beta_2.}\ (\Rightarrow\wedge)$$

Reading $\Rightarrow$ as $\leq$, these rules correspond to the quasi-equations

$$\begin{aligned} x_1 \leq y &\implies x_1 \wedge x_2 \leq y \\ x_2 \leq y &\implies x_1 \wedge x_2 \leq y \\ x \leq y_1 \ \&\ x \leq y_2 &\implies x \leq y_1 \wedge y_2. \end{aligned}$$

We also include a "cut rule" corresponding to the quasi-equation expressing the transitivity of $\leq$. The resulting proof system Lat is displayed in Figure 2.3.

EXAMPLE 2.2.1. Unary sequents corresponding to the commutativity of the meet operation can be derived in Lat as follows:

$$\dfrac{\dfrac{\dfrac{}{\beta \Rightarrow \beta}\,(\textsc{id})}{\alpha \wedge \beta \Rightarrow \beta}\,(\wedge\Rightarrow)_2 \qquad \dfrac{\dfrac{}{\alpha \Rightarrow \alpha}\,(\textsc{id})}{\alpha \wedge \beta \Rightarrow \alpha}\,(\wedge\Rightarrow)_1}{\alpha \wedge \beta \Rightarrow \beta \wedge \alpha.}\,(\Rightarrow\wedge)$$

Unary sequents corresponding to one half of the distributivity property for lattices can also be derived in Lat:

$$\dfrac{\dfrac{\dfrac{\dfrac{}{\alpha \Rightarrow \alpha}\,(\textsc{id})}{\alpha \Rightarrow \alpha \vee \beta}\,(\Rightarrow\vee)_1 \quad \dfrac{\dfrac{\dfrac{}{\beta \Rightarrow \beta}\,(\textsc{id})}{\beta \Rightarrow \alpha \vee \beta}\,(\Rightarrow\vee)_2}{\beta \wedge \gamma \Rightarrow \alpha \vee \beta}\,(\wedge\Rightarrow)_1}{\alpha \vee (\beta \wedge \gamma) \Rightarrow \alpha \vee \beta}\,(\vee\Rightarrow) \qquad \dfrac{\dfrac{\dfrac{}{\alpha \Rightarrow \alpha}\,(\textsc{id})}{\alpha \Rightarrow \alpha \vee \gamma}\,(\Rightarrow\vee)_1 \quad \dfrac{\dfrac{\dfrac{}{\gamma \Rightarrow \gamma}\,(\textsc{id})}{\gamma \Rightarrow \alpha \vee \gamma}\,(\Rightarrow\vee)_2}{\beta \wedge \gamma \Rightarrow \alpha \vee \gamma}\,(\wedge\Rightarrow)_2}{\alpha \vee (\beta \wedge \gamma) \Rightarrow \alpha \vee \gamma}\,(\vee\Rightarrow)}{\alpha \vee (\beta \wedge \gamma) \Rightarrow (\alpha \vee \beta) \wedge (\alpha \vee \gamma).}\,(\Rightarrow\wedge)$$

Lat is, as intended, a proof system for lattices in the sense that an inequation $\alpha \leq \beta$ is satisfied by all lattices if and only if the unary sequent $\alpha \Rightarrow \beta$ is derivable

Identity Axioms

$$\frac{}{\alpha \Rightarrow \alpha}\ (\text{ID})$$

Cut Rule

$$\frac{\alpha \Rightarrow \gamma \quad \gamma \Rightarrow \beta}{\alpha \Rightarrow \beta}\ (\text{CUT})$$

Left Operation Rules

$$\frac{\alpha_1 \Rightarrow \beta}{\alpha_1 \wedge \alpha_2 \Rightarrow \beta}\ (\wedge\Rightarrow)_1$$

$$\frac{\alpha_2 \Rightarrow \beta}{\alpha_1 \wedge \alpha_2 \Rightarrow \beta}\ (\wedge\Rightarrow)_2$$

$$\frac{\alpha_1 \Rightarrow \beta \quad \alpha_2 \Rightarrow \beta}{\alpha_1 \vee \alpha_2 \Rightarrow \beta}\ (\vee\Rightarrow)$$

Right Operation Rules

$$\frac{\alpha \Rightarrow \beta_1}{\alpha \Rightarrow \beta_1 \vee \beta_2}\ (\Rightarrow\vee)_1$$

$$\frac{\alpha \Rightarrow \beta_2}{\alpha \Rightarrow \beta_1 \vee \beta_2}\ (\Rightarrow\vee)_2$$

$$\frac{\alpha \Rightarrow \beta_1 \quad \alpha \Rightarrow \beta_2}{\alpha \Rightarrow \beta_1 \wedge \beta_2}\ (\Rightarrow\wedge)$$

FIGURE 2.3. The unary sequent calculus Lat

in Lat. The right-to-left direction is easy to check. Clearly, the identity axioms are satisfied by all lattices when we read $\Rightarrow$ as $\leq$, and it is not hard to check that if the premises of an instance of any rule of Lat are satisfied by all lattices, then so is the conclusion of that instance. Hence a simple induction on the height of a Lat-derivation establishes that $\vdash_{\mathsf{Lat}} \alpha \Rightarrow \beta$ implies $\vdash_{\mathcal{L}at} \alpha \leq \beta$, for any $\alpha, \beta \in \mathrm{Fm}$.

To establish the converse direction, we define a *Lindenbaum-Tarski* algebra (see the historical excursus for Chapter 3) based on derivability in Lat. Consider first the following binary relation Θ on the formula algebra **Fm**:

$$\langle \alpha, \beta \rangle \in \Theta \ :\Longleftrightarrow\ \vdash_{\mathsf{Lat}} \alpha \Rightarrow \beta \text{ and } \vdash_{\mathsf{Lat}} \beta \Rightarrow \alpha.$$

This relation is reflexive because of (ID), symmetric by definition, and transitive because of (CUT), and is therefore an equivalence relation on Fm. Moreover, it is compatible with the meet and join operations. Suppose that $\langle \alpha_1, \beta_1 \rangle \in \Theta$ and $\langle \alpha_2, \beta_2 \rangle \in \Theta$. Then, in particular, $\vdash_{\mathsf{Lat}} \alpha_1 \Rightarrow \beta_1$ and $\vdash_{\mathsf{Lat}} \alpha_2 \Rightarrow \beta_2$, and we obtain a Lat-derivation of $\alpha_1 \wedge \alpha_2 \Rightarrow \beta_1 \wedge \beta_2$ of the form

$$\dfrac{\dfrac{\dfrac{\vdots}{\alpha_1 \Rightarrow \beta_1}}{\alpha_1 \wedge \alpha_2 \Rightarrow \beta_1}\,(\wedge\Rightarrow)_1 \qquad \dfrac{\dfrac{\vdots}{\alpha_2 \Rightarrow \beta_2}}{\alpha_1 \wedge \alpha_2 \Rightarrow \beta_2}\,(\wedge\Rightarrow)_2}{\alpha_1 \wedge \alpha_2 \Rightarrow \beta_1 \wedge \beta_2.}\,(\Rightarrow\wedge)$$

Symmetrically, $\vdash_{\mathsf{Lat}} \beta_1 \wedge \beta_2 \Rightarrow \alpha_1 \wedge \alpha_2$, so $\langle \alpha_1 \wedge \alpha_2, \beta_1 \wedge \beta_2 \rangle \in \Theta$, and similar reasoning yields $\langle \alpha_1 \vee \alpha_2, \beta_1 \vee \beta_2 \rangle \in \Theta$.

Now let $\mathrm{Fm}/\Theta := \{[\alpha]_\Theta \mid \alpha \in \mathrm{Fm}\}$ be the set of equivalence classes of Θ. Then the quotient algebra $\mathbf{Fm}/\Theta = \langle \mathrm{Fm}/\Theta, \wedge^{\mathbf{Fm}/\Theta}, \vee^{\mathbf{Fm}/\Theta} \rangle$ with $[\alpha]_\Theta \wedge^{\mathbf{Fm}/\Theta} [\beta]_\Theta := [\alpha \wedge \beta]_\Theta$ and $[\alpha]_\Theta \vee^{\mathbf{Fm}/\Theta} [\beta]_\Theta := [\alpha \vee \beta]_\Theta$ is a lattice. (Indeed, it is the *free lattice* on countably infinitely many generators.) For example, $\wedge^{\mathbf{Fm}/\Theta}$ is commutative by virtue of the derivation of $\alpha \wedge \beta \Rightarrow \beta \wedge \alpha$ for arbitrary $\alpha, \beta \in \mathrm{Fm}$ given in Example 2.2.1. Observe also that in this lattice,

$$\begin{aligned}[\alpha]_\Theta \le [\beta]_\Theta &\iff [\alpha]_\Theta \wedge^{\mathbf{Fm}/\Theta} [\beta]_\Theta = [\alpha]_\Theta \\ &\iff [\alpha \wedge \beta]_\Theta = [\alpha]_\Theta \\ &\iff \langle \alpha \wedge \beta, \alpha \rangle \in \Theta \\ &\iff \vdash_{\mathsf{Lat}} \alpha \wedge \beta \Rightarrow \alpha \text{ and } \vdash_{\mathsf{Lat}} \alpha \Rightarrow \alpha \wedge \beta \\ &\iff \vdash_{\mathsf{Lat}} \alpha \Rightarrow \beta.\end{aligned}$$

Suppose finally that $\vdash_{\mathcal{L}at} \alpha \le \beta$. Then, in particular, $\pi_\Theta(\alpha) \le \pi_\Theta(\beta)$ in $\mathbf{Fm}/\Theta$, where $\pi_\Theta\colon \mathbf{Fm} \to \mathbf{Fm}/\Theta;\ \gamma \mapsto [\gamma]_\Theta$ is the canonical map. So $[\alpha]_\Theta \le [\beta]_\Theta$ in $\mathbf{Fm}/\Theta$, and, by the above equivalence, $\vdash_{\mathsf{Lat}} \alpha \Rightarrow \beta$. We have therefore established the following soundness and completeness result.

PROPOSITION 2.2.2. *A unary sequent $\alpha \Rightarrow \beta$ is derivable in* Lat *if and only if the inequation $\alpha \le \beta$ is satisfied by all lattices.*

Proof systems are usually designed to admit certain properties. Notice in particular that every rule of Lat except (CUT) has the *subformula property*: every formula occurring in the premises of an instance of such a rule occurs as a subformula of a formula in the conclusion. Derivations involving only these rules are much easier to investigate. Notably, it is easy to see that starting with any unary sequent and applying the rules of Lat except (CUT) backwards, we always terminate with inequations containing only variables.

It would be nice therefore to restrict our attention to "cut-free" derivations in Lat that make no use of the cut rule. This is possible so long as we are concerned only with inequations derivable from the empty set. For any proof system C containing a rule (CUT), let us define C° to be C without (CUT). We will say that C has *cut elimination* if whenever $\vdash_{\mathsf{C}} a$, also $\vdash_{\mathsf{C}^\circ} a$.[1]

PROPOSITION 2.2.3. Lat *has cut elimination.*

PROOF. It suffices to prove the following claim:

If $d_1 \vdash_{\mathsf{Lat}^\circ} \alpha \Rightarrow \gamma$ and $d_2 \vdash_{\mathsf{Lat}^\circ} \gamma \Rightarrow \beta$, then $\vdash_{\mathsf{Lat}^\circ} \alpha \Rightarrow \beta$.

We prove the claim by induction on $\mathrm{h}(d_1) + \mathrm{h}(d_2)$. For the base case, if $\mathrm{h}(d_1) = \mathrm{h}(d_2) = 1$, then both d_1 and d_2 consist of an instance of (ID), and $\alpha = \gamma = \beta$. Hence clearly $\vdash_{\mathsf{Lat}^\circ} \alpha \Rightarrow \beta$. For the inductive step there are two possibilities.

(a) d_1 or d_2 ends with an inference that decomposes α or β, respectively. Suppose, e.g., that $\alpha = \alpha_1 \vee \alpha_2$ and d_1 ends with

$$\dfrac{\dfrac{\vdots}{\alpha_1 \Rightarrow \gamma} \quad \dfrac{\vdots}{\alpha_2 \Rightarrow \gamma}}{\alpha_1 \vee \alpha_2 \Rightarrow \gamma.}\ (\vee\Rightarrow)$$

Let d_{11} and d_{12} be the derivations of $\alpha_1 \Rightarrow \gamma$ and $\alpha_2 \Rightarrow \gamma$, respectively. Since $\mathrm{h}(d_{11}) < \mathrm{h}(d_1)$ and $\mathrm{h}(d_{12}) < \mathrm{h}(d_1)$, applying the induction hypothesis twice yields $\vdash_{\mathsf{Lat}^\circ} \alpha_1 \Rightarrow \beta$ and $\vdash_{\mathsf{Lat}^\circ} \alpha_2 \Rightarrow \beta$. Hence, by an application of $(\vee\Rightarrow)$, also $\vdash_{\mathsf{Lat}^\circ} \alpha_1 \vee \alpha_2 \Rightarrow \beta$ as required. Other cases are very similar.

[1]This property might more accurately be called *cut admissibility*; in the field of proof theory, cut elimination for a calculus C requires an algorithm that given a derivation of a structure from $\emptyset$ in C, outputs a derivation of the structure from $\emptyset$ in C°.

(b) d_1 and d_2 both end with inferences that decompose γ. Suppose, e.g., that $\gamma = \gamma_1 \wedge \gamma_2$ and d_1 and d_2 end, for $i \in \{1, 2\}$, with

$$\frac{\dfrac{\vdots}{\alpha \Rightarrow \gamma_1} \quad \dfrac{\vdots}{\alpha \Rightarrow \gamma_2}}{\alpha \Rightarrow \gamma_1 \wedge \gamma_2}\,(\Rightarrow\wedge) \qquad \frac{\dfrac{\vdots}{\gamma_i \Rightarrow \beta}}{\gamma_1 \wedge \gamma_2 \Rightarrow \beta.}\,(\wedge\Rightarrow)_i$$

Let d_{1i} and d_{2i} be the derivations of $\alpha \Rightarrow \gamma_i$ and $\gamma_i \Rightarrow \beta$, respectively. Since $\mathrm{h}(d_{1i}) < \mathrm{h}(d_1)$ and $\mathrm{h}(d_{2i}) < \mathrm{h}(d_2)$, the induction hypothesis yields $\vdash_{\mathsf{Lat}^\circ} \alpha \Rightarrow \beta$ as required. The case $\gamma = \gamma_1 \vee \gamma_2$ is symmetrical. $\square$

A variety $\mathcal{V}$ is said to have a *decidable equational theory* if there exists an effective procedure for deciding if $\mathcal{V} \models \alpha \approx \beta$, for any $\alpha, \beta \in \mathrm{Fm}$. The following result is a consequence of Propositions 2.2.2 and 2.2.3 and the previously mentioned fact that proof search in the proof system Lat° is terminating.

COROLLARY 2.2.4. *The equational theory of lattices is decidable.*

We also obtain certain properties for lattices corresponding to the satisfaction of universal formulas in free lattices. For example, *Whitman's condition* states that

$$\vdash_{\mathcal{L}at} \alpha_1 \wedge \alpha_2 \leq \beta_1 \vee \beta_2 \implies \vdash_{\mathcal{L}at} \alpha_1 \wedge \alpha_2 \leq \beta_1 \text{ or } \vdash_{\mathcal{L}at} \alpha_1 \wedge \alpha_2 \leq \beta_2 \text{ or } \vdash_{\mathcal{L}at} \alpha_1 \leq \beta_1 \vee \beta_2 \text{ or } \vdash_{\mathcal{L}at} \alpha_2 \leq \beta_1 \vee \beta_2.$$

Since any Lat°-derivation of a unary sequent $\alpha_1 \wedge \alpha_2 \Rightarrow \beta_1 \vee \beta_2$ must end with an instance of one of the rules $(\wedge\Rightarrow)_1$, $(\wedge\Rightarrow)_2$, $(\Rightarrow\vee)_1$, or $(\Rightarrow\vee)_2$, there must also be a Lat°-derivation of $\alpha_1 \wedge \alpha_2 \Rightarrow \beta_1$, $\alpha_1 \wedge \alpha_2 \Rightarrow \beta_2$, $\alpha_1 \Rightarrow \beta_1 \vee \beta_2$, or $\alpha_2 \Rightarrow \beta_1 \vee \beta_2$.

2.3. The full Lambek calculus

Unary sequents provide a suitable framework for reasoning in the variety of lattices, but for residuated structures, more complex objects are required. Recall that a residuated pomonoid $\langle P, \cdot, \backslash, /, \mathrm{e}, \leq \rangle$ satisfies, for all $x, y, z \in P$,

$$y \leq x\backslash z \iff x \cdot y \leq z \iff x \leq z/y.$$

Reading from the center outwards, these equivalences suggest rules that remove residuals on the right of inequations by accumulating products on the left. Since $\cdot$ is associative, such products can be represented by finite sequences of formulas $\langle \alpha_1, \ldots, \alpha_n \rangle$, where the empty sequence represents e.

For any signature $\mathcal{L}$, let us denote finite (possibly empty) sequences of $\mathcal{L}$-formulas by metavariables Γ or Π and call such sequences of length at most 1, denoted by a metavariable Δ, *one-sequences*. We write Γ, Π for the concatenation of sequences and drop external brackets where convenient. An *$\mathcal{L}$-sequent* is an ordered pair, written $\Gamma \Rightarrow \Delta$, consisting of a finite sequence of $\mathcal{L}$-formulas Γ and a one-sequence of $\mathcal{L}$-formulas Δ.

Unless stated otherwise, let us assume for the rest of this chapter that we are working in the signature $\langle 2, 2, 2, 2, 0, 0 \rangle$ of pointed residuated lattices with binary operation symbols $\wedge$, $\vee$, $\cdot$, $\backslash$, $/$ and constant symbols e, f. Also, given $\star \in \{\wedge, \vee, \cdot\}$ and $\alpha_1, \ldots, \alpha_n \in \mathrm{Fm}$, we write $\alpha_1 \star \cdots \star \alpha_n$ to denote $\alpha_1 \star (\alpha_2 \star \cdots \star (\alpha_{n-1} \star \alpha_n) \cdots)$.

Reading the sequent arrow $\Rightarrow$ as $\leq$, and "," on the left as multiplication, one direction of the residuation property is expressed by the sequent rules

$$\frac{\alpha, \Gamma \Rightarrow \beta}{\Gamma \Rightarrow \alpha\backslash\beta}\ (\Rightarrow\backslash) \qquad \frac{\Gamma, \alpha \Rightarrow \beta}{\Gamma \Rightarrow \beta/\alpha.}\ (\Rightarrow/)$$

The intended interpretations of $\langle \alpha_1, \ldots, \alpha_n \rangle$ and $\langle\rangle$ on the left as $\alpha_1 \cdots \alpha_n$ and e, respectively, motivates the sequent rules

$$\frac{\Gamma_1, \alpha, \beta, \Gamma_2 \Rightarrow \Delta}{\Gamma_1, \alpha \cdot \beta, \Gamma_2 \Rightarrow \Delta}\ (\cdot\Rightarrow) \qquad \frac{\Gamma_1, \Gamma_2 \Rightarrow \Delta}{\Gamma_1, \mathrm{e}, \Gamma_2 \Rightarrow \Delta}\ (\mathrm{e}\Rightarrow) \qquad \frac{}{\Rightarrow \mathrm{e}.}\ (\Rightarrow\mathrm{e})$$

Similarly, the fact that every residuated lattice satisfies the quasi-equation

$$x_1 \leq y_1 \ \&\ x_2 \leq y_2 \implies x_1x_2 \leq y_1y_2$$

motivates the sequent rule

$$\frac{\Gamma_1 \Rightarrow \alpha \quad \Gamma_2 \Rightarrow \beta}{\Gamma_1, \Gamma_2 \Rightarrow \alpha \cdot \beta.}\ (\Rightarrow\cdot)$$

Sequent rules for the residuals on the left of the sequent arrow are a little harder to explain. Observe first that, since $x(x\backslash y) \leq y$ is satisfied by every residuated lattice, the same is true of the quasi-equation

$$z_2 \leq x \ \&\ y \leq z_1\backslash(w/z_3) \implies z_2(x\backslash y) \leq z_1\backslash(w/z_3),$$

and hence also, by residuation, the quasi-equation

$$z_2 \leq x \ \&\ z_1yz_3 \leq w \implies z_1z_2(x\backslash y)z_3 \leq w.$$

Similar reasoning for the other residual motivates the sequent rules

$$\frac{\Gamma_2 \Rightarrow \alpha \quad \Gamma_1, \beta, \Gamma_3 \Rightarrow \Delta}{\Gamma_1, \Gamma_2, \alpha\backslash\beta, \Gamma_3 \Rightarrow \Delta}\ (\backslash\Rightarrow) \qquad \frac{\Gamma_2 \Rightarrow \alpha \quad \Gamma_1, \beta, \Gamma_3 \Rightarrow \Delta}{\Gamma_1, \beta/\alpha, \Gamma_2, \Gamma_3 \Rightarrow \Delta.}\ (/\Rightarrow)$$

Transitivity is expressed by the cut rule for sequents

$$\frac{\Gamma_2 \Rightarrow \alpha \quad \Gamma_1, \alpha, \Gamma_3 \Rightarrow \Delta}{\Gamma_1, \Gamma_2, \Gamma_3 \Rightarrow \Delta.}\ (\textsc{cut})$$

These rules, together with identity axioms of the form $\alpha \Rightarrow \alpha$, form a sequent calculus for residuated pomonoids.

EXAMPLE 2.3.1. The following derivation for arbitrary formulas α, β, γ establishes that $\alpha\backslash(\beta/\gamma) \leq (\alpha\backslash\beta)/\gamma$ is satisfied by all residuated lattices:

$$\cfrac{\cfrac{\cfrac{\cfrac{}{\alpha \Rightarrow \alpha}\ (\textsc{id}) \quad \cfrac{\cfrac{}{\gamma \Rightarrow \gamma}\ (\textsc{id}) \quad \cfrac{}{\beta \Rightarrow \beta}\ (\textsc{id})}{\beta/\gamma, \gamma \Rightarrow \beta}\ (/\Rightarrow)}{\alpha, \alpha\backslash(\beta/\gamma), \gamma \Rightarrow \beta}\ (\backslash\Rightarrow)}{\alpha\backslash(\beta/\gamma), \gamma \Rightarrow \alpha\backslash\beta}\ (\Rightarrow\backslash)}{\alpha\backslash(\beta/\gamma) \Rightarrow (\alpha\backslash\beta)/\gamma.}\ (\Rightarrow/)$$

An interpretation for the empty sequence on the right side of sequents is provided by the following rules for the constant symbol f:

$$\frac{}{\mathrm{f} \Rightarrow}\ (\mathrm{f}\Rightarrow) \qquad \frac{\Gamma \Rightarrow}{\Gamma \Rightarrow \mathrm{f}.}\ (\Rightarrow\mathrm{f})$$

From an algebraic perspective, f has no defining properties for the class $\mathcal{RL}$ but can be used to define unary operations $x \mapsto x\backslash\mathrm{f}$ and $x \mapsto \mathrm{f}/x$. Notably, these operations

Identity Axioms

$$\frac{}{\alpha \Rightarrow \alpha}\ (\textsc{id})$$

Cut Rule

$$\frac{\Gamma_2 \Rightarrow \alpha \quad \Gamma_1, \alpha, \Gamma_3 \Rightarrow \Delta}{\Gamma_1, \Gamma_2, \Gamma_3 \Rightarrow \Delta}\ (\textsc{cut})$$

Left Operation Rules

$$\frac{\Gamma_1, \Gamma_2 \Rightarrow \Delta}{\Gamma_1, \mathrm{e}, \Gamma_2 \Rightarrow \Delta}\ (\mathrm{e}\Rightarrow)$$

$$\frac{}{\mathrm{f} \Rightarrow}\ (\mathrm{f}\Rightarrow)$$

$$\frac{\Gamma_2 \Rightarrow \alpha \quad \Gamma_1, \beta, \Gamma_3 \Rightarrow \Delta}{\Gamma_1, \beta/\alpha, \Gamma_2, \Gamma_3 \Rightarrow \Delta}\ (/\Rightarrow)$$

$$\frac{\Gamma_2 \Rightarrow \alpha \quad \Gamma_1, \beta, \Gamma_3 \Rightarrow \Delta}{\Gamma_1, \Gamma_2, \alpha\backslash\beta, \Gamma_3 \Rightarrow \Delta}\ (\backslash\Rightarrow)$$

$$\frac{\Gamma_1, \alpha, \beta, \Gamma_2 \Rightarrow \Delta}{\Gamma_1, \alpha\cdot\beta, \Gamma_2 \Rightarrow \Delta}\ (\cdot\Rightarrow)$$

$$\frac{\Gamma_1, \alpha, \Gamma_2 \Rightarrow \Delta}{\Gamma_1, \alpha\wedge\beta, \Gamma_2 \Rightarrow \Delta}\ (\wedge\Rightarrow)_1$$

$$\frac{\Gamma_1, \beta, \Gamma_2 \Rightarrow \Delta}{\Gamma_1, \alpha\wedge\beta, \Gamma_2 \Rightarrow \Delta}\ (\wedge\Rightarrow)_2$$

$$\frac{\Gamma_1, \alpha, \Gamma_2 \Rightarrow \Delta \quad \Gamma_1, \beta, \Gamma_2 \Rightarrow \Delta}{\Gamma_1, \alpha\vee\beta, \Gamma_2 \Rightarrow \Delta}\ (\vee\Rightarrow)$$

Right Operation Rules

$$\frac{}{\Rightarrow \mathrm{e}}\ (\Rightarrow\mathrm{e})$$

$$\frac{\Gamma \Rightarrow}{\Gamma \Rightarrow \mathrm{f}}\ (\Rightarrow\mathrm{f})$$

$$\frac{\Gamma, \alpha \Rightarrow \beta}{\Gamma \Rightarrow \beta/\alpha}\ (\Rightarrow/)$$

$$\frac{\alpha, \Gamma \Rightarrow \beta}{\Gamma \Rightarrow \alpha\backslash\beta}\ (\Rightarrow\backslash)$$

$$\frac{\Gamma_1 \Rightarrow \alpha \quad \Gamma_2 \Rightarrow \beta}{\Gamma_1, \Gamma_2 \Rightarrow \alpha\cdot\beta}\ (\Rightarrow\cdot)$$

$$\frac{\Gamma \Rightarrow \alpha}{\Gamma \Rightarrow \alpha\vee\beta}\ (\Rightarrow\vee)_1$$

$$\frac{\Gamma \Rightarrow \beta}{\Gamma \Rightarrow \alpha\vee\beta}\ (\Rightarrow\vee)_2$$

$$\frac{\Gamma \Rightarrow \alpha \quad \Gamma \Rightarrow \beta}{\Gamma \Rightarrow \alpha\wedge\beta}\ (\Rightarrow\wedge)$$

FIGURE 2.4. The full Lambek calculus FL

coincide with the negation $\neg$ in Boolean algebras and the inverse $^{-1}$ in ℓ-groups. In general, they are not involutions, but possess some properties of inverses; e.g., for an arbitrary formula α, we obtain derivations

$$\frac{\dfrac{\dfrac{}{\alpha \Rightarrow \alpha}\ (\textsc{id}) \quad \dfrac{}{\mathrm{f} \Rightarrow \mathrm{f}}\ (\textsc{id})}{\alpha, \alpha\backslash\mathrm{f} \Rightarrow \mathrm{f}}\ (\backslash\Rightarrow)}{\alpha \Rightarrow \mathrm{f}/(\alpha\backslash\mathrm{f})}\ (\Rightarrow/) \qquad \frac{\dfrac{\dfrac{}{\alpha \Rightarrow \alpha}\ (\textsc{id}) \quad \dfrac{}{\mathrm{f} \Rightarrow \mathrm{f}}\ (\textsc{id})}{\mathrm{f}/\alpha, \alpha \Rightarrow \mathrm{f}}\ (/\Rightarrow)}{\alpha \Rightarrow (\mathrm{f}/\alpha)\backslash\mathrm{f}.}\ (\Rightarrow\backslash)$$

To obtain genuine involutions, we generally require multiple-conclusion sequents where both the left and right sides of sequents consist of finite sequences of formulas. Multiple-conclusion sequent calculi are discussed in the historical excursus at the end of this chapter, but, for clarity of exposition, will not play a central role in the rest of the book.

Combining all of these rules with sequent versions of the rules of Lat for the meet and join operations, we obtain the *full Lambek calculus* FL presented in Figure 2.4. The sequent calculus RL is obtained by removing the rules for f.

EXAMPLE 2.3.2. Sequents corresponding to distributivity of products over binary joins are derivable in FL for arbitrary formulas α, β, γ as follows:

$$\dfrac{\dfrac{\dfrac{\dfrac{\dfrac{}{\alpha\Rightarrow\alpha}\,(\textsc{id}) \quad \dfrac{}{\beta\Rightarrow\beta}\,(\textsc{id})}{\alpha,\beta\Rightarrow\alpha\beta}\,(\Rightarrow\cdot)}{\alpha,\beta\Rightarrow\alpha\beta\vee\alpha\gamma}\,(\Rightarrow\vee)_1 \quad \dfrac{\dfrac{\dfrac{}{\alpha\Rightarrow\alpha}\,(\textsc{id}) \quad \dfrac{}{\gamma\Rightarrow\gamma}\,(\textsc{id})}{\alpha,\gamma\Rightarrow\alpha\gamma}\,(\Rightarrow\cdot)}{\alpha,\gamma\Rightarrow\alpha\beta\vee\alpha\gamma}\,(\Rightarrow\vee)_2}{\alpha,\beta\vee\gamma\Rightarrow\alpha\beta\vee\alpha\gamma}\,(\vee\Rightarrow)}{\alpha(\beta\vee\gamma)\Rightarrow\alpha\beta\vee\alpha\gamma}\,(\cdot\Rightarrow)$$

$$\dfrac{\dfrac{\dfrac{\dfrac{}{\alpha\Rightarrow\alpha}\,(\textsc{id}) \quad \dfrac{\dfrac{}{\beta\Rightarrow\beta}\,(\textsc{id})}{\beta\Rightarrow\beta\vee\gamma}\,(\Rightarrow\vee)_1}{\alpha,\beta\Rightarrow\alpha(\beta\vee\gamma)}\,(\Rightarrow\cdot)}{\alpha\beta\Rightarrow\alpha(\beta\vee\gamma)}\,(\cdot\Rightarrow) \quad \dfrac{\dfrac{\dfrac{}{\alpha\Rightarrow\alpha}\,(\textsc{id}) \quad \dfrac{\dfrac{}{\gamma\Rightarrow\gamma}\,(\textsc{id})}{\gamma\Rightarrow\beta\vee\gamma}\,(\Rightarrow\vee)_2}{\alpha,\gamma\Rightarrow\alpha(\beta\vee\gamma)}\,(\Rightarrow\cdot)}{\alpha\gamma\Rightarrow\alpha(\beta\vee\gamma)}\,(\cdot\Rightarrow)}{\alpha\beta\vee\alpha\gamma\Rightarrow\alpha(\beta\vee\gamma).}\,(\vee\Rightarrow)$$

Despite the name, FL is not as "full" as we might like. In particular, there also exist notable cases (such as fragments of linear logic) where two constant symbols are not enough. To deal with bounded structures, we can add constant symbols $\bot$ and $\top$ to the signature and introduce the sequent rules

$$\dfrac{}{\Gamma_1,\bot,\Gamma_2\Rightarrow\Delta}\,(\bot\Rightarrow) \qquad \dfrac{}{\Gamma\Rightarrow\top.}\,(\Rightarrow\top)$$

In sequent calculi for varieties of integral bounded residuated lattices, $\bot$ and $\top$ are inter-derivable with f and e, respectively. On the other hand, in varieties related to groups, e and f are often identified. We therefore restrict our attention in this book to FL and pointed residuated lattices, leaving implicit the fact that similar results may be obtained for calculi and algebras with different sets of constant symbols.

We now state the main theorem relating derivability in FL to satisfaction in pointed residuated lattices, noting that in the next chapter we will establish an even closer relationship between these two notions. For any sequent $\Gamma\Rightarrow\Delta$, let

$$\prod\Gamma := \begin{cases}\alpha_1\cdots\alpha_n & \text{if } \Gamma=\langle\alpha_1,\dots,\alpha_n\rangle;\\ \mathrm{e} & \text{if } \Gamma=\langle\rangle;\end{cases} \quad \text{and} \quad \sum\Delta := \begin{cases}\alpha & \text{if } \Delta=\langle\alpha\rangle;\\ \mathrm{f} & \text{if } \Delta=\langle\rangle.\end{cases}$$

THEOREM 2.3.3. *A sequent $\Gamma\Rightarrow\Delta$ is derivable in* FL *if and only if $\prod\Gamma\le\sum\Delta$ is satisfied by all pointed residuated lattices.*

PROOF. The left-to-right direction is established by a straightforward induction on the height of a derivation of a sequent $\Gamma\Rightarrow\Delta$ in FL, noting that, by design, whenever the premises of an instance of a rule of FL are satisfied by all pointed residuated lattices, so is the conclusion. The right-to-left direction proceeds, as in the lattice setting, by constructing a Lindenbaum-Tarski algebra for FL. That is, we define a binary relation Θ on Fm by

$$\langle\alpha,\beta\rangle\in\Theta \;:\Longleftrightarrow\; \vdash_{\mathsf{FL}}\alpha\Rightarrow\beta \text{ and } \vdash_{\mathsf{FL}}\beta\Rightarrow\alpha.$$

The relation is reflexive because of (ID), symmetric by definition, and transitive because of (CUT), and hence an equivalence relation on Fm. To show that it is compatible with the operations of pointed residuated lattices requires suitable derivations in FL. For example, the claim that $\langle\alpha_1, \beta_1\rangle \in \Theta$ and $\langle\alpha_2, \beta_2\rangle \in \Theta$ implies $\langle\alpha_1\backslash\alpha_2, \beta_1\backslash\beta_2\rangle \in \Theta$ is justified by the construction of derivations

$$\dfrac{\dfrac{\dfrac{\vdots}{\beta_1 \Rightarrow \alpha_1} \quad \dfrac{\vdots}{\alpha_2 \Rightarrow \beta_2}}{\beta_1, \alpha_1\backslash\alpha_2 \Rightarrow \beta_2}\,(\backslash\Rightarrow)}{\alpha_1\backslash\alpha_2 \Rightarrow \beta_1\backslash\beta_2}\,(\Rightarrow\backslash) \qquad \dfrac{\dfrac{\dfrac{\vdots}{\alpha_1 \Rightarrow \beta_1} \quad \dfrac{\vdots}{\beta_2 \Rightarrow \alpha_2}}{\alpha_1, \beta_1\backslash\beta_2 \Rightarrow \alpha_2}\,(\backslash\Rightarrow)}{\beta_1\backslash\beta_2 \Rightarrow \alpha_1\backslash\alpha_2.}\,(\Rightarrow\backslash)$$

We are then able to show — again by providing suitable derivations in FL — that the quotient algebra $\mathbf{Fm}/\Theta$ is a pointed residuated lattice. In particular, it follows using Example 2.3.2 that the equation (RL1) in Proposition 1.4.6 is satisfied by $\mathbf{Fm}/\Theta$, and the following derivation for arbitrary formulas α, β, γ establishes that this algebra also satisfies the equation (RL3):

$$\dfrac{\dfrac{\dfrac{\dfrac{}{\alpha \Rightarrow \alpha}\,(\textsc{id}) \quad \dfrac{}{\beta \Rightarrow \beta}\,(\textsc{id})}{\alpha, \alpha\backslash\beta \Rightarrow \beta}\,(\backslash\Rightarrow)}{\alpha, \alpha\backslash\beta \Rightarrow \beta \vee \gamma}\,(\Rightarrow\vee)_1}{\alpha\backslash\beta \Rightarrow \alpha\backslash(\beta \vee \gamma).}\,(\Rightarrow\backslash)$$

Moreover, reasoning exactly as for the calculus Lat, it follows that

$$[\alpha]_\Theta \le [\beta]_\Theta \iff \vdash_{\mathsf{FL}} \alpha \Rightarrow \beta.$$

Suppose finally that $\prod\Gamma \le \sum\Delta$ is satisfied by all pointed residuated lattices. Then $\pi_\Theta(\prod\Gamma) \le \pi_\Theta(\sum\Delta)$ in $\mathbf{Fm}/\Theta$, where $\pi_\Theta \colon \mathbf{Fm} \to \mathbf{Fm}/\Theta;\ \gamma \mapsto [\gamma]_\Theta$ is the canonical map. Hence $[\prod\Gamma]_\Theta \le [\sum\Delta]_\Theta$ in $\mathbf{Fm}/\Theta$, and, by the above equivalence, $\prod\Gamma \Rightarrow \sum\Delta$ is derivable in FL. But also $\Gamma \Rightarrow \prod\Gamma$ and $\sum\Delta \Rightarrow \Delta$ are derivable in FL using $(\Rightarrow\cdot)$, $(\Rightarrow\mathrm{e})$, $(\mathrm{f}\Rightarrow)$, and $(\textsc{id})$, and hence, using $(\textsc{cut})$, so is $\Gamma \Rightarrow \Delta$. $\square$

The real virtue of the Full Lambek calculus, and the source of its most important applications, is the fact that any sequent that is derivable in this system is derivable without using the cut rule. Below we give a constructive proof of this fact, following almost exactly the same steps as in the case of Lat, except that now — and this is a cause for concern in such proofs — there are many more cases to consider.

THEOREM 2.3.4. FL *has cut elimination.*

PROOF. It suffices to prove the following claim:

If $d_1 \vdash_{\mathsf{FL}^\circ} \Gamma_2 \Rightarrow \alpha$ and $d_2 \vdash_{\mathsf{FL}^\circ} \Gamma_1, \alpha, \Gamma_3 \Rightarrow \Delta$, then $\vdash_{\mathsf{FL}^\circ} \Gamma_1, \Gamma_2, \Gamma_3 \Rightarrow \Delta$.

We prove the claim by induction on the ordered pair $\langle \mathrm{cp}(\alpha), \mathrm{h}(d_1) + \mathrm{h}(d_2)\rangle \in \mathbb{N}^2$ in the lexicographic product of two copies of $\langle \mathbb{N}, \le\rangle$ (see Example 1.1.1).

For the base case, suppose that $\mathrm{h}(d_1) = \mathrm{h}(d_2) = 1$. Then either d_1 or d_2 must consist of an instance $\alpha \Rightarrow \alpha$ of $(\textsc{id})$, and the claim follows immediately.

For the inductive step, there are two possibilities.

(a) Suppose that one of the derivations d_1 or d_2 ends with an instance of a rule that decomposes a formula in Γ_1, Γ_2, Γ_3, or Δ. There are a large number of cases to check. Consider first the case where Δ is $\gamma_1\backslash\gamma_2$ and d_2 ends with

$$\dfrac{\dfrac{\vdots}{\gamma_1, \Gamma_1, \alpha, \Gamma_3 \Rightarrow \gamma_2}}{\Gamma_1, \alpha, \Gamma_3 \Rightarrow \gamma_1\backslash\gamma_2.}\ (\Rightarrow\backslash)$$

Let d_2' be the FL°-derivation of $\gamma_1, \Gamma_1, \alpha, \Gamma_3 \Rightarrow \gamma_2$. Since $\mathrm{h}(d_2') < \mathrm{h}(d_2)$, by the induction hypothesis, $\vdash_{\mathsf{FL}^\circ} \gamma_1, \Gamma_1, \Gamma_2, \Gamma_3 \Rightarrow \gamma_2$, and an application of $(\Rightarrow\backslash)$ yields $\vdash_{\mathsf{FL}^\circ} \Gamma_1, \Gamma_2, \Gamma_3 \Rightarrow \gamma_1\backslash\gamma_2$ as required.

Now consider the case where Γ_1 is $\Gamma_{11}, \Gamma_{12}, \gamma_1\backslash\gamma_2, \Gamma_{13}$ and d_2 ends with

$$\dfrac{\dfrac{\vdots}{\Gamma_{12} \Rightarrow \gamma_1} \quad \dfrac{\vdots}{\Gamma_{11}, \gamma_2, \Gamma_{13}, \alpha, \Gamma_3 \Rightarrow \Delta}}{\Gamma_{11}, \Gamma_{12}, \gamma_1\backslash\gamma_2, \Gamma_{13}, \alpha, \Gamma_3 \Rightarrow \Delta.}\ (\backslash\Rightarrow)$$

Let d_2' be the FL°-derivation of $\Gamma_{11}, \gamma_2, \Gamma_{12}, \alpha, \Gamma_3 \Rightarrow \Delta$. Since $\mathrm{h}(d_2') < \mathrm{h}(d_2)$, by the induction hypothesis, $\vdash_{\mathsf{FL}^\circ} \Gamma_{11}, \gamma_2, \Gamma_{12}, \Gamma_2, \Gamma_3 \Rightarrow \Delta$, and an application of $(\backslash\Rightarrow)$ yields $\vdash_{\mathsf{FL}^\circ} \Gamma_{11}, \gamma_1\backslash\gamma_2, \Gamma_{12}, \Gamma_2, \Gamma_3 \Rightarrow \Delta$ as required.

We have only begun to scratch the surface of possible cases, but the same pattern of reasoning can be repeated. Consider an instance of any operation rule of FL. If the formula α is not decomposed in this instance and occurs on the left, then it can be uniformly replaced by Γ_2 and we obtain another instance of the same rule. Similarly, if α is not decomposed and occurs on the right, then by replacing α with Δ and placing Γ_1 and Γ_3 appropriately on the left, we obtain another instance of the rule.

(b) Suppose that both d_1 and d_2 end with instances of rules that decompose α. Consider first the case where α is $\alpha_1\backslash\alpha_2$ and Γ_1 is Γ_{11}, Γ_{12}, and d_1 and d_2 end with

$$\dfrac{\dfrac{\vdots}{\alpha_1, \Gamma_2 \Rightarrow \alpha_2}}{\Gamma_2 \Rightarrow \alpha_1\backslash\alpha_2}\ (\Rightarrow\backslash) \qquad \dfrac{\dfrac{\vdots}{\Gamma_{12} \Rightarrow \alpha_1} \quad \dfrac{\vdots}{\Gamma_{11}, \alpha_2, \Gamma_3 \Rightarrow \Delta}}{\Gamma_{11}, \Gamma_{12}, \alpha_1\backslash\alpha_2, \Gamma_3 \Rightarrow \Delta.}\ (\backslash\Rightarrow)$$

Let d_1' be the FL°-derivation ending with $\alpha_1, \Gamma_2 \Rightarrow \alpha_2$ and let d_{21} and d_{22} be the FL°-derivations ending with $\Gamma_{12} \Rightarrow \alpha_1$ and $\Gamma_{11}, \alpha_2, \Gamma_3 \Rightarrow \Delta$, respectively. Since $\mathrm{cp}(\alpha_1) < \mathrm{cp}(\alpha_1\backslash\alpha_2)$, applying the induction hypothesis to d_{21} and d_1' yields an FL°-derivation d_3 of $\Gamma_{12}, \Gamma_2 \Rightarrow \alpha_2$. But also $\mathrm{cp}(\alpha_2) < \mathrm{cp}(\alpha_1\backslash\alpha_2)$, so a further application of the induction hypothesis, this time to d_3 and d_{22}, yields $\vdash_{\mathsf{FL}^\circ} \Gamma_{11}, \Gamma_{12}, \Gamma_2, \Gamma_3 \Rightarrow \Delta$ as required.

Now suppose that α is $\alpha_1 \vee \alpha_2$ and d_1 and d_2 end with (for $i \in \{1, 2\}$)

$$\dfrac{\dfrac{\vdots}{\Gamma_2 \Rightarrow \alpha_i}}{\Gamma_2 \Rightarrow \alpha_1 \vee \alpha_2}\ (\Rightarrow\vee)_i \qquad \dfrac{\dfrac{\vdots}{\Gamma_1, \alpha_1, \Gamma_3 \Rightarrow \Delta} \quad \dfrac{\vdots}{\Gamma_1, \alpha_2, \Gamma_3 \Rightarrow \Delta}}{\Gamma_1, \alpha_1 \vee \alpha_2, \Gamma_3 \Rightarrow \Delta.}\ (\vee\Rightarrow)$$

Let d_1' and d_{2i} be the FL°-derivations ending with $\Gamma_2 \Rightarrow \alpha_i$ and $\Gamma_1, \alpha_i, \Gamma_3 \Rightarrow \Delta$, respectively. Since $\mathrm{cp}(\alpha_i) < \mathrm{cp}(\alpha_1 \vee \alpha_2)$, the induction hypothesis applied to d_1' and d_{2i} yields $\vdash_{\mathsf{FL}^\circ} \Gamma_1, \Gamma_2, \Gamma_3 \Rightarrow \Delta$ as required.

Of course, there are again many more cases to consider, depending on the form of α, but these follow a common pattern and are left as instructive exercises for the reader. □

The decidability of the equational theory of pointed residuated lattices is a direct consequence of cut elimination for FL.[2] Combining Theorems 2.3.3 and 2.3.4, it follows that $\alpha \leq \beta$ is satisfied by all pointed residuated lattices if and only if the sequent $\alpha \Rightarrow \beta$ is derivable in the cut-free calculus FL°. To search for a possible FL°-derivation, we try to match the sequent with the conclusion of an instance of one of the rules, a well-defined and finite procedure. If there is no such match, then we stop and the sequent is not derivable. If there is at least one match, then we choose one of the finitely many options and continue with the corresponding instances of the premises. Since the premises always contain fewer occurrences of operation symbols than the conclusion, this process, which involves backtracking to consider alternative matches, terminates.

COROLLARY 2.3.5. *The equational theory of pointed residuated lattices is decidable.*

Cut elimination for FL also allows us to restrict derivations to sequents based on a subsignature $\mathcal{L}'$ without affecting derivability; that is, an $\mathcal{L}'$-sequent is FL-derivable if and only if it is derivable in the proof system FL restricted to $\mathcal{L}'$-sequents. This observation can be used to obtain conservative extension results. For example, considering the system RL restricted to sequents with a non-empty right side, it follows that a residuated lattice equation is satisfied by all residuated lattices if and only if it is satisfied by all pointed residuated lattices. Similarly, considering RL restricted to unary sequents and lattice formulas, it follows that a lattice equation is satisfied by all lattices if and only if it is satisfied by all residuated lattices.

2.4. Adding structural rules

Just as a variety of pointed residuated lattices is defined relative to $\mathcal{PRL}$ by a set of additional equations, so a proof system for such a variety can be defined by adding suitable rules to FL. Indeed, there is an easy way to achieve this: just add sequents corresponding to the additional equations as axioms. That is, for a variety defined relative to $\mathcal{PRL}$ by a set of equations E, extend FL with axioms $\sigma(\alpha) \Rightarrow \sigma(\beta)$ and $\sigma(\beta) \Rightarrow \sigma(\alpha)$, for each $\alpha \approx \beta \in E$ and substitution σ on $\mathbf{Fm}$. In general, however, this new system will not have cut elimination, and we therefore take a different approach. Equations of a suitably "simple" form are converted into *structural rules* that manipulate the structure of sequents without referring to particular formulas.

Let us first consider some fundamental examples.

- The *exchange rule* corresponds to the commutativity law $x \cdot y \approx y \cdot x$ and permits the swapping of (sequences of) formulas in sequents:

$$\frac{\Gamma_1, \Pi_1, \Pi_2, \Gamma_2 \Rightarrow \Delta}{\Gamma_1, \Pi_2, \Pi_1, \Gamma_2 \Rightarrow \Delta.}\ (\text{EL})$$

[2]Decidability of the equational theory of a class of algebras does not always follow directly from the existence of a cut-free sequent calculus for the class, however; perhaps more work is required to obtain a decision procedure or the calculus does not support such a procedure.

We can use this rule to establish the inter-derivability of the residual operations:

$$\dfrac{\dfrac{\dfrac{\dfrac{}{\alpha\Rightarrow\alpha}\,(\textsc{id}) \qquad \dfrac{}{\beta\Rightarrow\beta}\,(\textsc{id})}{\alpha,\alpha\backslash\beta\Rightarrow\beta}\,(\backslash\Rightarrow)}{\alpha\backslash\beta,\alpha\Rightarrow\beta}\,(\textsc{el})}{\alpha\backslash\beta\Rightarrow\beta/\alpha}\,(\Rightarrow/) \qquad \dfrac{\dfrac{\dfrac{\dfrac{}{\alpha\Rightarrow\alpha}\,(\textsc{id}) \qquad \dfrac{}{\beta\Rightarrow\beta}\,(\textsc{id})}{\beta/\alpha,\alpha\Rightarrow\beta}\,(/\Rightarrow)}{\alpha,\beta/\alpha\Rightarrow\beta}\,(\textsc{el})}{\beta/\alpha\Rightarrow\alpha\backslash\beta.}\,(\Rightarrow\backslash)$$

In the presence of the exchange rule, we therefore use just the left residual and write $\alpha\to\beta$ instead of $\alpha\backslash\beta$.

- *Weakening rules* add finite sequences of formulas on the left or right of a sequent:

$$\dfrac{\Gamma_1,\Gamma_2\Rightarrow\Delta}{\Gamma_1,\Pi,\Gamma_2\Rightarrow\Delta}\,(\textsc{wl}) \qquad \dfrac{\Pi\Rightarrow}{\Gamma_1,\Pi,\Gamma_2\Rightarrow\Delta.}\,(\textsc{wr})$$

In the presence of these rules, e and f play the same role as $\top$ and $\bot$; that is, the following derivations become available:

$$\dfrac{\dfrac{}{\Rightarrow\mathrm{e}}\,(\Rightarrow\mathrm{e})}{\Gamma\Rightarrow\mathrm{e}}\,(\textsc{wl}) \qquad \dfrac{\dfrac{}{\mathrm{f}\Rightarrow}\,(\mathrm{f}\Rightarrow)}{\Gamma_1,\mathrm{f},\Gamma_2\Rightarrow\Delta.}\,(\textsc{wr})$$

Other forms of weakening include the *mix rule*[3] and *empty axiom*

$$\dfrac{\Pi\Rightarrow \qquad \Gamma_1,\Gamma_2\Rightarrow\Delta}{\Gamma_1,\Pi,\Gamma_2\Rightarrow\Delta}\,(\textsc{mix}) \qquad \dfrac{}{\Rightarrow}\,,(\textsc{empty})$$

which can be used to relate the constant symbols e and f as follows:

$$\dfrac{\dfrac{}{\mathrm{f}\Rightarrow}\,(\mathrm{f}\Rightarrow) \qquad \dfrac{}{\Rightarrow\mathrm{e}}\,(\Rightarrow\mathrm{e})}{\mathrm{f}\Rightarrow\mathrm{e}}\,(\textsc{mix}) \qquad \dfrac{\dfrac{\dfrac{}{\Rightarrow}\,(\textsc{empty})}{\mathrm{e}\Rightarrow}\,(\mathrm{e}\Rightarrow)}{\mathrm{e}\Rightarrow\mathrm{f}.}\,(\Rightarrow\mathrm{f})$$

- The *contraction rule* combines consecutive copies of finite sequences of formulas:

$$\dfrac{\Gamma_1,\Pi,\Pi,\Gamma_2\Rightarrow\Delta}{\Gamma_1,\Pi,\Gamma_2\Rightarrow\Delta.}\,(\textsc{cl})$$

It can be used to derive sequents corresponding to the "square-increasing" law $x\le x\cdot x$ as follows:

$$\dfrac{\dfrac{\dfrac{}{\alpha\Rightarrow\alpha}\,(\textsc{id}) \qquad \dfrac{}{\alpha\Rightarrow\alpha}\,(\textsc{id})}{\alpha,\alpha\Rightarrow\alpha\cdot\alpha}\,(\Rightarrow\cdot)}{\alpha\Rightarrow\alpha\cdot\alpha.}\,(\textsc{cl})$$

For the "square-decreasing" law $x\cdot x\le x$, we can add the *mingle rule*

$$\dfrac{\Gamma_1,\Pi_1,\Gamma_2\Rightarrow\Delta \quad \Gamma_1,\Pi_2,\Gamma_2\Rightarrow\Delta}{\Gamma_1,\Pi_1,\Pi_2,\Gamma_2\Rightarrow\Delta,}\,(\textsc{mingle})$$

obtaining derivations of the form

$$\dfrac{\dfrac{\dfrac{}{\alpha\Rightarrow\alpha}\,(\textsc{id}) \qquad \dfrac{}{\alpha\Rightarrow\alpha}\,(\textsc{id})}{\alpha,\alpha\Rightarrow\alpha}\,(\textsc{mingle})}{\alpha\cdot\alpha\Rightarrow\alpha.}\,(\cdot\Rightarrow)$$

[3] This rule is not to be confused with the "mix rule" used by Gentzen to prove cut elimination for sequent calculi for (first-order) classical and intuitionistic logics.

The exchange, weakening, and contraction rules have been sufficiently well-studied to generate a useful notation. Namely, FL_X for $X \subseteq \{\mathsf{e}, \mathsf{w}, \mathsf{c}\}$ denotes FL extended with the exchange rule for $\mathsf{e} \in X$, both weakening rules for $\mathsf{w} \in X$, and the contraction rule for $\mathsf{c} \in X$. As explained in the historical excursus, several of these systems are well-known in their own right. In particular, $\mathsf{FL}_{\mathsf{ewc}}$ is a calculus for intuitionistic propositional logic, while FL_{e} is also known as multiplicative additive intuitionistic linear logic without additive constants.

Many other structural rules have appeared in the literature with a wide range of motivations and applications. Typically, these rules, including those introduced above, have two key properties: they are structural and analytic. We will call a sequent rule *structural* if it can be presented schematically as

$$\frac{\Psi_1 \Rightarrow \Xi_1 \quad \dots \quad \Psi_n \Rightarrow \Xi_n}{\Psi_0 \Rightarrow \Xi_0,}$$

where each Ψ_i is a finite sequence of metavariables for finite sequences of formulas and each Ξ_i is empty or a metavariable for a one-sequence of formulas. We call the rule *analytic* if every metavariable occurring in a Ψ_i or Ξ_i occurs also in Ψ_0 or Ξ_0.

EXAMPLE 2.4.1. The following "cyclicity" rule is both analytic and structural:

$$\frac{\Pi, \Gamma \Rightarrow}{\Gamma, \Pi \Rightarrow .}\ (\textsc{cycle})$$

This rule permits the identification of the two "negations" $x\backslash \mathrm{f}$ and f/x, i.e., we obtain the following derivations for any formula α:

$$\dfrac{\dfrac{\dfrac{\dfrac{\dfrac{}{\alpha \Rightarrow \alpha}\,(\textsc{id}) \quad \dfrac{}{\mathrm{f} \Rightarrow}\,(\mathrm{f}\Rightarrow)}{\alpha, \alpha\backslash \mathrm{f} \Rightarrow}\,(\backslash\Rightarrow)}{\alpha\backslash\mathrm{f}, \alpha \Rightarrow}\,(\textsc{cycle})}{\alpha\backslash \mathrm{f},\alpha\Rightarrow \mathrm{f}}\,(\Rightarrow\mathrm{f})}{\alpha\backslash\mathrm{f}\Rightarrow \mathrm{f}/\alpha}\,(\Rightarrow/) \qquad \dfrac{\dfrac{\dfrac{\dfrac{\dfrac{}{\alpha \Rightarrow \alpha}\,(\textsc{id}) \quad \dfrac{}{\mathrm{f} \Rightarrow}\,(\mathrm{f}\Rightarrow)}{\mathrm{f}/\alpha, \alpha \Rightarrow}\,(/\Rightarrow)}{\alpha, \mathrm{f}/\alpha \Rightarrow}\,(\textsc{cycle})}{\alpha, \mathrm{f}/\alpha\Rightarrow \mathrm{f}}\,(\Rightarrow\mathrm{f})}{\mathrm{f}/\alpha\Rightarrow \alpha\backslash\mathrm{f}.}\,(\Rightarrow\backslash)$$

In the presence of (MIX) and (EMPTY), which identify the constant symbols e and f, this rule can be used to derive the e-cyclicity equation $x\backslash \mathrm{e} \approx \mathrm{e}/x$.

Unfortunately, not all analytic structural rules preserve cut elimination when added to FL. We therefore introduce a further natural restriction, calling an analytic structural sequent rule *simple* if it can be presented schematically as

$$\frac{\Gamma, \Psi_1, \Pi \Rightarrow \Delta \quad \dots \quad \Gamma, \Psi_n, \Pi \Rightarrow \Delta}{\Gamma, \Psi_0, \Pi \Rightarrow \Delta,}$$

where Γ and Π are metavariables for finite sequences of formulas, Δ is a metavariable for a one-sequence of formulas, Ψ_0 is a finite sequence of distinct metavariables for finite sequences of formulas (different to Γ, Π, and Δ), and $\Psi_1, \dots, \Psi_n$ are finite sequences of metavariables occurring in Ψ_0.

EXAMPLE 2.4.2. (EL), (WL), (CL), and (MINGLE) are simple structural rules, whereas (WR), (MIX), (EMPTY), and (CYCLE) are analytic but not simple since they contain instances with an empty sequence on the right; the latter four rules also preserve cut elimination when added to FL, but the characterization is a little more delicate and will be omitted here.

We will prove below that adding any set of simple structural rules to FL yields a proof system that has cut elimination. First, however, we describe the varieties of pointed residuated lattices that are characterized by these systems.

Let us call an equation *simple* if it has the form

$$\alpha \le \beta_1 \vee \cdots \vee \beta_n,$$

where each of $\alpha, \beta_1, \ldots, \beta_n$ is either e or a product of variables, α contains at most one occurrence of any variable, and all variables occurring in $\beta_1, \ldots, \beta_n$ occur in α.

The relationship between simple equations and simple rules is easy to describe. Let $\Psi(\mathrm{e})$ be the empty sequence and define $\Psi(x_1 \cdots x_m)$, for (not necessarily distinct) variables $x_1, \ldots, x_m$, to be the sequence of metavariables $\Gamma_{x_1}, \ldots, \Gamma_{x_m}$. Given a simple equation ε of the form $\alpha \le \beta_1 \vee \cdots \vee \beta_n$, the simple rule $r(\varepsilon)$ is defined schematically by

$$\frac{\Gamma, \Psi(\beta_1), \Pi \Rightarrow \Delta \quad \ldots \quad \Gamma, \Psi(\beta_n), \Pi \Rightarrow \Delta}{\Gamma, \Psi(\alpha), \Pi \Rightarrow \Delta.}$$

Clearly, any simple rule may be expressed as $r(\varepsilon)$ for some simple equation ε.

Observe that for any simple equation ε of the form $\alpha \le \beta_1 \vee \cdots \vee \beta_n$, the sequent $\alpha \Rightarrow \beta_1 \vee \cdots \vee \beta_n$ is derivable in $\mathsf{FL} + r(\varepsilon)$: reasoning backwards, apply $(\cdot \Rightarrow)$ repeatedly to α, followed by an application of $r(\varepsilon)$, and then apply $(\Rightarrow \vee)_1$ or $(\Rightarrow\vee)_2$ as appropriate in each branch, and finally $(\Rightarrow\cdot)$, $(\textsc{id})$, and $(\Rightarrow\mathrm{e})$. Moreover, every pointed residuated lattice satisfying ε, satisfies also the quasi-equation

$$x\beta_1 y \le z \;\&\; \cdots \;\&\; x\beta_n y \le z \implies x\alpha y \le z,$$

where x, y, z are variables not occurring in $\alpha, \beta_1, \ldots, \beta_n$.

Example 2.4.3. The simple equation $\varepsilon = x_1x_2 \le x_1^2 \vee x_2^2$ yields the simple rule

$$\frac{\Gamma, \Gamma_1, \Gamma_1, \Pi \Rightarrow \Delta \quad \Gamma, \Gamma_2, \Gamma_2, \Pi \Rightarrow \Delta}{\Gamma, \Gamma_1, \Gamma_2, \Pi \Rightarrow \Delta,}\; r(\varepsilon)$$

and we obtain a corresponding derivation

$$\dfrac{\dfrac{\dfrac{\dfrac{\dfrac{}{x_1 \Rightarrow x_1}(\textsc{id}) \quad \dfrac{}{x_1 \Rightarrow x_1}(\textsc{id})}{x_1, x_1 \Rightarrow x_1^2}(\Rightarrow\cdot)}{x_1, x_1 \Rightarrow x_1^2 \vee x_2^2}(\Rightarrow\vee)_1 \quad \dfrac{\dfrac{\dfrac{}{x_2 \Rightarrow x_2}(\textsc{id}) \quad \dfrac{}{x_2 \Rightarrow x_2}(\textsc{id})}{x_2, x_2 \Rightarrow x_2^2}(\Rightarrow\cdot)}{x_2, x_2 \Rightarrow x_1^2 \vee x_2^2}(\Rightarrow\vee)_2}{x_1, x_2 \Rightarrow x_1^2 \vee x_2^2}r(\varepsilon)}{x_1x_2 \Rightarrow x_1^2 \vee x_2^2.}(\cdot\Rightarrow)$$

The following correspondence between a variety defined relative to $\mathcal{PRL}$ by a set of simple equations E and the proof system $\mathsf{FL}(E) := \mathsf{FL} + \{r(\varepsilon) \mid \varepsilon \in E\}$ is obtained by making suitable modifications to the proof of Theorem 2.3.3.

Theorem 2.4.4. *Let E be a set of simple equations. A sequent $\Gamma \Rightarrow \Delta$ is derivable in $\mathsf{FL}(E)$ if and only if the equation $\prod\Gamma \le \sum\Delta$ is satisfied by all pointed residuated lattices that satisfy E.*

The class of simple equations is more expressive than might first appear. Let us call sets of equations E_1 and E_2 *equivalent* in a variety $\mathcal{V}$ if each algebra in $\mathcal{V}$ that satisfies E_1 also satisfies E_2, and vice versa, that is, if E_1 and E_2 define the same variety relative to $\mathcal{V}$. Let us also call a formula or equation in which only operation symbols from a set S occur, an *S-formula* or *S-equation*, respectively.

PROPOSITION 2.4.5. *Every $\{\vee, \cdot, \mathrm{e}\}$-equation is equivalent in $\mathcal{PRL}$ to a finite set of simple equations or defines the trivial variety relative to $\mathcal{PRL}$.*

PROOF. By distributing products over joins, any $\{\vee, \cdot, \mathrm{e}\}$-equation is equivalent in $\mathcal{PRL}$ to an equation of the form

$$\alpha_1 \vee \cdots \vee \alpha_n \approx \beta_1 \vee \cdots \vee \beta_m,$$

for some $\{\cdot, \mathrm{e}\}$-formulas $\alpha_1, \ldots, \alpha_n, \beta_1, \ldots, \beta_m$. Such an equation is then equivalent in $\mathcal{PRL}$ to the set of equations

$$\{\alpha_i \leq \beta_1 \vee \cdots \vee \beta_m \mid i \in \{1, \ldots, n\}\} \cup \{\beta_j \leq \alpha_1 \vee \cdots \vee \alpha_n \mid j \in \{1, \ldots, m\}\}.$$

We may therefore restrict our attention to an equation in this set of the form

$$\alpha \leq \beta_1 \vee \cdots \vee \beta_m. \tag{2.1}$$

If (2.1) is simple, we are done. Otherwise, suppose first that some variable x occurs $k > 1$ times in α. We replace x uniformly in (2.1) by $x_1 \vee \cdots \vee x_k$, for some variables $x_1, \ldots, x_k$ not occurring in(2.1), and distribute products over joins to obtain an equation

$$\alpha'_1 \vee \cdots \vee \alpha'_r \leq \beta'_1 \vee \cdots \vee \beta'_s. \tag{2.2}$$

Let α' be one of the α'_i where each x_j occurs only once, and consider

$$\alpha' \leq \beta'_1 \vee \cdots \vee \beta'_s. \tag{2.3}$$

We claim that (2.1) and (2.3) are equivalent in $\mathcal{PRL}$. If $\mathbf{L} \in \mathcal{PRL}$ satisfies (2.1), then it follows, substituting x with $x_1 \vee \cdots \vee x_k$, that $\mathbf{L}$ satisfies (2.2) and so also (2.3). Conversely, if $\mathbf{L} \in \mathcal{PRL}$ satisfies (2.3), then, substituting each x_j with x, it follows that $\mathbf{L}$ satisfies (2.1). Hence, since α' has fewer repeated occurrences of variables, we can iterate this last step and assume now without loss of generality that the formula α in (2.1) contains no repeated occurrences of variables.

Suppose next that in (2.1) at least one of $\beta_1, \ldots, \beta_m$ contains a variable y that does not occur in α. Suppose without loss of generality (introducing extra occurrences of e if necessary) that, for each $j \in \{1, \ldots, m\}$,

$$\beta_j = \beta_{j,1} y \beta_{j,2} y \cdots \beta_{j,r_j} y \beta_{j,r_j+1},$$

where $r_j \in \mathbb{N}$ and y does not occur in $\beta_{j,1}, \ldots, \beta_{j,r_j+1}$. Now let z be a new variable that does not occur in (2.1) and, for each $j \in \{1, \ldots, m\}$, let β_j^* be the result of substituting y in β_j with

$$\textstyle\bigwedge_{i=1}^{m}((\beta_{i,1} \backslash (z / \beta_{i,2})) \wedge (\mathrm{e} / \beta_{i,3}) \wedge \cdots \wedge (\mathrm{e} / \beta_{i,r_i+1})).$$

Observe that if y occurs in β_j, then $r_j > 0$ and $\mathcal{PRL} \models \beta_j^* \leq z$.

Suppose first that y occurs in each of $\beta_1, \ldots, \beta_m$. If $\mathbf{L} \in \mathcal{PRL}$ satisfies (2.1), then $\mathbf{L} \models \alpha \leq \beta_1^* \vee \cdots \vee \beta_m^*$, so $\mathbf{L} \models \alpha \leq z$ and, substituting z with $u \wedge (u \backslash \mathrm{e})$ and all other variables with e, it follows that $\mathbf{L} \models \mathrm{e} \approx u$. Hence (2.1) defines the trivial variety relative to $\mathcal{PRL}$.

Now suppose without loss of generality that y occurs in each of $\beta_1, \ldots, \beta_l$ but not in $\beta_{l+1}, \ldots, \beta_m$, for some $l \in \{1, \ldots, m-1\}$. Then $\mathcal{PRL} \models \beta_1^* \vee \cdots \vee \beta_l^* \leq z$. Hence, if $\mathbf{L} \in \mathcal{PRL}$ satisfies (2.1), then $\mathbf{L} \models \alpha \leq z \vee \beta_{l+1} \vee \cdots \vee \beta_m$ and, substituting z with

β_{l+1}, also $\mathbf{L} \models \alpha \leq \beta_{l+1} \vee \cdots \vee \beta_m$. Clearly, if $\mathbf{L} \in \mathcal{PRL}$ satisfies $\alpha \leq \beta_{l+1} \vee \cdots \vee \beta_m$, then $\mathbf{L}$ satisfies (2.1). That is, (2.1) and $\alpha \leq \beta_{l+1} \vee \cdots \vee \beta_m$ are equivalent in $\mathcal{PRL}$. Hence, iterating this process, either (2.1) defines the trivial variety relative to $\mathcal{PRL}$ or we obtain a simple equation that is equivalent to (2.1) in $\mathcal{PRL}$. □

EXAMPLE 2.4.6. The "knotted" equation $x^2 \leq x^3$ is not simple, but is equivalent in $\mathcal{PRL}$ to $(x_1 \vee x_2)^2 \leq (x_1 \vee x_2)^3$ and hence, following the procedure described above, to the simple equation

$$x_1x_2 \leq x_1^3 \vee x_1^2x_2 \vee x_1x_2x_1 \vee x_1x_2^2 \vee x_2x_1^2 \vee x_2x_1x_2 \vee x_2^2x_1 \vee x_2^3.$$

The corresponding simple rule is then

$$\frac{\{\Gamma, \Gamma_i, \Gamma_j, \Gamma_k, \Pi \Rightarrow \Delta \mid \{i,j,k\} \subseteq \{1,2\}\}}{\Gamma, \Gamma_1, \Gamma_2, \Pi \Rightarrow \Delta.}$$

We provide now an algebraic proof of the fact that any extension of FL with simple rules has cut elimination. Let $\mathbf{Fm}^*$ be the free monoid generated by the set Fm, i.e., the set Fm^* of finite sequences of formulas with concatenation as monoid multiplication and the empty sequence $\langle\rangle$ as the neutral element. We denote the set of one-sequences of formulas by Fm^1, recalling that a sequence of length 1 is often denoted by its unique member.

Let us fix a set of simple equations E. For $\Gamma_1, \Gamma_2 \in \mathrm{Fm}^*$ and $\Delta \in \mathrm{Fm}^1$, we define

$$[\Gamma_1, \Gamma_2, \Delta]_E := \{\Pi \in \mathrm{Fm}^* \mid \vdash_{\mathsf{FL}(E)^\circ} \Gamma_1, \Pi, \Gamma_2 \Rightarrow \Delta\},$$

and collect these sets of sequences of formulas in the set

$$\mathcal{D} := \{[\Gamma_1, \Gamma_2, \Delta]_E \mid \Gamma_1, \Gamma_2 \in \mathrm{Fm}^* \text{ and } \Delta \in \mathrm{Fm}^1\}.$$

We obtain a nucleus on the pointed residuated lattice (see Example 1.4.3) $\wp(\mathbf{Fm}^*) = \langle \wp(\mathrm{Fm}^*), \cap, \cup, \cdot, \backslash, /, \{\langle\rangle\}, \{\mathrm{f}\}\rangle$ by defining, for each $X \subseteq \mathrm{Fm}^*$,

$$\uparrow^{\mathcal{D}} X := \{[\Gamma_1, \Gamma_2, \Delta]_E \in \mathcal{D} \mid X \subseteq [\Gamma_1, \Gamma_2, \Delta]_E\} \quad \text{and} \quad \gamma_E(X) := \bigcap \uparrow^{\mathcal{D}} X.$$

We leave it as an exercise for the reader to check that γ_E is indeed a nucleus on $\wp(\mathbf{Fm}^*)$ and hence, by Lemma 1.5.4 (see also Example 1.5.7), that

$$\mathbf{L}_E = \langle \wp(\mathrm{Fm}^*)_{\gamma_E}, \cap, \cup_{\gamma_E}, \circ_{\gamma_E}, \backslash, /, \mathrm{e}_{\gamma_E}, \mathrm{f}_{\gamma_E}\rangle$$

is a pointed residuated lattice with $X \cup_{\gamma_E} Y := \gamma_E(X \cup Y)$, $X \circ_{\gamma_E} Y := \gamma_E(X \cdot Y)$, $\mathrm{e}_{\gamma_E} := \gamma_E(\{\langle\rangle\})$, and $\mathrm{f}_{\gamma_E} := [\langle\rangle, \langle\rangle, \langle\rangle]_E$.

LEMMA 2.4.7. *$\mathbf{L}_E$ satisfies E.*

PROOF. Observe first that, for any $X_1, \ldots, X_n, Y \in \wp(\mathrm{Fm}^*)_{\gamma_E}$,

$$\begin{aligned}
X_1 \circ_{\gamma_E} \cdots \circ_{\gamma_E} X_n \subseteq Y &\iff X_1 \cdots X_n \subseteq Y \\
&\iff \Pi_1, \ldots, \Pi_n \in [\Gamma_1, \Gamma_2, \Delta]_E, \text{ for all} \\
&\qquad \Pi_1 \in X_1, \ldots, \Pi_n \in X_n \text{ and } [\Gamma_1, \Gamma_2, \Delta]_E \in \uparrow^{\mathcal{D}} Y \\
&\iff \vdash_{\mathsf{FL}(E)^\circ} \Gamma_1, \Pi_1, \ldots, \Pi_n, \Gamma_2 \Rightarrow \Delta, \text{ for all} \\
&\qquad \Pi_1 \in X_1, \ldots, \Pi_n \in X_n \text{ and } [\Gamma_1, \Gamma_2, \Delta]_E \in \uparrow^{\mathcal{D}} Y.
\end{aligned}$$

Now consider any simple equation $\varepsilon = \alpha \leq \beta_1 \vee \cdots \vee \beta_m$ in E, with $\alpha = x_1 \cdots x_n$ and $\beta_j = x_{N(j,1)} \cdots x_{N(j,k_j)}$, for each $j \in \{1, \ldots, m\}$. We show that $\mathbf{L}_E$ satisfies the quasi-equation $\beta_1 \leq y \ \& \ \cdots \ \& \ \beta_m \leq y \implies \alpha \leq y$ and hence, substituting $\beta_1 \vee \cdots \vee \beta_m$ for y, also $\alpha \leq \beta_1 \vee \cdots \vee \beta_m$.

Let $X_1, \ldots, X_n, Y \in \wp(\mathrm{Fm}^*)_{\gamma_E}$ and suppose that, for each $j \in \{1, \ldots, m\}$,

$$X_{N(j,1)} \circ_{\gamma_E} \cdots \circ_{\gamma_E} X_{N(j,k_j)} \subseteq Y.$$

Using the equivalence at the beginning of this proof, for any $\Pi_1 \in X_1, \ldots, \Pi_n \in X_n$ and $[\Gamma_1, \Gamma_2, \Delta]_E \in \uparrow^{\mathcal{D}} Y$, for each $j \in \{1, \ldots, m\}$,

$$\vdash_{\mathsf{FL}(E)^\circ} \Gamma_1, \Pi_{N(j,1)}, \ldots, \Pi_{N(j,k_j)}, \Gamma_2 \Rightarrow \Delta.$$

Hence, using $r(\varepsilon)$, for any $\Pi_1 \in X_1, \ldots, \Pi_n \in X_n$ and $[\Gamma_1, \Gamma_2, \Delta]_E \in \uparrow^{\mathcal{D}} Y$,

$$\vdash_{\mathsf{FL}(E)^\circ} \Gamma_1, \Pi_1, \ldots, \Pi_n, \Gamma_2 \Rightarrow \Delta.$$

So, by the equivalence at the beginning of this proof, $X_1 \circ_{\gamma_E} \cdots \circ_{\gamma_E} X_n \subseteq Y$. □

Let us write $[\alpha]_E$ to denote $[\langle\rangle, \langle\rangle, \alpha]_E$ for $\alpha \in \mathrm{Fm}$, and define a homomorphism $\varphi_E \colon \mathbf{Fm} \to \mathbf{L}_E$ by fixing $\varphi_E(x) = [x]_E$, for each variable x.

LEMMA 2.4.8. *Every* $\alpha \in \mathrm{Fm}$ *satisfies* $\alpha \in \varphi_E(\alpha) \subseteq [\alpha]_E$.

PROOF. We proceed by induction on $\mathrm{cp}(\alpha)$. For each variable x, by definition, $\varphi_E(x) = [x]_E$, and, since $\vdash_{\mathsf{FL}(E)^\circ} x \Rightarrow x$, also $x \in \varphi_E(x)$. Moreover, $\mathrm{e} \in \varphi_E(\mathrm{e}) = \gamma_E(\{\langle\rangle\}) \subseteq [\mathrm{e}]_E$, using $(\mathrm{e}\Rightarrow)$ and $(\Rightarrow\mathrm{e})$, and $\mathrm{f} \in \varphi_E(\mathrm{f}) = [\langle\rangle, \langle\rangle, \langle\rangle]_E \subseteq [\mathrm{f}]_E$, using $(\mathrm{f}\Rightarrow)$ and $(\Rightarrow\mathrm{f})$.

For the inductive step, we assume $\alpha \in \varphi_E(\alpha) \subseteq [\alpha]_E$ and $\beta \in \varphi_E(\beta) \subseteq [\beta]_E$ and consider the following cases, leaving the rest as instructive exercises for the reader:

($\cdot$) Let $\Pi \in \varphi_E(\alpha) \cdot \varphi_E(\beta)$. Then $\Pi = \Pi_1, \Pi_2$ with $\Pi_1 \in [\alpha]_E$ and $\Pi_2 \in [\beta]_E$, that is, $\vdash_{\mathsf{FL}(E)^\circ} \Pi_1 \Rightarrow \alpha$ and $\vdash_{\mathsf{FL}(E)^\circ} \Pi_2 \Rightarrow \beta$. Using $(\Rightarrow\cdot)$, we obtain $\vdash_{\mathsf{FL}(E)^\circ} \Pi_1, \Pi_2 \Rightarrow \alpha \cdot \beta$, that is, $\Pi_1, \Pi_2 \in [\alpha \cdot \beta]_E$. So $\varphi_E(\alpha) \cdot \varphi_E(\beta) \subseteq [\alpha \cdot \beta]_E$ and, since $[\alpha \cdot \beta]_E$ is γ_E-closed, also $\varphi_E(\alpha \cdot \beta) = \gamma_E(\varphi_E(\alpha) \cdot \varphi_E(\beta)) \subseteq [\alpha \cdot \beta]_E$. Suppose now that $\varphi_E(\alpha) \cdot \varphi_E(\beta) \subseteq [\Gamma_1, \Gamma_2, \Delta]_E$. Then $\langle \alpha, \beta \rangle \in [\Gamma_1, \Gamma_2, \Delta]_E$, that is, $\vdash_{\mathsf{FL}(E)^\circ} \Gamma_1, \alpha, \beta, \Gamma_2 \Rightarrow \Delta$. Using $(\cdot\Rightarrow)$, we obtain $\vdash_{\mathsf{FL}(E)^\circ} \Gamma_1, \alpha \cdot \beta, \Gamma_2 \Rightarrow \Delta$, that is, $\alpha \cdot \beta \in [\Gamma_1, \Gamma_2, \Delta]_E$. Hence $\alpha \cdot \beta \in \gamma_E(\varphi_E(\alpha) \cdot \varphi_E(\beta))$.

($\backslash$) Let $\Pi \in \varphi_E(\alpha) \backslash \varphi_E(\beta)$. Then $\alpha, \Pi \in \varphi_E(\alpha) \cdot \Pi \subseteq \varphi_E(\beta) \subseteq [\beta]_E$, that is, $\vdash_{\mathsf{FL}(E)^\circ} \alpha, \Pi \Rightarrow \beta$. Using $(\Rightarrow\backslash)$, we obtain $\vdash_{\mathsf{FL}(E)^\circ} \Pi \Rightarrow \alpha \backslash \beta$, that is, $\Pi \in [\alpha \backslash \beta]$. So $\varphi_E(\alpha \backslash \beta) = \varphi_E(\alpha) \backslash \varphi_E(\beta) \subseteq [\alpha \backslash \beta]_E$. Consider now any $\Gamma_2 \in \varphi_E(\alpha)$. Then $\Gamma_2 \in [\alpha]_E$, that is, $\vdash_{\mathsf{FL}(E)^\circ} \Gamma_2 \Rightarrow \alpha$. Let $\varphi_E(\beta) \subseteq [\Gamma_1, \Gamma_3, \Delta]_E$. Then $\vdash_{\mathsf{FL}(E)^\circ} \Gamma_1, \beta, \Gamma_3 \Rightarrow \Delta$. Using $(\backslash\Rightarrow)$, we obtain $\vdash_{\mathsf{FL}(E)^\circ} \Gamma_1, \Gamma_2, \alpha \backslash \beta, \Gamma_3 \Rightarrow \Delta$, that is, $\Gamma_2, \alpha \backslash \beta \in [\Gamma_1, \Gamma_3, \Delta]_E$. Hence $\Gamma_2, \alpha \backslash \beta \in \varphi_E(\beta)$. But then $\varphi_E(\alpha)(\alpha \backslash \beta) \subseteq \varphi_E(\beta)$, and so $\alpha \backslash \beta \in \varphi_E(\alpha) \backslash \varphi_E(\beta)$.

($\vee$) Let $\Pi \in \varphi_E(\alpha) \cup \varphi_E(\beta)$. Then $\Pi \in [\alpha]_E$ or $\Pi \in [\beta]_E$, that is, $\vdash_{\mathsf{FL}(E)^\circ} \Pi \Rightarrow \alpha$ or $\vdash_{\mathsf{FL}(E)^\circ} \Pi \Rightarrow \beta$. Using $(\Rightarrow\vee)_1$ or $(\Rightarrow\vee)_2$, we obtain $\vdash_{\mathsf{FL}(E)^\circ} \Pi \Rightarrow \alpha \vee \beta$, that is, $\Pi \in [\alpha \vee \beta]_E$. So $\varphi_E(\alpha) \cup \varphi_E(\beta) \subseteq [\alpha \vee \beta]_E$ and, since $[\alpha \vee \beta]_E$ is γ_E-closed, also $\varphi_E(\alpha \vee \beta) = \gamma_E(\varphi_E(\alpha) \cup \varphi_E(\beta)) \subseteq [\alpha \vee \beta]_E$. Suppose now that $\varphi_E(\alpha) \cup \varphi_E(\beta) \subseteq [\Gamma_1, \Gamma_2, \Delta]_E$. Then $\alpha, \beta \in [\Gamma_1, \Gamma_2, \Delta]_E$, that is, $\vdash_{\mathsf{FL}(E)^\circ} \Gamma_1, \alpha, \Gamma_2 \Rightarrow \Delta$ and $\vdash_{\mathsf{FL}(E)^\circ} \Gamma_1, \beta, \Gamma_2 \Rightarrow \Delta$. Using $(\vee\Rightarrow)$, we obtain $\vdash_{\mathsf{FL}(E)^\circ} \Gamma_1, \alpha \vee \beta, \Gamma_2 \Rightarrow \Delta$, that is, $\alpha \vee \beta \in [\Gamma_1, \Gamma_2, \Delta]_E$. Hence $\alpha \vee \beta \in \gamma_E(\varphi_E(\alpha) \cup \varphi_E(\beta))$. □

THEOREM 2.4.9. *Let E be a set of simple equations. The following statements are equivalent for any sequent $\Gamma \Rightarrow \Delta$:*

(1) $\prod\Gamma \leq \sum\Delta$ *is satisfied by all pointed residuated lattices that satisfy E.*
(2) $\prod\Gamma \leq \sum\Delta$ *is satisfied by $\mathbf{L}_E$.*
(3) $\Gamma \Rightarrow \Delta$ *is derivable in $\mathsf{FL}(E)^\circ$.*
(4) $\Gamma \Rightarrow \Delta$ *is derivable in $\mathsf{FL}(E)$.*

In particular, $\mathsf{FL}(E)$ has cut elimination.

PROOF.

(1) $\Rightarrow$ (2) Follows directly from Lemma 2.4.7.

(2) $\Rightarrow$ (3) Suppose that $\prod\Gamma \leq \sum\Delta$ is satisfied by $\mathbf{L}_E$. Consider first $\Gamma = \langle\beta_1, \ldots, \beta_n\rangle$ and $\Delta = \langle\alpha\rangle$. Then Lemma 2.4.8 yields $\beta_i \in \varphi_E(\beta_i)$, for each $i \in \{1, \ldots, n\}$ and $\varphi_E(\alpha) \subseteq [\alpha]_E$. Hence, since φ_E is a homomorphism from $\mathbf{Fm}$ to $\mathbf{L}_E$,

$$\begin{aligned}
\langle\beta_1, \ldots, \beta_n\rangle &\in \varphi_E(\beta_1) \cdots \varphi_E(\beta_n) \\
&\subseteq \gamma_E(\varphi_E(\beta_1) \cdots \varphi_E(\beta_n)) \\
&= \varphi_E(\beta_1 \cdots \beta_n) \\
&\subseteq \varphi_E(\alpha) \\
&\subseteq [\alpha]_E.
\end{aligned}$$

For $\Gamma = \langle\rangle$ and $\Delta = \langle\alpha\rangle$, it follows that $\langle\rangle \in \gamma_E(\{\langle\rangle\}) = \varphi_E(\mathrm{e}) \subseteq \varphi_E(\alpha) \subseteq [\alpha]_E$; for $\Gamma = \langle\beta_1, \ldots, \beta_n\rangle$ and $\Delta = \langle\rangle$, it follows that $\langle\beta_1, \ldots, \beta_n\rangle \in \varphi_E(\mathrm{f}) = [\langle\rangle, \langle\rangle, \langle\rangle]_E$; and for $\Gamma = \Delta = \langle\rangle$, it follows that $\langle\rangle \in [\langle\rangle, \langle\rangle, \langle\rangle]_E$. Hence in all cases, $\vdash_{\mathsf{FL}(E)^\circ} \Gamma \Rightarrow \Delta$.

(3) $\Rightarrow$ (4) Immediate.

(4) $\Rightarrow$ (1) Follows directly from Theorem 2.4.4. $\square$

In commutative settings, it can be convenient (e.g., to avoid lots of cases in proofs) to define proof systems based on sequents built from finite multisets, rather than sequences, of formulas. Formally, a *finite multiset* is an ordered pair $\langle S, \varphi\rangle$ consisting of a set S and a function $\varphi\colon S \to \mathbb{N}$ such that $\{a \in S \mid \varphi(a) > 0\}$ is finite, often denoted by a list containing $\varphi(a)$ occurrences of each element $a \in S$ between square brackets [and]. We call a finite multiset $\langle S, \varphi\rangle$ a *one-multiset* if there exists an element $a \in S$ such that $\varphi(a) \leq 1$ and $\varphi(b) = 0$, for all $b \in S \backslash \{a\}$. The multiset sum of finite multisets $\langle S, \varphi_1\rangle$ and $\langle S, \varphi_2\rangle$ is the finite multiset $\langle S, \varphi_1\rangle \uplus \langle S, \varphi_2\rangle := \langle S, \varphi\rangle$, where $\varphi(x) := \varphi_1(x) + \varphi_2(x)$.

Let us reuse the metavariables Γ and Π to denote arbitrary finite multisets of formulas, and Δ to denote a one-multiset of formulas, relying on context to avoid ambiguity. Following convention, we often write the finite multiset sum $\Gamma \uplus \Pi$ of Γ and Π as Γ, Π, removing brackets where appropriate.

A *multiset-sequent* is an ordered pair consisting of a finite multiset Γ and one-multiset Δ of formulas, written $\Gamma \Rightarrow \Delta$. Let us also harmlessly associate every finite multiset $\Gamma = [\alpha_1, \ldots, \alpha_n]$ of formulas with a fixed sequence $\Gamma^s := \langle\alpha_1, \ldots, \alpha_n\rangle$; we may then define $\prod\Gamma := \prod\Gamma^s$ and, for a one-multiset Δ of formulas, $\sum\Delta := \sum\Delta^s$.

Identity Axioms

$$\frac{}{\alpha \Rightarrow \alpha}\ (\textsc{id})$$

Cut Rule

$$\frac{\Gamma_2 \Rightarrow \alpha \quad \Gamma_1, \alpha \Rightarrow \Delta}{\Gamma_1, \Gamma_2 \Rightarrow \Delta}\ (\textsc{cut})$$

Left Operation Rules

$$\frac{\Gamma \Rightarrow \Delta}{\Gamma, \mathrm{e} \Rightarrow \Delta}\ (\mathrm{e}\Rightarrow)$$

$$\frac{}{\mathrm{f} \Rightarrow}\ (\mathrm{f}\Rightarrow)$$

$$\frac{\Gamma_2 \Rightarrow \alpha \quad \Gamma_1, \beta \Rightarrow \Delta}{\Gamma_1, \Gamma_2, \alpha \to \beta \Rightarrow \Delta}\ (\to\Rightarrow)$$

$$\frac{\Gamma, \alpha, \beta \Rightarrow \Delta}{\Gamma, \alpha \cdot \beta \Rightarrow \Delta}\ (\cdot\Rightarrow)$$

$$\frac{\Gamma, \alpha \Rightarrow \Delta}{\Gamma, \alpha \wedge \beta \Rightarrow \Delta}\ (\wedge\Rightarrow)_1$$

$$\frac{\Gamma, \beta \Rightarrow \Delta}{\Gamma, \alpha \wedge \beta \Rightarrow \Delta}\ (\wedge\Rightarrow)_2$$

$$\frac{\Gamma, \alpha \Rightarrow \Delta \quad \Gamma, \beta \Rightarrow \Delta}{\Gamma, \alpha \vee \beta \Rightarrow \Delta}\ (\vee\Rightarrow)$$

Right Operation Rules

$$\frac{}{\Rightarrow \mathrm{e}}\ (\Rightarrow\mathrm{e})$$

$$\frac{\Gamma \Rightarrow}{\Gamma \Rightarrow \mathrm{f}}\ (\Rightarrow\mathrm{f})$$

$$\frac{\Gamma, \alpha \Rightarrow \beta}{\Gamma \Rightarrow \alpha \to \beta}\ (\Rightarrow\to)$$

$$\frac{\Gamma_1 \Rightarrow \alpha \quad \Gamma_2 \Rightarrow \beta}{\Gamma_1, \Gamma_2 \Rightarrow \alpha \cdot \beta}\ (\Rightarrow\cdot)$$

$$\frac{\Gamma \Rightarrow \alpha}{\Gamma \Rightarrow \alpha \vee \beta}\ (\Rightarrow\vee)_1$$

$$\frac{\Gamma \Rightarrow \beta}{\Gamma \Rightarrow \alpha \vee \beta}\ (\Rightarrow\vee)_2$$

$$\frac{\Gamma \Rightarrow \alpha \quad \Gamma \Rightarrow \beta}{\Gamma \Rightarrow \alpha \wedge \beta}\ (\Rightarrow\wedge)$$

FIGURE 2.5. The multiset-sequent calculus CFL

By reinterpreting the rules of FL, we obtain the multiset-sequent calculus CFL presented in Figure 2.5. It is then straightforward to prove the following soundness, completeness, and cut elimination results, either directly or using the exchange rules to translate between derivations in CFL and $\mathsf{FL_e}$.

COROLLARY 2.4.10.

(a) *A multiset-sequent* $\Gamma \Rightarrow \Delta$ *is derivable in* CFL *if and only if* $\prod\Gamma \leq \sum\Delta$ *is satisfied by all commutative pointed residuated lattices.*

(b) CFL *has cut elimination.*

2.5. Hypersequent calculi

Sequent calculi provide a suitable environment for defining analytic proof systems for a broad spectrum of classes of algebras. However, for certain important cases — notably, varieties of semilinear residuated lattices — sequents are not enough. For example, the equation $\mathrm{e} \leq (x\backslash y) \vee (y\backslash x)$ is not simple and cannot be transformed into a simple analytic structural rule; to obtain an analytic calculus, we should adapt either the operation rules or the sequent framework itself.

Consider the sequent counterpart $\Rightarrow (x\backslash y) \vee (y\backslash x)$ of the above equation. In such cases, it would be helpful to represent the choice between the two sequents obtained via $(\Rightarrow\vee)_1$ and $(\Rightarrow\vee)_2$ "in parallel" as

$$\Rightarrow x\backslash y \quad \text{“or”} \quad \Rightarrow y\backslash x.$$

Hypersequents provide a suitable framework for representing such choices.

Let us focus here on the commutative case, writing $\to$ instead of $\backslash$ and $/$, and abbreviating "multiset-sequent" to "sequent". A *(multiset-)hypersequent* is a finite multiset of sequents $S_1, \ldots, S_n$, written

$$S_1 \mid \cdots \mid S_n.$$

We denote arbitrary hypersequents using the metavariables $\mathcal{G}$ and $\mathcal{H}$.

A ready supply of hypersequent rules is obtained directly from sequent rules. Given a sequent rule r, the *hypersequent version* of r consists of all inferences of the form

$$\frac{\mathcal{G} \mid S_1 \ \ldots\ \mathcal{G} \mid S_n}{\mathcal{G} \mid S,}$$

where $S_1, \ldots, S_n \ /\ S$ is any instance of r and $\mathcal{G}$ is any hypersequent. Similarly, given a sequent calculus C, we refer to the hypersequent calculus consisting of the hypersequent versions of the rules of C as the *hypersequent version* of C.

EXAMPLE 2.5.1. The hypersequent versions of the implication rules $(\to\Rightarrow)$ and $(\Rightarrow\to)$ of CFL (see Figure 2.5) are

$$\frac{\mathcal{G} \mid \Gamma_2 \Rightarrow \alpha \quad \mathcal{G} \mid \Gamma_1, \beta \Rightarrow \Delta}{\mathcal{G} \mid \Gamma_1, \Gamma_2, \alpha \to \beta \Rightarrow \Delta}\ (\to\Rightarrow) \qquad \frac{\mathcal{G} \mid \Gamma, \alpha \Rightarrow \beta}{\mathcal{G} \mid \Gamma \Rightarrow \alpha \to \beta,}\ (\Rightarrow\to)$$

and the hypersequent versions of the cut rule of this calculus is

$$\frac{\mathcal{G} \mid \Gamma_2 \Rightarrow \alpha \quad \mathcal{G} \mid \Gamma_1, \alpha \Rightarrow \Delta}{\mathcal{G} \mid \Gamma_1, \Gamma_2 \Rightarrow \Delta.}\ (\text{CUT})$$

We also make use of rules that relate only to hypersequents. In particular, the following "external structural rules" allow sequents to be weakened and contracted within a hypersequent:

$$\frac{\mathcal{G}}{\mathcal{G} \mid \mathcal{H}}\ (\text{EW}) \qquad \frac{\mathcal{G} \mid \mathcal{H} \mid \mathcal{H}}{\mathcal{G} \mid \mathcal{H}.}\ (\text{EC})$$

Essentially, these extra rules characterize "|" as a join-like disjunction: reasoning backwards, "external weakening" (EW) removes sequents and "external contraction" (EC) duplicates them. They provide greater flexibility for recording choices in a derivation. In particular, the rules for $\wedge$ on the left and those for $\vee$ on the right can be combined in the rules

$$\frac{\mathcal{G} \mid \Gamma, \alpha \Rightarrow \Delta \mid \Gamma, \beta \Rightarrow \Delta}{\mathcal{G} \mid \Gamma, \alpha \wedge \beta \Rightarrow \Delta}\ (\wedge\Rightarrow) \qquad \frac{\mathcal{G} \mid \Gamma \Rightarrow \alpha \mid \Gamma \Rightarrow \beta}{\mathcal{G} \mid \Gamma \Rightarrow \alpha \vee \beta.}\ (\Rightarrow\vee)$$

The following derivations show that $(\Rightarrow\vee)$ is inter-derivable with $(\Rightarrow\vee)_1$ and $(\Rightarrow\vee)_2$ using (EW) and (EC) (with $i \in \{1,2\}$):

$$\dfrac{\dfrac{\dfrac{\mathcal{G} \mid \Gamma \Rightarrow \alpha_1 \mid \Gamma \Rightarrow \alpha_2}{\mathcal{G} \mid \Gamma \Rightarrow \alpha_1 \mid \Gamma \Rightarrow \alpha_1 \vee \alpha_2}\,(\Rightarrow\vee)_2}{\mathcal{G} \mid \Gamma \Rightarrow \alpha_1 \vee \alpha_2 \mid \Gamma \Rightarrow \alpha_1 \vee \alpha_2}\,(\Rightarrow\vee)_1}{\mathcal{G} \mid \Gamma \Rightarrow \alpha_1 \vee \alpha_2}\,(\textsc{ec}) \qquad \dfrac{\dfrac{\mathcal{G} \mid \Gamma \Rightarrow \alpha_i}{\mathcal{G} \mid \Gamma \Rightarrow \alpha_1 \mid \Gamma \Rightarrow \alpha_2}\,(\textsc{ew})}{\mathcal{G} \mid \Gamma \Rightarrow \alpha_1 \vee \alpha_2.}\,(\Rightarrow\vee)$$

Similarly, $(\wedge\Rightarrow)$ is inter-derivable with $(\wedge\Rightarrow)_1$ and $(\wedge\Rightarrow)_2$ using (EW) and (EC).

On their own, the external weakening and external contraction rules do not add much to the expressive power of hypersequent calculi. Indeed, it is easy to prove that a hypersequent $S_1 \mid \cdots \mid S_n$ is derivable in the hypersequent version of a sequent calculus C extended with (EW) and (EC) if and only if S_i is derivable in C, for some $i \in \{1, \ldots, n\}$.

To reach beyond the scope of sequent calculi, we require structural rules that allow sequents to interact within a hypersequent. Perhaps the most influential example of such a rule is the following *communication rule*:

$$\dfrac{\mathcal{G} \mid \Gamma_1, \Gamma_2 \Rightarrow \Delta_1 \quad \mathcal{G} \mid \Pi_1, \Pi_2 \Rightarrow \Delta_2}{\mathcal{G} \mid \Gamma_1, \Pi_1 \Rightarrow \Delta_1 \mid \Gamma_2, \Pi_2 \Rightarrow \Delta_2.}\,(\textsc{com})$$

EXAMPLE 2.5.2. The communication rule is best understood by considering again the equation $\mathrm{e} \leq (x \to y) \vee (y \to x)$, which can now be derived as follows:

$$\dfrac{\dfrac{\dfrac{\dfrac{\dfrac{\dfrac{\dfrac{}{x \Rightarrow x}\,(\textsc{id}) \quad \dfrac{}{y \Rightarrow y}\,(\textsc{id})}{x \Rightarrow y \mid y \Rightarrow x}\,(\textsc{com})}{x \Rightarrow y \mid \;\Rightarrow y \to x}\,(\Rightarrow\to)}{\Rightarrow x \to y \mid \;\Rightarrow y \to x}\,(\Rightarrow\to)}{\Rightarrow x \to y \mid \;\Rightarrow (x \to y) \vee (y \to x)}\,(\Rightarrow\vee)_2}{\Rightarrow (x \to y) \vee (y \to x) \mid \;\Rightarrow (x \to y) \vee (y \to x)}\,(\Rightarrow\vee)_1}{\Rightarrow (x \to y) \vee (y \to x).}\,(\textsc{ec})$$

Note that the hypersequent $x \Rightarrow y \mid y \Rightarrow x$ two lines down may be understood as the "hypersequent translation" of $(x \to y) \vee (y \to x)$.

Let CSemFL be the hypersequent version of CFL presented in Figure 2.5 extended with (EW), (EC), and (COM). This calculus serves as a proof-theoretic counterpart to the variety $\mathcal{CSemPRL}$ of commutative, semilinear pointed residuated lattices, defined relative to $\mathcal{CPRL}$ by the commutative prelinearity law

$$((x \to y) \wedge \mathrm{e}) \vee ((y \to x) \wedge \mathrm{e}) \approx \mathrm{e}.$$

THEOREM 2.5.3. *A non-empty hypersequent $\Gamma_1 \Rightarrow \Delta_1 \mid \cdots \mid \Gamma_n \Rightarrow \Delta_n$ is derivable in* CSemFL *if and only if the equation $\mathrm{e} \leq (\prod\Gamma_1 \to \sum\Delta_1) \vee \cdots \vee (\prod\Gamma_n \to \sum\Delta_n)$ is satisfied by all commutative, semilinear pointed residuated lattices.*

PROOF. The left-to-right direction is established by an induction on the height of a derivation in CSemFL. The right-to-left direction is established via a Lindenbaum-Tarski algebra construction as in the proof of Theorem 2.3.3. □

For hypersequent calculi, cut elimination is complicated by the fact that whole sequents may be contracted using (EC). This means that a cut formula occurring in the premises of an application of (CUT) may appear in several sequents in a hypersequent occurring higher up in the derivation, e.g.,

$$\dfrac{\dfrac{\dfrac{\vdots}{\Gamma, \alpha \Rightarrow \Delta \mid \Gamma, \alpha \Rightarrow \Delta}}{\Gamma, \alpha \Rightarrow \Delta}\,(\text{EC}) \qquad \dfrac{\vdots}{\Pi \Rightarrow \alpha}}{\Gamma, \Pi \Rightarrow \Delta.}\,(\text{CUT})$$

To cope with this situation, we eliminate more general versions of (CUT) that perform multiple cuts in different sequents. We prove here the concrete case of cut elimination for CSemFL, referring to the bibliographical remarks for (general, uniform) cut elimination proofs for other hypersequent calculi.

Recall that CSemFL° denotes the calculus CSemFL without (CUT). For a finite multiset Γ, let $\Gamma^0 := [\,]$ and $\Gamma^{n+1} := \Gamma^n \uplus \Gamma$, for each $n \in \mathbb{N}$. We also often write $[\Gamma_i \Rightarrow \Delta_i]_{i=1}^n$ to denote the hypersequent $\Gamma_1 \Rightarrow \Delta_1 \mid \cdots \mid \Gamma_n \Rightarrow \Delta_n$.

THEOREM 2.5.4. CSemFL *has cut elimination.*

PROOF. It suffices to prove the following claims:

(I) If $d_1 \vdash_{\mathsf{CSemFL}^\circ} [\Gamma_i, [\alpha]^{\lambda_i} \Rightarrow \Delta_i]_{i=1}^n$ and $d_2 \vdash_{\mathsf{CSemFL}^\circ} \mathcal{H} \mid \Pi \Rightarrow \alpha$, then

$$\vdash_{\mathsf{CSemFL}^\circ} \mathcal{H} \mid [\Gamma_i, \Pi^{\lambda_i} \Rightarrow \Delta_i]_{i=1}^n.$$

(II) If $d_1 \vdash_{\mathsf{CSemFL}^\circ} \mathcal{G} \mid [\Gamma_i \Rightarrow \alpha]_{i=1}^n$ and $d_2 \vdash_{\mathsf{CSemFL}^\circ} \mathcal{H} \mid \Pi, \alpha \Rightarrow \Delta$, then

$$\vdash_{\mathsf{CSemFL}^\circ} \mathcal{G} \mid \mathcal{H} \mid [\Gamma_i, \Pi \Rightarrow \Delta]_{i=1}^n.$$

We prove (I) and (II) simultaneously by induction on $\langle \mathrm{cp}(\alpha), P(d_2), \mathrm{h}(d_1)\rangle \in \mathbb{N}^3$ in the lexicographic product of three copies of $\langle \mathbb{N}, \leq\rangle$, where $P(d_2)$ is 0 if d_2 ends with an operation rule applied to the distinguished occurrence of α, and 1 otherwise.

We consider the last step in the derivation d_1. Suppose first that this step is not an application of an operation rule that introduces a distinguished occurrence of α. We consider some representative cases, leaving the remainder to the reader.

(a) If d_1 ends with an instance of (ID) of the form $\mathcal{G}' \mid \beta \Rightarrow \beta$, where β is not a distinguished occurrence of α, then the concluding hypersequent for both (I) and (II) contains $\beta \Rightarrow \beta$ and is derivable using (ID). If d_1 ends with an instance of (ID) of the form $\alpha \Rightarrow \alpha$ for a distinguished occurrence of α on the left for (I) or on the right for (II), then the concluding hypersequent contains $\mathcal{H} \mid \Pi \Rightarrow \alpha$ or $\mathcal{H} \mid \Pi, \alpha \Rightarrow \Delta$, respectively, and is derivable by extending d_2 with an application of (EW).

(b) Suppose now that d_1 ends with an instance of an operation rule that does not introduce a distinguished occurrence of α. Consider, for example, claim (I) where Γ_1 is $\Gamma_{11} \uplus \Gamma_{12}$, $\mathcal{G}'$ is $[\Gamma_i, [\alpha]^{\lambda_i} \Rightarrow \Delta_i]_{i=2}^n$, and d_1 ends with

$$\dfrac{\dfrac{\vdots}{\mathcal{G}' \mid \Gamma_{11}, [\alpha]^{\mu} \Rightarrow \gamma_1} \quad \dfrac{\vdots}{\mathcal{G}' \mid \Gamma_{12}, [\alpha]^{\lambda_1 - \mu} \Rightarrow \gamma_2}}{\mathcal{G}' \mid \Gamma_{11}, \Gamma_{12}, [\alpha]^{\lambda_1} \Rightarrow \gamma_1 \cdot \gamma_2.}\,(\Rightarrow\cdot)$$

By two applications of the induction hypothesis,

$$\vdash_{\mathsf{CSemFL}^\circ} \mathcal{H}' \mid \Gamma_{11}, \Pi^{\mu} \Rightarrow \gamma_1 \quad \text{and} \quad \vdash_{\mathsf{CSemFL}^\circ} \mathcal{H}' \mid \Gamma_{12}, \Pi^{\lambda_1-\mu} \Rightarrow \gamma_2,$$

where $\mathcal{H}'$ is $\mathcal{H} \mid [\Gamma_i, \Pi^{\lambda_i} \Rightarrow \Delta_i]_{i=2}^n$, and an application of $(\Rightarrow\cdot)$ yields

$$\vdash_{\mathsf{CSemFL}^\circ} \mathcal{H}' \mid \Gamma_{11}, \Gamma_{12}, \Pi^{\lambda_1} \Rightarrow \gamma_1 \cdot \gamma_2.$$

Other cases follow the same general pattern.

(c) Suppose that d_1 ends with an application of a structural rule. The cases of (EW) and (EC) are immediate by virtue of the very general induction hypothesis. Consider claim (I) where Γ_1 is $\Gamma_{11} \uplus \Gamma_{12}$, Γ_2 is $\Gamma_{21} \uplus \Gamma_{22}$, $\mathcal{G}'$ is $[\Gamma_i, [\alpha]^{\lambda_i} \Rightarrow \Delta_i]_{i=3}^n$, and d_1 ends with

$$\dfrac{\dfrac{\vdots}{\mathcal{G}' \mid \Gamma_{11}, \Gamma_{21}, [\alpha]^{\mu} \Rightarrow \Delta_1} \quad \dfrac{\vdots}{\mathcal{G}' \mid \Gamma_{12}, \Gamma_{22}, [\alpha]^{\lambda_1+\lambda_2-\mu} \Rightarrow \Delta_2}}{\mathcal{G}' \mid \Gamma_{11}, \Gamma_{12}, [\alpha]^{\lambda_1} \Rightarrow \Delta_1 \mid \Gamma_{21}, \Gamma_{22}, [\alpha]^{\lambda_2} \Rightarrow \Delta_2.}\ (\textsc{com})$$

By two applications of the induction hypothesis,

$$\vdash_{\mathsf{CSemFL}^\circ} \mathcal{H}' \mid \Gamma_{11}, \Gamma_{21}, \Pi^{\mu} \Rightarrow \Delta_1 \quad \text{and} \quad \vdash_{\mathsf{CSemFL}^\circ} \mathcal{H}' \mid \Gamma_{12}, \Gamma_{22}, \Pi^{\lambda_1+\lambda_2-\mu} \Rightarrow \Delta_2,$$

where $\mathcal{H}'$ is $\mathcal{H} \mid [\Gamma_i, \Pi^{\lambda_i} \Rightarrow \Delta_i]_{i=3}^n$, and an application of (COM) yields

$$\vdash_{\mathsf{CSemFL}^\circ} \mathcal{H}' \mid \Gamma_{11}, \Gamma_{12},, \Pi^{\lambda_1} \Rightarrow \Delta_1 \mid \Gamma_{21}, \Gamma_{22}, \Pi^{\lambda_2} \Rightarrow \Delta_2.$$

Now consider the case where the last step in d_1 is an application of an operation rule that introduces one of the distinguished occurrences of α. Suppose, for example, that α is $\beta \to \gamma$, Γ_1 is $\Gamma_{11} \uplus \Gamma_{12}$, $\mathcal{G}'$ is $[\Gamma_i, [\alpha]^{\lambda_i} \Rightarrow \Delta_i]_{i=2}^n$, and d_1 ends with

$$\dfrac{\dfrac{\vdots}{\mathcal{G}' \mid \Gamma_{11}, [\alpha]^{\mu} \Rightarrow \beta} \quad \dfrac{\vdots}{\mathcal{G}' \mid \Gamma_{12}, [\alpha]^{\lambda_1-1-\mu}, \gamma \Rightarrow \Delta_1}}{\mathcal{G}' \mid \Gamma_{11}, \Gamma_{12}, [\alpha]^{\lambda_1-1}, \beta \to \gamma \Rightarrow \Delta_1.}\ (\to\Rightarrow)$$

Two applications of the induction hypothesis yield

$$d_3 \vdash_{\mathsf{CSemFL}^\circ} \mathcal{H}' \mid \Gamma_{11}, \Pi^{\mu} \Rightarrow \beta \quad \text{and} \quad d_4 \vdash_{\mathsf{CSemFL}^\circ} \mathcal{H}' \mid \Gamma_{12}, \Pi^{\lambda_1-1-\mu}, \gamma \Rightarrow \Delta_1,$$

where $\mathcal{H}'$ is $\mathcal{H} \mid [\Gamma_i, \Pi^{\lambda_i} \Rightarrow \Delta_i]_{i=2}^n$. We consider the derivation d_5 ending with

$$\dfrac{\dfrac{d_3 \vdots}{\mathcal{H}' \mid \Gamma_{11}, \Pi^{\mu} \Rightarrow \beta} \quad \dfrac{d_4 \vdots}{\mathcal{H}' \mid \Gamma_{12}, \Pi^{\lambda_1-1-\mu}, \gamma \Rightarrow \Delta_1}}{\mathcal{H}' \mid \Gamma_{11}, \Gamma_{12}, \Pi^{\lambda_1-1}, \beta \to \gamma \Rightarrow \Delta_1,}\ (\to\Rightarrow)$$

and identify two possibilities.

(a) Suppose first that $P(d_2) = 1$. Then $\langle \mathrm{cp}(\alpha), 0, \mathrm{h}(d_2)\rangle < \langle \mathrm{cp}(\alpha), 1, \mathrm{h}(d_1)\rangle$ and applying claim (II) of the induction hypothesis to d_2 and d_5 yields

$$\vdash_{\mathsf{CSemFL}^\circ} \mathcal{H}' \mid \mathcal{H} \mid \Gamma_{11}, \Gamma_{12}, \Pi^{\lambda_1} \Rightarrow \Delta_1.$$

The desired result follows by a further application of (EC).

(b) Now suppose that $P(d_2) = 0$. Then d_2 must have the form

$$\dfrac{\dfrac{d_6 \vdots}{\mathcal{H} \mid \Pi, \beta \Rightarrow \gamma}}{\mathcal{H} \mid \Pi \Rightarrow \beta \to \gamma.}\ (\Rightarrow\to)$$

Since $\langle \mathrm{cp}(\beta), p, \mathrm{h}(d_3)\rangle < \langle \mathrm{cp}(\alpha), 0, \mathrm{h}(d_1)\rangle$, for any $p \in \{0,1\}$, applying claim (I) of the induction hypothesis to d_3 and d_6 yields

$$d_7 \vdash_{\mathsf{CSemFL}^\circ} \mathcal{H} \mid \mathcal{H}' \mid \Gamma_{11}, \Pi^{\mu+1} \Rightarrow \gamma.$$

But then also $\langle \mathrm{cp}(\gamma), p, \mathrm{h}(d_4)\rangle < \langle \mathrm{cp}(\alpha), 0, \mathrm{h}(d_1)\rangle$, for any $p \in \{0,1\}$, so, by claim (I) of the induction hypothesis applied to d_4 and d_7,

$$\vdash_{\mathsf{CSemFL}^\circ} \mathcal{H} \mid \mathcal{H}' \mid \mathcal{H}' \mid \Gamma_{11}, \Gamma_{12}, \Pi^{\lambda_1} \Rightarrow \Delta_1.$$

The desired result follows by a further application of (EC).

The many other cases follow the same pattern. □

Unlike sequent calculi such as FL or CFL, cut elimination for CSemFL does not yield (at least directly) decidability for a corresponding equational theory. There is no obvious way to control applications of the external contraction rule as we proceed upwards in a derivation. Indeed, decidability of the equational theory of commutative, semilinear pointed residuated lattices is an open problem. It is therefore reasonable to ask if such calculi have any practical value. Fortunately, the answer is yes; as will be shown in Chapter 5, we can use CSemFL to prove that $\mathcal{CSemRL}$ is generated as a quasivariety by its dense totally ordered members.

Finally, note that a hypersequent calculus for a variety of commutative, semilinear pointed residuated lattices defined relative to $\mathcal{CSemPRL}$ by simple equations can be obtained by extending CSemFL with hypersequent versions of simple rules. Cut elimination for these calculi is proved either syntactically, by supplementing the proof of Theorem 2.5.4 with extra cases, or algebraically, by adapting the proof of Theorem 2.4.9 to deal with hypersequents. These methods extend, in fact, to a much broader class of hypersequent rules that — like the communication rule — allow interaction between sequents. However, this class does not include rules that correspond to the key equations defining varieties of cancellative or divisible residuated lattices. (See the bibliographical remarks for further details.)

2.6. Historical excursus

In this section, we trace the main sources of residuation from a proof-theoretic perspective, beginning with Gerhard Gentzen's pioneering work on sequent calculi for classical logic and intuitionistic logic, then turning our attention to the origins of substructural logics in linguistics, set theory, philosophy, and computer science.

2.6.1. Gentzen's sequent calculus. Analytic proof systems have, arguably, an official birth date: 1935, the year that Gentzen published his *Untersuchungen über das logische Schließen* [**155**]. In this milestone work, Gentzen introduced

(i) the general format of sequent (as well as natural deduction) calculi;
(ii) the sequent calculus LK for first-order classical logic;
(iii) the sequent calculus LJ for first-order intuitionistic logic, the first calculus with a restriction on structural rules;
(iv) the method of cut elimination, applied to both LJ and LK.[4]

Identity Axioms

$$\frac{}{\alpha \Rightarrow \alpha}\ (\text{ID})$$

Cut Rule

$$\frac{\Gamma_1, \alpha \Rightarrow \Delta_1 \quad \Gamma_2 \Rightarrow \alpha, \Delta_2}{\Gamma_1, \Gamma_2 \Rightarrow \Delta_1, \Delta_2}\ (\text{CUT})$$

Left Operation Rules

$$\frac{}{\Gamma, 0 \Rightarrow \Delta}\ (0\Rightarrow)$$

$$\frac{\Gamma \Rightarrow \alpha, \Delta}{\Gamma, \neg\alpha \Rightarrow \Delta}\ (\neg\Rightarrow)$$

$$\frac{\Gamma_1 \Rightarrow \alpha, \Delta_1 \quad \Gamma_2, \beta \Rightarrow \Delta_2}{\Gamma_1, \Gamma_2, \alpha \to \beta \Rightarrow \Delta_1, \Delta_2}\ (\to\Rightarrow)$$

$$\frac{\Gamma, \alpha \Rightarrow \Delta}{\Gamma, \alpha \wedge \beta \Rightarrow \Delta}\ (\wedge\Rightarrow)_1$$

$$\frac{\Gamma, \beta \Rightarrow \Delta}{\Gamma, \alpha \wedge \beta \Rightarrow \Delta}\ (\wedge\Rightarrow)_2$$

$$\frac{\Gamma, \alpha \Rightarrow \Delta \quad \Gamma, \beta \Rightarrow \Delta}{\Gamma, \alpha \vee \beta \Rightarrow \Delta}\ (\vee\Rightarrow)$$

Right Operation Rules

$$\frac{}{\Gamma \Rightarrow 1, \Delta}\ (\Rightarrow 1)$$

$$\frac{\Gamma, \alpha \Rightarrow \Delta}{\Gamma \Rightarrow \neg\alpha, \Delta}\ (\Rightarrow\neg)$$

$$\frac{\Gamma, \alpha \Rightarrow \beta, \Delta}{\Gamma \Rightarrow \alpha \to \beta, \Delta}\ (\Rightarrow\to)$$

$$\frac{\Gamma \Rightarrow \alpha, \Delta}{\Gamma \Rightarrow \alpha \vee \beta, \Delta}\ (\Rightarrow\vee)_1$$

$$\frac{\Gamma \Rightarrow \beta, \Delta}{\Gamma \Rightarrow \alpha \vee \beta, \Delta}\ (\Rightarrow\vee)_2$$

$$\frac{\Gamma \Rightarrow \alpha, \Delta \quad \Gamma \Rightarrow \beta, \Delta}{\Gamma \Rightarrow \alpha \wedge \beta, \Delta}\ (\Rightarrow\wedge)$$

Left Structural Rules

$$\frac{\Gamma \Rightarrow \Delta}{\Gamma, \alpha \Rightarrow \Delta}\ (\text{WL})$$

$$\frac{\Gamma, \alpha, \alpha \Rightarrow \Delta}{\Gamma, \alpha \Rightarrow \Delta}\ (\text{CL})$$

$$\frac{\Gamma_1, \beta, \alpha, \Gamma_2 \Rightarrow \Delta}{\Gamma_1, \alpha, \beta, \Gamma_2 \Rightarrow \Delta}\ (\text{EL})$$

Right Structural Rules

$$\frac{\Gamma \Rightarrow \Delta}{\Gamma \Rightarrow \alpha, \Delta}\ (\text{WR})$$

$$\frac{\Gamma \Rightarrow \alpha, \alpha, \Delta}{\Gamma \Rightarrow \alpha, \Delta}\ (\text{CR})$$

$$\frac{\Gamma \Rightarrow \Delta_1, \beta, \alpha, \Delta_2}{\Gamma \Rightarrow \Delta_1, \alpha, \beta, \Delta_2}\ (\text{ER})$$

FIGURE 2.6. The sequent calculus GCL

The propositional part GCL of Gentzen's calculus LK (extended with rules for the constant symbols 0 and 1) is presented in Figure 2.6, using "multiple-conclusion" sequents $\Gamma \Rightarrow \Delta$, where both Γ and Δ are finite sequences of formulas over the signature $\langle 2, 2, 2, 1, 0, 0\rangle$ with operation symbols $\wedge$, $\vee$, $\to$, $\neg$, 1, and 0. The calculus LJ is the same as LK except that the rules are restricted to "single-conclusion" sequents $\Gamma \Rightarrow \Delta$ where Δ contains at most one formula. This means in particular that LJ has no right exchange or right contraction rules, and the right weakening

[4]Note that Stanisław Jaśkowski and Paul Hertz introduced, prior to Gentzen, systems similar to natural deduction calculi and sequent calculi, respectively; see, e.g., **[200]** and **[234]**.

rule is applied only to sequents with an empty conclusion. In a sense, then, LJ may be called the first *substructural* sequent calculus.[5]

Gentzen's aim was not to provide an algorithmic presentation of the equations satisfied by some class of algebras. Rather, he was interested in regimenting the behavior of logical connectives (and quantifiers) in mathematical proofs, reading the symbols $\wedge$, $\vee$, $\to$, and $\neg$ in Figure 2.6 as conjunction, disjunction, implication, and negation, respectively. Indeed, Gentzen explicitly introduced natural deduction calculi for intuitionistic and classical first-order logic to "set up a formal system which comes as close as possible to actual reasoning" [**155**, p. 68], capturing features of informal mathematical proofs such as the use of assumptions. His calculi were later taken as the basis for a philosophical account of the meaning of connectives and quantifiers known as *logical inferentialism* [**108**]. Whereas in Hilbert systems, the behavior of connectives and quantifiers is governed by axioms in which other logical operators may occur, in Gentzen's natural deduction and sequent calculi, their inferential role is separately determined by specific rules; in a sense, the meaning of each connective can be read off these rules, with no reference to external models.

Gentzen himself displayed only a cursory interest in these philosophical issues. He touched fleetingly on the meaning of connectives with respect to natural deduction, but assigned sequent calculi only an instrumental role as a suitable framework for proving his cut elimination "Hauptsatz" for LK and LJ. There are clear similarities between the two formats, however. The distinction in natural deduction between introduction and elimination rules is paralleled in sequent calculi by the distinction between right and left introduction rules. For example, the rules $(\to\Rightarrow)$ and $(\Rightarrow\to)$ in Figure 2.6 correspond to implication elimination and introduction, respectively, in natural deduction. On the other hand, the structural rules of LK and LJ have no natural deduction analogues; their role is played by corresponding design choices concerning the discharge of assumptions.

Further refinements of the sequent calculus approach were also motivated mostly by logical, rather than algebraic, considerations. In particular, the development of sequent calculi found (much later) to correspond to varieties of (pointed) residuated lattices was an outcome of investigations into the effects of *removing* structural rules from LK or LJ. One such effect became evident early on. Oiva Ketonen suggested in [**216**] a new version of LK with the aim of making its rules invertible. Ketonen's version contains the following left conjunction and right disjunction rules:

$$\frac{\Gamma, \alpha, \beta \Rightarrow \Delta}{\Gamma, \alpha \wedge \beta \Rightarrow \Delta}\ (\wedge\Rightarrow)' \qquad \frac{\Gamma \Rightarrow \alpha, \beta, \Delta}{\Gamma \Rightarrow \alpha \vee \beta, \Delta.}\ (\Rightarrow\vee)'$$

Notably, these rules are inter-derivable with $(\wedge\Rightarrow)_1, (\wedge\Rightarrow)_2$ and $(\Rightarrow\vee)_1, (\Rightarrow\vee)_2$, respectively, using the structural rules of weakening and contraction.

Similarly, Haskell B. Curry [**93**] (and others) considered the following variants of the rules $(\vee\Rightarrow)$ and $(\Rightarrow\wedge)$:

$$\frac{\Gamma_1, \alpha \Rightarrow \Delta_1 \quad \Gamma_2, \beta \Rightarrow \Delta_2}{\Gamma_1, \Gamma_2, \alpha \vee \beta \Rightarrow \Delta_1, \Delta_2}\ (\vee\Rightarrow)' \qquad \frac{\Gamma_1 \Rightarrow \alpha, \Delta_1 \quad \Gamma_2 \Rightarrow \beta, \Delta_2}{\Gamma_1, \Gamma_2 \Rightarrow \alpha \wedge \beta, \Delta_1, \Delta_2.}\ (\Rightarrow\wedge)'$$

[5]The first use of the expression "substructural logic" can be traced back to a suggestion of Kosta Došen and Peter Schroeder-Heister at the Tübingen conference on "Logics with restricted structural rules" in 1990 [**105**].

Unlike $(\vee\Rightarrow)$ and $(\Rightarrow\wedge)$, these rules are context-free (the contexts of the premises do not need to match). However, they are inter-derivable with $(\vee\Rightarrow)$ and $(\Rightarrow\wedge)$, respectively, again using the structural rules of contraction and weakening (see, e.g., [**303**, pp. 11–15] for details).

The inter-derivability of rules using the structural rules of LK allows some freedom as to the choice of left and right rules for conjunction and disjunction connectives. Suppose, however, that we remove weakening or contraction rules from LK. Then the proofs of inter-derivability for Gentzen's rules with Ketonen and Curry's rules are no longer valid. Hence, if the operation rules for a connective are understood to define their meaning, we no longer have, for example, a single conjunction but *two*: a lattice conjunction $\wedge$ that retains the properties of a meet but has no residual, and a monoidal conjunction $\cdot$ that loses idempotency but still forms a residuated pair with a corresponding implication.

2.6.2. The Lambek calculus. Chronologically, substructural sequent calculi were first applied in the field of *linguistics*. Joachim Lambek [**231**] made use of a substructural sequent calculus — now known as the *Lambek calculus* — to represent transformations on syntactic types of a formal grammar. Lambek's approach built on earlier work on *categorial grammar* in the 1930s by Kazimierz Ajdukiewicz, who aimed to develop an analysis of natural language by assigning syntactic types to linguistic expressions that describe their syntactic roles (e.g., verb, noun phrase, verb phrase, sentence). A naive approach to this task would consist of listing a number of lexical atoms (e.g., Joan, smiles, charmingly) and a number of mutually unrelated types (e.g., NP = noun phrase; V = verb; Adv = adverb; VP = verb phrase; S = sentence), and then tagging each lexical atom with the appropriate type (e.g., Joan: NP, smiles: V, charmingly: Adv).

Ajdukiewicz understood, however, that the stock of basic grammatical categories can be substantially reduced by the use of *type-forming operations* $\backslash$ and $/$, where an expression v has type $\alpha\backslash\beta$ (respectively, β/α) if whenever the expression v' has type α, then the expression $v'v$ (respectively, vv') has type β. Indeed, the whole apparatus of categorial grammar can then be constructed out of just two basic types, n (noun) and s (sentence). For example, in English, John works is a sentence, but works John is not. The intransitive verb works has type $n\backslash s$: applied to the right of an expression of type n, it yields an expression of type s. On the other hand, the adjective poor has type n/n: applied to the left of an expression of type n, it yields another expression of type n (a complex noun phrase). We may write these transformations as $n, n\backslash s \Rightarrow s$ and $n/n, n \Rightarrow n$, respectively. More generally, the following transformations are permissible for types α, β in Ajdukiewicz's categorial grammar:

$$\alpha, \alpha\backslash\beta \Rightarrow \beta \quad \text{and} \quad \beta/\alpha, \alpha \Rightarrow \beta.$$

Lambek extended the deductive power of categorial grammar by defining a sequent calculus for permissible transformations on types, introducing a new type-forming operation $\cdot$ such that v has type $\alpha \cdot \beta$ whenever $v = v'v''$, with v' of type α and v'' of type β, and admitting, in addition to modus ponens, patterns of *hypothetical reasoning* corresponding to right introduction rules for $\backslash$ and $/$.

Extending Lambek's sequent calculus with suitable rules for the lattice operation symbols $\wedge$ and $\vee$, and constant symbols e and f, gives the *full Lambek calculus* FL (see Figure 2.4), defined explicitly by Hiroakira Ono in [**293**].

2.6.3. Many-valued logics. The first examples of logics failing to admit the structural rule of contraction occurred in the setting of *many-valued logics*. In the 1920s, Jan Łukasiewicz introduced logics with truth values $0, \frac{1}{n-1}, \ldots, \frac{n-2}{n-1}, 1$, for each $n > 2$, and an infinite-valued logic Ł over the closed real unit interval $[0,1]$, where 0 represents absolute falsity, 1 represents absolute truth, and values between 0 and 1 can be thought of as intermediate "degrees of truth" [**239, 240**]. The truth functions corresponding to negation and implication are

$$\neg x := 1 - x \quad \text{and} \quad x \to y := \min(1, 1 - x + y).$$

In classical logic, there are several equivalent ways to define disjunction via negation and implication; for example, $\alpha \vee \beta$ is classically equivalent to both $\neg\alpha \to \beta$ and $(\alpha \to \beta) \to \beta$. However, in Ł, truth functions for two distinct disjunctions emerge:

$$x \oplus y := \neg x \to y = \min(1, x + y) \quad \text{and} \quad x \vee y := (x \to y) \to y = \max(x, y).$$

Similarly, there are two distinct truth functions for conjunction:

$$x \cdot y := \neg(x \to \neg y) = \max(0, x + y - 1) \quad \text{and} \quad x \wedge y := \neg(\neg x \vee \neg y) = \min(x, y).$$

A completeness proof for an axiomatization for Ł conjectured by Łukasiewicz [**240**] was obtained by Mordchaj Wajsberg in the 1930s, but published proofs appeared only in the late 1950s: a syntactic proof by Alan Rose and J. Barkley Rosser [**328**] and an algebraic proof (introducing MV-algebras as algebraic semantics for Ł; see Example 1.4.12) by Chen Chung Chang [**67**].

In subsequent years, the study of the infinite-valued logic Ł became ever more intertwined with research into *fuzzy logics*. Following the approach promoted by Petr Hájek in his influential monograph [**174**], the conjunction connective $\cdot$ of a fuzzy logic is interpreted by a continuous t-norm — a continuous, commutative, associative, and increasing binary function on $[0,1]$ with neutral element 1 — and the implication connective $\to$ is interpreted by its residual (see Example 1.4.15). The constant symbol f is interpreted by 0 and the definable connectives e, $\wedge$, and $\vee$ are interpreted by 1, min, and max, respectively. Fundamental examples of continuous t-norms are the Łukasiewicz t-norm $x \star_{\text{Ł}} y := \max(0, x + y - 1)$, the Gödel t-norm $x \star_{\text{G}} y := \min(x, y)$, and the product t-norm $x \star_{\text{P}} y := xy$ (the ordinary product of real numbers), which give rise, respectively, to Łukasiewicz logic Ł, Gödel-Dummett logic G [**107, 167**], and product logic P [**176**].

Common generalizations of the three main continuous t-norm logics include, in chronological order, Hájek's *basic fuzzy logic* BL [**174**], *monoidal t-norm based logic* MTL [**121**], and *uninorm logic* UL [**255**]. Algebraic semantics for these logics are provided by varieties of commutative, semilinear pointed residuated lattices, while their central role as fuzzy logics is justified by the fact that BL is the logic of continuous t-norms [**74**], MTL is the logic of residuated t-norms [**202**], and UL is the logic of residuated uninorms [**255**]. Algebraically, this means that the corresponding varieties are generated as quasivarieties by their "standard" members with lattice reduct $\langle [0,1], \min, \max \rangle$.

Although there was some awareness by the 1980s that substructural logics and fuzzy logics form part of a common hierarchy [**61**, **293**], a full understanding of the latter as genuine substructural logics was hindered by the lack of a good proof theory. The crucial missing ingredient, the notion of a hypersequent, was introduced by Arnon Avron in 1987 [**10**], leading in subsequent years to hypersequent calculi for G [**12**], MTL [**14**], Ł [**258**], P [**257**], UL [**255**], and many other fuzzy logics (see [**259**] for details).[6] Critically, all these calculi can be obtained by removing structural rules from suitable hypersequent calculi for classical logic.

2.6.4. Set theory. A further intriguing motivation for dropping the contraction rules arose from the realization that the formula $(\alpha \to (\alpha \to \beta)) \to (\alpha \to \beta)$, known as the *law of contraction*, plays a crucial role in formalizing set-theoretic paradoxes. At the turn of the twentieth century, Bertrand Russell showed that naive set theory yields, via unrestricted comprehension, a formula that is provably equivalent to its own negation. This produces a contradiction if the underlying logic proves the law of excluded middle and certain other classical principles. Some mathematicians, notably L.E.J. Brouwer, therefore began to nurture the belief that an intuitionistically correct set theory would escape this paradox. In fact, as shown by the following variant of an argument originally devised in 1942 by Curry [**92**] for illative combinatory logic, something akin to Russell's paradox follows also from intuitionistically acceptable principles. Let α be any sentence in the language of the theory, and let $C := \{x \mid (x \in x) \to \alpha\}$. Then we may reason as follows:

1.	$(C \in C) \to ((C \in C) \to \alpha)$	definition of C
2.	$((C \in C) \to \alpha) \to (C \in C)$	definition of C
3.	$((C \in C) \to ((C \in C) \to \alpha)) \to ((C \in C) \to \alpha)$	law of contraction
4.	$(C \in C) \to \alpha$	$1, 3$ modus ponens
5.	$C \in C$	$2, 4$ modus ponens
6.	α	$4, 5$ modus ponens.

Hence any set theory which contains an unrestricted comprehension axiom and the (intuitionistically correct) law of contraction is bound to be trivial. From a sequent calculus perspective, however, the law of excluded middle and the law of contraction are equally vicious: the former requires an application of right contraction, and the latter, an application of left contraction. Intuitionistic logic is not substructural enough to accommodate a set theory with unrestricted comprehension.

In a series of papers in the late 1950s and early 1960s (see, e.g., [**345**]), Thoralf Skolem explored the possibility of reconstructing set theory on a non-classical logical basis. In particular, Skolem considered Łukasiewicz logic Ł a plausible candidate for this reconstruction. Russell's paradoxical sentence is equivalent to its own negation, but this causes no problem in Ł, where any formula with a truth value of 0.5 has this property. Moreover, the above-mentioned contraction-related principles do not hold in this logic. On the other hand, weakening, which does hold in Ł, also plays a role in producing the set-theoretic paradoxes. Indeed, a result by V.N. Grišin [**172**] showed that if the extensionality axioms are added to classical logic without contraction (a weaker system than Ł), contraction can be recovered, so the system extended with comprehension is again trivial. Subsequent research

[6] Similar structures had also appeared in earlier papers by Mints [**267**] and Pottinger [**309**], but with the aim of providing cut-free calculi for modal logics.

on logical bases for naive set theories has therefore focused on even weaker systems. It would be misleading, however, to assume that the Curry paradox depends on the behavior of any particular connective, like implication. More recent investigations in set theory and formal theories of truth (where the Curry argument can be recast as a variant of the Liar) have yielded purely structural versions of the paradox, where no logical connective seems to play a role [**21**, **184**]. Crucially, applications of the structural rule of contraction are still needed for their proofs to go through.

2.6.5. Relevance logics. A significant source of substructural logics lacking weakening rules may be found in the philosophy of logic and the long-standing debate over *entailment*. Modern modal logic originated with C.I. Lewis' critical attitude towards Russell's formulation of classical propositional logic and what he called its "counterintuitive and repugnant to common sense" "paradoxes of material implication" such as the laws of *a fortiori* and *ex absurdo quodlibet*

$$\alpha \to (\beta \to \alpha) \quad \text{and} \quad \neg\alpha \to (\alpha \to \beta).$$

In a series of works culminating in the monograph *Symbolic Logic*, coauthored with C.H. Langford [**235**], Lewis introduced calculi of *strict implication* to describe a tighter notion of implication intended to be true whenever it is impossible that the antecedent holds while the consequent does not. These calculi avoid paradoxes of material implication, but derive "paradoxes of strict implication" such as

$$\alpha \to (\beta \vee \neg\beta) \quad \text{and} \quad (\alpha \wedge \neg\alpha) \to \beta.$$

The Lewis-Langford analysis was deemed inadequate by many commentators for failing to take into account the *relevance* connection, absent in these two formulas, between the antecedent and consequent of a logical implication. On the other hand, Lewis showed that these paradoxes can be obtained using just a few seemingly unobjectionable modes of inference [**235**, Ch. 8]. E.g., for *ex absurdo quodlibet*, we may reason as follows:

1.	$\alpha \wedge \neg\alpha$	assumption
2.	α	1, simplification: from $\alpha \wedge \beta$ derive α
3.	$\neg\alpha$	1, simplification: from $\alpha \wedge \beta$ derive β
4.	$\alpha \vee \beta$	2, addition: from α derive $\alpha \vee \beta$
5.	β	3, 4, disjunctive syllogism: from $\neg\alpha, \alpha \vee \beta$ derive β.

It is therefore incumbent on a relevant logician to explain what is wrong with this reasoning. Although several responses to Lewis were already proposed in the 1930s, 1940s, and 1950s, the most successful came in the late 1950s from Alan R. Anderson and Nuel D. Belnap, who, developing ideas of Wilhelm Ackermann [**1**], introduced the systems of *relevant logic* E and R (see [**5**, **6**]). Anderson and Belnap argued that the argument by Lewis is a *fallacy of ambiguity*: it equivocates over the meaning of disjunction. There is no single disjunction that allows both addition and the disjunctive syllogism; rather, there is an *intensional* relevant disjunction for which the disjunctive syllogism holds, but not addition, and an *extensional* truth-functional disjunction for which addition holds, but not the disjunctive syllogism. Although the connection was only implicit in Anderson and Belnap's papers, the extensional "or" roughly corresponds to a lattice disjunction, while the intensional "or" corresponds to the De Morgan dual of a monoid operation.

The placing of Anderson and Belnap's R within the framework of substructural logics is most evident for its implication-negation fragment, which corresponds to the implication-negation fragment of LK without weakening [**5**]. However, the full logic corresponding to the propositional part of LK without weakening is not R, but rather a weaker logic LR, studied intensively by Robert K. Meyer in his PhD thesis [**264**] that fails to derive the critical half of the distributivity law

$$(\alpha \wedge (\beta \vee \gamma)) \to ((\alpha \wedge \beta) \vee (\alpha \wedge \gamma)).$$

Meyer considered an extension of LR with a special rule to recover distributivity. Another route to recovering this property without re-introducing weakening was proposed independently by J. Michael Dunn [**110**] and Grigori Mints [**268**] based on a generalization of sequents that incorporates two structure-building operations: a semi-colon, interpreted as the intensional conjunction, and a comma, interpreted as the lattice conjunction. More precisely, a *DM-structure* can be defined inductively by stating that (i) all formulas and the empty set are DM-structures; and (ii) if X and Y are DM-structures, then X, Y and $X; Y$ are DM-structures.

A *DM-sequent* is an ordered pair consisting of a DM-structure X and a formula α, written $X \Rightarrow \alpha$. Substructures are also defined inductively by (i) X is a substructure of X; (ii) any substructure of X or Y is a substructure of X, Y and $X; Y$. Denoting by $X[Y]$ an occurrence of a substructure Y in X, we can then transfer operation rules from sequents to DM-sequents; e.g., for the implication rules, we obtain

$$\frac{Y \Rightarrow \alpha \quad X[\beta] \Rightarrow \gamma}{X[Y; \alpha \to \beta] \Rightarrow \gamma}\ (\to\Rightarrow) \qquad \frac{X; \alpha \Rightarrow \beta}{X \Rightarrow \alpha \to \beta.}\ (\Rightarrow\to)$$

Critically, this format allows the behavior of the comma and semi-colon to be governed by different structural rules. In particular, we can permit weakening for the comma (but not the semi-colon) via the rule

$$\frac{X[Y] \Rightarrow \alpha}{X[Y, Z] \Rightarrow \alpha.}\ (\textsc{wl})$$

In this way, we obtain a calculus that has cut elimination for the positive (i.e., negation-free) fragment of R that is substructural with respect to a DM-sequent calculus for classical logic. This perspective — that a logic can be understood, proof-theoretically, as a substructural logic in a suitably general framework — finds its fullest expression in the *display logic* formalism introduced by Belnap in [**23**]. In this setting, vast families of substructural (and other) logics, including the full logic R, can be presented as display logic calculi that have — on the basis of a uniform theorem — cut elimination. The price to be paid for this superior flexibility is the introduction of further structure-building operations. These can be awkward both proof-theoretically for establishing properties such as decidability and interpolation, and also algebraically, since these operations may not correspond to operations definable in the original signature.

An algebraic semantics for R with constants consisting of algebras term-equivalent to distributive, commutative involutive residuated lattices satisfying the equation $x \leq x{\cdot}x$ was provided by Dunn in 1966 [**109**]. More general residuated structures for relevant logics were later studied in detail by this author in the framework of *gaggle theory* [**111–113**]. In the early 1970s, an operational semantics for the implication fragment of R was introduced by Alasdair Urquhart [**358**], and a series of papers by Meyer and Richard Routley provided a Kripke-style relational semantics for R and an array of kindred systems admitting distributivity (see [**6**] and [**330**, vol. 1]). This semantic thread, to which Larissa Maksimova [**245**] and Kit Fine [**127**] also made significant contributions, triggered in turn the introduction of further relevant systems, mostly weaker than E or R, motivated by natural semantical conditions.

2.6.6. Linear logic. Further motivation for dropping structural rules arose from the constructive perspective on logic. Since Heyting, partisans of this approach have focused on the notion of *proof*, stressing, however, that what really matters is not whether a given formula is provable, but how it is proved. A formula may therefore be identified with the set of its proofs, so that a proof of α from the assumptions $\alpha_1, \dots, \alpha_n$ — seen as a method for converting proofs of $\alpha_1, \dots, \alpha_n$ into a proof of α — amounts to a function $\varphi(x_1, \dots, x_n)$ that associates to elements $a_1 \in \alpha_1, \dots, a_n \in \alpha_n$ the element $\varphi(a_1, \dots, a_n) \in \alpha$.

This idea is already implicit in the Brouwer-Heyting-Kolmogorov interpretation of intuitionistic logic. In the 1960s, William Howard [**196**] added to this interpretation the identification of intuitionistic natural deduction proofs with terms of the typed lambda calculus. A proof of the formula α is associated with a term of type α, and the computational content of the inference rules in the $\{\wedge, \to\}$-fragment of the intuitionistic natural deduction calculus can then be described as follows:

(i) if t and s are terms of type α and type β, respectively, then $\langle t, s\rangle$ (the *pairing* of t and s) is a term of type $\alpha \wedge \beta$;

(ii) if t is a term of type $\alpha\wedge\beta$, then $\pi_1(t)$ and $\pi_2(t)$ (the first and second *projections* of t) are terms of type α and β, respectively;

(iii) if x is a variable of type α and t is a term of type β, then $\lambda x.t$ (the *abstraction* of t with respect to x) is a term of type $\alpha \to \beta$;

(iv) if t and s are terms of type $\alpha \to \beta$ and type α, respectively, then ts (the *application* of t to s) is a term of type β.

The resulting correspondence between intuitionistic natural deduction proofs and terms in the lambda calculus with projection and pairing functors can be seen as a fully-fledged isomorphism (and is indeed referred to as the *Curry-Howard isomorphism*) in that there is a perfect match between the notions of conversion, normality, and reduction introduced in the two frameworks.

In the light of the Curry-Howard isomorphism, it became clear that the problem of finding a "semantics of proofs" for a given constructive logic and the problem of providing a lambda calculus (or, for that matter, functional programming) with a semantical interpretation are two sides of the same coin. In Dana Scott's *domain theory*, a first attempt to accomplish this task, a type α was interpreted by means of a particular topological space. An alternative interpretation was provided in the 1980s by Jean-Yves Girard [**163**] who introduced the notion of a *coherent space*: a set A equipped with a reflexive and symmetric relation $R^{\mathbf{A}}$, called the *coherence relation* of the space. A type (*alias* formula) α is associated with a coherent space $\mathbf{A} = \langle A, R^{\mathbf{A}}\rangle$, and a term t of type α (*alias* a proof of α) with a *clique* of $\mathbf{A}$, i.e., with a subset $B \subseteq A$ of pairwise coherent elements of A. Compound formulas and their proofs are interpreted using more complex coherent spaces. This semantics of proofs allows for substructural distinctions; for example,

(i) $\mathbf{A} \cdot \mathbf{B}$ is a coherent space with universe $A \times B$ and coherence relation defined for $a, a' \in A$ and $b, b' \in B$ by

$$\langle a, b\rangle R^{\mathbf{A} \cdot \mathbf{B}} \langle a', b'\rangle \;:\Longleftrightarrow\; aR^{\mathbf{A}}a' \text{ and } bR^{\mathbf{B}}b';$$

(ii) $\mathbf{A} \wedge \mathbf{B}$ is a coherent space with universe $\{\langle a, 0\rangle \mid a \in A\} \cup \{\langle b, 1\rangle \mid b \in B\}$ and coherence relation defined for $a, a' \in A$ and $b, b' \in B$ by $\langle a, 0\rangle R^{\mathbf{A} \wedge \mathbf{B}} \langle b, 1\rangle$,

$$\langle a, 0\rangle R^{\mathbf{A} \wedge \mathbf{B}} \langle a', 0\rangle \;:\Longleftrightarrow\; aR^{\mathbf{A}}a', \quad \text{and} \quad \langle b, 1\rangle R^{\mathbf{A} \wedge \mathbf{B}} \langle b', 1\rangle \;:\Longleftrightarrow\; bR^{\mathbf{B}}b'.$$

Omitting details, let us just mention that Girard also introduces a coherent space $\mathbf{A} \to \mathbf{B}$ corresponding to intuitionistic implication and a coherent space $\mathbf{A} \multimap \mathbf{B}$ corresponding to a new kind of implication, which he calls *linear* implication. If we let $!\mathbf{A}$ be the space whose universe is the set of finite cliques of $\mathbf{A}$, and whose coherence relation is given by

$$cR^{!\mathbf{A}}c' \;:\Longleftrightarrow\; \text{there exists a clique } c'' \text{ of } \mathbf{A} \text{ satisfying } c, c' \subseteq c'',$$

then we get the fundamental property that the space $\mathbf{A} \to \mathbf{B}$ is isomorphic to the space $!\mathbf{A} \multimap \mathbf{B}$; that is, the semantics of coherent spaces yields a decomposition of intuitionistic implication into the new linear implication $\multimap$ and unary operation $!$.

How can we make intuitive sense of this *linear logic*? One option is to view formulas as concrete resources that once consumed in a deduction, cannot be reused, but allow also formulas of the form $!\alpha$ to represent "ideal" resources that can be reused at will. Hence, while the availability of an intuitionistic implication $\alpha \to \beta$ means that by using as many α's as we want we can get one β, the availability of a linear implication $\alpha \multimap \beta$ expresses the fact that using just one α we can get one β, an interpretation that fits perfectly with the coherent space isomorphism mentioned above. We can also view the other compound formulas of linear logic as concrete resources; for example, $\alpha \cdot \beta$ expresses the availability of *both* resource α and resource β, while $\alpha \wedge \beta$ expresses the availability of *any* one of these resources.

In his seminal 1987 paper, Girard introduced a sequent calculus for this new logic that corresponds (with a quite different syntax) to the propositional part of LK without weakening and contraction, together with rules for alternative conjunction, disjunction, and truth constants, and extended with the following rules for the unary connective $!$ (defining also $?\alpha := \neg!\neg\alpha$):

$$\frac{\Gamma, \alpha \Rightarrow \Delta}{\Gamma, !\alpha \Rightarrow \Delta}\,(!\Rightarrow) \qquad \frac{!\Gamma \Rightarrow \alpha}{!\Gamma \Rightarrow !\alpha}\,(\Rightarrow!) \qquad \frac{\Gamma \Rightarrow \Delta}{\Gamma, !\alpha \Rightarrow \Delta}\,(!\mathrm{w}) \qquad \frac{\Gamma, !\alpha, !\alpha \Rightarrow \Delta}{\Gamma, !\alpha \Rightarrow \Delta,}\,(!\mathrm{c})$$

where $!\Gamma$ is obtained by prefixing $!$ to all formulas in Γ.

The importance of the exponentials is explained by the fact that Girard was not interested in setting up a logic weaker than classical or intuitionistic logic; rather, he wanted a logic that permits a better analysis of proofs via a stricter control of structural rules. Exponentials are there precisely to recapture the deductive power of weakening and contraction, an aim that is attained to some extent by showing that both classical logic and intuitionistic logic can be embedded into linear logic.

2.7. Bibliographical remarks

For general introductions to proof theory, focusing mostly on Hilbert systems and sequent calculi for classical and intuitionistic logic, we refer to [**199**, **285**, **353**, **357**].

Detailed presentations of proof systems for substructural logics, overlapping in some aspects with the development in this chapter, may be found in [**28, 145, 303, 327**], while fuzzy logics and hypersequent calculi are the main focus of [**259**].

The calculus for lattices presented in Figure 2.3 has several antecedents. It is implicit in Whitman's work on free lattices [**370**], which establishes decidability of the equational theory of lattices, but was first defined explicitly by Schulte Mönting in [**343**]. Decidability of the universal theory of lattices was proved earlier using quite different techniques by Skolem [**346**].

Our account of the relationship between the full Lambek calculus and the variety of pointed residuated lattices is based largely on the work of Ono [**293, 294, 296, 297**]. The algebraic cut elimination proof for extensions of FL with simple rules — notably, Lemma 2.4.8 — originated in Okada's papers [**288–290**], and was further developed in [**22, 70, 291, 356**]. These methods have led to a general algebraic approach to proof theory based on the notion of a residuated frame [**143, 144**]; in particular, the definition of a simple equation and Proposition 2.4.5 (fixing a small gap in the proof) may be found in [**143**].

The hypersequent calculus CSemFL for commutative, semilinear pointed residuated lattices is taken from [**255**]. A uniform syntactic proof of cut elimination for a broad family of hypersequent calculi extending CSemFL with structural rules was provided in [**259**], and an algebraic approach was developed in [**71**].

Varieties of cancellative or divisible (pointed) residuated lattices do not fit into these general frameworks, but "non-standard" proof systems have been obtained for significant cases. In particular, an analytic multiple-conclusion (with finite multisets of formulas on both sides of a sequent) hypersequent calculus was defined for the variety of Abelian ℓ-groups in [**258**], using an implication left rule

$$\frac{\mathcal{G} \mid \Gamma, \beta \Rightarrow \alpha, \Delta}{\mathcal{G} \mid \Gamma, \alpha \to \beta \Rightarrow \Delta.} \ (\to\Rightarrow)$$

Similar analytic calculi were obtained for the variety of MV-algebras and other varieties related to fuzzy logics in [**257–259**], and, in [**146**], an analytic one-sided (with sequences of formulas as sequents) hypersequent calculus was provided for the variety of ℓ-groups.

CHAPTER 3

Consequence relations

Derivability in a proof system C yields a natural notion of consequence; that is, the ordered pairs $\langle X, a\rangle$ such that a is derivable from X in C form a *consequence relation.* Such relations also arise naturally in other settings. Notably, associated with any variety $\mathcal{V}$ is a family of *equational* consequence relations that can be described via congruences of $\mathcal{V}$-free algebras and — as will be explored in Chapter 8 — provide "bridges" between its algebraic and logical properties.

The aim of this chapter is to provide a general abstract framework for the study of consequence relations arising in algebra and logic. In particular, we will see that consequence relations may be *equivalent* in the sense that there exist well-behaved translations between their consequences. Most pertinent to the topic of this book, consequence relations for varieties of pointed residuated lattices are equivalent to consequence relations based on extensions of the full Lambek calculus. In later chapters, we may therefore focus on varieties and use proof systems when available, safe in the knowledge that relevant results transfer also to substructural logics.

3.1. Abstract consequence relations

In order to define a relation describing logical consequence, we must first decide what kind of entities such a relation relates. As in the preceding chapter on proof systems, we give a very simple answer. We do not care if we are reasoning about formulas or more complicated objects such as equations, sequents, or hypersequents; we just suppose that these objects form a non-empty set A. We then assume that the consequences of each $X \subseteq A$ form a set $\gamma(X) \subseteq A$ with the following natural properties, satisfied for all $X, Y \subseteq A$:

(i) $X \subseteq \gamma(X)$: all members of X are consequences of X;

(ii) $\gamma(X) = \gamma(\gamma(X))$: any consequence of the consequences of X is a consequence of X;

(iii) if $X \subseteq Y$, then $\gamma(X) \subseteq \gamma(Y)$: any consequence of X is a consequence of any superset of X.

That is, γ is a closure operator on $\langle \wp(A), \subseteq\rangle$ (see Section 1.3), called a *consequence operator* on A.

Alternatively, we may collect together all ordered pairs $\langle X, a\rangle$ consisting of a subset $X \subseteq A$ and a consequence $a \in A$ of X. An *abstract consequence relation* (or *consequence relation* for short) over A is a relation $\vdash \subseteq \wp(A) \times A$ satisfying the following properties, for all $a \in A$ and $X, Y \subseteq A$:

(i) if $a \in X$, then $X \vdash a$ *(reflexivity)*;

(ii) if $X \vdash a$ and $X \subseteq Y$, then $Y \vdash a$ *(monotonicity)*;

(iii) if $X \vdash a$ and $Y \vdash b$, for every $b \in X$, then $Y \vdash a$ *(transitivity)*.

Following convention, we write $X \vdash Y$ if $X \vdash a$, for all $a \in Y$, and $\vdash a$ if $\emptyset \vdash a$. In the latter case, a is said to be a *theorem* of $\vdash$. We also write $X \dashv\vdash Y$ if both $X \vdash Y$ and $Y \vdash X$ hold.

PROPOSITION 3.1.1. *Let A be any non-empty set. Then there exists a bijective correspondence between the set of consequence operators γ on A and the set of consequence relations $\vdash$ over A. More precisely, if $\vdash$ is a consequence relation over A, then the following map is a consequence operator on A:*

$$\gamma_{\vdash} : \wp(A) \to \wp(A); \quad X \mapsto \{a \in A \mid X \vdash a\}.$$

Conversely, if γ is a consequence operator on A, then the following condition defines a consequence relation $\vdash_{\gamma} \subseteq \wp(A) \times A$ over A:

$$X \vdash_{\gamma} a \;:\Longleftrightarrow\; a \in \gamma(X).$$

Moreover, $\gamma_{\vdash_{\gamma}} = \gamma$ and $\vdash_{\gamma_{\vdash}} = \vdash$.

PROOF. Let $\vdash$ be any consequence relation over A. Then $\gamma_{\vdash}$ is enlarging (since $\vdash$ is reflexive), idempotent (since $\vdash$ is transitive), and an order-endomorphism of $\langle \wp(A), \subseteq \rangle$ (since $\vdash$ is monotonic). A similar appeal to the three properties of any consequence operator γ shows that $\vdash_{\gamma}$ satisfies the conditions for a consequence relation. Finally, for any consequence operator γ on A and $X \subseteq A$,

$$\gamma_{\vdash_{\gamma}}(X) = \{a \in A \mid X \vdash_{\gamma} a\} = \{a \in A \mid a \in \gamma(X)\} = \gamma(X),$$

and, for any consequence relation $\vdash$ over A and $X \cup \{a\} \subseteq A$,

$$X \vdash_{\gamma_{\vdash}} a \iff a \in \gamma_{\vdash}(X) \iff X \vdash a.$$

That is, $\gamma_{\vdash_{\gamma}} = \gamma$ and $\vdash_{\gamma_{\vdash}} = \vdash$. □

A subset X of a non-empty set A is called a *theory* of a consequence relation $\vdash$ over A if $X \vdash a$ implies $a \in X$. The theories of $\vdash$ are the *closed subsets* of A under $\gamma_{\vdash}$; by Proposition 1.3.6, they form a complete lattice $\langle \gamma_{\vdash}[\wp(A)], \subseteq \rangle$ with operations $\cap$ and $\vee$, where $X \vee Y = \gamma_{\vdash}(X \cup Y)$.

No restriction is made here on the cardinalities of the sets occurring in consequences, which may be infinite as well as finite. However, since logical deductions generally proceed from finitely many premises, we often consider consequence operators γ on A that satisfy, for any $X \subseteq A$,

$$\gamma(X) = \bigcup \{\gamma(Y) \mid Y \subseteq X \text{ and } Y \text{ is finite}\},$$

which holds if and only if, for any $X \cup \{a\} \subseteq A$,

$$X \vdash_{\gamma} a \iff Y \vdash_{\gamma} a, \text{ for some finite } Y \subseteq X.$$

In this case, we call γ and the associated consequence relation $\vdash_{\gamma}$ *finitary*.

Proof systems provide an important source of finitary consequence relations.

PROPOSITION 3.1.2. *Let $\mathsf{C} = \langle A, R \rangle$ be any proof system. Then $\vdash_{\mathsf{C}}$ is a finitary consequence relation over A.*

PROOF. Let $X, Y \subseteq A$ and $a \in A$. For reflexivity, if $a \in X$, then a single node labeled with a is a derivation for $X \vdash_{\mathsf{C}} a$. For monotonicity, suppose that $X \vdash_{\mathsf{C}} a$ and $X \subseteq Y$. Clearly, any derivation for $X \vdash_{\mathsf{C}} a$ is also a derivation for $Y \vdash_{\mathsf{C}} a$. For transitivity, suppose that $Y \vdash_{\mathsf{C}} a$ and $X \vdash_{\mathsf{C}} b$ for every $b \in Y$. We can perform a surgery on the derivation for $Y \vdash_{\mathsf{C}} a$ by extending each leaf labeled with a $b \in Y$ with a derivation for $X \vdash_{\mathsf{C}} b$. The result is a derivation for $X \vdash_{\mathsf{C}} a$. Finally, if $X \vdash_{\mathsf{C}} a$, then let Y be the finite set of members of X occurring in a derivation for $X \vdash_{\mathsf{C}} a$. Clearly, $Y \vdash_{\mathsf{C}} a$. Hence $\vdash_{\mathsf{C}}$ is a finitary consequence relation over A. □

One of the distinctive properties of logical consequence is its formal character: what follows logically from a set of premises can be inferred from the premises purely by virtue of their logical form. "It's sunny and it's windy" does not entail "It's sunny" because of the actual weather conditions or the meanings of the words "sunny" or "windy"; the entailment is preserved whenever we replace these words by terms in the appropriate grammatical categories.

To give a proper account of this feature, we consider operations σ acting on members of a non-empty set A by uniformly substituting their non-logical components in such a way as to leave their logical form unchanged. It is natural also to assume that such actions σ_1, σ_2 can be composed as $\sigma_1 \circ \sigma_2$ and allow an identity action 1 that leaves members of A unaltered. The formal character of logical consequence therefore translates into the requirement that, for any action σ, if a is a logical consequence of X, then $\sigma(a)$ is a logical consequence of the set $\sigma[X] = \{\sigma(b) \mid b \in X\}$.

More formally, let $\mathbf{M} = \langle M, \circ, 1 \rangle$ be any monoid. An $\mathbf{M}$-*set* $\mathbf{A} = \langle A, \cdot \rangle$ consists of a non-empty set A and a binary operation $\cdot \colon M \times A \to A$ satisfying, for all $\sigma_1, \sigma_2 \in M$ and $a \in A$,

(i) $1 \cdot a = a$;

(ii) $(\sigma_1 \circ \sigma_2) \cdot a = \sigma_1 \cdot (\sigma_2 \cdot a)$.

Considering the elements of M as "actions" on the set A, we adopt the more intuitive notation $\sigma(a)$ for $\sigma \cdot a$, recalling also that we often omit the monoid multiplication symbol when writing products.

A consequence operator γ on a non-empty set A is said to be *action-invariant* on an $\mathbf{M}$-set $\mathbf{A} = \langle A, \cdot \rangle$ if, for any $\sigma \in M$ and $X \subseteq A$,

$$\sigma[\gamma(X)] \subseteq \gamma(\sigma[X]).$$

It follows that γ is action-invariant if and only if, for any $\sigma \in M$ and $X \cup \{a\} \subseteq A$,

$$X \vdash_{\gamma} a \implies \sigma[X] \vdash_{\gamma} \sigma(a).$$

We therefore also call any consequence relation $\vdash$ over A that satisfies this property *action-invariant* over $\mathbf{A}$.

Proof systems often, but not always, yield action-invariant consequence relations. Let $\mathbf{A}$ be an $\mathbf{M}$-set for some monoid $\mathbf{M}$. A rule r for A is called $\mathbf{A}$-*schematic* if for every instance $a_1, \dots, a_n \,/\, a$ of r and $\sigma \in M$, also $\sigma(a_1), \dots, \sigma(a_n) \,/\, \sigma(a)$ is an instance of r. A proof system is called $\mathbf{A}$-*schematic* (or just *schematic* if $\mathbf{A}$ is clear from the context) if all of its rules are $\mathbf{A}$-schematic. It is easy to see that each $\mathbf{A}$-schematic proof system gives rise to an action-invariant consequence relation; however, this can also be the case for a proof system that is not $\mathbf{A}$-schematic.

The class of consequence relations is sufficiently broad to encompass a diverse range of examples. For most consequence relations used in logic, however, the objects in the set A are constructed from the members of a formula algebra $\mathbf{Fm}_{\mathcal{L}}(X)$ of signature $\mathcal{L}$ over a set of variables X, recalling that we omit the subscript if the signature is clear from the context and write just $\mathbf{Fm}$ when X is a fixed countably infinite set. In particular, A could be the set of $\mathcal{L}$-formulas, $\mathcal{L}$-equations, (unary) $\mathcal{L}$-sequents, or $\mathcal{L}$-hypersequents.

Suppose first that the set A is Fm. Substitutions on Fm, endomorphisms of $\mathbf{Fm}$, form a monoid $\mathbf{End}$ under function composition, and the binary operation $\cdot\colon \mathrm{End} \times \mathrm{Fm} \to \mathrm{Fm}$ can be defined as substitution application. A consequence relation over Fm that is action-invariant over the $\mathbf{End}$-set $\langle \mathrm{Fm}, \cdot \rangle$ is typically called *substitution-invariant* over $\mathbf{Fm}$.[1] A *(propositional) logic* may then be defined as a pair $\mathrm{L} = \langle \mathbf{Fm}, \vdash_{\mathrm{L}} \rangle$ where $\vdash_{\mathrm{L}}$ is a substitution-invariant consequence relation over $\mathbf{Fm}$.

EXAMPLE 3.1.3. Classical propositional logic can be defined as $\mathrm{CL} := \langle \mathbf{Fm}, \vdash_{\mathsf{HCL}} \rangle$, where $\mathbf{Fm}$ is the formula algebra of signature $\langle 2, 2, 2, 1, 0, 0 \rangle$ with connectives $\wedge$, $\vee$, $\to$, $\neg$, 1, and 0, and HCL is the Hilbert system presented in Figure 2.1.

By Proposition 3.1.2, every schematic Hilbert system gives rise to a finitary logic. In fact, the converse is also true: it is easy to check that every finitary logic $\mathrm{L} = \langle \mathbf{Fm}, \vdash_{\mathrm{L}} \rangle$ arises from derivability in a schematic Hilbert system containing a schematic inference rule $\langle \{\alpha_1, \dots, \alpha_n\}, \alpha \rangle$, for all $\alpha_1, \dots, \alpha_n, \alpha \in \mathrm{Fm}$ satisfying $\{\alpha_1, \dots, \alpha_n\} \vdash_{\mathrm{L}} \alpha$. This result is known as the *Łoś-Suszko theorem.*

Sequent and hypersequent calculi provide a further important source of finitary, substitution-invariant consequence relations, where substitutions on Fm extend to these settings in the obvious way. Let σ be any substitution on Fm. For any finite, possibly empty, sequence or multiset Γ of $\mathcal{L}$-formulas $\alpha_1, \dots, \alpha_n$, we let $\sigma(\Gamma)$ be the sequence or multiset of $\mathcal{L}$-formulas $\sigma(\alpha_1), \dots, \sigma(\alpha_n)$. Then, for any (multiset) $\mathcal{L}$-sequent $S = \Gamma \Rightarrow \Delta$ and $\mathcal{L}$-hypersequent $\mathcal{G} = S_1 \mid \cdots \mid S_m$, we let $\sigma(S) := \sigma(\Gamma) \Rightarrow \sigma(\Delta)$ and $\sigma(\mathcal{G}) := \sigma(S_1) \mid \cdots \mid \sigma(S_n)$, respectively. Substitutions extend to multiple-conclusion sequents analogously. As in the case of formulas, we may then speak of substitution-invariant consequence relations over sets of sequents or hypersequents. By Proposition 3.1.2, every schematic sequent or hypersequent calculus — in particular, the systems Lat, FL, CFL, and CSemFL introduced in Chapter 2 — gives rise to finitary, substitution-invariant consequence relations.

EXAMPLE 3.1.4. The sequent calculus GCL for classical propositional logic displayed in Figure 2.6 is defined over the set of multiple-conclusion sequents $\Gamma \Rightarrow \Delta$, where Γ and Δ are finite sequences of formulas over the signature $\langle 2, 2, 2, 1, 0, 0 \rangle$ with connectives $\wedge$, $\vee$, $\to$, $\neg$, 1, and 0. Since the rules of GCL are schematic, it follows that $\vdash_{\mathsf{GCL}}$ is a finitary, substitution-invariant consequence relation.

It is worth remarking at this juncture that a sequent calculus and its cut-free variant typically give rise to different consequence relations. For example,

$$\{x \Rightarrow y, y \Rightarrow z\} \vdash_{\mathsf{FL}} x \Rightarrow z, \quad \text{but} \quad \{x \Rightarrow y, y \Rightarrow z\} \nvdash_{\mathsf{FL}^\circ} x \Rightarrow z.$$

The cut elimination result for FL ensures only that these two consequence relations have the same *theorems.*

---

[1] The term "structural" is also widely used, but would be somewhat misleading in the context of this book, since the property is unrelated to the concept of a structural rule.

3.2. Equational consequence relations

Formulas are the basic statements of logic, whereas, in algebra, this role is typically taken by *equations*. In this section, we show that a natural family of "equational" consequence relations can be associated with any class of algebras of the same signature, and that when this class is a variety, these consequence relations can be expressed via congruences of free algebras for the class. For convenience, we recall certain relevant definitions from universal algebra along the way, referring the reader to Appendix B for further details.

Let us fix an arbitrary signature $\mathcal{L}$ for the remainder of this section, writing $\mathcal{L}_n$ to denote the set of operation symbols of $\mathcal{L}$ of arity $n \in \mathbb{N}$. Also, let X be a set of variables such that $X \cup \mathcal{L}_0 \neq \emptyset$, ensuring that the formula algebra $\mathbf{Fm}(X)$ exists. Recall that $\mathrm{Eq}_{\mathcal{L}}(X)$ denotes the set of $\mathcal{L}$-equations (ordered pairs of $\mathcal{L}$-formulas) over a set of variables X, and that we write $\mathrm{Eq}(X)$ when $\mathcal{L}$ is clear from the context, or just Eq when X is a fixed countably infinite set of variables.

The monoid of substitutions $\mathbf{End}(\mathbf{Fm}(X))$ on $\mathbf{Fm}(X)$ acts on $\mathrm{Eq}(X)$ by mapping an $\mathcal{L}$-equation $\alpha \approx \beta$ to $\sigma \cdot (\alpha \approx \beta) := \sigma(\alpha) \approx \sigma(\beta)$, for each $\sigma \in \mathrm{End}(\mathbf{Fm}(X))$. A consequence relation over $\mathrm{Eq}(X)$ is called *substitution-invariant* over $\mathbf{Eq}(X) := \mathbf{Fm}(X)^2$ if it is action-invariant over the $\mathbf{End}(\mathbf{Fm}(X))$-set $\langle \mathrm{Eq}(X), \cdot \rangle$.

For any map $\varphi\colon A \to B$, let $\ker(\varphi) := \{\langle a_1, a_2\rangle \mid \varphi(a_1) = \varphi(a_2)\}$ and let $\hom(\mathbf{A}, \mathbf{B})$ denote the set of all homomorphisms from an $\mathcal{L}$-algebra $\mathbf{A}$ to an $\mathcal{L}$-algebra $\mathbf{B}$, noting that $\ker(\varphi) \in \mathrm{Con}(\mathbf{A})$, for any $\varphi \in \hom(\mathbf{A}, \mathbf{B})$.

Given any class $\mathcal{K}$ of $\mathcal{L}$-algebras, we define for $\Sigma \cup \{\varepsilon\} \subseteq \mathrm{Eq}(X)$,

$$\begin{aligned}\Sigma \vdash_{\mathcal{K}}^{X} \varepsilon \;:\Longleftrightarrow\; & \text{for all } \mathbf{A} \in \mathcal{K} \text{ and } \varphi \in \hom(\mathbf{Fm}(X), \mathbf{A}),\\ & \Sigma \subseteq \ker(\varphi) \implies \varepsilon \in \ker(\varphi).\end{aligned}$$

Note that the consequences of $\vdash_{\mathcal{K}}^{X}$ from *finite* sets of $\mathcal{L}$-equations may be understood as the $\mathcal{L}$-quasi-equations that are satisfied by $\mathcal{K}$; that is, $\{\varepsilon_1, \dots, \varepsilon_n\} \vdash_{\mathcal{K}}^{X} \varepsilon$ if and only if $\mathcal{K} \models \varepsilon_1 \,\&\, \cdots \,\&\, \varepsilon_n \implies \varepsilon$.

The following proposition allows us to refer to $\vdash_{\mathcal{K}}^{X}$ as the *equational consequence relation* of $\mathcal{K}$ over $\mathrm{Eq}(X)$.

PROPOSITION 3.2.1. *Let $\mathcal{K}$ be any class of $\mathcal{L}$-algebras and let X be a set such that $X \cup \mathcal{L}_0 \neq \emptyset$. Then $\vdash_{\mathcal{K}}^{X}$ is a substitution-invariant consequence relation over $\mathbf{Eq}(X)$.*

PROOF. Reflexivity and monotonicity follow immediately from the definition. For transitivity, suppose that $\Sigma \vdash_{\mathcal{K}}^{X} \varepsilon$ and $\Pi \vdash_{\mathcal{K}}^{X} \Sigma$. Let $\mathbf{A} \in \mathcal{K}$ and $\varphi \in \hom(\mathbf{Fm}(X), \mathbf{A})$. If $\Pi \subseteq \ker(\varphi)$, then $\Sigma \subseteq \ker(\varphi)$ and hence $\varepsilon \in \ker(\varphi)$. That is, $\Pi \vdash_{\mathcal{K}}^{X} \varepsilon$. So $\vdash_{\mathcal{K}}^{X}$ is a consequence relation over $\mathrm{Eq}(X)$. Finally, for substitution-invariance, suppose that $\Sigma \vdash_{\mathcal{K}}^{X} \varepsilon$ and let $\sigma \in \mathrm{End}(\mathbf{Fm}(X))$. For any $\mathbf{A} \in \mathcal{K}$ and $\varphi \in \hom(\mathbf{Fm}(X), \mathbf{A})$, also $\varphi \circ \sigma \in \hom(\mathbf{Fm}(X), \mathbf{A})$, so

$$\sigma[\Sigma] \subseteq \ker(\varphi) \implies \Sigma \subseteq \ker(\varphi \circ \sigma) \implies \varepsilon \in \ker(\varphi \circ \sigma) \implies \sigma(\varepsilon) \in \ker(\varphi).$$

That is, $\sigma[\Sigma] \vdash_{\mathcal{K}}^{X} \sigma(\varepsilon)$ as required. □

Different classes of $\mathcal{L}$-algebras can give rise to the same equational consequence relations. A precise characterization of this situation is obtained using well-known results from universal algebra (see Appendix B). For a class of $\mathcal{L}$-algebras $\mathcal{K}$, let $\mathbb{I}(\mathcal{K})$, $\mathbb{H}(\mathcal{K})$, $\mathbb{S}(\mathcal{K})$, $\mathbb{P}(\mathcal{K})$, and $\mathbb{P}_U(\mathcal{K})$ denote the classes of isomorphic copies of algebras in $\mathcal{K}$, homomorphic images of algebras in $\mathcal{K}$, subalgebras of algebras in $\mathcal{K}$, direct products of algebras in $\mathcal{K}$, and ultraproducts of algebras in $\mathcal{K}$, respectively.

PROPOSITION 3.2.2. *Let $\mathcal{K}_1$ and $\mathcal{K}_2$ be any classes of $\mathcal{L}$-algebras. Then $\vdash^X_{\mathcal{K}_1} = \vdash^X_{\mathcal{K}_2}$, for every set X such that $X \cup \mathcal{L}_0 \neq \emptyset$, if and only if $\mathbb{ISP}(\mathcal{K}_1) = \mathbb{ISP}(\mathcal{K}_2)$.*

PROOF. Suppose first that $\vdash^X_{\mathcal{K}_1} = \vdash^X_{\mathcal{K}_2}$, for every set X such that $X \cup \mathcal{L}_0 \neq \emptyset$. To prove that $\mathbb{ISP}(\mathcal{K}_1) = \mathbb{ISP}(\mathcal{K}_2)$, it suffices, by symmetry and the fact that $\mathbb{PS}(\mathcal{K}) \subseteq \mathbb{SP}(\mathcal{K})$, for any class $\mathcal{K}$, to show that $\mathcal{K}_1 \subseteq \mathbb{ISP}(\mathcal{K}_2)$. Given $\mathbf{A} \in \mathcal{K}_1$, let $\psi\colon \mathbf{Fm}(A) \to \mathbf{A}$ be the unique surjective homomorphism extending the identity map on A, and let $\Sigma := \ker(\psi)$. By the Homomorphism theorem (Corollary B.3), $\mathbf{A} \cong \mathbf{Fm}(A)/\Sigma$. For each $\varepsilon \in \mathrm{Eq}(A)\backslash\Sigma$, clearly $\Sigma \not\vdash^A_{\mathcal{K}_1} \varepsilon$, and hence, by assumption, $\Sigma \not\vdash^A_{\mathcal{K}_2} \varepsilon$. So, for each $\varepsilon \in \mathrm{Eq}(A)\backslash\Sigma$, there exist $\mathbf{B}_\varepsilon \in \mathcal{K}_2$ and a homomorphism $\varphi_\varepsilon\colon \mathbf{Fm}(A) \to \mathbf{B}_\varepsilon$ such that $\Sigma \subseteq \ker(\varphi_\varepsilon)$ and $\varepsilon \notin \ker(\varphi_\varepsilon)$. Let φ be the induced homomorphism from $\mathbf{Fm}(A)$ into the direct product of the family of $\mathcal{L}$-algebras $\{\mathbf{B}_\varepsilon \mid \varepsilon \in \mathrm{Eq}(A)\backslash\Sigma\} \subseteq \mathcal{K}_2$. Then $\ker(\varphi) = \Sigma$ and the Homomorphism theorem (Corollary B.3) yields $\mathbf{Fm}(A)/\Sigma \in \mathbb{ISP}(\mathcal{K}_2)$. Hence also $\mathbf{A} \in \mathbb{ISP}(\mathcal{K}_2)$.

Now suppose that $\mathbb{ISP}(\mathcal{K}_1) = \mathbb{ISP}(\mathcal{K}_2)$ and let X be any set such that $X \cup \mathcal{L}_0 \neq \emptyset$. To obtain $\vdash^X_{\mathcal{K}_1} = \vdash^X_{\mathcal{K}_2}$, it suffices to observe, leaving the details to the reader, that $\vdash^X_{\mathcal{K}} = \vdash^X_{\mathbb{I}(\mathcal{K})} = \vdash^X_{\mathbb{S}(\mathcal{K})} = \vdash^X_{\mathbb{P}(\mathcal{K})}$, for any class of $\mathcal{L}$-algebras $\mathcal{K}$. $\square$

EXAMPLE 3.2.3. By Stone's representation theorem, every Boolean algebra embeds into a power of the two-element Boolean algebra $\mathbf{2}$; that is, $\mathcal{BA} = \mathbb{ISP}(\mathbf{2})$. Hence $\vdash^X_{\mathcal{BA}} = \vdash^X_{\{\mathbf{2}\}}$, for any set X.

Similarly, we obtain a precise characterization of the case where different classes of $\mathcal{L}$-algebras give rise to equational consequence relations with the same theorems.

PROPOSITION 3.2.4. *Let $\mathcal{K}_1$ and $\mathcal{K}_2$ be any classes of $\mathcal{L}$-algebras. Then $\vdash^X_{\mathcal{K}_1}$ and $\vdash^X_{\mathcal{K}_2}$ have the same theorems, for every set X such that $X \cup \mathcal{L}_0 \neq \emptyset$, if and only if $\mathbb{HSP}(\mathcal{K}_1) = \mathbb{HSP}(\mathcal{K}_2)$.*

PROOF. Suppose first that $\vdash^X_{\mathcal{K}_1}$ and $\vdash^X_{\mathcal{K}_2}$ have the same theorems, for every set X such that $X \cup \mathcal{L}_0 \neq \emptyset$. By the HSP theorem (Theorem B.12), there exists for $\mathbb{HSP}(\mathcal{K}_1)$ an equational basis $\Sigma \subseteq \mathrm{Eq}(X)$ for some set X satisfying $X \cup \mathcal{L}_0 \neq \emptyset$. Clearly, $\vdash^X_{\mathcal{K}_1} \Sigma$, and hence, by assumption, $\vdash^X_{\mathcal{K}_2} \Sigma$ and $\mathcal{K}_2 \subseteq \mathbb{HSP}(\mathcal{K}_1)$. Reasoning symmetrically, $\mathcal{K}_1 \subseteq \mathbb{HSP}(\mathcal{K}_2)$ and it follows easily that $\mathbb{HSP}(\mathcal{K}_1) = \mathbb{HSP}(\mathcal{K}_2)$.

Suppose now that $\mathbb{HSP}(\mathcal{K}_1) = \mathbb{HSP}(\mathcal{K}_2)$ and let X be any set such that $X \cup \mathcal{L}_0 \neq \emptyset$. If $\vdash^X_{\mathcal{K}_1} \varepsilon$, then, since $\mathcal{K}_1$ and $\mathbb{HSP}(\mathcal{K}_1)$ satisfy the same $\mathcal{L}$-equations, $\vdash^X_{\mathbb{HSP}(\mathcal{K}_1)} \varepsilon$, and so, by assumption, $\vdash^X_{\mathbb{HSP}(\mathcal{K}_2)} \varepsilon$ and $\vdash^X_{\mathcal{K}_2} \varepsilon$. The opposite direction is symmetrical. $\square$

We can also give a precise characterization of the case where a class of $\mathcal{L}$-algebras yields finitary equational consequence relations.

PROPOSITION 3.2.5. *Let $\mathcal{K}$ be any class of $\mathcal{L}$-algebras. Then $\vdash^X_{\mathcal{K}}$ is finitary, for every set X such that $X \cup \mathcal{L}_0 \neq \emptyset$, if and only if $\mathbb{ISPP}_U(\mathcal{K}) = \mathbb{ISP}(\mathcal{K})$.*

PROOF. Suppose first that $\vdash^X_{\mathcal{K}}$ is finitary, for every set X such that $X \cup \mathcal{L}_0 \neq \emptyset$. Then, for any set X such that $X \cup \mathcal{L}_0 \neq \emptyset$, since $\mathcal{K}$ and $\mathbb{P}_U(\mathcal{K})$ satisfy the same $\mathcal{L}$-quasi-equations and $\vdash^X_{\mathcal{K}}$ is finitary, $\vdash^X_{\mathcal{K}} = \vdash^X_{\mathbb{P}_U(\mathcal{K})}$. Hence $\mathbb{ISPP}_U(\mathcal{K}) = \mathbb{ISP}(\mathcal{K})$, by Proposition 3.2.2.

Suppose now that $\vdash^X_{\mathcal{K}}$ is not finitary, for some set X such that $X \cup \mathcal{L}_0 \neq \emptyset$. Then there exists $\Sigma \cup \{\varepsilon\} \subseteq \mathrm{Eq}(X)$ such that $\Sigma \vdash^X_{\mathcal{K}} \varepsilon$, but $\Pi \nvdash^X_{\mathcal{K}} \varepsilon$ for every finite subset $\Pi \subseteq \Sigma$. Let I be the set of finite subsets of Σ and let $\Pi^* := \{\Pi' \in I \mid \Pi \subseteq \Pi'\}$, for each $\Pi \in I$. Then no finite intersection of elements of $\{\Pi^* \mid \Pi \in I\}$ is empty and, by the Ultrafilter lemma (Lemma B.14), there exists an ultrafilter U on I extending this set. For each $\Pi \in I$, there exist $\mathbf{A}_\Pi \in \mathcal{K}$ and $\varphi_\Pi \in \mathrm{hom}(\mathbf{Fm}(X), \mathbf{A}_\Pi)$ such that $\Pi \subseteq \ker(\varphi_\Pi)$, but $\varepsilon \notin \ker(\varphi_\Pi)$. So $\{\delta\}^* \subseteq \{\Pi \in I \mid \Pi \vdash^X_{\mathcal{K}} \delta\}$, for each $\delta \in \Sigma$, and, since U is closed upwards, $\{\Pi \in I \mid \Pi \vdash^X_{\mathcal{K}} \delta\} \in U$. The family $\{\varphi_\Pi\}_{\Pi \in I}$ induces a homomorphism φ from $\mathbf{Fm}(X)$ to the ultraproduct $\prod_{\Pi \in I} \mathbf{A}_\Pi / U \in \mathbb{P}_U(\mathcal{K})$. Then $\Sigma \subseteq \ker(\varphi)$, but $\varepsilon \notin \ker(\varphi)$, since $\{\Pi \in I \mid \Pi \vdash^X_{\mathcal{K}} \varepsilon\} = \emptyset \notin U$. Hence $\Sigma \nvdash^X_{\mathbb{P}_U(\mathcal{K})} \varepsilon$ and Proposition 3.2.2 yields $\mathbb{ISPP}_U(\mathcal{K}) \neq \mathbb{ISP}(\mathcal{K})$. $\square$

Since quasivarieties are closed under taking ultraproducts, Proposition 3.2.5 yields the following refinement of Proposition 3.2.1 for these classes.

COROLLARY 3.2.6. *Let $\mathcal{K}$ be a quasivariety and let X be a set such that $X \cup \mathcal{L}_0 \neq \emptyset$. Then $\vdash^X_{\mathcal{K}}$ is a finitary, substitution-invariant consequence relation over* $\mathbf{Eq}(X)$.

Let us turn our attention now to an interpretation of equational consequence for varieties, expressed in terms of congruences of free algebras, that has proved to be very useful in relating various logical and algebraic properties (see Chapter 8).

Given any variety $\mathcal{V}$ of $\mathcal{L}$-algebras and set X such that $X \cup \mathcal{L}_0 \neq \emptyset$, we consider the following congruence of the formula algebra $\mathbf{Fm}(X)$:

$$\Theta_{\mathcal{V}}(X) := \bigcap\{\Theta \in \mathrm{Con}(\mathbf{Fm}(X)) \mid \mathbf{Fm}(X)/\Theta \in \mathcal{V}\}.$$

Let $\bar{\alpha} := [\alpha]_{\Theta_{\mathcal{V}}(X)} = \{\beta \in \mathrm{Fm} \mid \langle \alpha, \beta \rangle \in \Theta_{\mathcal{V}}(X)\}$ for $\alpha \in \mathrm{Fm}(X)$, $\overline{T} := \{\bar{\alpha} \mid \alpha \in T\}$ for $T \subseteq \mathrm{Fm}(X)$, and $\overline{\Sigma} := \{\langle \bar{\alpha}, \bar{\beta} \rangle \mid \alpha \approx \beta \in \Sigma\}$ for $\Sigma \subseteq \mathrm{Eq}(X)$.

The unique (up to isomorphism) *$\mathcal{V}$-free algebra over $\overline{X}$* is identified with the quotient $\mathbf{F}_{\mathcal{V}}(X) := \mathbf{Fm}(X)/\Theta_{\mathcal{V}}(X)$, noting in particular that $\mathbf{F}_{\mathcal{V}}(X)$ is generated by $\overline{X}$, and if $\mathcal{V}$ contains a non-trivial algebra, then $|X| = |\overline{X}|$. When the variety $\mathcal{V}$ is clear from the context, we omit the subscript and write just $\mathbf{F}(X)$.

The following result describes the relationship between the equational consequence relations of $\mathcal{V}$ and the congruences of $\mathcal{V}$-free algebras.

PROPOSITION 3.2.7. *Let $\mathcal{V}$ be any variety of $\mathcal{L}$-algebras and let X be a set such that $X \cup \mathcal{L}_0 \neq \emptyset$. The following statements are equivalent, for any $\Sigma \cup \{\varepsilon\} \subseteq \mathrm{Eq}(X)$:*

(1) $\Sigma \vdash^X_{\mathcal{V}} \varepsilon$.
(2) $\varepsilon \in \mathrm{Cg}^{\mathbf{Fm}(X)}(\Sigma) \vee \Theta_{\mathcal{V}}(X)$.
(3) $\bar{\varepsilon} \in \mathrm{Cg}^{\mathbf{F}(X)}(\overline{\Sigma})$.

PROOF. Let $\Psi := \mathrm{Cg}^{\mathbf{Fm}(X)}(\Sigma) \vee \Theta_{\mathcal{V}}(X)$, noting that $\mathbf{Fm}(X)/\Psi \in \mathcal{V}$.

(1) ⇒ (2) Suppose that $\Sigma \vdash_{\mathcal{V}}^{X} \varepsilon$ and consider the canonical map

$$\pi_{\Psi}\colon \mathbf{Fm}(X) \to \mathbf{Fm}(X)/\Psi; \quad \alpha \mapsto [\alpha]_{\Psi}.$$

Since $\Sigma \vdash_{\mathcal{V}}^{X} \varepsilon$ and $\Sigma \subseteq \ker(\pi_{\Psi}) = \Psi$, also $\varepsilon \in \ker(\pi_{\Psi}) = \Psi$.

(2) ⇒ (1) Suppose that $\varepsilon \in \Psi$. For any $\mathbf{A} \in \mathcal{V}$ and $\varphi \in \hom(\mathbf{Fm}(X), \mathbf{A})$ such that $\Sigma \subseteq \ker(\varphi)$, clearly, $\mathrm{Cg}^{\mathbf{Fm}(X)}(\Sigma) \subseteq \ker(\varphi)$ and $\Theta_{\mathcal{V}}(X) \subseteq \ker(\varphi)$, so $\varepsilon \in \Psi \subseteq \ker(\varphi)$. That is, $\Sigma \vdash_{\mathcal{V}}^{X} \varepsilon$.

(2) ⇔ (3) An application of the Correspondence theorem (Theorem B.4.(d)), using the canonical map $\pi_{\Theta_{\mathcal{V}}(X)}\colon \mathbf{Fm}(X) \to \mathbf{F}(X); \alpha \mapsto \bar{\alpha}$, yields $\Psi = \pi_{\Theta_{\mathcal{V}}(X)}^{-1}[\mathrm{Cg}^{\mathbf{F}(X)}(\bar{\Sigma})]$ and hence $\varepsilon \in \Psi$ if and only if $\bar{\varepsilon} \in \mathrm{Cg}^{\mathbf{F}(X)}(\bar{\Sigma})$. □

Since $\mathrm{Cg}^{\mathbf{A}}$ is an algebraic closure operator for any algebra $\mathbf{A}$, condition (3) of Proposition 3.2.7 may also be written as

$$\bar{\varepsilon} \in \mathrm{Cg}^{\mathbf{F}(X)}(\bar{\Sigma}) = \bigcup\{\mathrm{Cg}^{\mathbf{F}(X)}(\bar{\Pi}) \mid \Pi \subseteq \Sigma \text{ and } \Pi \text{ is finite}\}.$$

Hence $\Sigma \vdash_{\mathcal{V}}^{X} \varepsilon$ if and only if $\Pi \vdash_{\mathcal{V}}^{X} \varepsilon$, for some finite $\Pi \subseteq \Sigma$. That is, $\vdash_{\mathcal{V}}^{X}$ is finitary, as already observed in Corollary 3.2.6.

Observe next that $\Theta_{\mathcal{V}}(X) = \Theta_{\mathcal{V}}(Y) \cap \mathrm{Fm}(X)^2$, for any set Y such that $X \subseteq Y$, and hence, for convenience, we may assume that $\mathbf{F}(X)$ is a subalgebra of $\mathbf{F}(Y)$. Similarly, any congruence Θ of $\mathbf{F}(X)$ generates a congruence of $\mathbf{F}(Y)$ that restricts to Θ on $\mathbf{F}(X)$, i.e., $\mathrm{Cg}^{\mathbf{F}(Y)}(\Theta) \cap \mathrm{F}(X)^2 = \Theta$ (Lemma B.10). Hence, it follows from Proposition 3.2.7 that the equational consequence relation of $\mathcal{V}$ over $\mathrm{Eq}(Y)$ restricts to the equational consequence relation of $\mathcal{V}$ over $\mathrm{Eq}(X)$.

COROLLARY 3.2.8. *Let $\mathcal{V}$ be any variety of $\mathcal{L}$-algebras and let X be a set such that $X \cup \mathcal{L}_0 \neq \emptyset$. For any $\Sigma \cup \{\varepsilon\} \subseteq \mathrm{Eq}(X)$ and set Y such that $X \subseteq Y$,*

$$\Sigma \vdash_{\mathcal{V}}^{X} \varepsilon \iff \bar{\varepsilon} \in \mathrm{Cg}^{\mathbf{F}(X)}(\bar{\Sigma}) \iff \bar{\varepsilon} \in \mathrm{Cg}^{\mathbf{F}(Y)}(\bar{\Sigma}) \iff \Sigma \vdash_{\mathcal{V}}^{Y} \varepsilon.$$

In light of this fact, we may unambiguously write $\Sigma \vdash_{\mathcal{V}} \varepsilon$ to denote that $\Sigma \vdash_{\mathcal{V}}^{X} \varepsilon$ for *any* set X satisfying $\Sigma \cup \{\varepsilon\} \subseteq \mathrm{Eq}(X)$. Note also that, combining Corollary 3.2.8 with Proposition 3.2.5, it follows that the consequence relation of $\mathcal{V}$ over a countably infinite set fixes the consequence relations of $\mathcal{V}$ over any set.

Let us conclude this section by considering a further well-known notion of equational consequence, focusing on the case where X is a countably infinite set. Given any $\mathcal{L}$-algebra $\mathbf{A}$, let

$$\mathrm{Id}_{\mathbf{A}} := \bigcap\{\ker(\varphi) \mid \varphi \in \hom(\mathbf{Fm}, \mathbf{A})\} = \{\varepsilon \in \mathrm{Eq} \mid \vdash_{\mathbf{A}} \varepsilon\},$$

and, for any class of $\mathcal{L}$-algebras $\mathcal{K}$ and $\Sigma \cup \{\varepsilon\} \subseteq \mathrm{Eq}$, define

$$\Sigma \vdash_{\mathcal{K}}^{\mathrm{FI}} \varepsilon :\iff (\Sigma \subseteq \mathrm{Id}_{\mathbf{A}} \implies \varepsilon \in \mathrm{Id}_{\mathbf{A}}, \text{ for all } \mathbf{A} \in \mathcal{K}).$$

It is not hard to check that $\vdash_{\mathcal{K}}^{\mathrm{FI}}$ is a substitution-invariant consequence relation. Moreover, when $\mathcal{K}$ is the class of *all* $\mathcal{L}$-algebras, this is precisely the consequence relation of equational logic defined in Section 2.1.

To obtain an analogue of Proposition 3.2.7 for this alternate definition of equational consequence, we recall a further definition from universal algebra. Call a congruence Θ of an algebra $\mathbf{A}$ *fully invariant* if for every endomorphism φ of $\mathbf{A}$, whenever

$\langle a, b\rangle \in \Theta$, also $\langle \varphi(a), \varphi(b)\rangle \in \Theta$. The smallest fully invariant congruence of $\mathbf{A}$ containing the set $R \subseteq A^2$ is denoted by $\mathrm{Cg}^{\mathbf{A}}_{\mathrm{FI}}(R)$.

PROPOSITION 3.2.9. *Let $\mathcal{V}$ be any variety. The following statements are equivalent, for any $\Sigma \cup \{\varepsilon\} \subseteq \mathrm{Eq}$:*

(1) $\Sigma \vdash^{\mathrm{FI}}_{\mathcal{V}} \varepsilon$.
(2) $\varepsilon \in \mathrm{Cg}^{\mathbf{Fm}}_{\mathrm{FI}}(\Sigma) \vee \Theta_{\mathcal{V}}$.
(3) $\bar{\varepsilon} \in \mathrm{Cg}^{\mathbf{F}}_{\mathrm{FI}}(\bar{\Sigma})$.

PROOF. We show first that $\Theta = \mathrm{Id}_{\mathbf{Fm}/\Theta}$, for any fully invariant congruence Θ of $\mathbf{Fm}$. Let $\pi_\Theta\colon \mathbf{Fm} \to \mathbf{Fm}/\Theta;\ \alpha \mapsto [\alpha]_\Theta$ be the canonical map. By definition, $\vdash_{\mathbf{Fm}/\Theta} \mathrm{Id}_{\mathbf{Fm}/\Theta}$, so $\mathrm{Id}_{\mathbf{Fm}/\Theta} \subseteq \ker(\pi_\Theta) = \Theta$. Now consider any $\langle \alpha, \beta\rangle \in \Theta$ and $\varphi \in \hom(\mathbf{Fm}, \mathbf{Fm}/\Theta)$ satisfying $\varphi = \pi_\Theta \circ \sigma_\varphi$, for some substitution $\sigma_\varphi\colon \mathbf{Fm} \to \mathbf{Fm}$. Since Θ is fully invariant, $\langle \sigma_\varphi(\alpha), \sigma_\varphi(\beta)\rangle \in \Theta$, and hence

$$\varphi(\alpha) = \pi_\Theta(\sigma_\varphi(\alpha)) = [\sigma_\varphi(\alpha)]_\Theta = [\sigma_\varphi(\beta)]_\Theta = \pi_\Theta(\sigma_\varphi(\beta)) = \varphi(\beta).$$

That is, $\langle \alpha, \beta\rangle \in \ker(\varphi)$. So $\langle \alpha, \beta\rangle \in \mathrm{Id}_{\mathbf{Fm}/\Theta}$ as required.

Now, let Θ be the fully invariant congruence $\mathrm{Cg}^{\mathbf{Fm}}_{\mathrm{FI}}(\Sigma) \vee \Theta_{\mathcal{V}}$. If $\Sigma \vdash^{\mathrm{FI}}_{\mathcal{V}} \varepsilon$, then, since $\Sigma \subseteq \Theta$ and $\mathbf{Fm}/\Theta \in \mathcal{V}$, the above claim yields $\varepsilon \in \Theta$. Conversely, suppose that $\varepsilon \in \Theta$ and consider any $\mathbf{A} \in \mathcal{V}$ such that $\Sigma \subseteq \mathrm{Id}_{\mathbf{A}}$. Since $\mathrm{Id}_{\mathbf{A}}$ is a fully invariant congruence of $\mathbf{Fm}$, we obtain $\varepsilon \in \Theta = \mathrm{Cg}^{\mathbf{Fm}}_{\mathrm{FI}}(\Sigma) \vee \Theta_{\mathcal{V}} \subseteq \mathrm{Id}_{\mathbf{A}}$. So $\Sigma \vdash^{\mathrm{FI}}_{\mathcal{V}} \varepsilon$. The equivalence of (2) and (3) then follows as in the proof of Proposition 3.2.7. □

3.3. Equivalence of consequence relations

The abstract notion of consequence relation presented in Section 3.1 is, for many purposes, too fine-grained to be of much use. Two consequence relations, on the same or distinct sets, might differ, not because they describe intrinsically different consequences, but rather because they exhibit these consequences in different guises. In such cases, we would like to say that these consequence relations are *equivalent*.

Our definition of equivalence of consequence relations will proceed in two steps. Note first that, for any sets A, B, every map $\varphi\colon A \to \wp(B)$ extends to a map between the complete lattices $\langle \wp(A), \subseteq\rangle$ and $\langle \wp(B), \subseteq\rangle$,

$$\varphi\colon \wp(A) \to \wp(B); \quad X \mapsto \bigcup\{\varphi(a) \mid a \in X\},$$

that preserves arbitrary joins and is therefore, by Lemma 1.2.9, residuated.

Let $\vdash_1$ and $\vdash_2$ be consequence relations over non-empty sets A_1 and A_2, respectively. We say that $\vdash_1$ and $\vdash_2$ are *similar* if there exist maps, referred to as *transformers*, $\tau\colon A_1 \to \wp(A_2)$ and $\rho\colon A_2 \to \wp(A_1)$ satisfying the following conditions for all $X \cup \{a\} \subseteq A_1$ and $Y \cup \{b\} \subseteq A_2$:

(S1) $X \vdash_1 a \iff \tau(X) \vdash_2 \tau(a)$;
(S2) $Y \vdash_2 b \iff \rho(Y) \vdash_1 \rho(b)$;
(S3) $\{a\} \dashv\vdash_1 \rho(\tau(a))$;
(S4) $\{b\} \dashv\vdash_2 \tau(\rho(b))$.

In other words, $\vdash_1$ and $\vdash_2$ are similar when

(i) $\vdash_1$ is faithfully interpretable via the map τ into $\vdash_2$ (S1);

(ii) $\vdash_2$ is faithfully interpretable via the map ρ into $\vdash_1$ (S2);

(iii) ρ and τ are mutually inverse up to inter-derivability (S3), (S4).

The next lemma simplifies this definition.

LEMMA 3.3.1. *The consequence relations $\vdash_1$ and $\vdash_2$ are similar if and only if there exist maps $\tau\colon A_1 \to \wp(A_2)$ and $\rho\colon A_2 \to \wp(A_1)$ satisfying either (S1) and (S4), or (S2) and (S3).*

PROOF. Suppose that the maps $\tau\colon A_1 \to \wp(A_2)$ and $\rho\colon A_2 \to \wp(A_1)$ satisfy (S1) and (S4). Then, for any $Y \cup \{b\} \subseteq A_2$,

$$Y \vdash_2 b \overset{(S4)}{\iff} \tau(\rho(Y)) \vdash_2 \tau(\rho(b)) \overset{(S1)}{\iff} \rho(Y) \vdash_1 \rho(b).$$

That is, (S2) is satisfied. Consider next any $a \in X$. Then $\tau(a) \dashv\vdash_2 \tau(\rho(\tau(a)))$, using (S4), so $\rho(\tau(a)) \vdash_1 a$ and $\{a\} \vdash_1 \rho(\tau(a))$, using (S1). That is, (S3) is also satisfied. The case where τ and ρ satisfy (S2) and (S3) is symmetrical. □

The similarity of two consequence relations $\vdash_1$ and $\vdash_2$ can also be formulated via conditions on the corresponding consequence operators $\gamma_1 := \gamma_{\vdash_1}$ and $\gamma_2 := \gamma_{\vdash_2}$. That is, it can be shown that $\vdash_1$ and $\vdash_2$ are similar if and only if there exist maps $\tau\colon A_1 \to \wp(A_2)$ and $\rho\colon A_2 \to \wp(A_1)$ (extending as described above to maps $\tau\colon \wp(A_1) \to \wp(A_2)$ and $\rho\colon \wp(A_2) \to \wp(A_1)$) satisfying

$$\begin{aligned} &\textit{either} \quad \gamma_1 = \bar{\tau}\gamma_2\tau \text{ and } \gamma_2\tau\rho = \gamma_2 \\ &\textit{or} \quad \gamma_2 = \bar{\rho}\gamma_1\rho \text{ and } \gamma_1\rho\tau = \gamma_1, \end{aligned}$$

where $\bar{\tau}$ denotes the residual of τ restricted to $\gamma_2[\wp(A_2)]$ and $\bar{\rho}$ denotes the residual of ρ restricted to $\gamma_1[\wp(A_1)]$. In this case, it follows that $\bar{\tau}$ is an isomorphism betwen the lattices $\langle \gamma_2[\wp(A_2)], \subseteq \rangle$ and $\langle \gamma_1[\wp(A_1)], \subseteq \rangle$ of theories of $\vdash_2$ and $\vdash_1$, respectively, with inverse $\bar{\rho}$, as depicted in the following diagram:

$$\begin{array}{ccccc} \wp(A_2) & \xrightarrow{\ \rho\ } & \wp(A_1) & \xrightarrow{\ \tau\ } & \wp(A_2) \\ \downarrow{\scriptstyle \gamma_2} & & \downarrow{\scriptstyle \gamma_1} & & \downarrow{\scriptstyle \gamma_2} \\ \gamma_2[\wp(A_2)] & \underset{\bar{\rho}}{\twoheadleftarrow} & \gamma_1[\wp(A_1)] & \underset{\bar{\tau}}{\twoheadleftarrow} & \gamma_2[\wp(A_2)]. \end{array}$$

Similarity clearly enjoys the properties of an equivalence relation on the class of all consequence relations. Strengthening this notion to accommodate an appropriate action-invariance property requires some additional work, however. Suppose that $\langle A, \cdot \rangle$ is an $\mathbf{M}$-set, for some monoid $\mathbf{M} = \langle M, \circ, 1 \rangle$, and consider the power set monoid $\wp(\mathbf{M}) = \langle \wp(M), \circ, \{1\} \rangle$, where, for $S_1, S_2 \subseteq M$,

$$S_1 \circ S_2 := \{\sigma_1 \circ \sigma_2 \mid \sigma_1 \in S_1 \text{ and } \sigma_2 \in S_2\}.$$

Then $\langle \wp(A), \cdot \rangle$ is a $\wp(\mathbf{M})$-set, where $\cdot\colon \wp(M) \times \wp(A) \to \wp(A)$ is defined, for $S \subseteq M$ and $X \subseteq A$, by

$$S \cdot X := \{\sigma \cdot a \mid \sigma \in S \text{ and } a \in X\}.$$

Note that $\cdot$ preserves arbitrary joins; that is, for any $S \subseteq M$ and $W \subseteq \wp(A)$,

$$S \cdot \bigcup W = \bigcup \{S \cdot X \mid X \in W\}.$$

Suppose now that $\langle A_1, \cdot_1 \rangle$ and $\langle A_2, \cdot_2 \rangle$ are both **M**-sets. A map $\varphi \colon \wp(A_1) \to \wp(A_2)$ is called *action-invariant* if, for all $S \subseteq M$ and $X \subseteq A_1$,

$$S \cdot_2 \varphi(X) = \varphi(S \cdot_1 X).$$

Having set the stage, we are now ready to present our definition of equivalence. Let $\vdash_1$ and $\vdash_2$ be action-invariant consequence relations over **M**-sets $\langle A_1, \cdot_1 \rangle$ and $\langle A_2, \cdot_2 \rangle$, respectively. We say that $\vdash_1$ and $\vdash_2$ are *equivalent* if the following two conditions are satisfied:

(E1) $\vdash_1$ and $\vdash_2$ are similar, with transformers τ and ρ;

(E2) τ and ρ are action-invariant.

Some special cases of this definition have played a notable role in the literature. Consider first the case where

(i) $\mathrm{L} = \langle \mathbf{Fm}, \vdash_1 \rangle$ is a logic of signature $\mathcal{L}$;

(ii) $\vdash_2$ is the equational consequence relation of a class $\mathcal{K}$ of $\mathcal{L}$-algebras over Eq.

If $\vdash_1$ and $\vdash_2$ are equivalent, then L is said to be *algebraizable* with *equivalent algebraic semantics* $\mathcal{K}$, and τ and ρ are referred to as *formula-equation* and *equation-formula transformers*, respectively. These transformers need not be unique, even up to equivalence in L. Also, $\mathcal{K}$ might not be the unique equivalent algebraic semantics for L; for example, the equational consequence relations of the two-element Boolean algebra and the variety of all Boolean algebras coincide. Nevertheless, any two equivalent algebraic semantics for a finitary logic L both generate, by Propositions 3.2.2 and 3.2.5, a quasivariety that is also an equivalent algebraic semantics for L and can therefore be called the equivalent quasivariety semantics for L. Let us also remark that it is not uncommon to find distinct algebraizable logics with the same equivalent algebraic semantics (see, e.g., Example 3.3.4). Of course, if these equivalences share the same formula-equation transformer, then they must coincide.

Consider next the case where GL is a sequent calculus such that the consequence relation $\vdash_{\mathsf{GL}}$ over the set of $\mathcal{L}$-sequents is equivalent to a consequence relation $\vdash_{\mathrm{L}}$ over the set of $\mathcal{L}$-formulas with $\tau(\alpha) = \{\Rightarrow \alpha\}$. Then (S1) means that $\vdash_{\mathrm{L}}$ is faithfully interpretable into the *external consequence relation* of GL: the consequence relation $\vdash^{e}_{\mathsf{GL}}$ on formulas defined by

$$T \vdash^{e}_{\mathsf{GL}} \alpha \;:\Longleftrightarrow\; \{\Rightarrow \beta \mid \beta \in T\} \vdash_{\mathsf{GL}} \Rightarrow \alpha.$$

Below we consider some pertinent examples of equivalence between consequence relations, with the twofold aim of, firstly, combining some threads interspersed throughout the previous two chapters and, secondly, illustrating how to shuttle back and forth between formulas, equations, and sequents. In each case, we confine ourselves to discussing similarity; the additional requirement needed to confirm equivalence does not create any difficulty in these contexts.

EXAMPLE 3.3.2. Let us sketch first a proof of the well-known fact that classical propositional logic $\mathrm{CL} := \langle \mathbf{Fm}, \vdash_{\mathsf{HCL}} \rangle$, introduced via the Hilbert system HCL in Example 3.1.3, is both algebraizable with the variety of Boolean algebras (in a suitable signature) as equivalent algebraic semantics, and equivalent also to the consequence relation on multiple-conclusion sequents defined via the calculus GCL in Example 3.1.4.

Observe first that to apply the definition of algebraizability, Boolean algebras must be considered as algebras of the same signature as CL. For convenience, let $\mathcal{B}ool$ denote the variety of algebras in this signature with a Boolean algebra reduct that satisfies also the equation $x \to y \approx \neg x \vee y$. We will show that the consequence relations $\vdash_{\mathsf{HCL}}$, $\vdash_{\mathcal{B}ool}$, and $\vdash_{\mathsf{GCL}}$ are all equivalent.

Let us first establish the equivalence of $\vdash_{\mathsf{HCL}}$ and $\vdash_{\mathcal{B}ool}$. By Lemma 3.3.1, it suffices to show that the following maps satisfy conditions (S1) and (S4):

$\tau\colon \mathrm{Fm} \to \wp(\mathrm{Eq});\ \alpha \mapsto \{\alpha \approx 1\}$ and $\rho\colon \mathrm{Eq} \to \wp(\mathrm{Fm});\ \alpha \approx \beta \mapsto \{\alpha \to \beta, \beta \to \alpha\}$.

Checking (S1) amounts to establishing that, for all $T \cup \{\alpha\} \subseteq \mathrm{Fm}$,

$$T \vdash_{\mathsf{HCL}} \alpha \iff \{\gamma \approx 1 \mid \gamma \in T\} \vdash_{\mathcal{B}ool} \alpha \approx 1.$$

The left-to-right direction is established by a straightforward induction on the length of an HCL-derivation of α from T, making use of standard properties of Boolean algebras. The converse direction is established via the Lindenbaum-Tarski algebra construction encountered earlier in our proofs of Proposition 2.2.2 and Theorem 2.3.3. We assume $T \nvdash_{\mathsf{HCL}} \alpha$ and prove that $\{\gamma \approx 1 \mid \gamma \in T\} \nvdash_{\mathcal{B}ool} \alpha \approx 1$ by choosing a suitable $\mathbf{A} \in \mathcal{B}ool$ and homomorphism $\varphi \in \mathrm{hom}(\mathbf{Fm}, \mathbf{A})$ such that $\varphi(\gamma) = 1^{\mathbf{A}}$, for all $\gamma \in T$, but $\varphi(\alpha) \neq 1^{\mathbf{A}}$. This $\mathbf{A}$ is defined as $\mathbf{Fm}/\Theta_T$, where $\Theta_T := \{\langle \alpha, \beta \rangle \in \mathrm{Fm}^2 \mid T \vdash_{\mathsf{HCL}} \{\alpha \to \beta, \beta \to \alpha\}\}$, showing, by heavy recourse to properties of derivability in HCL, that Θ_T is a congruence of $\mathbf{Fm}$ and that $\mathbf{Fm}/\Theta_T \in \mathcal{B}ool$. The homomorphism φ is then the canonical map $\pi_{\Theta_T}\colon \mathbf{Fm} \to \mathbf{Fm}/\Theta_T;\ \gamma \mapsto [\gamma]_{\Theta_T}$.

For (S4), we simply observe that, for any $\mathbf{A} \in \mathcal{B}ool$ and $\varphi \in \mathrm{hom}(\mathbf{Fm}, \mathbf{A})$,

$$\varphi(\alpha) = \varphi(\beta) \iff \varphi(\alpha \to \beta) = \varphi(\beta \to \alpha) = 1^{\mathbf{A}},$$

and hence, as required,

$$\{\alpha \approx \beta\} \dashv\vdash_{\mathcal{B}ool} \{\alpha \to \beta \approx 1, \beta \to \alpha \approx 1\}.$$

Next, we establish the equivalence of $\vdash_{\mathsf{HCL}}$ and $\vdash_{\mathsf{GCL}}$. For any $\alpha \in \mathrm{Fm}$ and finite sequences Γ, Δ of formulas, let $\tau(\alpha) := \{\Rightarrow \alpha\}$ and $\rho(\Gamma \Rightarrow \Delta) := \{\bigwedge\Gamma \to \bigvee\Delta\}$, where $\bigwedge\langle \alpha_1, \dots, \alpha_n \rangle := \alpha_1 \wedge \dots \wedge \alpha_n$ $(n \in \mathbb{N}^{>0})$, $\bigvee\langle \beta_1, \dots, \beta_m \rangle := \beta_1 \vee \dots \vee \beta_m$ $(m \in \mathbb{N}^{>0})$, $\bigwedge\langle\rangle := 1$, and $\bigvee\langle\rangle := 0$.

In this case, we establish the left-to-right directions of (S1) and (S2), and both directions of (S3) and (S4). These conditions imply the right-to-left direction of (S1): just observe that if $\tau(T) \vdash_{\mathsf{GCL}} \tau(\alpha)$, then, by the left-to-right direction of (S2), also $\rho(\tau(T)) \vdash_{\mathsf{HCL}} \rho(\tau(\alpha))$, and repeated applications of (S3) yield $T \vdash_{\mathsf{HCL}} \alpha$.

For the left-to-right direction of (S1), we may assume that the set of premises is finite (since any consequence relation obtained from a proof system is finitary) and show that, for all $\gamma_1, \dots, \gamma_n, \alpha \in \mathrm{Fm}$,

$$\{\gamma_1, \dots, \gamma_n\} \vdash_{\mathsf{HCL}} \alpha \implies \{\Rightarrow \gamma_1, \dots, \Rightarrow \gamma_n\} \vdash_{\mathsf{GCL}} \Rightarrow \alpha.$$

This is proved by induction on the length of an HCL-derivation of α from $\gamma_1, \dots, \gamma_n$. It is easily checked that $\Rightarrow \alpha$ is GCL-derivable for every axiom α of HCL, and, clearly, each $\Rightarrow \gamma_i$ is GCL-derivable from $\{\Rightarrow \gamma_1, \dots, \Rightarrow \gamma_n\}$. Moreover, the only inference rule of HCL, modus ponens, preserves GCL-derivability. An analogous induction on the height of a GCL-derivation of a sequent $\Gamma \Rightarrow \Delta$ from a set of sequents S suffices for the left-to-right direction of (S2).

(S3) is trivial, because $\{\alpha\} \dashv\vdash_{\mathsf{HCL}} \{1 \to \alpha\}$. For (S4), we need to show that

$$\{\Gamma \Rightarrow \Delta\} \dashv\vdash_{\mathsf{GCL}} \{\Rightarrow \bigwedge\Gamma \to \bigvee\Delta\}.$$

For the left-to-right direction, we observe that $\Rightarrow \bigwedge\Gamma \to \bigvee\Delta$ is derivable from $\Gamma \Rightarrow \Delta$ by applying $(\Rightarrow\to)$, $(\wedge\Rightarrow)_1$, $(\wedge\Rightarrow)_2$, $(\Rightarrow\vee)_1$, and $(\Rightarrow\vee)_2$ together with the structural rules and, for the cases where Γ or Δ is empty, cuts with the derivable sequents $\Rightarrow 1$ and $0 \Rightarrow$. For the right-to-left direction, observe that applications of $(\to\Rightarrow)$, $(\vee \Rightarrow)$, $(\Rightarrow \wedge)$, $(0 \Rightarrow)$, $(\Rightarrow 1)$, and (ID) yield $\vdash_{\mathsf{GCL}} \Gamma, \bigwedge\Gamma \to \bigvee\Delta \Rightarrow \Delta$. An application of (CUT) with this sequent and $\Rightarrow \bigwedge\Gamma \to \bigvee\Delta$ then produces a GCL-derivation of $\Gamma \Rightarrow \Delta$ from $\Rightarrow \bigwedge\Gamma \to \bigvee\Delta$ as required.

EXAMPLE 3.3.3. In Chapter 2, we justified the claim that Lat is a calculus for equational reasoning in the variety $\mathcal{L}at$ of lattices using Proposition 2.2.2. In fact, as we now show, this justification extends to the equivalence of the consequence relations $\vdash_{\mathsf{Lat}}$ and $\vdash_{\mathcal{L}at}$. For $\alpha, \beta \in \mathrm{Fm}$, let

$$\tau(\alpha \Rightarrow \beta) := \{\alpha \wedge \beta \approx \alpha\} \quad \text{and} \quad \rho(\alpha \approx \beta) := \{\alpha \Rightarrow \beta, \beta \Rightarrow \alpha\}.$$

By Lemma 3.3.1, it suffices to show that τ and ρ satisfy (S1) and (S4). For (S4), it is easily verified that

$$\{\alpha \approx \beta\} \dashv\vdash_{\mathcal{L}at} \{\alpha \wedge \beta \approx \alpha, \beta \wedge \alpha \approx \beta\}.$$

For (S1), we need to show that, for any set of unary $\mathcal{L}$-sequents $S \cup \{\alpha \Rightarrow \beta\}$, defining $S^{\leq} := \{\alpha' \wedge \beta' \approx \alpha' \mid \alpha' \Rightarrow \beta' \in S\}$,

$$S \vdash_{\mathsf{Lat}} \alpha \Rightarrow \beta \iff S^{\leq} \vdash_{\mathcal{L}at} \alpha \wedge \beta \approx \alpha.$$

Since the proof is very similar to the proof of Proposition 2.2.2, which establishes this correspondence for the case where S is empty, we confine ourselves here to a few brief remarks.

The left-to-right direction is established by an easy induction on the height of a Lat-derivation of $\alpha \Rightarrow \beta$ from S. For the right-to-left direction, let S be any set of unary $\mathcal{L}$-sequents. We define a binary relation Θ_S on Fm by

$$\langle \alpha, \beta \rangle \in \Theta_S \;:\iff\; S \vdash_{\mathsf{Lat}} \{\alpha \Rightarrow \beta, \beta \Rightarrow \alpha\},$$

and show that Θ_S is a congruence of $\mathbf{Fm}$, and $\mathbf{Fm}/\Theta_S \in \mathcal{L}at$, observing (as in the proof of Proposition 2.2.2) that, for any $\alpha, \beta \in \mathrm{Fm}$,

$$[\alpha]_{\Theta_S} \leq^{\mathbf{Fm}/\Theta_S} [\beta]_{\Theta_S} \iff S \vdash_{\mathsf{Lat}} \alpha \Rightarrow \beta.$$

Finally, we suppose that $S^{\leq} \vdash_{\mathcal{L}at} \alpha \wedge \beta \approx \alpha$ and let $\pi_{\Theta_S} : \mathbf{Fm} \to \mathbf{Fm}/\Theta_S;\ \gamma \mapsto [\gamma]_{\Theta_S}$ be the canonical map. Using the above equivalence, $\Sigma^{\leq} \subseteq \ker(\pi_{\Theta_S})$, and hence $\langle \alpha \wedge \beta, \alpha \rangle \in \ker(\pi_{\Theta_S})$, yielding $S \vdash_{\mathsf{Lat}} \alpha \Rightarrow \beta$.

EXAMPLE 3.3.4. For a natural example of two different algebraizable logics with the same equivalent algebraic semantics, consider the ordered group of integers $\mathbf{Z} = \langle \mathbb{Z}, \min, \max, +, -, 0 \rangle$ and maps τ_1 and τ_2 defined for $\alpha \in \mathrm{Fm}$ by

$$\tau_1(\alpha) := \{\alpha \approx \mathrm{e}\} \quad \text{and} \quad \tau_2(\alpha) := \{a \wedge \mathrm{e} \approx \mathrm{e}\}.$$

Defining now for $T \cup \{\alpha\} \subseteq \mathrm{Fm}$,

$$T \vdash_1 \alpha :\Longleftrightarrow \tau_1(T) \vdash_{\mathbf{Z}} \tau_1(\alpha) \quad \text{and} \quad T \vdash_2 \alpha :\Longleftrightarrow \tau_2(T) \vdash_{\mathbf{Z}} \tau_2(\alpha),$$

we obtain logics $\mathrm{L}_1 = \langle \mathrm{Fm}, \vdash_1 \rangle$ and $\mathrm{L}_2 = \langle \mathrm{Fm}, \vdash_2 \rangle$ that are both algebraizable with $\{\mathbf{Z}\}$ as an equivalent algebraic semantics, but are distinct; e.g.,

$$\{x\} \nvdash_1 x \vee y \text{ and } \{x, y\} \vdash_1 x^{-1}y, \text{ but } \{x\} \vdash_2 x \vee y \text{ and } \{x, y\} \nvdash_2 x^{-1}y.$$

3.4. Residuated lattices and the full Lambek calculus

Example 3.3.2 shows that classical propositional logic, the logic of Boolean algebras, can be presented in three alternative but equivalent formats: as a consequence relation on formulas, sequents, or equations. We show here that the same is true for the logic of pointed residuated lattices. The consequence relation on formulas of this logic emerges, however, not from a Hilbert system — although such a system can certainly be defined — but rather as the external consequence relation of the full Lambek calculus FL.

THEOREM 3.4.1. *The following consequence relations are equivalent:*

1. $\vdash_{\mathsf{FL}}$, *the derivability relation of the calculus* FL *(see Figure 2.4);*
2. $\vdash_{\mathcal{PRL}}$, *the equational consequence relation of the variety* $\mathcal{PRL}$ *over* Fm*;*
3. $\vdash^{e}_{\mathsf{FL}}$, *the external consequence relation of the calculus* FL.

PROOF. We first establish the equivalence of $\vdash_{\mathsf{FL}}$ and $\vdash_{\mathcal{PRL}}$. Suitable transformers are defined by setting for any sequent $\Gamma \Rightarrow \Delta$ and $\alpha, \beta \in \mathrm{Fm}$,

$$\tau(\Gamma \Rightarrow \Delta) := \{\textstyle\prod\Gamma \leq \sum\Delta\} \quad \text{and} \quad \rho(\alpha \approx \beta) := \{\alpha \Rightarrow \beta, \beta \Rightarrow \alpha\},$$

recalling that

$$\textstyle\prod\Gamma := \begin{cases} \alpha_1 \cdots \alpha_n & \text{if } \Gamma = \langle \alpha_1, \ldots, \alpha_n \rangle; \\ \mathrm{e} & \text{if } \Gamma = \langle\rangle; \end{cases} \quad \text{and} \quad \textstyle\sum\Delta := \begin{cases} \alpha & \text{if } \Delta = \langle \alpha \rangle; \\ \mathrm{f} & \text{if } \Delta = \langle\rangle. \end{cases}$$

Again, it suffices to verify conditions (S1) and (S4) in the definition of similarity. For (S1), we prove that $S \vdash_{\mathsf{FL}} \Gamma \Rightarrow \Delta$ if and only if $\tau(S) \vdash_{\mathcal{PRL}} \tau(\Gamma \Rightarrow \Delta)$, for any set of sequents $S \cup \{\Gamma \Rightarrow \Delta\}$, that is, letting $S^{\leq} := \{\prod\Gamma' \leq \sum\Delta' \mid \Gamma' \Rightarrow \Delta' \in S\}$,

$$S \vdash_{\mathsf{FL}} \Gamma \Rightarrow \Delta \iff S^{\leq} \vdash_{\mathcal{PRL}} \textstyle\prod\Gamma \leq \sum\Delta.$$

A standard induction on the height of an FL-derivation of $\Gamma \Rightarrow \Delta$ from S, as in the proof of Theorem 2.3.3, suffices for the left-to-right direction. If $\Gamma \Rightarrow \Delta$ is an axiom, then it is an instance of (ID), ($\Rightarrow$e), or (f$\Rightarrow$) and is translated by τ into an equation of the form $\alpha \leq \alpha$ that is satisfied by all pointed residuated lattices. If $\Gamma \Rightarrow \Delta$ is in S, the claim is trivial. To illustrate the inductive step, suppose that $\Gamma \Rightarrow \alpha\backslash\beta$ has been derived from S in FL via a final application of ($\Rightarrow\backslash$) to the premise $\alpha, \Gamma \Rightarrow \beta$. Then, by the induction hypothesis,

$$S^{\leq} \vdash_{\mathcal{PRL}} \alpha \cdot \textstyle\prod\Gamma \leq \beta.$$

If $S^{\leq} \subseteq \ker(\varphi)$, for some $\mathbf{L} \in \mathcal{PRL}$ and $\varphi \in \hom(\mathbf{Fm}, \mathbf{L})$, then

$$\varphi(\alpha \cdot \textstyle\prod\Gamma) = \varphi(\alpha) \cdot^{\mathbf{L}} \varphi(\prod\Gamma) \leq^{\mathbf{L}} \varphi(\beta),$$

and, by residuation, $\varphi(\prod\Gamma) \leq^{\mathbf{L}} \varphi(\alpha)\backslash^{\mathbf{L}}\varphi(\beta)$, yielding $\prod\Gamma \leq \alpha\backslash\beta \in \ker(\varphi)$. So

$$S^{\leq} \vdash_{\mathcal{PRL}} \prod\Gamma \leq \alpha\backslash\beta.$$

For the right-to-left direction, we suppose that $S \nvdash_{\mathsf{FL}} \Gamma \Rightarrow \Delta$, and prove that $S^{\leq} \nvdash_{\mathcal{PRL}} \prod\Gamma \leq \sum\Delta$. Let Θ_S be the binary relation on Fm defined by

$$\langle \alpha, \beta \rangle \in \Theta_S \;:\Longleftrightarrow\; S \vdash_{\mathsf{FL}} \{\alpha \Rightarrow \beta, \beta \Rightarrow \alpha\}.$$

It follows that

(a) Θ_S is a congruence of **Fm**;
(b) the quotient $\mathbf{Fm}/\Theta_S$ is a pointed residuated lattice.

The proof of (a) proceeds exactly as in the proof of Theorem 2.3.3; indeed, the relation Θ in that proof is just the special case of Θ_S where $S = \emptyset$. The proof of (b) is also a slight variant of the corresponding proof given in Theorem 2.3.3, allowing also derivations from sequents in S in FL. Observe in particular that

$$\begin{aligned} [\alpha]_{\Theta_S} \leq^{\mathbf{Fm}/\Theta_S} [\beta]_{\Theta_S} &\iff [\alpha]_{\Theta_S} \wedge [\beta]_{\Theta_S} = [\alpha]_{\Theta_S} \\ &\iff \langle \alpha \wedge \beta, \alpha \rangle \in \Theta_S \\ &\iff S \vdash_{\mathsf{FL}} \{\alpha \wedge \beta \Rightarrow \alpha, \alpha \Rightarrow \alpha \wedge \beta\} \\ &\iff S \vdash_{\mathsf{FL}} \alpha \Rightarrow \beta. \end{aligned}$$

Consider the canonical map $\pi_{\Theta_S} \colon \mathbf{Fm} \to \mathbf{Fm}/\Theta_S;\ \gamma \mapsto [\gamma]_{\Theta_S}$. Given $\Gamma' \Rightarrow \Delta' \in S$, it follows using $(\cdot \Rightarrow)$, $(\mathrm{e} \Rightarrow)$, and $(\Rightarrow \mathrm{f})$ that $S \vdash_{\mathsf{FL}} \prod\Gamma' \Rightarrow \sum\Delta'$ and hence, by the above equivalence, $[\prod\Gamma']_{\Theta_S} \leq^{\mathbf{Fm}/\Theta_S} [\sum\Delta']_{\Theta_S}$. Moreover, using the fact that $\vdash_{\mathsf{FL}} \Gamma \Rightarrow \prod\Gamma$ and $\vdash_{\mathsf{FL}} \sum\Delta \Rightarrow \Delta$, it follows from the assumption $S \nvdash_{\mathsf{FL}} \Gamma \Rightarrow \Delta$ that also $S \nvdash_{\mathsf{FL}} \prod\Gamma \Rightarrow \sum\Delta$. Hence, by the above equivalence, $[\prod\Gamma]_{\Theta_S} \nleq^{\mathbf{Fm}/\Theta_S} [\sum\Delta]_{\Theta_S}$. So we may conclude that $S^{\leq} \nvdash_{\mathcal{PRL}} \prod\Gamma \leq \sum\Delta$, completing the proof of (S1).

For (S4), it suffices to observe that $\tau(\rho(\alpha \approx \beta)) = \{\alpha \wedge \beta \approx \alpha, \beta \wedge \alpha \approx \beta\}$ and

$$\{\alpha \approx \beta\} \dashv\vdash_{\mathcal{PRL}} \{\alpha \wedge \beta \approx \alpha, \beta \wedge \alpha \approx \beta\}.$$

The equivalence of $\vdash_{\mathsf{FL}}$ and $\vdash^{\mathrm{e}}_{\mathsf{FL}}$ via the transformers

$$\tau(\Gamma \Rightarrow \Delta) := \prod\Gamma\backslash\sum\Delta \quad \text{and} \quad \rho(\alpha) := \{\ \Rightarrow \alpha\}$$

is left as an easy exercise. □

Certain aspects of these equivalences extend to all varieties of pointed residuated lattices. Let $\mathcal{V}$ be any such variety and define, for any $T \cup \{\alpha\} \subseteq \mathrm{Fm}$,

$$T \vdash_{\mathcal{V}} \alpha \;:\Longleftrightarrow\; \{\mathrm{e} \leq \beta \mid \beta \in T\} \vdash_{\mathcal{V}} \mathrm{e} \leq \alpha.$$

We use $\vdash_{\mathcal{V}}$ here to denote a consequence relation for $\mathcal{V}$ over *formulas*, rather than the equational consequence relation of $\mathcal{V}$ over Eq, relying on context to distinguish the two meanings.

The proof of the following result is straightforward.

PROPOSITION 3.4.2. *Let $\mathcal{V}$ be any variety of pointed residuated lattices. Then the equational consequence relation of $\mathcal{V}$ over* Eq *and the consequence relation $\vdash_{\mathcal{V}}$ over* Fm *are equivalent.*

If $\mathcal{V}$ is the variety of all pointed residuated lattices, then $\vdash_{\mathcal{V}}$ is just the external consequence relation $\vdash^{e}_{\mathsf{FL}}$ of the full Lambek calculus. Moreover, for any variety $\mathcal{V}$ of pointed residuated lattices defined relative to $\mathcal{PRL}$ by additional equations, $\vdash_{\mathcal{V}}$ will be the external consequence relation of the full Lambek calculus extended with axioms corresponding to these equations.

3.5. Historical excursus

Both the abstract treatment of consequence relations and algebraic perspective on logic presented in this chapter may be found in the pioneering work of Alfred Tarski. In this section, we trace the historical development of the ideas feeding into this work and subsequent developments relevant to this book.

3.5.1. Bernard Bolzano. Bernard Bolzano brought reflection on the logical foundations of mathematics to unprecedented levels of awareness and depth. A mathematician by trade, he sought from the beginning of his career (e.g., in his 1810 work *Beyträge zu einer begründeteren Darstellung der Mathematik*) to establish the foundations of the mathematical disciplines on firmer ground. In accordance with the time-honored tradition of "doctrine of method" followed by French seventeenth century theorists (Pascal, Descartes, Arnauld) and more recent adherents (Crusius, Lambert, Kant), Bolzano believed that logic serves as a preliminary methodological framework for stating rules that properly found these disciplines.

The second volume of his monumental work *Wissenschaftslehre* (1837) contains a detailed development of a powerful logical system, including an analysis of logical consequence, viewed by some as an anticipation of Tarski (see **[62]** for details). In particular, he considers separately the relations of *derivability* (*Ableitbarkeit*) and *consecution* (*Abfolge*) among propositions.

Derivability for Bolzano is, surprisingly perhaps, multiple-conclusion: propositions $q_1,\dots,q_m$ are said to be derivable from propositions $p_1,\dots,p_n$ with respect to component concepts $a_1,\dots,a_r$ if and only if

(i) there exists a sequence of concepts $\langle b_1,\dots,b_r\rangle$ such that, for each $i\in\{1,\dots,n\}$, the result $p_i(a_1/b_1,\dots,a_r/b_r)$ of uniformly substituting each occurrence of a_j by b_j in p_i is true (*compatibility clause*);

(ii) every such sequence of concepts $\langle b_1,\dots,b_r\rangle$ yields also a true conclusion $q_i(a_1/b_1,\dots,a_r/b_r)$, for each $i\in\{1,\dots,m\}$.

This notion is intended to formalize several features Bolzano requires of "good" mathematical proofs: the compatibility clause reflects his desire, inherited from Arnauld and Kant, to assign direct proofs a higher status than ex absurdo proofs, while the fact that derivability is relativized to a sequence of components — possibly a *proper* subsequence of the sequence of all non-logical concepts contained in the propositions at issue — is an attempt to describe an *enthymematic* consequence relation broad enough to encompass standard forms of mathematical reasoning.

Bolzano's consecution relation is even more interesting, since it goes some way towards investigating a stronger *causal* concept of consequence. To illustrate the difference between the two notions, consider the following propositions:

(A) In Rome the temperature is higher in August than in January.

(B) In Rome the mercury columns of thermometers are higher in August than in January.

(A) and (B) are derivable from one another (with respect to component concepts "Rome", "August", "January"), but only (A) causally implies (B), not vice versa. According to Bolzano, only the best mathematical proofs are built from causal inferences; in other cases, there is at most a "transfer of evidence" from the premises to the conclusions. Consecution is not formally defined by Bolzano; rather it is characterized via a list of properties, among which occurs the condition: if p and q are mathematical propositions and p causally implies q, then p cannot be "more complex" than q. In one notable passage of his *Wissenschaftslehre* (§§ 216-221), Bolzano considers in detail the formal structure of proof trees explicitly taken as mirroring the structure of actual mathematical proofs, and viewed as bottom-up proof processes of "going from a consequence up to its reasons": nodes in the trees correspond to mathematical propositions, and arcs to instances of the consecution relation. Taking advantage of the non-decreasing complexity condition, he tries to show that such proof trees are analytic and hence finite. This account is correct only under further assumptions, but nevertheless provides one of the earliest and clearest examples of a formal analysis of the logical structure of mathematical proofs.

3.5.2. The early tradition in algebraic logic. Bolzano's insights wielded only a weak influence on subsequent generations of logicians. Despite this, an undercurrent in central European philosophy can be traced that leads from Bolzano to Tarski via authors such as Robert Zimmermann, Franz Brentano, and Kazimierz Twardowski, who passed some of Bolzano's ideas on to Tarski's advisor Stanisław Leśniewski [**344**]. Tarski's development of algebraic logic may be viewed as a result of combining ideas from the Bolzano school with methods from the algebra of logic, established by George Boole and his followers.

According to some scholars, Boole's *Mathematical Analysis of Logic*, published in 1847, and the expanded 1854 version *An Investigation of the Laws of Thought*, mark the official birth of modern mathematical logic. In fact, Boole did not depart from tradition as radically as it might seem in these works, because one of his main concerns — as for his contemporary Augustus De Morgan — was providing an algebraic treatment of Aristotelian syllogistic logic, which occupies about one third of the first work. Boole started with the assumption that ordinary logic is concerned with assertions about *classes* of objects. He then translated the latter into equations in the language of classes. His approach contained numerous errors, partly due to his insistence that the algebra of logic should behave like ordinary algebra, but offered significant new perspectives, as noted already in Section 1.6.4.

Stanley Jevons, Charles Sanders Peirce, and Ernst Schröder proposed a number of improvements and modifications to the work of Boole. Jevons, in particular, was dissatisfied with Boole's choice of primitive set-theoretical operations. He did not like the fact that aggregation was a partial operation, undefined on pairs of classes with non-empty intersection. In his 1864 work *Pure Logic*, he suggested a variant of Boole's calculus which he proudly advertised as based only on "processes of self-evident meaning and force". He viewed $+$ as a total operation making sense for any pair of classes, and essentially corresponding to set-theoretic union. He also showed that the expressions used remained interpretable throughout the intermediate steps of his calculations, overcoming one of the drawbacks of Boole's work [**177**].

The most significant contribution to logic of the versatile Peirce — philosopher, mathematician, and authority on several of the pure and applied sciences — is usually considered to be the foundation of the algebra of relations. However, in the 12,000 pages of his published work — rising to an astounding 90,000 if his unpublished manuscripts are taken into account— much more can be found. He investigated the laws of propositional logic, discovering that the usual propositional connectives are all definable using the single connective NAND. He also introduced quantifiers, although, unlike Frege, he did not propose an axiomatic calculus for quantified logic, and conceived of complex graphs for representing logical syntax in two or even three dimensions.

Let us also note that, contrary to the dominant paradigm of Hilbertian formalism, adherents to the algebra of logic tradition inherited from Boole a markedly anti-formalistic stance on logic and mathematics: mathematical language was seen as a system of *interpreted* symbols, and semantic notions like validity and satisfiability were accorded priority over syntactic notions. This view had some unfortunate consequences. Thoralf Skolem, for example, came close to proving the completeness theorem for first-order logic, but refrained from providing an explicit formulation because he viewed the consistency of a system as equivalent *by definition* to its satisfiability (see [**59**] for details).

3.5.3. Lindenbaum and Tarski. In the early decades of the twentieth century, the interests of the dominant tradition in logic — whose main heralds were Gottlob Frege, Bertrand Russell, and David Hilbert — revolved around the foundations of mathematics, while the themes cherished by algebraic logic disappeared from the agenda of most leading researchers. Curiously, however, it was work within the Fregean tradition that triggered a revival of algebraic logic and, ultimately, a confluence of the two streams. In the 1920s and early 1930s, while Hilbert was perfecting Frege's axiomatic calculus — resulting in the standard presentation of first-order logic presented in his 1939 work *Grundlagen der Mathematik* with Paul Bernays — Arend Heyting, Emil Post, Jan Łukasiewicz, and C.I. Lewis introduced first Hilbert-style calculi for some non-classical (intuitionistic, many-valued, and modal) propositional logics. Classical logic therefore had by the 1930s a number of competitors, each intended to capture a different concept of *logical consequence.* The question then arose as to what these concepts should have in common. In answering this question, Adolf Lindenbaum and Tarski showed how to associate in a canonical way, a (propositional) logical calculus and its corresponding consequence relation with a class of algebras.

Many of the concepts and results presented in this chapter originated in the work of Lindenbaum or Tarski. Consequence operators are already present in Tarski's 1930 paper *On some concepts of metamathematics*, although their introduction certainly predated this work. The Lindenbaum-Tarski algebra construction first appears in Tarski's 1935 paper *Foundations of the calculus of systems*, where, however, there is a lot of unfamiliar terminology, by present-day standards, to wade through to recognize the core ingredients of the method. (The treatment in J.C.C. McKinsey's 1941 article *Solution of the decision problem for Lewis* S2 *and* S4 *systems with an application to topology* [**248**], on the other hand, is closer to the development found in contemporary textbooks.) It is a matter of debate as to whether Lindenbaum contributed in any way to this concept (see [**129**, **269**]). No trace of a contribution can be found in Lindenbaum's published output (see [**378**]). It should however be noted that, within the Warsaw circle of logicians gravitating around Łukasiewicz during the 1920s and 1930s, there was a tradition of collaboration and coauthorship, and that the results included in a particular paper were not always to be credited solely to the authors. For example, an editor's footnote to Chapter 4 of [**355**], reproducing the Łukasiewicz-Tarski paper *Investigations into the sentential calculus* (1930), warns us that the article

> does not contain results discovered by both the authors jointly, but is a compilation of theorems and concepts belonging to five different persons. [...] Theorem 3, for instance, is not a theorem of Łukasiewicz and Tarski, but a Theorem of Lindenbaum.

This historical context lends plausibility to a remark of Stanisław Surma [**349**, p. 724], who reconstructs the birth of Lindenbaum-Tarski algebras as follows[2]:

> This idea was first made known [by Lindenbaum] in the academic year 1926/27 to Professor J. Łukasiewicz's Seminar on Mathematical Logic held at Warsaw University. [...It] was again presented by Lindenbaum in his address to the First Congress of Polish Mathematicians held in Lvov in September, 1927, but it was not included in the summary of the addresses in the Congress Proceedings [...] The Lindenbaum method has never been published by its author [...] during his lifetime, and it has survived solely through the oral tradition of the Warsaw School of Mathematics.

The main part of the proof outlined in Example 3.3.2 may also be found, essentially, in Tarski's *Foundations of the calculus of systems*. In fact, Tarski demonstrated a weaker kind of correspondence between Boolean algebras and classical propositional logic. If $\mathrm{L} = \langle \mathbf{Fm}, \vdash_{\mathrm{L}} \rangle$ is a logic in the signature $\mathcal{L}$, and $\mathcal{K}$ is a class of $\mathcal{L}$-algebras, we say that $\mathcal{K}$ is an *algebraic semantics* for L if, for some formula-equation transformer τ, condition (S1) in the definition of equivalence holds true between $\vdash_{\mathrm{L}}$ and $\vdash_{\mathcal{K}}$. In other words, we do not assume that there is an equation-formula transformer ρ to act as an inverse of τ. The concept of an algebraic semantics is weaker than the notion of algebraizability in at least two respects:

- First, the same logic can have more than one algebraic semantics and there can be different logics with the same algebraic semantics. For example, by Glivenko's theorem [**166**], CL admits not only $\mathcal{B}ool$ as an algebraic semantics, but also (by choosing $\tau(\alpha) = \{\neg\neg\alpha \approx 1\}$) the class $\mathcal{HA}$ of Heyting algebras [**39**], which forms an algebraic semantics for intuitionistic logic IL.

[2]According to Josep Maria Font [**129**], the credit should be divided more or less as follows: it was Lindenbaum who came up with the idea of a matrix on the formula algebra of the signature of classical propositional logic, while the idea of factoring this algebra modulo the relation of provable equivalence in a theory is Tarski's.

- Second, the relation between a logic L and an algebraic semantics $\mathcal{K}$ for L is asymmetric. The transformer τ provides the class $\mathcal{K}$ with the expressive resources to indicate when a given *formula* is valid in L. On the other hand, the logic L need not have the expressive resources to indicate when a given *equation* is satisfied by $\mathcal{K}$. For example, there is no way in CL to express, by means of a condition involving a set of formulas, when it is the case that an equation $\alpha \approx \beta$ is satisfied by its algebraic semantics $\mathcal{HA}$. This contrasts with the more familiar algebraic semantics of CL, where an equation $\alpha \approx \beta$ is satisfied by $\mathcal{B}ool$ just in case $\vdash_{\mathsf{HCL}} \alpha \to \beta$ and $\vdash_{\mathsf{HCL}} \beta \to \alpha$.

3.5.4. Tarski's legacy. Lindenbaum's premature death in 1941 prevented him from exerting an even greater impact on the development of logic. Tarski, on the other hand, permanently settled in the United States in 1939, establishing in Berkeley what would soon become the leading research group in algebraic logic worldwide. Among Tarski's PhD students were such prominent figures in logic and algebra as Bjarni Jónsson, C.C. Chang, Richard Montague, Donald Monk, and many others, and Tarski's charismatic personality also deeply affected researchers who were not his students such as Leon Henkin, Dana Scott, and Ralph McKenzie.

Tarski's groundbreaking contributions spanned several areas of mathematics and philosophy, including geometry and the philosophy of language. Together with his students and collaborators, he founded the field of *model theory*, giving the first rigorous definitions of satisfaction, truth, and logical consequence for first-order languages, introducing fundamental concepts such as elementary equivalence and elementary extension, generalizing the Löwenheim-Skolem theorem to arbitrary infinite cardinalities, and proving some of the earliest preservation results [**97**, **363**].

In the field of algebraic logic, Tarski either started or substantially developed at least three threads whose investigation became the *leitmotiv* of the discipline from the 1950s to the 1970s.

Algebraic semantics for first-order logic. In his 1936 paper *On the concept of logical consequence*, Tarski laid the foundations of the semantics for first-order logic. Although the insights contained in this work prompted Henkin to provide what is nowadays the standard textbook completeness proof, the approach fails to identify an algebraic counterpart of first-order logic analogous to that of Boolean algebras with respect to classical propositional logic. The decisive step in that direction was made by Tarski, with his students Louise Chin and Frederick Thompson, in the late 1940s and early 1950s, with the introduction of the concept of a *cylindric algebra*.

Let κ be any ordinal. An algebra $\langle A, \wedge, \vee, \neg, 0, 1, \{c_\lambda\}_{\lambda \in \kappa}, \{d_{\lambda,\mu}\}_{\lambda,\mu \in \kappa}\rangle$ is called a *cylindric algebra of dimension* κ if $\langle A, \wedge, \vee, \neg, 0, 1\rangle$ is a Boolean algebra, each c_λ is a unary operation, each $d_{\lambda,\mu}$ is a member of A, and the following equations are satisfied, for all $\lambda, \mu, \lambda_1, \lambda_2 \in \kappa$ with $\mu \notin \{\lambda, \lambda_1, \lambda_2\}$:

$$
\begin{array}{ll}
c_\lambda(0) \approx 0 & d_{\lambda,\lambda} \approx 1 \\
x \leq c_\lambda(x) & d_{\lambda_1,\lambda_2} \approx c_\mu(d_{\lambda_1,\mu} \wedge d_{\mu,\lambda_2}) \\
c_\lambda(x \wedge c_\lambda(y)) \approx c_\lambda(x) \wedge c_\lambda(y) & c_\kappa(d_{\lambda,\mu} \wedge x) \wedge c_\lambda(d_{\lambda,\mu} \wedge \neg x) \approx 0 \\
c_\lambda(c_\mu(x)) \approx c_\mu(c_\lambda(x)). &
\end{array}
$$

This definition is motivated by cylindric algebras of the form

$$\langle F/\Theta_T, \wedge, \vee, \neg, 0, 1, \{c_\lambda\}_{\lambda\in\kappa}, \{d_{\lambda,\mu}\}_{\lambda,\mu\in\kappa}\rangle,$$

where F is the set of formulas of a first-order language $\mathcal{L}$ over the set of variables $\{x_\lambda\}_{\lambda\in\kappa}$, $T \subseteq F$ is a first-order theory, $\langle\alpha, \beta\rangle \in \Theta_T$ if and only if $\alpha \leftrightarrow \beta$ is a classical consequence of T, $c_\lambda(\alpha) = [(\exists x_\lambda)\alpha]_{\Theta_T}$, and $d_{\lambda,\mu} = [x_\lambda \approx x_\mu]_{\Theta_T}$. Boolean algebras of sets form cylindric algebras of sets when equipped with suitable constants (the *diagonal sets*) and unary operations (the *cylindrification* operations), but unlike the pure Boolean algebra case, not every cylindric algebra is representable as a subdirect product of such set-based algebras. Indeed, for $\kappa \geq 3$, the class of representable cylindric algebras of dimension κ is not finitely axiomatizable over the class of all cylindric algebras of the same dimension, while for $\kappa = 2$, two additional equations suffice. The number 3 is also decisive for the decidability problem: for $\kappa \geq 3$, the equational theories of cylindric algebras and representable cylindric algebras of dimension κ are undecidable, while for $\kappa = 2$, they are both decidable (see **[182]**).

Relation algebras. In the late 1940s and early 1950s, Jónsson and Tarski revived one of the traditional themes of algebraic logic, namely, the calculus of relations. They considered Boolean algebras of binary relations on a set X, under the familiar operations of intersection, union, and complement, with $\emptyset$ and X^2 as designated elements, and expanded these algebras by a binary operation $\circ$ (*relational product*) and a unary operation $^\smile$ (*converse*) defined by

$$R \circ S := \{\langle a, b\rangle \in X^2 \mid \langle a, x\rangle \in R \text{ and } \langle x, b\rangle \in S, \text{ for some } x \in X\}$$
$$R^\smile := \{\langle a, b\rangle \in X^2 \mid \langle b, a\rangle \in R\}.$$

They designated the identity relation $1'$ and introduced the more general notion of a *relation algebra*, a Boolean algebra with additional operations $\circ$, $^\smile$, and $1'$ that satisfies certain equations. A relation algebra is *representable* if it is a subdirect product of algebras of binary relations over a set. Similarly to the case of cylindric algebras, not every relation algebra is representable **[241]**; however, the class of representable relation algebras is a variety **[354]**. Tarski and Jónsson proved that the equational theories of relation algebras and representable relation algebras are undecidable; for a proof and a comprehensive account of the theory of relation algebras, see **[243]**.

Matrix semantics. A (*logical*) *matrix* is an ordered pair $\mathbf{M} = \langle\mathbf{A}, F\rangle$ consisting of an $\mathcal{L}$-algebra $\mathbf{A}$ and a set of *designated elements* $F \subseteq A$. Given $T \cup \{\alpha\} \subseteq \mathrm{Fm}$, let $T \vdash_{\mathbf{M}} \alpha$ denote that, for all $\varphi \in \hom(\mathbf{Fm}, \mathbf{A})$, whenever $\varphi[T] \subseteq F$, also $\varphi(\alpha) \in F$. Then any class $\mathcal{M}$ of matrices of the same signature yields a consequence relation over Fm defined by $T \vdash_{\mathcal{M}} \alpha$ if and only if $T \vdash_{\mathbf{M}} \alpha$, for all $\mathbf{M} \in \mathcal{M}$. Conversely, given any logic L, let $\mathcal{M}(\mathrm{L})$ denote the class of all matrices of the form $\langle\mathbf{Fm}, T\rangle$, where T is a theory of $\vdash_{\mathrm{L}}$.

Logical matrices were an important tool for many logicians already in the 1920s — including those in the Polish school and others such as Bernays, Lewis, and Post — and were employed either to establish the independence of particular axioms in some Hilbert-style calculus, or to show the deductive equivalence of calculi with different signatures. Łukasiewicz, Tarski, Lindenbaum, and other pre-war Warsaw logicians fine-tuned this technique; however, matrices were first used to study sets of formulas rather than consequence relations, and attention was often restricted to matrices which validated certain specific rules of inference such as modus ponens [**97**].

After World War Two, algebraic logic rapidly developed in Poland. Jerzy Łoś, Roman Suszko, Helena Rasiowa, Roman Sikorski, and Riszard Wójcicki kept the flag flying throughout the 1950s and 1960s, developing in detail the theory of logical matrices initiated within the Łukasiewicz circle [**35**]. Łoś and Suszko [**238**] were the first to apply matrices to the investigation of consequence relations. Although it had been noticed earlier that logical matrices provide a tool to define logics that might fail to be finitely axiomatizable, they showed that any logic defined by a finite matrix is finitary, and found a necessary and sufficient condition for a logic L to admit a matrix $\mathbf{M}$ such that $\vdash_{\mathrm{L}} = \vdash_{\mathbf{M}}$, proving also the Łoś-Suszko theorem mentioned in Section 3.1. Later, Wojcicki [**372**] proved that $\vdash_{\mathrm{L}} = \vdash_{\mathcal{M}(\mathrm{L})}$, for any logic L. For an extensive treatment of matrix semantics, see, e.g., [**95**] or [**131**].

3.5.5. Algebraizable logics and beyond. One of the most influential books for the development of algebraic logic was Rasiowa's *An Algebraic Approach to Non-classical Logic* [**323**], which was published in 1974, but contains ideas and methods appearing already in Rasiowa's publications from the early 1950s onwards [**130**]. In this work, Rasiowa generalized Tarski's algebraic completeness proof for classical propositional logic to the class of all logics $\mathrm{L} = \langle \mathbf{Fm}, \vdash_{\mathrm{L}} \rangle$, based on a signature containing a binary operation symbol $\to$, such that the following conditions are satisfied for every theory T of $\vdash_{\mathrm{L}}$:

(I1) the relation $\leq_T = \{\langle \alpha, \beta \rangle \in \mathrm{Fm}^2 \mid \alpha \to \beta \in T\}$ is reflexive and transitive;

(I2) the equivalence relation induced by $\leq_T$ is a congruence of $\mathbf{Fm}$;

(I3) T is an order-filter of $\langle \mathrm{Fm}, \leq_T \rangle$;

(I4) $\alpha \in T$ and $\beta \in \mathrm{Fm}$ imply that $\beta \leq_T \alpha$.

Rasiowa's investigation of these so-called *implicative logics* and their corresponding algebraic models is remarkable in that it extends Tarskian methods to a whole class of logical systems (including classical and intuitionistic logics), paving the way for the even more abstract approaches that followed.

A further major breakthrough in the discipline may be dated to 1989 and the publication of Wim Blok and Don Pigozzi's monograph on algebraizable logics [**36**]. The notion of an *algebraizable logic*, introduced in [**36**] and recalled in Section 3.3 above, avoids the flaws of the weaker concept of an algebraic semantics of a logic. In particular:

- Ambiguity is drastically reduced. A finitary algebraizable logic can have more than one equivalent algebraic semantics, but all such classes of algebras generate the same quasivariety; moreover, although there are still different logics with the same equivalent algebraic semantics, they must have a different formula-equation transformer τ.
- Asymmetry disappears. The requirement that there be an equation-formula transformer ρ inverting τ entails that the equational consequence relation of the equivalent quasivariety semantics of an algebraizable logic L is fully interpretable in the consequence relation of L.

Blok and Pigozzi's volume is an in-depth investigation of algebraizable logics and some other classes of logics of particular significance. In addition to the results already mentioned, its achievements include:

(i) a thorough investigation of the *Leibniz congruence* $\Omega(\mathbf{M})$ associated to a logical matrix $\mathbf{M} = \langle \mathbf{A}, F \rangle$: the largest congruence Θ of $\mathbf{A}$ for which F is a union of Θ-congruence classes (the function Ω is called a *Leibniz operator*);

(ii) an algorithmic axiomatization of the equivalent quasivariety semantics of a given algebraizable logic;

(iii) a proof that the lattices of theories of an algebraizable logic and the equational consequence relation of its equivalent quasivariety semantics are isomorphic;

(iv) two intrinsic characterizations of algebraizability, one with a "syntactic" flavor and another that appeals to properties of the Leibniz operator.

The largest class of logics that is tractable with the methods introduced by Blok and Pigozzi is the class of *protoalgebraic logics* [**95**], which, however, fails to encompass some important and well-motivated logics. The Barcelona school of algebraic logic, led by Josep Maria Font and Ramon Jansana, developed a theory of *generalized matrices* [**132**] with the goal of extending the abstract investigation of logics beyond the protoalgebraic case. Other recent developments that expand the Blok-Pigozzi paradigm in different directions are the theories of *order algebraizability* [**78, 322**], *reverse algebraization* [**37**], and *truth-equational logics* [**277, 320**].

3.6. Bibliographical remarks

The approach followed in Sections 3.1 and 3.3 originated in the work of Blok and Jónsson [**34**] and was further developed in [**133, 140, 153, 158**]. These papers build on earlier studies of consequence relations on formulas and algebraizability initiated by Blok and Pigozzi in their influential monograph [**36**]. For further details on the Lindenbaum-Tarski method and abstract algebraic logic, we refer to [**79, 80, 131**].

The account of equational consequence relations given in Section 3.2 is based on [**256**], which systematizes results presented in various forms in the literature on abstract algebraic logic and universal algebra. In particular, Propositions 3.2.2 and 3.2.5 follow from results of Mal'cev [**247**], while Proposition 3.2.7, describing the correspondence between equational consequence in varieties and congruences of free algebras, may be found in [**308**]. Equational logic and its relationship with fully invariant congruences was introduced by Birkhoff [**29**].

The logics L_1 and L_2 presented in Example 3.3.4 are known, respectively, as the *logic of equilibrium*, introduced in [**154**], and *Abelian logic*, introduced independently in [**266**] and [**61**]; a detailed comparison of these logics can be found in [**305**]. The equivalence of the second and third consequence relations listed in Theorem 3.4.1 has been part of the folklore of the subject for some time, but was first clearly stated and proved in detail in [**148**].

Correspondences between sequent calculi and equational consequence relations were first considered in [**316, 325**]; the account in Section 3.3 is essentially based on [**319**]. For the concept of the external consequence relation of a sequent calculus, we refer to [**11**]. Let us also remark that interesting and somewhat unexpected equivalence results can be proved by focusing on proof systems whose deductive strength is intermediate between GCL and its cut-free version GCL°. Suppose for example that we add to the $\langle \wedge, \vee, \neg \rangle$-fragment of GCL°, formulated with invertible rules, the *inverses* of all the logical rules, yielding a hybrid system $\mathsf{GCL}^\circ_{\mathsf{inv}}$ with left and right *elimination* rules in addition to the usual introduction rules. The resulting consequence relation $\vdash_{\mathsf{GCL}^\circ_{\mathsf{inv}}}$ is then equivalent (see [**20**, **312**, **317**]) to the consequence relation of Priest's Logic of Paradox LP [**314**], a three-valued paraconsistent logic. This approach develops ideas already present in embryo in Girard's three-valued semantics for the classical sequent calculus without cut [**164**].

CHAPTER 4

Structure theory

In this chapter, we lay the groundwork for an approach to the study of residuated structures based on the guiding principles and motivations of the *Conrad Program* for ℓ-groups: a series of papers by Paul Conrad in the 1960s [**84**–**87**] that investigate the structure of ℓ-groups primarily via lattice-theoretic properties of their lattices of convex ℓ-subgroups. As we show here, large parts of this program extend profitably to the setting of e-cyclic residuated lattices, a class that covers the vast majority of the "algebras of logic" studied in this book. Subsequent chapters will provide further compelling evidence of the significance of ℓ-groups in the study of these algebras; in particular, we will see in Chapters 6 and 7 that important classes of residuated lattices may be viewed as classes of ℓ-groups with a nucleus or co-nucleus.

Sections 4.1 and 4.2 focus on the structure of lattices of convex subuniverses of e-cyclic residuated lattices, demonstrating that many remarkable properties that hold in the ℓ-group case extend to this more general setting. We then prove in Section 4.3 that the variety of residuated lattices is e-*regular*: every congruence of a residuated lattice is determined by its equivalence class of e. We also show that these equivalence classes admit an elegant description as *normal* convex subuniverses, where "normality" appropriately generalizes the group-theoretic notion, and hence that the congruence lattice of a residuated lattice is isomorphic to the lattice of its normal convex subuniverses. In Section 4.4, we describe the generation of normal convex subuniverses in residuated lattices.

4.1. Convex subuniverses

In this section, we study the structure of residuated lattices that are e-*cyclic*, that is, residuated lattices that satisfy the equation $x\backslash \mathrm{e} \approx \mathrm{e}/x$. The main result is Theorem 4.1.7, which establishes that the lattice of convex subuniverses of such a residuated lattice is an algebraic distributive lattice whose compact elements — the principal convex subuniverses — form a sublattice. We also provide in Lemma 4.1.2 an element-wise description of the convex subuniverse generated by an arbitrary subset of an e-cyclic residuated lattice.

Let $\mathbf{P} = \langle P, \leq \rangle$ be a poset. A subset $C \subseteq P$ is said to be *convex* in $\mathbf{P}$ if whenever $x, y \in C$, $z \in P$, and $x \leq z \leq y$, then $z \in C$.

Let $\mathbf{L}$ be any residuated lattice. A subuniverse of $\mathbf{L}$ that is convex in the lattice reduct of $\mathbf{L}$ is called a *convex subuniverse* of $\mathbf{L}$. It is easily confirmed that the intersection of any family of convex subuniverses of $\mathbf{L}$ is again a convex subuniverse of $\mathbf{L}$, and that the same holds for the union of any up-directed family of convex subuniverses of $\mathbf{L}$. The set $\mathcal{C}(\mathbf{L})$ of convex subuniverses of $\mathbf{L}$ ordered by set-inclusion therefore forms an algebraic closure system of $\langle \wp(L), \subseteq \rangle$ (in particular, an algebraic lattice) that, for convenience, we also denote by $\mathcal{C}(\mathbf{L})$.

For any $S \subseteq L$, let $\mathrm{C}[S]$ (or $\mathrm{C}_{\mathbf{L}}[S]$ whenever reference to $\mathbf{L}$ is needed) denote $\bigcap\{H \in \mathcal{C}(\mathbf{L}) \mid S \subseteq H\}$, the *convex subuniverse of* $\mathbf{L}$ *generated by* S. We call $H \in \mathcal{C}(\mathbf{L})$ *finitely generated* if $H = \mathrm{C}[S]$, for some finite $S \subseteq L$, and *principal* if $H = \mathrm{C}[a] := \mathrm{C}[\{a\}]$, for some $a \in L$.

The following definition of the *absolute value* of an element $x \in L$ — identified here with a negative element, contrary to the usual ℓ-group definition — will simplify our presentation:

$$|x| := x \wedge (\mathrm{e}/x) \wedge \mathrm{e}.$$

For $S \subseteq L$, we let $|S| := \{|x| \mid x \in S\}$. We also say that $x, y \in L$ are *disjoint*, denoted by $x \perp y$, if $|x| \vee |y| = \mathrm{e}$.

The proof of the following lemma is routine and left to the reader.

LEMMA 4.1.1. *Let* $\mathbf{L}$ *be a residuated lattice and let* $x \in L$ *and* $y \in L^-$.

(a) $x \leq \mathrm{e}$ *if and only if* $|x| = x$.
(b) $|x| \leq x \leq |x|\backslash \mathrm{e}$.
(c) $|x| = \mathrm{e}$ *if and only if* $x = \mathrm{e}$.
(d) $y \leq x \leq y\backslash \mathrm{e}$ *if and only if* $y \leq |x|$.
(e) *If* $H \in \mathcal{C}(\mathbf{L})$, *then* $x \in H$ *if and only if* $|x| \in H$.

For any residuated lattice $\mathbf{L}$ and $S \subseteq L$, let $\langle S \rangle$ denote the universe of the submonoid of $\langle L, \cdot\, \mathrm{e} \rangle$ generated by S, that is,

$$\langle S \rangle := \{x_1 \cdots x_n \mid x_1, \ldots, x_n \in S\},$$

recalling that $x_1 \cdots x_n$ is e for $n = 0$. In the case where $\mathbf{L}$ is e-cyclic, we obtain the following useful description of the generation of convex subuniverses.

LEMMA 4.1.2. *Let* $\mathbf{L}$ *be an* e*-cyclic residuated lattice. For any* $S \subseteq L$,

$$\begin{aligned} \mathrm{C}[S] = \mathrm{C}[|S|] &= \{x \in L \mid h \leq x \leq h\backslash \mathrm{e}, \textit{ for some } h \in \langle |S| \rangle\} \\ &= \{x \in L \mid h \leq |x|, \textit{ for some } h \in \langle |S| \rangle\}. \end{aligned}$$

PROOF. Let $H := \{x \in L \mid h \leq x \leq h\backslash \mathrm{e}$, for some $h \in \langle |S| \rangle\}$. The definition of the absolute value of an element and Lemma 4.1.1.(b) yield $H \subseteq \mathrm{C}[S] = \mathrm{C}[|S|]$. Hence, taking into account Lemma 4.1.1.(d), it suffices to prove that H is a convex subuniverse of $\mathbf{L}$ that contains $|S|$.

If $x \in |S|$, then $x \leq x \leq x\backslash \mathrm{e}$, so $|S| \subseteq H$. Now consider $x, y \in H$ and $h, k \in \langle |S| \rangle$ such that $h \leq x \leq h\backslash \mathrm{e}$ and $k \leq y \leq k\backslash \mathrm{e}$. Observe first that, since $h, k \leq \mathrm{e}$,

$$(hk)(kh) \leq hk \leq xy \leq (h\backslash \mathrm{e})(k\backslash \mathrm{e}) \leq kh\backslash \mathrm{e} \leq (hk)(kh)\backslash \mathrm{e}.$$

Hence $xy \in H$, and H is closed under multiplication. Moreover,

$$hk \leq h \wedge k \leq x \wedge y \leq x \vee y \leq (h\backslash \mathrm{e}) \vee (k\backslash \mathrm{e}) \leq hk\backslash \mathrm{e}.$$

Hence $x \wedge y, x \vee y \in H$, and H is closed under the lattice operations.

We show next that H is closed under the residual operations. Since $\mathbf{L}$ is e-cyclic, $y \leq k\backslash \mathrm{e} = \mathrm{e}/k$, so $ykh \leq h \leq x$ and $hkh \leq kh \leq y\backslash x$. But also $k \leq y$ and $x \leq h\backslash \mathrm{e}$, so, using Propositions 1.4.2 and 1.4.4,

$$hkh \leq y\backslash x \leq k\backslash x \leq k\backslash(h\backslash \mathrm{e}) = hk\backslash \mathrm{e} \leq hkh\backslash \mathrm{e}.$$

Hence $y\backslash x \in H$. The proof that $x/y \in H$ is symmetric, since by e-cyclicity, $H = \{x \in L \mid h \leq x \leq \mathrm{e}/h, \text{ for some } h \in \langle|S|\rangle\}$.

Finally, for convexity, if $z \in L$ and $x \leq z \leq y$, then also $z \in H$, since

$$hk \leq h \leq x \leq z \leq y \leq k\backslash \mathrm{e} \leq hk\backslash \mathrm{e}. \qquad \square$$

COROLLARY 4.1.3. *Let $\mathbf{L}$ be an e-cyclic residuated lattice. For any $a \in L$,*

$$\begin{aligned} \mathrm{C}[a] = \mathrm{C}[|a|] &= \{x \in L \mid |a|^n \leq x \leq |a|^n\backslash \mathrm{e}, \textit{ for some } n \in \mathbb{N}\} \\ &= \{x \in L \mid |a|^n \leq |x|, \textit{ for some } n \in \mathbb{N}\}. \end{aligned}$$

Lemma 4.1.2 also implies that a convex subuniverse is completely determined by its negative cone.

COROLLARY 4.1.4. *Let $\mathbf{L}$ be an e-cyclic residuated lattice. Then $H = \mathrm{C}[H^-]$, for any $H \in \mathcal{C}(\mathbf{L})$.*

The next simple auxiliary result will be used quite often in calculations.

LEMMA 4.1.5. *For any residuated lattice $\mathbf{L}$ and $x_1, \ldots, x_n, y \in L^-$,*

$$(x_1 \vee y) \cdots (x_n \vee y) \leq (x_1 \cdots x_n) \vee y.$$

PROOF. For $n = 2$, observe that, for any $x_1, x_2, y \in L^-$,

$$(x_1 \vee y)(x_2 \vee y) = x_1x_2 \vee x_1y \vee yx_2 \vee y^2 \leq x_1x_2 \vee y.$$

The general case follows by an easy induction. $\square$

The following lemma shows that the poset $\mathcal{K}(\mathcal{C}(\mathbf{L}))$ of *compact* (equivalently, *finitely generated*) convex subuniverses of an arbitrary e-cyclic residuated lattice $\mathbf{L}$ ordered by inclusion is a sublattice of $\mathcal{C}(\mathbf{L})$.

LEMMA 4.1.6. *Let $\mathbf{L}$ be an e-cyclic residuated lattice. For any $a, b \in L$,*

$$\mathrm{C}[a] \cap \mathrm{C}[b] = \mathrm{C}[|a| \vee |b|] \quad \textit{and} \quad \mathrm{C}[a] \vee \mathrm{C}[b] = \mathrm{C}[|a| \wedge |b|] = \mathrm{C}[|a||b|].$$

Moreover, $\mathcal{K}(\mathcal{C}(\mathbf{L}))$ is a sublattice of $\mathcal{C}(\mathbf{L})$ consisting of all the principal convex subuniverses of $\mathbf{L}$.

PROOF. In view of Corollary 4.1.4, it suffices to establish the two equalities in the lemma for any $a, b \in L^-$.

For the first equality, note that $a, b \leq a \vee b \leq \mathrm{e}$, so $\mathrm{C}[a \vee b] \subseteq \mathrm{C}[a] \cap \mathrm{C}[b]$. To establish the reverse inclusion, consider any $x \in \mathrm{C}[a] \cap \mathrm{C}[b]$. By Corollary 4.1.3, there exist $m, n \in \mathbb{N}$ such that $a^m \leq x \leq a^m\backslash \mathrm{e}$ and $b^n \leq x \leq b^n\backslash \mathrm{e}$. Successive applications of Lemma 4.1.5 yield $(a \vee b)^{mn} \leq (a^m \vee b)^n \leq a^m \vee b^n$, and hence

$$(a \vee b)^{mn} \leq a^m \vee b^n \leq x \leq (a^m\backslash \mathrm{e}) \wedge (b^n\backslash \mathrm{e}) = (a^m \vee b^n)\backslash \mathrm{e} \leq (a \vee b)^{mn}\backslash \mathrm{e}.$$

So $x \in \mathrm{C}[a \vee b]$, by Corollary 4.1.3, showing that $\mathrm{C}[a] \cap \mathrm{C}[b] \subseteq \mathrm{C}[a \vee b]$.

For the second equality, note that the inequalities $ab \leq a \wedge b \leq a, b \leq \mathrm{e}$ imply that $\mathrm{C}[a] \vee \mathrm{C}[b] \subseteq \mathrm{C}[a \wedge b] \subseteq \mathrm{C}[ab]$. But also $ab \in \mathrm{C}[a] \vee \mathrm{C}[b]$, so $\mathrm{C}[ab] \subseteq \mathrm{C}[a] \vee \mathrm{C}[b]$.

It follows immediately that $\mathcal{K}(\mathcal{C}(\mathbf{L}))$ is a sublattice of $\mathcal{C}(\mathbf{L})$ consisting of all the principal convex subuniverses of $\mathbf{L}$. □

We now present the main result of this section.

THEOREM 4.1.7. *Let* $\mathbf{L}$ *be an* e*-cyclic residuated lattice. Then* $\mathcal{C}(\mathbf{L})$ *is an algebraic distributive lattice.*

PROOF. Let $H, J, K \in \mathcal{C}(\mathbf{L})$. We prove that $H \cap (J \vee K) \subseteq (H \cap J) \vee (H \cap K)$, since the other inclusion holds trivially. Let $x \in H \cap (J \vee K)$. Then $x \in H$ and there exist $a_1, \ldots, a_n \in J^- \cup K^-$ such that $a_1 \cdots a_n \leq |x|$. By Lemma 4.1.5,

$$(a_1 \vee |x|) \cdots (a_n \vee |x|) \leq (a_1 \cdots a_n) \vee |x| = |x|.$$

Since $|x| \in L^-$, each element $a_i \vee |x|$ belongs to $(H \cap J)^-$ when $a_i \in J^-$, or to $(H \cap K)^-$ when $a_i \in K^-$. Hence $|x| \in \mathrm{C}[(H \cap J) \cup (H \cap K)] = (H \cap J) \vee (H \cap K)$, and Lemma 4.1.1.(b) guarantees that also $x \in (H \cap J) \vee (H \cap K)$ as required. □

A convex subuniverse of an e-cyclic residuated lattice $\mathbf{L}$ can be restricted to a convex subuniverse of any subalgebra $\mathbf{H}$ of $\mathbf{L}$. Observe that the order-homomorphisms

$$\varphi\colon \mathcal{C}(\mathbf{H}) \to \mathcal{C}(\mathbf{L});\ J \mapsto \mathrm{C}_{\mathbf{L}}[J] \quad \text{and} \quad \psi\colon \mathcal{C}(\mathbf{L}) \to \mathcal{C}(\mathbf{H});\ K \mapsto K \cap H$$

form a residuated pair $\langle \varphi, \psi \rangle$, since, for all $J \in \mathcal{C}(\mathbf{H})$ and $K \in \mathcal{C}(\mathbf{L})$,

$$J \subseteq \psi(K) = K \cap H \iff J \subseteq K \iff \varphi(J) = \mathrm{C}_{\mathbf{L}}[J] \subseteq K.$$

Given any $J \in \mathcal{C}(\mathbf{H})$, defining $S := \{x \in L \mid a \leq x \leq \mathrm{e}, \text{ for some } a \in J\}$ and $K := \mathrm{C}_{\mathbf{L}}[S]$ gives $S = K^-$, by Lemma 4.1.2, and $\psi(K) = K \cap H = J$. So ψ is surjective and, since $\langle \varphi, \psi \rangle$ is a residuated pair, φ is injective, by Lemma 1.2.4.(c). These observations imply the following result, which may be viewed as an analogue of the congruence extension property for convex subuniverses.

PROPOSITION 4.1.8. *Let* $\mathbf{H}$ *be any subalgebra of an* e*-cyclic residuated lattice* $\mathbf{L}$ *and let* $J \in \mathcal{C}(\mathbf{H})$. *Then* $J = \mathrm{C}_{\mathbf{L}}[J] \cap H$.

Similarly, the lattice structure of the convex subuniverses of an integral residuated lattice is inherited by the convex subuniverses of its images under certain nuclei. Let us call a nucleus γ on a residuated lattice $\mathbf{L}$ *dense* if $\langle L, \cdot, \mathrm{e} \rangle$ is generated by L_γ as a monoid, i.e., $\langle L_\gamma \rangle = L$.

LEMMA 4.1.9. *Let* γ *be a dense nucleus on an integral residuated lattice* $\mathbf{L}$. *The lattices* $\mathcal{C}(\mathbf{L})$ *and* $\mathcal{C}(\mathbf{L}_\gamma)$ *are isomorphic with an isomorphism given by the mutually inverse maps* $\varphi\colon \mathcal{C}(\mathbf{L}_\gamma) \to \mathcal{C}(\mathbf{L});\ J \mapsto \mathrm{C}_{\mathbf{L}}[J]$ *and* $\psi\colon \mathcal{C}(\mathbf{L}) \to \mathcal{C}(\mathbf{L}_\gamma);\ K \mapsto K \cap L_\gamma$.

PROOF. We check first that $K \cap L_\gamma \in \mathcal{C}(\mathbf{L}_\gamma)$, for any $K \in \mathcal{C}(\mathbf{L})$. Clearly, it is closed under $\wedge$, $\backslash$, and $/$, and contains $\gamma(\mathrm{e}) = \mathrm{e}$. Also, for any $\gamma(a), \gamma(b) \in K \cap L_\gamma$, clearly $\gamma(a) \circ_\gamma \gamma(b), \gamma(a) \vee_\gamma \gamma(b) \in L_\gamma$ and, since $\gamma(a)\gamma(b) \leq \gamma(a) \circ_\gamma \gamma(b) \leq \gamma(a) \vee_\gamma \gamma(b) \leq \mathrm{e}$, by convexity, $\gamma(a) \circ_\gamma \gamma(b), \gamma(a) \vee_\gamma \gamma(b) \in K$. The fact that $\langle \varphi, \psi \rangle$ is a residuated pair follows as in the case considered above for subalgebras of $\mathbf{L}$. Hence, to establish that φ is an isomorphism with inverse ψ, it suffices, by Lemma 1.2.4, to show that φ is a bijection.

First, consider any $H_1, H_2 \in \mathcal{C}(\mathbf{L}_\gamma)$ with $\varphi(H_1) = \varphi(H_2)$. Let $a \in H_1$. Then $a \in \mathrm{C}_\mathbf{L}[H_1] = \mathrm{C}_\mathbf{L}[H_2]$, and, since $\mathbf{L}$ is integral, there exist $g_1, \ldots, g_n \in H_2$ such that $g_1 \cdots g_n \leq a$. But then $g_1 \circ_\gamma \cdots \circ_\gamma g_n = \gamma(g_1 \cdots g_n) \leq \gamma(a) = a$, which implies that $a \in H_2$. Hence $H_1 \subseteq H_2$. Symmetrically, $H_1 = H_2$, so φ is injective. We establish surjectivity by showing that $\mathrm{C}_\mathbf{L}[K \cap L_\gamma] = K$, for all $K \in \mathcal{C}(\mathbf{L})$. For the non-trivial direction, let $a \in K$. Since γ is dense, there exist $g_1, \ldots, g_n \in L_\gamma$ such that $a = g_1 \cdots g_n$. But then, by integrality and convexity, $g_1, \ldots, g_n \in K$, so $a \in \mathrm{C}_\mathbf{L}[K \cap L_\gamma]$. □

4.2. Polars and prime convex subuniverses

In this section, we study the prime (meet-irreducible) elements and polars (pseudo-complements of elements) of $\mathcal{C}(\mathbf{L})$ for an arbitrary e-cyclic residuated lattice $\mathbf{L}$. In particular, it is shown in Lemma 4.2.7 that if $\mathbf{L}$ satisfies a certain prelinearity law, then a convex subuniverse of $\mathbf{L}$ is prime exactly when the convex subuniverses containing it form a chain under set-inclusion. A description of *minimal*, prime convex subuniverses in terms of polars is provided by Proposition 4.2.13.

Let $\mathbf{L}$ be any e-cyclic residuated lattice. Then $\mathcal{C}(\mathbf{L})$ is an algebraic distributive lattice (Theorem 4.1.7) and hence, by Proposition 1.3.8.(b), a *frame*; that is, for any $H \in \mathcal{C}(\mathbf{L})$ and $\{K_i \mid i \in I\} \subseteq \mathcal{C}(\mathbf{L})$,

$$H \cap \bigvee\{K_i \mid i \in I\} = \bigvee\{H \cap K_i \mid i \in I\}.$$

Moreover, the binary meet operation of $\mathcal{C}(\mathbf{L})$ is residuated (see Example 1.4.11): for all $H, K \in \mathcal{C}(\mathbf{L})$, there exists in $\mathcal{C}(\mathbf{L})$ the *relative pseudo-complement* of H with respect to K,

$$H \to K := \max\{J \in \mathcal{C}(\mathbf{L}) \mid H \cap J \subseteq K\}.$$

We also call $H^\perp := H \to \{\mathrm{e}\}$ the *pseudo-complement* of H, and sometimes refer to $\mathcal{C}(\mathbf{L})$ as a *Brouwerian algebra*, by which we mean $\mathcal{C}(\mathbf{L})$ equipped with the additional operations $\to$ and $\top := L$.

The next lemma gives an element-wise description of relative pseudo-complements in terms of absolute values.

LEMMA 4.2.1. *For any* e-*cyclic residuated lattice* $\mathbf{L}$ *and* $H, K \in \mathcal{C}(\mathbf{L})$,

$$H \to K = \{a \in L \mid |a| \vee |x| \in K, \textit{ for all } x \in H\};$$
$$H^\perp = \{a \in L \mid a \perp x, \textit{ for all } x \in H\}.$$

PROOF. The proof is a direct consequence of Lemmas 4.1.1.(e) and 4.1.6. Just note that, for any $H, K \in \mathcal{C}(\mathbf{L})$ and $a \in L$,

$$\begin{aligned} a \in H \to K &\iff \mathrm{C}[a] \subseteq H \to K \\ &\iff \mathrm{C}[a] \cap H \subseteq K \\ &\iff \mathrm{C}[a] \cap \mathrm{C}[x] \subseteq K, \text{ for all } x \in H \\ &\iff \mathrm{C}[|a| \vee |x|] \subseteq K, \text{ for all } x \in H \\ &\iff |a| \vee |x| \in K, \text{ for all } x \in H. \end{aligned}$$

The description of $H^\perp$ is now clear. □

Observe next that, for any e-cyclic residuated lattice $\mathbf{L}$, the map $^{\perp}\colon \mathcal{C}(\mathbf{L}) \to \mathcal{C}(\mathbf{L})$ satisfies, for all $H, K \in \mathcal{C}(\mathbf{L})$,

$$H \subseteq K^{\perp} \iff K \subseteq H^{\perp}.$$

That is, $^{\perp}\colon \mathcal{C}(\mathbf{L}) \to \mathcal{C}(\mathbf{L})^{\partial}$ is residuated and its own residual. Hence, by Lemma 1.2.8, it converts arbitrary joins in $\mathcal{C}(\mathbf{L})$ to intersections. Moreover, using Lemma 1.2.4 and Proposition 1.3.4, it follows that $H^{\perp\perp\perp} = H^{\perp}$, for each $H \in \mathcal{C}(\mathbf{L})$, and that the map $^{\perp\perp}$ is a closure operator on $\mathcal{C}(\mathbf{L})$. Indeed, since $H^{\perp\perp} \cap K^{\perp\perp} \subseteq (H \cap K)^{\perp\perp}$, for all $H, K \in \mathcal{C}(\mathbf{L})$, this map is a nucleus on the Brouwerian algebra $\mathcal{C}(\mathbf{L})$. The following result is then a direct consequence of Proposition 1.5.9.

PROPOSITION 4.2.2. *Let $\mathbf{L}$ be an* e-*cyclic residuated lattice. Then*

$$\langle \mathcal{C}(\mathbf{L})_{\perp\perp}, \cap, \vee^{\perp\perp}, {}^{\perp}, \{\mathrm{e}\}, L \rangle$$

is a complete Boolean algebra, where $H \vee^{\perp\perp} K := (H \vee K)^{\perp\perp} = (H^{\perp} \cap K^{\perp})^{\perp}$.

Let $\mathbf{L}$ be an e-cyclic residuated lattice. We call $H \in \mathcal{C}(\mathbf{L})$ a *polar* of $\mathbf{L}$ if $H^{\perp\perp} = H$, i.e., if $H \in \mathcal{C}(\mathbf{L})_{\perp\perp}$. It is then straightforward to show that for any $A \subseteq L$, the set $\{b \in L \mid a \perp b, \text{ for all } a \in A\}$ is a polar of $\mathbf{L}$. Hence we may define, for any $a \in L$,

$$a^{\perp} := \{b \in L \mid a \perp b\},$$

called a *principal polar* of $\mathbf{L}$, and for any $A \subseteq L$,

$$A^{\perp} := \bigcap \{a^{\perp} \mid a \in A\} = \{b \in L \mid a \perp b, \text{ for all } a \in A\}.$$

We call $H \in \mathcal{C}(\mathbf{L})$ *prime* if it is meet-irreducible in $\mathcal{C}(\mathbf{L})$, that is, if $J \cap K = H$ implies $J = H$ or $K = H$, for any $J, K \in \mathcal{C}(\mathbf{L})$. Since $\mathcal{C}(\mathbf{L})$ is distributive, $H \in \mathcal{C}(\mathbf{L})$ is prime if and only if it is meet-prime in $\mathcal{C}(\mathbf{L})$, that is, if $J \cap K \subseteq H$ implies $J \subseteq H$ or $K \subseteq H$, for any $J, K \in \mathcal{C}(\mathbf{L})$.

The following lemma provides an alternative characterization of the prime convex subuniverses of an e-cyclic residuated lattice.

LEMMA 4.2.3. *Let $\mathbf{L}$ be any* e-*cyclic residuated lattice and let $H \in \mathcal{C}(\mathbf{L})$. The following statements are equivalent:*

(1) *H is a prime convex subuniverse of $\mathbf{L}$.*
(2) *For all $a, b \in L$, if $|a| \vee |b| \in H$, then $a \in H$ or $b \in H$.*

PROOF.

(1) $\Rightarrow$ (2) Suppose that H is prime, and consider any $a, b \in L$ such that $|a| \vee |b| \in H$. Then $\mathrm{C}[a] \cap \mathrm{C}[b] = \mathrm{C}[|a| \vee |b|] \subseteq H$, by Lemma 4.1.6, and, by assumption, $\mathrm{C}[a] \subseteq H$ or $\mathrm{C}[b] \subseteq H$. Hence $a \in H$ or $b \in H$.

(2) $\Rightarrow$ (1) Suppose that H satisfies (2), and consider any $J, K \in \mathcal{C}(\mathbf{L})$ such that $J \cap K \subseteq H$. Assume towards a contradiction that $J \not\subseteq H$ and $K \not\subseteq H$, i.e., there exist $a \in K \backslash H$ and $b \in J \backslash H$. Then $\mathrm{C}[|a| \vee |b|] = \mathrm{C}[a] \cap \mathrm{C}[b] \subseteq J \cap K \subseteq H$, by Lemma 4.1.6, so $|a| \vee |b| \in H$. But then $a \in H$ or $b \in H$, by (2), a contradiction. $\square$

Several related properties, playing a key role in the structure theory of residuated lattices, are referred to in the literature as *prelinearity laws.* Let us call a residuated lattice $\mathbf{L}$ *prelinear*[1] if it satisfies the equation

$$((x\backslash y)\wedge \mathrm{e})\vee((y\backslash x)\wedge \mathrm{e})\approx \mathrm{e}.$$

The next result provides equivalent characterizations of prelinearity for residuated lattices that satisfy a weak form of the distributivity law.

PROPOSITION 4.2.4. *A residuated lattice satisfying* $\mathrm{e}\wedge(y\vee z)\approx(\mathrm{e}\wedge y)\vee(\mathrm{e}\wedge z)$ *is prelinear if and only if it satisfies any one of the following equations:*

$$\begin{array}{ll}(\text{PL1}) & \mathrm{e}\leq(x\backslash y)\vee(y\backslash x)\\ (\text{PL2}) & (y\wedge z)\backslash x\approx(y\backslash x)\vee(z\backslash x)\\ (\text{PL3}) & x\backslash(y\vee z)\approx(x\backslash y)\vee(x\backslash z).\end{array}$$

PROOF. Let $\mathbf{L}$ be a residuated lattice satisfying $\mathrm{e}\wedge(y\vee z)\approx(\mathrm{e}\wedge y)\vee(\mathrm{e}\wedge z)$. Clearly, $\mathbf{L}$ is prelinear if and only if it satisfies (PL1). The remaining equivalences are established as follows:

(PL1) $\Rightarrow$ (PL2) Suppose that $\mathbf{L}$ satisfies (PL1). Every residuated lattice satisfies $(y\backslash x)\vee(z\backslash x)\leq(y\wedge z)\backslash x$, so it suffices to establish the converse. Let $a,b,c\in L$. Clearly, $b(\mathrm{e}\wedge(b\backslash c))\leq b$ and $b(\mathrm{e}\wedge(b\backslash c))\leq c$, so $b(\mathrm{e}\wedge(b\backslash c))\leq b\wedge c$. Hence $b(\mathrm{e}\wedge(b\backslash c))((b\wedge c)\backslash a)\leq a$ and it follows then that $(\mathrm{e}\wedge(b\backslash c))((b\wedge c)\backslash a)\leq b\backslash a$, $(\mathrm{e}\wedge(b\backslash c))((b\wedge c)\backslash a)\leq(b\backslash a)\vee(c\backslash a)$, and $\mathrm{e}\wedge(b\backslash c)\leq[(b\backslash a)\vee(c\backslash a)]/((b\wedge c)\backslash a)$. Likewise, $\mathrm{e}\wedge(c\backslash b)\leq[(b\backslash a)\vee(c\backslash a)]/((b\wedge c)\backslash a)$, so, using (PL1),

$$\mathrm{e}=\mathrm{e}\wedge[(b\backslash c)\vee(c\backslash b)]=[\mathrm{e}\wedge(b\backslash c)]\vee[\mathrm{e}\wedge(c\backslash b)]\leq[(b\backslash a)\vee(c\backslash a)]/[(b\wedge c)\backslash a].$$

Finally, by residuation, $(b\wedge c)\backslash a\leq(b\backslash a)\vee(c\backslash a)$ as required.

(PL2) $\Rightarrow$ (PL1) Suppose that $\mathbf{L}$ satisfies (PL2). Then, for all $a,b,c\in L$,

$$\begin{aligned}(a\backslash b)\vee(b\backslash a)&\geq[(a\backslash b)\wedge\mathrm{e}]\vee[(b\backslash a)\wedge\mathrm{e}]\\&=[a\backslash(a\wedge b)\wedge\mathrm{e}]\vee[b\backslash(a\wedge b)\wedge\mathrm{e}]\\&=[(a\backslash(a\wedge b))\vee(b\backslash(a\wedge b))]\wedge\mathrm{e}\\&=[(a\wedge b)\backslash(a\wedge b)]\wedge\mathrm{e}\\&=\mathrm{e}.\end{aligned}$$

(PL1) $\Leftrightarrow$ (PL3) Very similar to the proof that (PL1) $\Leftrightarrow$ (PL2). $\square$

The equation (PL3) — or dually, the equation (PL2) — suffices, in conjunction with one further equation, to ensure distributivity.

PROPOSITION 4.2.5. *Let* $\mathbf{L}$ *be a residuated lattice satisfying* $x\backslash(y\vee z)\approx(x\backslash y)\vee(x\backslash z)$ *and* $x\backslash x\approx\mathrm{e}$ *(e.g., any* ℓ*-group). Then* $\mathbf{L}$ *is distributive.*

[1]Such residuated lattices could also be called *left prelinear* and residuated lattices satisfying the dual equation *right prelinear*. Although these equations define different varieties, virtually all the properties considered in this book hold for (left) prelinear residuated lattices if and only if they hold for right prelinear residuated lattices.

PROOF. Let $a, b \in L$ with $a \leq b$ and consider the map $\varphi\colon L \to L$; $x \mapsto a(b\backslash x)$. Clearly, $\varphi(b) = a(b\backslash b) = a\mathrm{e} = a$ and $\varphi(x) = a(b\backslash x) \leq a(a\backslash x) \leq x$, for all $x \in L$. Moreover, φ is a join-homomorphism, since, for any $x, y \in L$,

$$\varphi(x \vee y) = a(b\backslash(x \vee y)) = a((b\backslash x) \vee (b\backslash y)) = a(b\backslash x) \vee a(b\backslash y) = \varphi(x) \vee \varphi(y).$$

Hence, by Proposition 1.1.10, the lattice reduct of $\mathbf{L}$ is distributive. □

COROLLARY 4.2.6. *Let* $\mathbf{L}$ *be a prelinear residuated lattice satisfying* $x\backslash x \approx \mathrm{e}$ *and* $\mathrm{e} \wedge (y \vee z) \approx (\mathrm{e} \wedge y) \vee (\mathrm{e} \wedge z)$. *Then* $\mathbf{L}$ *is distributive.*

The next lemma provides a characterization of the prime convex subuniverses of a prelinear, e-cyclic residuated lattice.

LEMMA 4.2.7. *Let* $\mathbf{L}$ *be a prelinear,* e*-cyclic residuated lattice and let* $H \in \mathcal{C}(\mathbf{L})$. *The following statements are equivalent:*

(1) H *is a prime convex subuniverse of* $\mathbf{L}$.
(2) *For all* $a, b \in L$, *if* $|a| \vee |b| \in H$, *then* $a \in H$ *or* $b \in H$.
(3) *For all* $a, b \in L$, *if* $a \perp b$, *then* $a \in H$ *or* $b \in H$.
(4) *For all* $a, b \in L$, *either* $(a\backslash b) \wedge \mathrm{e} \in H$ *or* $(b\backslash a) \wedge \mathrm{e} \in H$.
(5) $\langle \{K \in \mathcal{C}(\mathbf{L}) \mid H \subseteq K\}, \subseteq \rangle$ *is a chain.*

PROOF.

(1) $\Rightarrow$ (2) $\Rightarrow$ (3) Follows directly from Lemma 4.2.3.

(3) $\Rightarrow$ (4) Follows immediately using prelinearity.

(4) $\Rightarrow$ (5) Suppose that H satisfies (4), and consider any $J, K \in \mathcal{C}(\mathbf{L})$ with $H \subseteq J$ and $H \subseteq K$. Assume towards a contradiction that J and K are incomparable. Then, by Corollary 4.1.4, also J^- and K^- are incomparable, so there exist $a \in J^-\backslash K^-$ and $b \in K^-\backslash J^-$. By (4), either $(a\backslash b) \wedge \mathrm{e} \in H$ or $(b\backslash a) \wedge \mathrm{e} \in H$. However, if $(a\backslash b) \wedge \mathrm{e} \in H \subseteq J$, then $a((a\backslash b) \wedge \mathrm{e}) \in J$ and hence, since $a((a\backslash b) \wedge \mathrm{e}) \leq b \leq \mathrm{e}$, also $b \in J$, a contradiction. Analogously, if $(b\backslash a) \wedge \mathrm{e} \in H \subseteq K$, then $b((b\backslash a) \wedge \mathrm{e}) \in K$ and it follows that $a \in K$, a contradiction.

(5) $\Rightarrow$ (1) Obvious. □

COROLLARY 4.2.8. *Let* $\mathbf{L}$ *be a prelinear,* e*-cyclic residuated lattice. If* $\{\mathrm{e}\}$ *is a prime convex subuniverse of* $\mathbf{L}$ *(or, equivalently,* $\mathcal{C}(\mathbf{L})$ *is totally ordered), then* $\mathbf{L}$ *is totally ordered.*

PROOF. Suppose that $\{\mathrm{e}\}$ is a prime convex subuniverse of $\mathbf{L}$, or, equivalently, by Lemma 4.2.7, that $\mathcal{C}(\mathbf{L})$ is totally ordered. Consider any $a, b \in L$. By Lemma 4.2.7, either $(a\backslash b) \wedge \mathrm{e} = \mathrm{e}$ or $(b\backslash a) \wedge \mathrm{e} = \mathrm{e}$. If, $(a\backslash b) \wedge \mathrm{e} = \mathrm{e}$, then $\mathrm{e} \leq a\backslash b$ and so $a \leq b$. Likewise, if $(b\backslash a) \wedge \mathrm{e} = \mathrm{e}$, then $\mathrm{e} \leq b\backslash a$ and so $b \leq a$. □

A convex subuniverse H of a residuated lattice $\mathbf{L}$ is completely meet-irreducible in $\mathcal{C}(\mathbf{L})$ if whenever $\{K_i\}_{i \in I}$ is a family of convex subuniverses of $\mathbf{L}$ and $H = \bigcap_{i \in I} K_i$, then $H = K_i$, for some $i \in I$. Every such convex subuniverse has a unique cover in $\mathcal{C}(\mathbf{L})$: the intersection of all the convex subuniverses that properly contain it. Moreover, in the case where $\mathbf{L}$ is e-cyclic, it follows from Proposition 1.3.8 that every member of $\mathcal{C}(\mathbf{L})$ is a meet of completely meet-irreducible members of $\mathcal{C}(\mathbf{L})$, and hence of prime convex subuniverses of $\mathbf{L}$.

We close this section by discussing the relationship between principal polars and minimal, prime convex subuniverses of an e-cyclic residuated lattice. It has been proved for ℓ-groups, MV-algebras, and other more general classes of (bounded) residuated lattices, that if H is a minimal, prime convex subuniverse (often called a filter or prefilter) of such an algebra, then $H = \bigcup\{x^{\perp} \mid x \notin H\}$. However, as we show below, this is a result about algebraic distributive lattices, and has little to do with particular properties of the algebras in question.

For the rest of this section, let $\mathbf{A}$ denote any non-trivial, algebraic distributive lattice with least element $\bot$ and greatest element $\top$, such that the poset $\mathcal{K}(\mathbf{A})$ of compact elements of $\mathbf{A}$ forms a sublattice. We let $a \to b$ denote the relative pseudo-complement of a relative to b, and $\neg a$ the pseudo-complement $a \to \bot$ of a.

LEMMA 4.2.9. *Let $p \neq \top$ be a meet-irreducible element of $\mathbf{A}$, let $a \in \mathcal{K}(\mathbf{A})$, and let $b \in A$ be such that $a \to b \leq p$. Then there exists a meet-irreducible element q such that $a \not\leq q$ and $b \leq q \leq p$.*

PROOF. Let $S := \{x \in {\uparrow}b \mid a \to x \leq p\}$. Then $S \neq \emptyset$, since $b \in S$, and an application of Zorn's lemma yields a maximal element q of S. Suppose towards a contradiction that q is not meet-irreducible. Then $x \not\leq q$, $y \not\leq q$, and $x \wedge y \leq q$, for some $x, y \in A$. So $a \to (q \vee x) \not\leq p$ and $a \to (q \vee y) \not\leq p$, by the maximality of q, and $c_1 \not\leq p$, $c_2 \not\leq p$, $c_1 \leq a \to (q \vee x)$, and $c_2 \leq a \to (q \vee y)$, for some $c_1, c_2 \in \mathcal{K}(\mathbf{A})$. By residuation, $a \wedge c_1 \leq q \vee x$ and $a \wedge c_2 \leq q \vee y$. So $a \wedge (c_1 \wedge c_2) \leq (q \vee x) \wedge (q \vee y) = q \vee (x \wedge y) = q$ and, by residuation again, $c_1 \wedge c_2 \leq a \to q$. But also, since $\mathcal{K}(\mathbf{A})$ is a lattice, $c_1 \wedge c_2 \in \mathcal{K}(\mathbf{A})$, and hence $c_1 \wedge c_2 \not\leq p$, yielding $a \to q \not\leq p$, which contradicts the fact that $q \in S$. So q is meet-irreducible. Finally, $a \to q \leq p$ implies that $b \leq q \leq p$ and, since $p \neq \top$, also $a \to q \neq \top$, yielding $a \not\leq q$. □

The minimal elements of the poset of meet-irreducible elements of $\mathbf{A}$ will be referred to as *minimal meet-irreducible* elements of $\mathbf{A}$. We state the following result for future reference.

LEMMA 4.2.10. *Every meet-irreducible element of $\mathbf{A}$ exceeds a minimal meet-irreducible element of $\mathbf{A}$. In particular, the meet of all minimal meet-irreducible elements of $\mathbf{A}$ is the least element $\bot$.*

PROOF. A direct application of Zorn's lemma implies that every meet-irreducible element exceeds a minimal meet-irreducible element. The second statement is then a direct consequence of Proposition 1.3.8. □

The next lemma supplies a useful description of minimal meet-irreducible elements.

LEMMA 4.2.11. *Let p be a meet-irreducible element of $\mathbf{A}$. The following statements are equivalent:*

(1) *p is a minimal meet-irreducible element of $\mathbf{A}$.*
(2) *$c \leq p$ implies $\neg c \not\leq p$, for all $c \in \mathcal{K}(\mathbf{A})$.*
(3) *$p = \bigvee\{\neg c \mid c \in \mathcal{K}(\mathbf{A})$ and $c \not\leq p\}$.*

PROOF.

(1) $\Rightarrow$ (2) Assume that (2) fails. Then there exists $c \in \mathcal{K}(\mathbf{A})$ such that $c, \neg c \leq p$. If $p = \top$, then, since $\mathbf{A}$ is non-trivial, it is clearly not a minimal meet-irreducible element of $\mathbf{A}$. Otherwise, by Lemma 4.2.9, with $b = \bot$, there exists a meet-irreducible element q such that $c \nleq q$ and $q \leq p$. However, since $c \leq p$, it follows that $q < p$, and hence p is not a minimal meet-irreducible element of $\mathbf{A}$.

(2) $\Rightarrow$ (3) Suppose that (2) holds and let $x := \bigvee\{\neg c \mid c \in \mathcal{K}(\mathbf{A}) \text{ and } c \nleq p\}$. Clearly, $x \leq p$. Now consider any $c \in \mathcal{K}(\mathbf{A})$ such that $c \leq p$. By (2), $\neg c \nleq p$, so there exists $d \in \mathcal{K}(\mathbf{A})$ such that $d \nleq p$ and $c \leq \neg d \leq x$. Hence $x = p$.

(3) $\Rightarrow$ (1) Suppose that (3) holds, but, towards a contradiction, $q < p$ for some meet-irreducible element q of $\mathbf{A}$. Consider any $t \in \mathcal{K}(\mathbf{A})$ such that $t \leq p$ and $t \nleq q$. Since $t \wedge \neg t = \bot \leq q$ and q is meet-irreducible, $\neg t \leq q < p$. So $t, \neg t \leq p$. By (3), $c_1, \ldots, c_n \nleq p$ and $t \leq \neg c_1 \vee \cdots \vee \neg c_n$, for some $c_1, \ldots, c_n \in \mathcal{K}(\mathbf{A})$. Hence

$$\neg\neg c_1 \wedge \cdots \wedge \neg\neg c_n = \neg(\neg c_1 \vee \cdots \vee \neg c_n) \leq \neg t \leq p.$$

But then $c_1 \wedge \cdots \wedge c_n \leq p$, a contradiction. □

Lemmas 4.2.10 and 4.2.11 yield the following results for e-cyclic residuated lattices.

PROPOSITION 4.2.12. *Every prime convex subuniverse of an* e*-cyclic residuated lattice* $\mathbf{L}$ *contains a minimal, prime convex subuniverse of* $\mathbf{L}$. *In particular, the intersection of all minimal, prime convex subuniverses of* $\mathbf{L}$ *is* $\{\mathrm{e}\}$.

PROPOSITION 4.2.13. *Let* $\mathbf{L}$ *be an* e*-cyclic residuated lattice and let* P *be any prime convex subuniverse of* $\mathbf{L}$. *The following statements are equivalent:*

(1) P *is a minimal, prime convex subuniverse of* $\mathbf{L}$.
(2) *For all* $x \in L$, *if* $x \in P$, *then* $x^{\perp} \nsubseteq P$.
(3) $P = \bigcup\{x^{\perp} \mid x \in L \backslash P\}$.

PROOF. The equivalence follows directly from Lemma 4.2.11, noting that in (3), the join is replaced by a union because the set $\{x^{\perp} \mid x \in L \backslash P\}$ is up-directed. □

We obtain a further result that has been proved in the literature for various classes of residuated lattices as a direct translation of the following lattice-theoretical fact.

LEMMA 4.2.14. *For any* $a \in A$,

$$\neg a = \bigwedge\{p \in A \mid p \text{ is minimal meet-irreducible and } a \nleq p\}.$$

PROOF. Let $a \in A$ and $M := \{p \in A \mid p \text{ is minimal meet-irreducible and } a \nleq p\}$. Then $\neg a \leq p$, since $a \wedge \neg a = \bot \leq p$, for any $p \in M$, so $\neg a \leq \bigwedge M$. To prove the reverse inequality, consider any $c \in A$ such that $c \nleq \neg a$, that is, $c \wedge a \neq \bot$. By Lemma 4.2.10, there exists a minimal meet-irreducible element $q \in A$ such that $c \wedge a \nleq q$. Then $c \nleq q$ and $a \nleq q$. It follows that $q \in M$ and hence $c \nleq \bigwedge M$. We have proved that $c \nleq \neg a$ implies $c \nleq \bigwedge M$, so $\bigwedge M \leq \neg a$. □

Specializing this lemma to e-cyclic residuated lattices yields the following result.

PROPOSITION 4.2.15. *Let* $\mathbf{L}$ *be an* e*-cyclic residuated lattice and* $H \in \mathcal{C}(\mathbf{L})$. *Then* $H^{\perp} = \bigcap\{P \in \mathcal{C}(\mathbf{L}) \mid P \text{ is a minimal, prime convex subuniverse of } \mathbf{L} \text{ and } H \nsubseteq P\}$.

4.3. Congruence relations

The main result of this section is Theorem 4.3.7, stating that the congruences of any residuated lattice are determined by its normal convex subuniverses. In particular, the variety $\mathcal{RL}$ of residuated lattices, and hence also the variety $\mathcal{PRL}$ of pointed residuated lattices, is e-*regular*: every congruence of a (pointed) residuated lattice is determined by the equivalence class of e. A more economical proof of e-regularity for $\mathcal{RL}$ could be given by observing that this is a *Mal'cev property*, meaning that it can be checked if a variety has the property by checking the satisfaction of certain quasi-equations involving finitely many terms (two, in this special case). However, a concrete description of the congruence classes of e, to be given in the next section, is essential for developing the structure theory of residuated lattices and its applications to substructural logics.

Let $\mathbf{L}$ be an arbitrary residuated lattice. For any $u, x \in L$, let

$$\lambda_u(x) := (u\backslash xu) \wedge \mathrm{e} \quad \text{and} \quad \rho_u(x) := (ux/u) \wedge \mathrm{e}.$$

We refer to λ_u and ρ_u as *left conjugation* and *right conjugation* by u. An *iterated conjugation* map is a composition of maps $\gamma = \gamma_1\gamma_2 \cdots \gamma_n$, where each γ_i is a right conjugation or a left conjugation by an element $u_i \in L$. The set of all iterated conjugation maps for $\mathbf{L}$ will be denoted by Γ.

A convex subuniverse H of $\mathbf{L}$ is said to be *normal* if $\lambda_u(x), \rho_u(x) \in H$, for all $x \in H$ and $u \in L$. It can easily be confirmed that the intersection of any family of normal convex subuniverses of $\mathbf{L}$ is again a normal convex subuniverse of $\mathbf{L}$, and that the same holds for the union of any up-directed family of normal convex subuniverses of $\mathbf{L}$. The set $\mathcal{NC}(\mathbf{L})$ of normal convex subuniverses of $\mathbf{L}$ ordered by set-inclusion therefore forms an algebraic closure system of $\langle \wp(L), \subseteq \rangle$ (in particular, an algebraic lattice) that, for convenience, we also denote by $\mathcal{NC}(\mathbf{L})$.

EXAMPLE 4.3.1. The residuated lattice described by the following Hasse diagram and multiplication table has non-normal convex subuniverses $\{\mathrm{e}, a\}$ and $\{\mathrm{e}, b\}$:

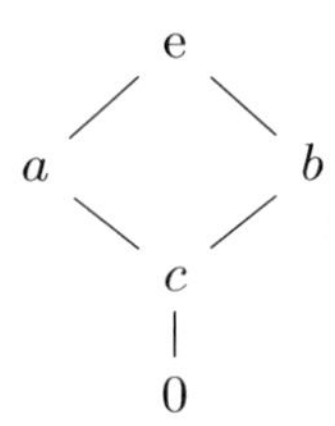

$\cdot$	0	a	b	c	e
0	0	0	0	0	0
a	0	a	0	0	a
b	0	c	b	c	b
c	0	c	0	0	c
e	0	a	b	c	e.

For example, $\lambda_c(a) = (c\backslash ac) \wedge \mathrm{e} = b \notin \{\mathrm{e}, a\}$.

LEMMA 4.3.2. *Let $\mathbf{L}$ be a residuated lattice and let $H \in \mathcal{C}(\mathbf{L})$. The following statements are equivalent:*

(1) *H is normal.*
(2) *H is closed under all iterated conjugation maps.*
(3) *For all $a, b \in L$, $(b\backslash a) \wedge \mathrm{e} \in H$ if and only if $(a/b) \wedge \mathrm{e} \in H$.*

PROOF.

(1) $\Leftrightarrow$ (2) Immediate from the definitions.

(1) $\Rightarrow$ (3) Suppose that H is normal and $(a/b) \wedge \mathrm{e} \in H$, for some $a, b \in L$. Since H is normal, $b\backslash[((a/b) \wedge \mathrm{e})b] \wedge \mathrm{e} \in H$. But $b\backslash[((a/b) \wedge \mathrm{e})b] \wedge \mathrm{e} \leq [b\backslash(a/b)b] \wedge \mathrm{e} \leq (b\backslash a) \wedge \mathrm{e} \leq \mathrm{e} \in H$, so $(b\backslash a) \wedge \mathrm{e} \in H$. The opposite direction follows by symmetry.

(3) $\Rightarrow$ (1) Suppose that (3) holds and consider any $a \in H$ and $b \in L$. Then $a \wedge \mathrm{e} \in H$ and $a \wedge \mathrm{e} \leq (ab/b) \wedge \mathrm{e} \leq \mathrm{e} \in H$, so $(ab/b) \wedge \mathrm{e} \in H$. Hence, by assumption, $(b\backslash ab) \wedge \mathrm{e} \in H$. Likewise, $(ba/b) \wedge \mathrm{e} \in H$. That is, H is normal. $\square$

Given any normal convex subuniverse H of a residuated lattice $\mathbf{L}$, let

$$\Theta_H := \{\langle a, b\rangle \in L^2 \mid (a\backslash b) \wedge (b\backslash a) \wedge \mathrm{e} \in H\}.$$

Note that $(a/b) \wedge (b/a) \wedge \mathrm{e} \in H$ if and only if $(a/b) \wedge \mathrm{e} \in H$ and $(b/a) \wedge \mathrm{e} \in H$, and a dual equivalence holds for the other residual. Hence Lemma 4.3.2 implies the following result.

LEMMA 4.3.3. *If H is a normal convex subuniverse of a residuated lattice $\mathbf{L}$, then $\Theta_H = \{\langle a, b\rangle \in L^2 \mid (a/b) \wedge (b/a) \wedge \mathrm{e} \in H\}$.*

We now prove a series of three lemmas establishing a correspondence between the normal convex subuniverses and congruences of a residuated lattice.

LEMMA 4.3.4. *If H is a normal convex subuniverse of a residuated lattice $\mathbf{L}$, then Θ_H is a congruence of $\mathbf{L}$.*

PROOF. Let H be a normal convex subuniverse of a residuated lattice $\mathbf{L}$. It is easily checked that Θ_H is an equivalence relation. To prove that it is a congruence, we check that it is compatible with the binary operations of $\mathbf{L}$.

Θ_H is compatible with multiplication.
Suppose that $a \;\Theta_H\; b$ and $c \in L$. Then $(a/b) \wedge \mathrm{e} \leq (ac/bc) \wedge \mathrm{e} \leq \mathrm{e}$. so $(ac/bc) \wedge \mathrm{e} \in H$. Similarly, $(bc/ac) \wedge \mathrm{e} \in H$, so $ac \;\Theta_H\; bc$. Also $\rho_c((a/b) \wedge \mathrm{e}) = (c[(a/b) \wedge \mathrm{e}]/c) \wedge \mathrm{e} \in H$, using the normality of H. But $\rho_c((a/b) \wedge \mathrm{e}) \leq [c(a/b)/c] \wedge \mathrm{e} \leq [(ca/b)/c] \wedge \mathrm{e} = (ca/cb) \wedge \mathrm{e} \leq \mathrm{e} \in H$, so $(ca/cb) \wedge \mathrm{e} \in H$. Similarly, $(cb/ca) \wedge \mathrm{e} \in H$, so $ca \;\Theta_H\; cb$.

Θ_H is compatible with the meet operation.
Suppose that $a \;\Theta_H\; b$ and $c \in L$. Set $r := (a/b) \wedge \mathrm{e}$. Then $r \leq a/b$ and $r \leq \mathrm{e}$ yield $rb \leq a$ and $rc \leq c$, respectively. So $r(b \wedge c) \leq rb \wedge rc \leq a \wedge c$. By residuation, $r \leq (a \wedge c)/(b \wedge c)$, and hence $r = (r \wedge \mathrm{e}) \leq [(a \wedge c)/(b \wedge c)] \wedge \mathrm{e} \leq \mathrm{e}$. So $[(a \wedge c)/(b \wedge c)] \wedge \mathrm{e} \in H$. Similarly, $[(b \wedge c)/(a \wedge c)] \wedge \mathrm{e} \in H$, so $(a \wedge c) \;\Theta_H\; (b \wedge c)$.

Θ_H is compatible with the join operation.
Analogous to the case of meet.

Θ_H is compatible with right division.
Suppose that $a \;\Theta_H\; b$ and $c \in L$. Then $(a/b) \wedge \mathrm{e} \leq [(a/c)/(b/c)] \wedge \mathrm{e} \leq \mathrm{e}$ yields $[(a/c)/(b/c)] \wedge \mathrm{e} \in H$. Similarly, $[(b/c)/(a/c)] \wedge \mathrm{e} \in H$, so $(a/c) \;\Theta_H\; (b/c)$. Moreover, $(a/b) \wedge \mathrm{e} \leq [(c/a)\backslash(c/b)] \wedge \mathrm{e} \leq \mathrm{e}$, so $[(c/a)\backslash(c/b)] \wedge \mathrm{e} \in H$, and, by Lemma 4.3.2, also $[(c/b)/(c/a)] \wedge \mathrm{e} \in H$. Similarly, $[(c/a)/(c/b)] \wedge \mathrm{e} \in H$, so $(c/a) \;\Theta_H\; (c/b)$.

Θ_H is compatible with left division.
Analogous to the case of right division. $\square$

LEMMA 4.3.5. *If Θ is a congruence of a residuated lattice $\mathbf{L}$, then $[\mathrm{e}]_\Theta = \{a \in L \mid a \;\Theta\; \mathrm{e}\}$ is a normal convex subuniverse of $\mathbf{L}$.*

PROOF. Let Θ be a congruence of a residuated lattice $\mathbf{L}$. It is easily checked that $[\mathrm{e}]_\Theta$ forms a subuniverse of $\mathbf{L}$. Convexity can also be checked directly, or is a consequence of the well-known fact that the equivalence classes of any lattice congruence are convex. Finally, let $a \in [\mathrm{e}]_\Theta$ and $c \in L$. Then $\lambda_c(a) \in [\mathrm{e}]_\Theta$, since

$$\lambda_c(a) = (c\backslash ac) \wedge \mathrm{e} \;\Theta\; (c\backslash \mathrm{e}c) \wedge \mathrm{e} = (c\backslash c) \wedge \mathrm{e} = \mathrm{e},$$

Likewise, $\rho_c(a) \in [\mathrm{e}]_\Theta$. So $[\mathrm{e}]_\Theta$ is a normal convex subuniverse of $\mathbf{L}$. □

LEMMA 4.3.6. *Let $\mathbf{L}$ be a residuated lattice and let $\Theta \in \mathrm{Con}(\mathbf{L})$. The following statements are equivalent:*

(1) $a \;\Theta\; b$.
(2) $[(a/b) \wedge \mathrm{e}] \;\Theta\; \mathrm{e}$ *and* $[(b/a) \wedge \mathrm{e}] \;\Theta\; \mathrm{e}$.
(3) $[(b\backslash a) \wedge \mathrm{e}] \;\Theta\; \mathrm{e}$ *and* $[(a\backslash b) \wedge \mathrm{e}] \;\Theta\; \mathrm{e}$.

PROOF.

(1) ⇒ (2) Suppose that $a \;\Theta\; b$. Then $(a/b) \;\Theta\; (b/b)$ and $(a/a) \;\Theta\; (b/a)$, and hence $[(a/b) \wedge \mathrm{e}] \;\Theta\; [(b/b) \wedge \mathrm{e}] = \mathrm{e}$ and $\mathrm{e} = [(a/a) \wedge \mathrm{e}] \;\Theta\; [(b/a) \wedge \mathrm{e}]$ as required.

(2) ⇒ (1) Suppose that $[(a/b) \wedge \mathrm{e}] \;\Theta\; \mathrm{e}$ and $[(b/a) \wedge \mathrm{e}] \;\Theta\; \mathrm{e}$. Setting $r := [(a/b) \wedge \mathrm{e}]b$ and $s := [(b/a) \wedge \mathrm{e}]a$, yields $r \;\Theta\; b$ and $s \;\Theta\; a$. Moreover, $r \leq (a/b)b \leq a$ and $s \leq (b/a)a \leq b$, so $r = (a \wedge r) \;\Theta\; (a \wedge b)$ and $s = (b \wedge s) \;\Theta\; (b \wedge a)$. Hence $b \;\Theta\; r \;\Theta\; (a \wedge b) \;\Theta\; s \;\Theta\; a$ as required.

(2) ⇔ (3) Follows from the normality of $[\mathrm{e}]_\Theta$. □

The preceding discussion leads to the main result of this section.

THEOREM 4.3.7. *Let $\mathbf{L}$ be any residuated lattice. The lattice of normal convex subuniverses of $\mathbf{L}$ is isomorphic to the congruence lattice of $\mathbf{L}$ and is therefore an algebraic distributive lattice. The isomorphism is given by the mutually inverse maps $H \mapsto \Theta_H$ and $\Theta \mapsto [\mathrm{e}]_\Theta$.*

PROOF. It follows from Lemmas 4.3.4 and 4.3.5 that $\Theta_H \in \mathrm{Con}(\mathbf{L})$ and $[\mathrm{e}]_\Theta \in \mathcal{NC}(\mathbf{L})$, for any $H \in \mathcal{NC}(\mathbf{L})$ and $\Theta \in \mathrm{Con}(\mathbf{L})$. Since the maps $H \mapsto \Theta_H$ and $\Theta \mapsto [\mathrm{e}]_\Theta$ are clearly order-preserving, it remains to prove that they are mutually inverse and hence lattice isomorphisms.

Observe first that $\Theta = \Theta_{[\mathrm{e}]_\Theta}$, for any $\Theta \in \mathrm{Con}(\mathbf{L})$, since, using Lemma 4.3.6,

$$\begin{aligned} a \;\Theta\; b &\iff ((a\backslash b) \wedge \mathrm{e}) \;\Theta\; \mathrm{e} \text{ and } ((b\backslash a) \wedge \mathrm{e}) \;\Theta\; \mathrm{e} \\ &\iff ((a\backslash b) \wedge \mathrm{e}) \in [\mathrm{e}]_\Theta \text{ and } ((b\backslash a) \wedge \mathrm{e}) \in [\mathrm{e}]_\Theta \\ &\iff a \;\Theta_{[\mathrm{e}]_\Theta}\; b. \end{aligned}$$

Next, observe that $H = [\mathrm{e}]_{\Theta_H}$, for any $H \in \mathcal{NC}(\mathbf{L})$, since

$$\begin{aligned} h \in H &\iff |h| \in H \\ &\iff (\mathrm{e}\backslash h) \wedge (h\backslash \mathrm{e}) \wedge \mathrm{e} \in H \\ &\iff h \;\Theta_H\; \mathrm{e} \\ &\iff h \in [\mathrm{e}]_{\Theta_H}. \end{aligned}$$

Finally, the fact that $\mathcal{NC}(\mathbf{L})$ is an algebraic distributive lattice is an immediate consequence of Proposition 1.4.7. □

By Lemma 4.3.2, every convex subuniverse of a commutative residuated lattice is normal, so in this case we may reformulate Theorem 4.3.7 as follows.

COROLLARY 4.3.8. *Let* $\mathbf{L}$ *be any commutative residuated lattice. The lattice of convex subuniverses of* $\mathbf{L}$ *is isomorphic to the congruence lattice* $\mathrm{Con}(\mathbf{L})$. *The isomorphism is given by the mutually inverse maps* $H \mapsto \Theta_H$ *and* $\Theta \mapsto [\mathrm{e}]_\Theta$.

Moreover, it follows from Proposition 4.1.8 that commutative residuated lattices enjoy a well-known algebraic property.

PROPOSITION 4.3.9. *For any subalgebra* $\mathbf{H}$ *of a commutative residuated lattice* $\mathbf{L}$ *and* $\Theta \in \mathrm{Con}(\mathbf{H})$, *there exists* $\Phi \in \mathrm{Con}(\mathbf{L})$ *such that* $\Theta = \Phi \cap H^2$. *That is, every commutative residuated lattice has the congruence extension property.*

Theorem 4.3.7 also allows us to reinterpret familiar theorems of universal algebra in the setting of normal convex subuniverses. For example, the fact that an algebra $\mathbf{A}$ is subdirectly irreducible if and only if Δ_A is completely meet-irreducible in $\mathrm{Con}(\mathbf{A})$ (Proposition B.6) yields the following characterization of subdirectly irreducible residuated lattices.

COROLLARY 4.3.10. *A residuated lattice* $\mathbf{L}$ *is subdirectly irreducible if and only if* $\{\mathrm{e}\}$ *is completely meet-irreducible in* $\mathcal{NC}(\mathbf{L})$.

Given a normal convex subuniverse H of a residuated lattice $\mathbf{L}$, we often denote the quotient algebra $\mathbf{L}/\Theta_H$ by $\mathbf{L}/H$. In particular, the Homomorphism theorem (Corollary B.3) for residuated lattices may be restated in the following form.

COROLLARY 4.3.11. *Let* $\mathbf{L}$ *and* $\mathbf{M}$ *be residuated lattices and let* $\varphi\colon \mathbf{L} \to \mathbf{M}$ *be a surjective homomorphism. Then* $H := \{x \in L \mid \varphi(x) = \mathrm{e}\}$ *is a normal convex subuniverse of* $\mathbf{L}$ *and there exists a unique isomorphism* $\psi\colon \mathbf{L}/H \to \mathbf{M}$ *such that* $\psi\pi_{\Theta_H} = \varphi$.

A reformulation of the Correspondence theorem (Theorem B.4) is left as an exercise for the reader. Instead, we show that an analogue of this theorem holds for convex subuniverses of e-cyclic residuated lattices.

PROPOSITION 4.3.12. *Let* $\mathbf{L}$ *and* $\mathbf{M}$ *be* e*-cyclic residuated lattices and let* $\varphi\colon \mathbf{L} \to \mathbf{M}$ *be a surjective homomorphism, defining* $H := \{x \in L \mid \varphi(x) = \mathrm{e}\}$. *Then the map* $\bar{\varphi}\colon [H, L] \to \mathcal{C}(\mathbf{M})$; $J \mapsto \varphi[J]$ *is a lattice isomorphism from the interval* $[H, L]$ *of* $\mathcal{C}(\mathbf{L})$ *to* $\mathcal{C}(\mathbf{M})$ *with inverse map* $\bar{\varphi}^{-1}\colon \mathcal{C}(\mathbf{M}) \to [H, L]$; $K \mapsto \varphi^{-1}[K]$.

PROOF. Most of the details of the proof are standard and are left to the reader. We only prove that $\bar{\varphi}$ is injective. Suppose that $\bar{\varphi}(J_1) = \bar{\varphi}(J_2)$, for some $J_1, J_2 \in [H, L]$. Let $a \in J_1^-$. Since $\bar{\varphi}(J_1) = \bar{\varphi}(J_2)$, there exists $b \in J_2^-$ such that $\varphi(a) = \varphi(b)$. It follows that $\varphi((b\backslash a) \wedge \mathrm{e}) = (\varphi(b)\backslash\varphi(a)) \wedge \mathrm{e} = \mathrm{e}$, and hence $(b\backslash a) \wedge \mathrm{e} \in H \subseteq J_2$. But then, since $b((b\backslash a) \wedge \mathrm{e}) \leq a \leq \mathrm{e}$, also $a \in J_2^-$. So $J_1^- \subseteq J_2^-$, and likewise $J_2^- \subseteq J_1^-$. We have shown that $J_1^- = J_2^-$, which implies that $J_1 = J_2$ by Corollary 4.1.4. □

We obtain a further useful description of the congruences of a residuated lattice by considering the *negative cones* of its normal convex subuniverses.

Let $\mathbf{L}$ be any residuated lattice. We call a subset $M \subseteq L^-$ that is convex in $\langle L^-, \leq \rangle$, contains e, and is closed under multiplication and all iterated conjugation maps by elements of $\mathbf{L}$ (not just $\mathbf{L}^-$) a *normal convex submonoid* of $\mathbf{L}^-$. Below we show that the poset $\mathcal{NCS}(\mathbf{L}^-)$ of normal convex submonoids of $\mathbf{L}^-$ ordered by inclusion is isomorphic to the lattice $\mathcal{NC}(\mathbf{L})$, and hence also to $\mathrm{Con}(\mathbf{L})$. In particular, normal convex subuniverses are determined by their negative cones.

First, given any $a, b \in L$, let $[a,b]_r := (ab/ba) \wedge \mathrm{e}$ and $[a,b]_l := (ba\backslash ab) \wedge \mathrm{e}$. The following lemma, which can be verified by straightforward calculation, provides a useful method for converting one of the division operations into its dual.

LEMMA 4.3.13. *Every residuated lattice satisfies the equations*

$$z/y \leq py\backslash z, \quad \text{where } p := [z/y, y]_r$$
$$x\backslash z \leq z/xq, \quad \text{where } q := [x, x\backslash z]_l.$$

We are now ready to establish the mentioned correspondence.

THEOREM 4.3.14. *Let $\mathbf{L}$ be a residuated lattice. If H is a normal convex subuniverse of $\mathbf{L}$, then $M_H := H^-$ is a normal convex submonoid of $\mathbf{L}^-$. Conversely, if M is a normal convex submonoid of $\mathbf{L}^-$, then $H_M := \{a \in L \mid s \leq a \leq s\backslash \mathrm{e}$, for some $s \in M\}$ is a normal convex subuniverse of $\mathbf{L}$. Moreover, the maps $H \mapsto M_H$ and $M \mapsto H_M$ are mutually inverse and establish a lattice isomorphism between $\mathcal{NC}(\mathbf{L})$ and $\mathcal{NCS}(\mathbf{L}^-)$.*

PROOF. It is easily verified that M_H is a normal convex submonoid of $\mathbf{L}^-$, for any normal convex subuniverse H of $\mathbf{L}$, so we just consider the converse. Let M be any normal convex submonoid of $\mathbf{L}^-$. To show that H_M is a normal convex subuniverse of $\mathbf{L}$, note first that the arguments for the convexity and closure of H_M under the multiplication and lattice operations are identical to the corresponding parts of the proof of Lemma 4.1.2. The normality of H_M is also immediate. However, Lemma 4.1.2 does not directly imply that H_M is a convex subuniverse of $\mathbf{L}$, as it assumes the e-cyclicity of $\mathbf{L}$. It remains therefore to verify that H_M is closed with respect to the division operations. Let $a, b \in H_M$. Then there exist $s, t \in M$ such that $s \leq a \leq s\backslash \mathrm{e}$ and $t \leq b \leq t\backslash \mathrm{e}$.

Closure under left division.
Observe first that $a\backslash b \leq s\backslash(t\backslash \mathrm{e}) = ts\backslash \mathrm{e}$. Now, $tsa \leq b$, since $t \leq b$ and $sa \leq \mathrm{e}$, and hence $ats(ats\backslash tsa) \leq tsa \leq b$ and $ts(ats\backslash tsa) \leq a\backslash b$. Setting $p := ats\backslash tsa$ and $q := ts(p \wedge \mathrm{e})$ yields $p \wedge \mathrm{e} = [ts, a]_l \in M$, so $q \in M$ and $q \leq tsp \leq a\backslash b$. Finally, setting $r := qts \in M$, it follows that $r \leq a\backslash b \leq r\backslash \mathrm{e}$, and therefore $a\backslash b \in H_M$.

Closure under right division.
Observe first that $stb \leq a$ and hence $st \leq a/b$, since $s \leq a$ and $tb \leq \mathrm{e}$. Moreover, Lemma 4.3.13 with $p := [(s\backslash \mathrm{e})/t, t]_r$ yields $a/b \leq (s\backslash \mathrm{e})/t \leq pt\backslash(s\backslash \mathrm{e}) = spt\backslash \mathrm{e}$. To see that $p \in M$, observe that

$$(((s\backslash \mathrm{e})/t)t/t((s\backslash \mathrm{e})/t)) \wedge \mathrm{e} \geq (((s\backslash \mathrm{e})/t)t/((s\backslash \mathrm{e})/t)) \wedge \mathrm{e},$$

and the claim follows from the normality and convexity of M. Finally, setting $r := (st)(spt) \in M$, it follows that $r \leq a/b \leq r\backslash \mathrm{e}$, and therefore $a/b \in H_M$.

The maps $H \mapsto M_H$ and $M \mapsto H_M$ are clearly order-preserving and mutually inverse and hence establish the desired lattice isomorphism. ☐

4.4. Normal convex subuniverse generation

The main result of this section is Theorem 4.4.3, which provides a description of the generation of normal convex submonoids and normal convex subuniverses in a residuated lattice. In addition to laying — along with Theorem 4.1.7 and Lemma 4.1.2 — the foundations for the structure theory of residuated lattices, this description yields a useful local deduction theorem for substructural logics.

First, we establish a useful property of left and right conjugation.

LEMMA 4.4.1. *Let* $\mathbf{L}$ *be a residuated lattice. For any* $a_1, \dots, a_n, b \in L$,

$$\prod_{j=1}^{n} \rho_b(a_j) \leq \rho_b(\prod_{j=1}^{n} a_j) \quad \textit{and} \quad \prod_{j=1}^{n} \lambda_b(a_j) \leq \lambda_b(\prod_{j=1}^{n} a_j).$$

PROOF. For left conjugation and $n = 2$, observe that, for any $a_1, a_2, b \in L$,

$$\begin{aligned} \rho_b(a_1)\rho_b(a_2) &= [(ba_1/b) \wedge \mathrm{e}][(ba_2/b) \wedge \mathrm{e}] \\ &\leq [(ba_1/b)(ba_2/b)] \wedge \mathrm{e} \\ &\leq [((ba_1/b)ba_2)/b] \wedge \mathrm{e} && \text{(by Proposition 1.4.4.(i))} \\ &\leq (ba_1a_2/b) \wedge \mathrm{e} \\ &= \rho_b(a_1a_2). \end{aligned}$$

The general case follows by an easy induction, and the case for right conjugation follows by symmetry. □

Given any residuated lattice $\mathbf{L}$, let $\mathrm{NCS}[T]$ denote the normal convex submonoid of $\mathbf{L}^-$ generated by $T \subseteq L^-$, and $\mathrm{NC}[S]$ the normal convex subuniverse of $\mathbf{L}$ generated by $S \subseteq L$. Recalling that Γ is the set of all iterated conjugation maps on $\mathbf{L}$, let also $\Gamma[T] := \{\gamma(a) \mid \gamma \in \Gamma \text{ and } a \in T\}$ for $T \subseteq L^-$.

LEMMA 4.4.2. *Let* $\mathbf{L}$ *be a residuated lattice. For any* $T \subseteq L^-$,

$$\mathrm{NCS}[T] = \{x \in L \mid y \leq x \leq \mathrm{e}, \textit{ for some } y \in \langle\Gamma[T]\rangle\}.$$

PROOF. Lemma 4.1.2, applied to the e-cyclic residuated lattice $\mathbf{L}^-$, implies that $\{x \in L \mid y \leq x \leq \mathrm{e}, \text{ for some } y \in \langle\Gamma[T]\rangle\}$ is the convex subuniverse $\mathrm{C}_{\mathbf{L}^-}[\Gamma[T]]$ of $\mathbf{L}^-$ generated by $\Gamma[T]$. Hence it suffices to observe, using Lemma 4.4.1, that $\mathrm{C}_{\mathbf{L}^-}[\Gamma[T]]$ is closed under left and right conjugation. □

Combining Theorem 4.3.14 and Lemma 4.4.2, we obtain the main result of this section.

THEOREM 4.4.3. *Let* $\mathbf{L}$ *be a residuated lattice. For any* $S \subseteq L$,

$$\begin{aligned} \mathrm{NC}[S] &= \{x \in L \mid y \leq x \leq y\backslash \mathrm{e}, \textit{ for some } y \in \langle\Gamma[|S|]\rangle\} \\ &= \{x \in L \mid y \leq |x|, \textit{ for some } y \in \langle\Gamma[|S|]\rangle\}. \end{aligned}$$

COROLLARY 4.4.4. *Let* $\mathbf{L}$ *be a commutative residuated lattice. For any* $S \subseteq L$,

$$\mathrm{NC}[S] = \{x \in L \mid \prod_{i=1}^{k} |y_i| \leq |x|, \textit{ for some } y_1, \dots, y_k \in S\}.$$

The proofs of the next two results are easy consequences of Theorem 4.4.3.

COROLLARY 4.4.5. *Let* **L** *be a residuated lattice. For any* $a \in L$,

$$\begin{aligned}\mathrm{NC}[a] &= \{x \in L \mid y \leq x \leq y\backslash \mathrm{e}, \textit{ for some } y \in \langle \Gamma[|a|]\rangle\} \\ &= \{x \in L \mid y \leq |x|, \textit{ for some } y \in \langle \Gamma[|a|]\rangle\}.\end{aligned}$$

COROLLARY 4.4.6. *For any elements* a, b *of a residuated lattice* **L**,

$$\mathrm{NC}[a] \cap \mathrm{NC}[b] \supseteq \mathrm{NC}[|a| \vee |b|] \quad \textit{and} \quad \mathrm{NC}[a] \vee \mathrm{NC}[b] = \mathrm{NC}[|a| \wedge |b|].$$

It is clear that the lattice $\mathcal{NC}(\mathbf{L})$ of normal convex subuniverses of a residuated lattice **L** forms a closure system of the lattice $\mathcal{C}(\mathbf{L})$ of convex subuniverses of **L**. Hence arbitrary meets in $\mathcal{NC}(\mathbf{L})$ are intersections and coincide with the meets in $\mathcal{C}(\mathbf{L})$. We show now that joins in $\mathcal{NC}(\mathbf{L})$ also coincide with the corresponding joins in $\mathcal{C}(\mathbf{L})$ when **L** is e-cyclic. In view of Theorem 4.4.3 and Lemma 4.1.2, e-cyclicity guarantees that the normal convex subuniverse $\mathrm{NC}[S]$ generated by $S \subseteq L$ is the convex subuniverse $\mathrm{C}[\Gamma[|S|]]$ generated by $\Gamma[|S|]$.

PROPOSITION 4.4.7. *Let* **L** *be an* e*-cyclic residuated lattice. Then* $\mathcal{NC}(\mathbf{L})$ *is a complete sublattice of* $\mathcal{C}(\mathbf{L})$.

PROOF. Let $\{H_i\}_{i \in I}$ be a family of normal convex subuniverses of **L**, and let $H := \bigcup_{i \in I} H_i$. Consider any $x \in \bigvee^{\mathcal{NC}(\mathbf{L})}\{H_i \mid i \in I\} = \mathrm{NC}[H] = \mathrm{NC}[|H|]$. It follows from Theorem 4.4.3 that $|x| \geq y$, for some $y \in \langle \Gamma[|H|]\rangle$. Since each H_i is a normal convex subuniverse,

$$\Gamma[|H|] = \bigcup_{i \in I} \Gamma[|H_i|] = \bigcup_{i \in I} |H_i| = |H|.$$

Hence $y \in \langle |H| \rangle$ and $x \in \mathrm{C}[H] = \bigvee^{\mathcal{C}(\mathbf{L})}\{H_i \mid i \in I\}$. The converse inclusion is clear, so the joins in $\mathcal{NC}(\mathbf{L})$ and $\mathcal{C}(\mathbf{L})$ coincide. □

Let **L** again be any residuated lattice and let $\gamma_{\mathcal{NC}(\mathbf{L})}$ denote the closure operator on $\mathcal{C}(\mathbf{L})$ corresponding to $\mathcal{NC}(\mathbf{L})$; that is, $\gamma_{\mathcal{NC}(\mathbf{L})}(H) = \mathrm{NC}[H]$ for $H \in \mathcal{C}(\mathbf{L})$. Some surprising consequences, to be discussed in the remainder of this section, occur when $\gamma_{\mathcal{NC}(\mathbf{L})}$ preserves finite intersections and is therefore a nucleus on the Brouwerian algebra $\mathcal{C}(\mathbf{L})$. This is trivially the case when **L** is commutative or, more generally, *Hamiltonian* — that is, when every convex subuniverse of **L** is normal — but these are not the only examples. In particular, it will be shown in Chapter 5 that $\gamma_{\mathcal{NC}(\mathbf{L})}$ is a nucleus whenever **L** is semilinear.

LEMMA 4.4.8. *Let* **L** *be an* e*-cyclic residuated lattice. The following statements are equivalent:*

(1) *The closure operator* $\gamma_{\mathcal{NC}(\mathbf{L})}$ *is a nucleus on the Brouwerian algebra* $\mathcal{C}(\mathbf{L})$.
(2) $\mathrm{NC}[a] \cap \mathrm{NC}[b] = \mathrm{NC}[|a| \vee |b|]$, *for all* $a, b \in L$.
(3) *The poset* $\mathcal{K}(\mathcal{NC}(\mathbf{L}))$ *of compact normal convex subuniverses of* **L** *consists of all principal normal convex subuniverses of* **L** *and is a sublattice of* $\mathcal{NC}(\mathbf{L})$.

PROOF. For convenience, let $\eta := \gamma_{\mathcal{NC}(\mathbf{L})}$.

(1) ⇒ (2) Suppose that (1) holds and let $a, b \in L$. Then $\mathrm{C}[a] \cap \mathrm{C}[b] = \mathrm{C}[|a| \vee |b|]$, by Lemma 4.1.6, and hence $\eta(\mathrm{C}[a]) \cap \eta(\mathrm{C}[b]) = \eta(\mathrm{C}[|a| \vee |b|])$, that is, $\mathrm{NC}[a] \cap \mathrm{NC}[b] = \mathrm{NC}[|a| \vee |b|]$.

(2) ⇒ (1) Suppose that (2) holds and let $H, K \in \mathcal{C}(\mathbf{L})$. Using again Lemma 4.1.6,

$$\begin{aligned}
\eta(H \cap K) &= \eta(\bigcup\{\mathrm{C}[a] \mid a \in H^-\} \cap \bigcup\{\mathrm{C}[b] \mid b \in K^-\}) \\
&= \eta(\bigcup\{\mathrm{C}[a] \cap \mathrm{C}[b] \mid a \in H^- \text{ and } b \in K^-\}) \\
&= \bigcup\{\eta(\mathrm{C}[|a| \vee |b|]) \mid a \in H^- \text{ and } b \in K^-\} \\
&= \bigcup\{\mathrm{NC}[|a| \vee |b|] \mid a \in H^- \text{ and } b \in K^-\} \\
&= \bigcup\{\mathrm{NC}[a] \cap \mathrm{NC}[b] \mid a \in H^- \text{ and } b \in K^-\} \\
&= \bigcup\{\mathrm{NC}[a] \mid a \in H^-\} \cap \bigcup\{\mathrm{NC}[b] \mid b \in K^-\} \\
&= \eta(H) \cap \eta(K).
\end{aligned}$$

(2) ⇔ (3) is an immediate consequence of Corollary 4.4.6. □

COROLLARY 4.4.9. *Let* $\mathbf{L}$ *be a prelinear,* e-*cyclic residuated lattice that satisfies the equivalent conditions of Lemma 4.4.8, and let* $H \in \mathcal{NC}(\mathbf{L})$. *The following statements are equivalent:*

(1) *H is a meet-prime element of $\mathcal{NC}(\mathbf{L})$.*
(2) *H is a prime convex subuniverse of $\mathbf{L}$.*
(3) *For all $a, b \in L$, if $|a| \vee |b| \in H$, then $a \in H$ or $b \in H$.*
(4) *For all $a, b \in L$, if $a \perp b$, then $a \in H$ or $b \in H$.*
(5) *$\langle\{K \in \mathcal{NC}(\mathbf{L}) \mid H \subseteq K\}, \subseteq\rangle$ is a chain.*

PROOF. By Lemma 4.2.7, it suffices to establish the equivalence of the first two conditions. Clearly, (2) implies (1). For the converse, consider any meet-prime element H of $\mathcal{NC}(\mathbf{L})$. Let $J, K \in \mathcal{C}(\mathbf{L})$ be such that $J \cap K \subseteq H$. Then $\gamma_{\mathcal{NC}(\mathbf{L})}(J) \cap \gamma_{\mathcal{NC}(\mathbf{L})}(K) \subseteq \gamma_{\mathcal{NC}(\mathbf{L})}(H) = H$. By assumption, $\gamma_{\mathcal{NC}(\mathbf{L})}(J) \subseteq H$ or $\gamma_{\mathcal{NC}(\mathbf{L})}(K) \subseteq H$, and therefore $J \subseteq H$ or $K \subseteq H$. So H is a prime convex subuniverse of $\mathbf{L}$. □

These results, describing the generation of normal convex subuniverses of residuated lattices, provide useful tools for studying consequence in varieties of these algebras. Recall first that, by Proposition 3.2.7, for any variety $\mathcal{V}$ and $\Sigma \cup \{\varepsilon\} \subseteq \mathrm{Eq}$,

$$\Sigma \vdash_{\mathcal{V}} \varepsilon \iff \bar{\varepsilon} \in \mathrm{Cg}^{\mathbf{F}_{\mathcal{V}}}(\bar{\Sigma}).$$

Now let $\mathcal{V}$ be a variety of residuated lattices. In this case, we obtain a description of consequence relations on formulas for $\mathcal{V}$ in terms of the generation of normal convex subuniverses of free algebras of $\mathcal{V}$. That is, for $T \cup \{\alpha\} \subseteq \mathrm{Fm}$,

$$\begin{aligned}
T \vdash_{\mathcal{V}} \alpha &\iff \{\mathrm{e} \leq \beta \mid \beta \in T\} \vdash_{\mathcal{V}} \mathrm{e} \leq \alpha \\
&\iff \langle \overline{\alpha \wedge \mathrm{e}}, \bar{\mathrm{e}}\rangle \in \mathrm{Cg}^{\mathbf{F}_{\mathcal{V}}}(\{\langle \overline{\beta \wedge \mathrm{e}}, \bar{\mathrm{e}}\rangle \mid \beta \in T\}) \\
&\iff \overline{\alpha \wedge \mathrm{e}} \in \mathrm{NC}[\{\overline{\beta \wedge \mathrm{e}} \mid \beta \in T\}] \\
&\iff \bar{\gamma} \leq \bar{\alpha}, \text{ for some } \bar{\gamma} \in \langle \Gamma(\{\overline{\beta \wedge \mathrm{e}} \mid \beta \in T\})\rangle.
\end{aligned}$$

If $\mathcal{V}$ is commutative, we obtain the following useful reformulation:

$$T \vdash_{\mathcal{V}} \alpha \iff \textstyle\prod_{i=1}^{k} \overline{\beta_i \wedge \mathrm{e}} \leq \bar{\alpha}, \text{ for some } k \in \mathbb{N} \text{ and } \beta_1, \ldots, \beta_k \in T.$$

In particular, this description yields a *local deduction theorem* relating consequences to implications between formulas in $\mathcal{V}$.

PROPOSITION 4.4.10. *Let $\mathcal{V}$ be a variety of commutative residuated lattices. Then, for any $T \cup \{\alpha, \beta\} \subseteq \mathrm{Fm}$,*

$$T \cup \{\alpha\} \vdash_{\mathcal{V}} \beta \iff T \vdash_{\mathcal{V}} (\alpha \wedge \mathrm{e})^n \to \beta, \textit{ for some } n \in \mathbb{N}.$$

PROOF. By the above reasoning, $T \cup \{\alpha\} \vdash_{\mathcal{V}} \beta$ if and only if

$$(\overline{\alpha \wedge \mathrm{e}})^n \textstyle\prod_{i=1}^k \overline{\alpha_i \wedge \mathrm{e}} \leq \overline{\beta}, \text{ for some } n, k \in \mathbb{N} \text{ and } \alpha_1, \ldots, \alpha_k \in T,$$

which, by residuation, holds if and only if

$$\textstyle\prod_{i=1}^k \overline{\alpha_i \wedge \mathrm{e}} \leq (\overline{\alpha \wedge \mathrm{e}})^n \to \overline{\beta}, \text{ for some } n, k \in \mathbb{N} \text{ and } \alpha_1, \ldots, \alpha_k \in T,$$

which holds if and only if $T \vdash_{\mathcal{V}} (\alpha \wedge \mathrm{e})^n \to \beta$, for some $n \in \mathbb{N}$. ☐

4.5. Bibliographical remarks

For the most important consequences of Conrad's approach to ℓ-groups, we refer to his papers [**84**–**87**], survey [**7**], and "Blue Notes" [**89**] (see also [**9**, **99**]).

Sections 4.1 and 4.2 are based on [**49**]. The development of the material of these sections is a natural extension of the techniques used to study lattices of normal convex subuniverses. Most of the results of Sections 4.3 and 4.4 first appeared in [**43**, **44**] (see also [**145**, **206**, **260**]). Proposition 4.4.7 was proved in [**49**]; it was also proved in this paper that for an e-cyclic residuated lattice $\mathbf{L}$, the lattice $\mathcal{C}(\mathbf{L})$ can be embedded (as a complete sublattice) into the congruence lattice of the lattice reduct of $\mathbf{L}$, giving an alternative proof that $\mathcal{C}(\mathbf{L})$ is distributive (Theorem 4.1.7). Lemma 4.4.8 and Corollary 4.4.9 have not appeared elsewhere.

Theorem 4.3.7 establishes a bijective correspondence between congruences of a residuated lattice $\mathbf{L}$ and its normal convex subuniverses. Another notion that may serve as a substitute for the concept of a normal convex subuniverse is that of a *deductive filter* of $\mathbf{L}$: a subset F of L satisfying

(i) ${\uparrow}\{\mathrm{e}\} \subseteq F$;
(ii) if $x, x \backslash y \in F$, then $y \in F$;
(iii) $x, y/x \in F$, then $y \in F$;
(iv) if $x, y \in F$, then $x \wedge y \in F$;
(v) if $x \in F$ and $y \in L$, then $y \backslash xy \in F$ and $yx/y \in F$.

It is easy to see that a non-empty subset of L is a deductive filter of $\mathbf{L}$ if and only if it is an order-filter that is closed under multiplication and conjugation. Moreover, $F \subseteq L$ is a deductive filter of $\mathbf{L}$ if and only if $F = {\uparrow}H$, for some normal convex subuniverse H of $\mathbf{L}$. The lattice of deductive filters of $\mathbf{L}$ is therefore isomorphic to the lattice of normal convex subuniverses of $\mathbf{L}$ via the mutually inverse maps $H \mapsto {\uparrow}H$ and $F \mapsto \langle F^- \rangle$. The logical significance of deductive filters lies in the fact that the consequence relation $\vdash_{\mathsf{RL}}$ coincides with the consequence relation given by logical matrices of the form $\langle \mathbf{L}, F \rangle$ such that $\mathbf{L}$ is a residuated lattice and F is a deductive filter of $\mathbf{L}$ [**148**].

In [**49**], it is proved that convex subuniverses of an e-cyclic residuated lattice $\mathbf{L}$ correspond to prefilters of $\mathbf{L}$ — also studied in dual form as ideals of DRℓ-monoids in [**224, 225, 227**]. A *prefilter* of $\mathbf{L}$ is a subset P of L that contains e, is closed under multiplication, and satisfies

(†) if $x \in P$ and $y \in L$, then $|x| \leq |y|$ implies $y \in P$.

Every convex subuniverse of $\mathbf{L}$ is a prefilter of $\mathbf{L}$, by Lemma 4.1.1.(e). For the converse, let P be any prefilter of $\mathbf{L}$. If $x \in \mathrm{C}[P]$, then, by Lemma 4.1.2, there exist $p_1, \ldots, p_n \in |P|$ such that $p_1 \cdots p_n \leq |x|$. From (†), it follows that $|P| \subseteq P$, and, since P is closed under multiplication, $p_1 \cdots p_n \in P$. Hence also $x \in P$. So $\mathrm{C}[P] = P$; that is, P is a convex subuniverse of $\mathbf{L}$.

Finally, let us mention some connections between the results of this chapter and relatively normal lattices. A lower-bounded distributive lattice $\mathbf{A} = \langle A, \wedge, \vee, \bot \rangle$ is said to be *relatively normal* if its prime lattice-ideals form a root system under set-inclusion. Relatively normal lattices form an important class of lattices, spanning dual relative Stone lattices, the lattice of cozero-sets of any topological space, and lattices of compact congruences for well-studied ordered algebraic structures such as semilinear ℓ-groups, Riesz spaces, and f-rings. (The reader is referred to [**181, 347**] for details and an extensive bibliography.) Lemma 4.2.7 implies that the lattice $\mathcal{K}(\mathcal{C}(\mathbf{L}))$ of principal convex subuniverses of any prelinear, e-cyclic residuated lattice is relatively normal. Likewise, Lemma 4.4.8 and Corollary 4.4.9 show that if $\mathbf{L}$ satisfies the assumptions of Corollary 4.4.9, then the lattice $\mathcal{K}(\mathcal{NC}(\mathbf{L}))$ of normal principal convex subuniverses of $\mathbf{L}$ is relatively normal. In Chapter 5, it will be shown that these assumptions are satisfied when $\mathbf{L}$ is semilinear.

Relative normality admits several useful reformulations [**274, 275**]. For any lower-bounded distributive lattice $\mathbf{A}$, the following statements are equivalent:

(1) $\mathbf{A}$ is relatively normal.

(2) Any pair of incomparable prime lattice-ideals of $\mathbf{A}$ have corresponding disjoint open neighborhoods in the Stone space of $\mathbf{A}$.

(3) The join of any pair of incomparable prime lattice-filters in $\mathbf{A}$ is A.

(4) For any $a, b \in A$, there exist $a', b' \in A$ with $a' \wedge b' = \bot$ and $a' \vee b = a \vee b = a \vee b'$.

Condition (4) plays a key role in the study of relatively normal lattices and can be used to provide an alternative proof of the relative normality of $\mathcal{K}(\mathcal{C}(\mathbf{L}))$ (see [**49**] and the proof of Proposition 4.2.13).

CHAPTER 5

Semilinearity and distributivity

Building on the general structure theory developed in Chapter 4, we turn our attention in this and the next two chapters to algebras that admit more concrete representations. In particular, we focus here on varieties of residuated lattices that can be represented in some fashion via *totally ordered* structures.

First, in Section 5.1, we provide equational bases for varieties consisting of algebras that are isomorphic to subdirect products of totally ordered residuated lattices. Such algebras are called *semilinear*[1] and include Abelian ℓ-groups as well as other structures relevant to the study of many-valued logics. In Section 5.2, we develop various algebraic and syntactic methods for determining whether or not a variety of semilinear residuated lattices is *densifiable*, that is, whether or not each of its totally ordered members embeds into one of its dense, totally ordered members.

In Section 5.3, we investigate the class of *ℓ-monoids*, encompassing residual-free reducts of residuated lattices in which multiplication distributes over binary meets as well as joins. We show that an ℓ-monoid is distributive as a lattice if and only if it embeds into an ℓ-monoid of order-endomorphisms of a chain, generalizing the well-known theorem of Charles Holland that every ℓ-group embeds into an ℓ-group of order-automorphisms of a chain. In Section 5.4, we use this characterization to obtain generation and decidability results for varieties of distributive ℓ-monoids and ℓ-groups.

5.1. Equational bases for semilinear varieties

A residuated lattice is called *semilinear* if it is isomorphic to a subdirect product of totally ordered residuated lattices. In particular, a semilinear residuated lattice is totally ordered if and only if it is finitely subdirectly irreducible, and every variety of semilinear residuated lattices is generated as a quasivariety by its totally ordered members. In this section, we provide equational bases for varieties of semilinear residuated lattices (Theorem 5.1.2), and characterize semilinear, e-cyclic residuated lattices in terms of their minimal, prime convex subuniverses and (principal) polars (Theorem 5.1.7).

For any variables u and x, let $\lambda_u(x) := (u\backslash xu) \wedge \mathrm{e}$ and $\rho_u(x) := (ux/u) \wedge \mathrm{e}$.

LEMMA 5.1.1. *Let* $\mathbf{L}$ *be any residuated lattice that satisfies the quasi-equation*

$$\textit{(DC)} \quad x \vee y \approx \mathrm{e} \implies \lambda_u(x) \vee \rho_v(y) \approx \mathrm{e}.$$

If $a_1, \ldots, a_n \in L$ *satisfy* $a_1 \vee \cdots \vee a_n = \mathrm{e}$, *then* $\mathrm{NC}[a_1] \cap \cdots \cap \mathrm{NC}[a_n] = \{\mathrm{e}\}$.

[1]The more traditional, but less descriptive, name for these algebras is *representable*.

PROOF. Suppose that $a_1, \dots, a_n \in L$ satisfy $a_1 \vee \cdots \vee a_n = \mathrm{e}$, and consider an element $c \in \mathrm{NC}[a_1] \cap \cdots \cap \mathrm{NC}[a_n]$. By Theorem 4.4.3, there exist conjugation maps $\gamma_1^{(1)}, \dots, \gamma_{k_1}^{(1)}, \dots, \gamma_1^{(n)}, \dots, \gamma_{k_n}^{(n)} \in \Gamma$ such that, for each $i \in \{1, \dots, n\}$,

$$a_i^* := \gamma_1^{(i)}(a_i) \cdots \gamma_{k_i}^{(i)}(a_i) \leq |c|.$$

Repeated applications of (DC) yield, for each $i \in \{1, \dots, n\}$ and $j \in \{1, \dots, k_i\}$,

$$a_1 \vee \cdots \vee a_{i-1} \vee \gamma_j^{(i)}(a_i) \vee a_{i+1} \vee \cdots \vee a_n = \mathrm{e}.$$

Hence, using the fact that $x \perp z$ and $y \perp z$ implies $xy \perp z$ in any residuated lattice (see Lemma 4.1.5), for each $i \in \{1, \dots, n\}$,

$$a_1 \vee \cdots \vee a_{i-1} \vee a_i^* \vee a_{i+1} \vee \cdots \vee a_n = \mathrm{e}.$$

But then, reasoning inductively,

$$\mathrm{e} = a_1^* \vee \cdots \vee a_n^* \leq |c| \leq \mathrm{e}.$$

So $c = \mathrm{e}$, by Lemma 4.1.1.(c). That is, $\mathrm{NC}[a_1] \cap \cdots \cap \mathrm{NC}[a_n] = \{\mathrm{e}\}$. □

The next result shows that prelinearity is a necessary condition for semilinearity; to see that it is not sufficient, consider any ℓ-group that is not semilinear, such as the ℓ-group of automorphisms of the real number line.

THEOREM 5.1.2. *The following statements are equivalent:*

(1) *$\mathcal{V}$ is a variety of semilinear residuated lattices.*

(2) *$\mathcal{V}$ is a variety of residuated lattices satisfying one of the following equations:*

$$\text{(SL1)} \quad \lambda_u((x \vee y)\backslash x) \vee \rho_v((x \vee y)\backslash y) \approx \mathrm{e}$$
$$\text{(SL2)} \quad \lambda_u((y\backslash x) \wedge \mathrm{e}) \vee \rho_v((x\backslash y) \wedge \mathrm{e}) \approx \mathrm{e}.$$

(3) *$\mathcal{V}$ is a variety of prelinear residuated lattices satisfying*

$$\text{(DC)} \quad x \vee y \approx \mathrm{e} \implies \lambda_u(x) \vee \rho_v(y) \approx \mathrm{e}.$$

PROOF.

(1) ⇒ (2) It is straightforward to check that (SL1) and (SL2) are satisfied by all totally ordered residuated lattices, and hence by all semilinear residuated lattices.

(2) ⇒ (3) Suppose that $\mathbf{L} \in \mathcal{V}$ satisfies (SL1). By substituting e for u and v in this equation, it can be checked that $\mathbf{L}$ satisfies $((x\backslash y) \wedge \mathrm{e}) \vee ((y\backslash x) \wedge \mathrm{e}) \approx \mathrm{e}$, and is therefore prelinear. Next, consider $a, b \in L$ such that $a \vee b = \mathrm{e}$. Note that $a = \mathrm{e}\backslash a = (a \vee b)\backslash a$, and likewise $b = (a \vee b)\backslash b$. Hence, by (SL1), for all $c, d \in L$,

$$\lambda_c(a) \vee \rho_d(b) = \lambda_c((a \vee b)\backslash a) \vee \rho_d((a \vee b)\backslash b)) = \mathrm{e}.$$

So $\mathbf{L}$ satisfies (DC). Moreover, every residuated lattice satisfies $(y\backslash x) \wedge \mathrm{e} \leq (x \vee y)\backslash x$ and $(x\backslash y) \wedge \mathrm{e} \leq (x \vee y)\backslash y$. Hence, if $\mathbf{L} \in \mathcal{V}$ satisfies (SL2), it also satisfies (SL1), by Proposition 1.4.2.

(3) ⇒ (1) Suppose that $\mathcal{V}$ is a variety of prelinear residuated lattices satisfying (DC). To show that $\mathcal{V}$ is semilinear, it suffices, by the Subdirect representation theorem (Theorem B.7), to prove that each subdirectly irreducible member $\mathbf{L}$ of $\mathcal{V}$ is totally ordered. Let $a, b \in L$. Since $\mathbf{L}$ is prelinear and satisfies (DC), by Lemma 5.1.1,

$$\mathrm{NC}[(a\backslash b) \wedge \mathrm{e}] \cap \mathrm{NC}[(b\backslash a) \wedge \mathrm{e}] = \{\mathrm{e}\}.$$

But $\mathbf{L}$ is subdirectly irreducible, so either $\mathrm{NC}[(a\backslash b)\wedge \mathrm{e}] = \{\mathrm{e}\}$ or $\mathrm{NC}[(b\backslash a)\wedge \mathrm{e}] = \{\mathrm{e}\}$, by Corollary 4.3.10. In the first case, $(a\backslash b) \wedge \mathrm{e} = \mathrm{e}$ and $a \leq b$, and in the second case, $(b\backslash a) \wedge \mathrm{e} = \mathrm{e}$ and $b \leq a$. Hence $\mathbf{L}$ is totally ordered. □

The theorem also yields a simplified equational characterization for commutative residuated lattices.

COROLLARY 5.1.3. *A commutative residuated lattice* $\mathbf{L}$ *is semilinear if and only if it satisfies the commutative prelinearity law* $((x \to y) \wedge \mathrm{e}) \vee ((y \to x) \wedge \mathrm{e}) \approx \mathrm{e}$.

Note further that (SL1) is equivalent in the setting of ℓ-groups to the equation $\mathrm{e} \leq x \vee y^{-1}x^{-1}y$, which defines the variety of semilinear ℓ-groups relative to $\mathcal{LG}$.

We turn our attention now to the structure of semilinear residuated lattices. Recall first from Corollary 4.4.6 that, for any residuated lattice $\mathbf{L}$ and elements $a, b \in L$, we have $\mathrm{NC}[a]\vee\mathrm{NC}[b] = \mathrm{NC}[|a|\wedge|b|]$, but, in general, only $\mathrm{NC}[a]\cap\mathrm{NC}[b] \supseteq \mathrm{NC}[|a|\vee|b|]$. By Lemma 4.1.6 and Corollary 4.3.8, when $\mathbf{L}$ is commutative, the latter inclusion becomes an equality and the poset of compact normal convex subuniverses of $\mathbf{L}$ is a sublattice of $\mathcal{NC}(\mathbf{L})$ that consists of all the principal normal convex subuniverses of $\mathbf{L}$. The next result shows that the same conclusion holds when $\mathbf{L}$ is semilinear.

PROPOSITION 5.1.4. *Let* $\mathbf{L}$ *be any semilinear residuated lattice. For any* $a, b \in L$,

$$\mathrm{NC}[a] \cap \mathrm{NC}[b] = \mathrm{NC}[|a| \vee |b|] \quad \textit{and} \quad \mathrm{NC}[a] \vee \mathrm{NC}[b] = \mathrm{NC}[|a| \wedge |b|] = \mathrm{NC}[|a||b|].$$

The poset $\mathcal{K}(\mathcal{NC}(\mathbf{L}))$ *of compact normal subuniverses of* $\mathbf{L}$ *therefore consists of all principal normal convex subuniverses of* $\mathbf{L}$ *and is a sublattice of* $\mathcal{NC}(\mathbf{L})$.

PROOF. It suffices to establish the two equalities, noting that the second can be proved as in Lemma 4.1.6. To prove the first equality, consider any $a, b \in L$. Since $\mathrm{NC}[a] = \mathrm{NC}[|a|]$ and $\mathrm{NC}[b] = \mathrm{NC}[|b|]$, by Corollary 4.1.3, we may assume that $a, b \in L^-$. The inclusion $\mathrm{NC}[a \vee b] \subseteq \mathrm{NC}[a] \cap \mathrm{NC}[b]$ is clear, so it remains to prove that $\mathrm{NC}[a] \cap \mathrm{NC}[b] \subseteq \mathrm{NC}[a \vee b]$. Suppose, without loss of generality, that $\mathbf{L}$ is a subdirect product of a family $\{\mathbf{L}_t\}_{t\in T}$ of totally ordered residuated lattices. Let $c \in ([\mathrm{NC}[a] \cap \mathrm{NC}[b])^-$. Then there exist iterated conjugation maps $\gamma_1, \ldots, \gamma_n, \delta_1, \ldots, \delta_m \in \Gamma$ such that $\prod_{j=1}^n \gamma_j(a) \leq c \leq \mathrm{e}$ and $\prod_{i=1}^m \delta_i(b) \leq c \leq \mathrm{e}$.

Fix an arbitrary $t \in T$ and suppose, without loss of generality, that $\pi_t(a) \geq \pi_t(b)$, where $\pi_t \colon \mathbf{L} \to \mathbf{L}_t$ is the usual projection map. Then

$$\begin{aligned} \mathrm{e} \geq \pi_t(c) &\geq \pi_t(\textstyle\prod_{j=1}^n \gamma_j(a)) \\ &= \textstyle\prod_{j=1}^n \pi_t(\gamma_j(a)) \\ &= \textstyle\prod_{j=1}^n \pi_t(\gamma_j(a \vee b)) \\ &\geq [\textstyle\prod_{j=1}^n \pi_t(\gamma_j(a \vee b))][\textstyle\prod_{i=1}^m \pi_t(\delta_i(a \vee b))] \\ &= \pi_t([\textstyle\prod_{j=1}^n (\gamma_j(a \vee b))][\textstyle\prod_{i=1}^m (\delta_i(a \vee b))]). \end{aligned}$$

It follows that $c \in \mathrm{NC}[a \vee b]$, since

$$\mathrm{e} \geq c \geq [\textstyle\prod_{j=1}^n (\gamma_j(a \vee b))][\textstyle\prod_{i=1}^m (\delta_i(a \vee b))].$$

Since a normal convex subuniverse is completely determined by its negative cone (Theorem 4.3.14), $\mathrm{NC}[a] \cap \mathrm{NC}[b] \subseteq \mathrm{NC}[a \vee b]$ as required. □

The reader may wonder at this point if every semilinear residuated lattice is e-cyclic. The following example shows that this is not the case.

EXAMPLE 5.1.5. Consider the four-element algebra $\mathbf{C}_4 = \langle\{\bot, a, \mathrm{e}, \top\}, \wedge, \vee, \cdot, \backslash, /, \mathrm{e}\rangle$ with a total order given by $\bot < a < \mathrm{e} < \top$ and multiplication table

$\cdot$	$\bot$	a	e	$\top$
$\bot$	$\bot$	$\bot$	$\bot$	$\bot$
a	$\bot$	a	a	$\top$
e	$\bot$	a	e	$\top$
$\top$	$\bot$	a	$\top$	$\top$.

It is easily checked that $\mathbf{C}_4$ is a totally ordered residuated lattice; however, $a\backslash\mathrm{e} = \mathrm{e}$ and $\mathrm{e}/a = \top$, so $\mathbf{C}_4$ is not e-cyclic.

Let us therefore consider in more detail the case where $\mathbf{L}$ is an e-cyclic, semilinear residuated lattice. Note first that by Lemma 4.4.8 and Proposition 5.1.4, the closure operator $\gamma_{\mathcal{NC}(\mathbf{L})}$ on the Brouwerian algebra $\mathcal{C}(\mathbf{L})$ of convex subuniverses of $\mathbf{L}$ is a nucleus (i.e., it preserves finite intersections). The next result, a direct consequence of Corollary 4.4.9 and Theorem 5.1.2, provides a characterization of the meet-prime elements of the lattice $\mathcal{NC}(\mathbf{L})$ of normal convex subuniverses of $\mathbf{L}$ in this setting.

COROLLARY 5.1.6. *Let $\mathbf{L}$ be any semilinear, e-cyclic residuated lattice and let $H \in \mathcal{NC}(\mathbf{L})$. The following statements are equivalent:*

(1) *H is a meet-prime element of $\mathcal{NC}(\mathbf{L})$.*
(2) *H is a prime convex subuniverse of $\mathbf{L}$.*
(3) *For all $a, b \in L$, if $|a| \vee |b| \in H$, then $a \in H$ or $b \in H$.*
(4) *For all $a, b \in L$, if $a \perp b$, then $a \in H$ or $b \in H$.*
(5) *$\langle\{K \in \mathcal{NC}(\mathbf{L}) \mid H \subseteq K\}, \subseteq\rangle$ is a chain.*

We now show that in the presence of e-cyclicity and prelinearity, semilinearity can be described by properties of the lattice of convex subuniverses.

THEOREM 5.1.7. *Let $\mathbf{L}$ be any prelinear, e-cyclic residuated lattice. The following statements are equivalent:*

(1) *$\mathbf{L}$ is semilinear.*
(2) *All (principal) polars of $\mathbf{L}$ are normal.*
(3) *All minimal, prime convex subuniverses of $\mathbf{L}$ are normal.*

PROOF.

(1) $\Rightarrow$ (2) Suppose that $\mathbf{L}$ is semilinear. Since every polar of $\mathbf{L}$ is an intersection of principal polars of $\mathbf{L}$, it suffices to show that $h^{\perp} = \{a \in L \mid a \perp h\}$ is normal, for any $h \in L$. Let $a \in h^{\perp}$ and $u \in L$. Then $a \perp h$ and, since $\mathbf{L}$ is semilinear, Theorem 5.1.2 yields $\lambda_u(|a|) \perp \rho_{\mathrm{e}}(|h|)$, and so $\lambda_u(|a|) \perp h$. Hence $\lambda_u(|a|) \in h^{\perp}$. But $|a| \leq a$ implies $\lambda_u(|a|) \leq \lambda_u(a) \leq \mathrm{e}$, so $\lambda_u(a) \in h^{\perp}$. Likewise, $\rho_u(a) \in h^{\perp}$. Hence $h^{\perp}$ is a normal convex subuniverse of $\mathbf{L}$.

(2) $\Rightarrow$ (3) Suppose that all (principal) polars of $\mathbf{L}$ are normal. Let $H \in \mathcal{C}(\mathbf{L})$ be any minimal, prime convex subuniverse of $\mathbf{L}$. Then $H = \bigcup\{h^{\perp} \mid h \in L\backslash H\}$, by Proposition 4.2.13. Since, by assumption, each $h^{\perp}$ is normal, so is H.

(3) $\Rightarrow$ (1) Let $\{H_i \mid i \in I\}$ be the set of minimal, prime convex subuniverses of $\mathbf{L}$. Suppose that H_i is normal, for each $i \in I$. By Lemma 4.2.7, the interval $[H_i, L]$ in $\mathcal{C}(\mathbf{L})$ is a chain, for each $i \in I$. Hence, by Proposition 4.3.12, for each $i \in I$, the lattice of convex subuniverses of $\mathbf{L}/H_i$ is a chain, and $\mathbf{L}/H_i$ is totally ordered, by Corollary 4.2.8. Since $\bigcap\{H_i \mid i \in I\} = \{\mathrm{e}\}$, by Proposition 4.2.12, it follows that $\mathbf{L}$ is isomorphic to a subdirect product of the family of totally ordered residuated lattices $\{\mathbf{L}/H_i\}_{i \in I}$, and is therefore semilinear. □

5.2. Densifiable varieties

Every variety of semilinear residuated lattices is generated as a quasivariety by its totally ordered members. Some of these varieties, however, are generated as a quasivariety also by their *dense* totally ordered members. In this section, we provide various algebraic and syntactic methods for identifying such varieties — Theorems 5.2.6, 5.2.7, and 5.2.8 — and apply them to some key examples.

First, we give an equivalent characterization of the varieties in question.

PROPOSITION 5.2.1. *A variety $\mathcal{V}$ of semilinear residuated lattices is generated as a quasivariety by its dense, totally ordered members if and only if every totally ordered member of $\mathcal{V}$ embeds into a dense, totally ordered member of $\mathcal{V}$.*

PROOF. Suppose first that every totally ordered member of $\mathcal{V}$ embeds into a dense, totally ordered member of $\mathcal{V}$. Every member of $\mathcal{V}$ is semilinear and hence embeds into a product of totally ordered members of $\mathcal{V}$, and therefore also, by assumption, into a product of dense, totally ordered members of $\mathcal{V}$. So $\mathcal{V}$ is generated as a quasivariety by its dense, totally ordered members.

Now suppose that $\mathcal{V}$ is generated as a quasivariety by its dense, totally ordered members. Let $\mathbf{L}$ be any totally ordered member of $\mathcal{V}$. It follows by the Fundamental theorem of ultraproducts (or can be confirmed directly) that the class of dense, totally ordered members of $\mathcal{V}$ is closed under taking ultraproducts. Hence $\mathbf{L}$ embeds into a product of dense, totally ordered residuated lattices. Since $\mathbf{L}$ is totally ordered, Δ_L is meet-irreducible in $\mathrm{Con}(\mathbf{L})$. So $\mathbf{L}$ embeds into a dense, totally ordered member of $\mathcal{V}$ by the Relativized Jónsson lemma (Theorem B.15). □

Let us call any variety of semilinear residuated lattices that satisfies one of the equivalent conditions in Proposition 5.2.1 *densifiable.*

EXAMPLE 5.2.2. By Corollary 5.1.3, the variety $\mathcal{RSA}$ of semilinear Brouwerian algebras, also known as *relative Stone algebras*, consists of all Brouwerian algebras satisfying $(x \to y) \vee (y \to x) \approx \mathrm{e}$. Consider any totally ordered member $\mathbf{L}$ of $\mathcal{RSA}$, observing that, for any $a, b \in L$, if $a \leq b$, then $a \to^{\mathbf{L}} b = \mathrm{e}$, otherwise $a \to^{\mathbf{L}} b = b$. Let $\mathbf{M}$ be the dense, totally ordered member of $\mathcal{RSA}$ obtained by extending the lexicographic product of $\langle L\backslash\{\mathrm{e}\}, \leq^{\mathbf{L}}\rangle$ and $\langle [0,1) \cap \mathbb{R}, \leq\rangle$ with a greatest element e, and, for $a, b \in M$, defining $a \to^{\mathbf{M}} b = \mathrm{e}$ if $a \leq b$, otherwise $a \to^{\mathbf{M}} b = b$. Clearly, $\mathbf{L}$ embeds into $\mathbf{M}$. So $\mathcal{RSA}$ is densifiable. On the other hand, any variety $\mathcal{V}$ of relative Stone algebras generated by a finite, totally ordered member $\mathbf{L}_n$ of $\mathcal{RSA}$ with $n > 1$ elements satisfies $\mathrm{e} \leq x_1 \vee (x_1 \to x_2) \vee (x_2 \to x_3) \vee \cdots \vee (x_{n-1} \to x_n)$, and cannot have a totally ordered member with more than n elements; in particular, $\mathbf{L}_n$ does not embed into any dense, totally ordered member of $\mathcal{V}$, so $\mathcal{V}$ is not densifiable.

The next result provides a useful tool for establishing failures of densifiability.

LEMMA 5.2.3.

(a) *Every dense, totally ordered residuated lattice is* e*-cyclic.*
(b) *Every densifiable variety of semilinear residuated lattices satisfies* e*-cyclicity.*

PROOF.

(a) Let $\mathbf{L}$ be any dense, totally ordered residuated lattice. Suppose towards a contradiction that $a\backslash \mathrm{e} > \mathrm{e}/a$, for some $a \in L$. Then, by residuation, $a > (a\backslash \mathrm{e})\backslash \mathrm{e}$, and, by assumption, $a > b > (a\backslash \mathrm{e})\backslash \mathrm{e}$, for some $b \in L$. By residuation again, $\mathrm{e} > a\backslash b$ and $(a\backslash \mathrm{e})b > \mathrm{e}$, yielding $(a\backslash \mathrm{e})b > a\backslash b$ and $a(a\backslash \mathrm{e})b > b$. But $\mathrm{e} \geq a(a\backslash \mathrm{e})$, so $b = \mathrm{e}b \geq a(a\backslash \mathrm{e})b > b$, a contradiction. Hence $a\backslash \mathrm{e} \leq \mathrm{e}/a$ and, by symmetry, $a\backslash \mathrm{e} \geq \mathrm{e}/a$. So $\mathbf{L}$ is e-cyclic.

(b) It suffices to recall that a densifiable variety of semilinear residuated lattices is generated as a quasivariety by its dense, totally ordered members. □

Since the four-element totally ordered residuated lattice presented in Example 5.1.5 is not e-cyclic, we obtain the following negative result.

COROLLARY 5.2.4. *The variety of semilinear residuated lattices is not densifiable.*

We now provide a useful necessary and sufficient "one-step" condition for a variety of semilinear residuated lattices to be densifiable.

LEMMA 5.2.5. *Let $\mathcal{V}$ be any variety of semilinear residuated lattices. Then $\mathcal{V}$ is densifiable if and only if, for every countable, totally ordered member $\mathbf{L}$ of $\mathcal{V}$ and any $a, b \in L$ satisfying $a < b$, there exist a countable, totally ordered member $\mathbf{M}$ of $\mathcal{V}$ and $c \in M$ such that $\mathbf{L}$ is a subalgebra of $\mathbf{M}$ and $a < c < b$.*

PROOF. For the left-to-right direction, suppose that $\mathcal{V}$ is densifiable. Let $\mathbf{L}$ be any countable, totally ordered member of $\mathcal{V}$ and consider any $a, b \in L$ satisfying $a < b$. Since $\mathcal{V}$ is densifiable, $\mathbf{L}$ is a subalgebra of some dense, totally ordered member $\mathbf{K}$ of $\mathcal{V}$, and there exists $c \in K$ such that $a < c < b$. Hence, the subalgebra $\mathbf{M}$ of $\mathbf{K}$ generated by $L \cup \{c\}$ is a countable, totally ordered member of $\mathcal{V}$ such that $\mathbf{L}$ is a subalgebra of $\mathbf{M}$ and $c \in M$ satisfies $a < c < b$.

For the right-to-left direction, suppose that for every countable, totally ordered member $\mathbf{L}$ of $\mathcal{V}$ and any $a, b \in L$ satisfying $a < b$, there exist a countable, totally ordered member $\mathbf{M}$ of $\mathcal{V}$ and $c \in M$ such that $\mathbf{L}$ is a subalgebra of $\mathbf{M}$ and $a < c < b$. We show that every countable, totally ordered member $\mathbf{L}$ of $\mathcal{V}$ is a subalgebra of some dense, totally ordered member $\mathbf{M}$ of $\mathcal{V}$. The result then follows using the fact that every totally ordered member of $\mathcal{V}$ embeds into an ultraproduct of countable, totally ordered members of $\mathcal{V}$ and hence into an ultraproduct of dense, totally ordered members of $\mathcal{V}$, which is then also a dense, totally ordered member of $\mathcal{V}$.

Let $\mathbf{L}_0 := \mathbf{L}$. We define, for each $n \in \mathbb{N}$, a countable, totally ordered member $\mathbf{L}_{n+1}$ of $\mathcal{V}$ such that $\mathbf{L}_n$ is a subalgebra of $\mathbf{L}_{n+1}$ and, for any $a, b \in L_n$ satisfying $a < b$, there exists $c \in L_{n+1}$ such that $a < c < b$. Enumerate all pairs $a_i, b_i \in L_n$ satisfying $a_i < b_i$, for each $i \in \mathbb{N}$. Let $\mathbf{L}_n^0 := \mathbf{L}_n$ and, inductively, using the assumption, let

$\mathbf{L}_n^{i+1}$, for each $i \in \mathbb{N}$, be a countable, totally ordered member of $\mathcal{V}$ such that $\mathbf{L}_n^i$ is a subalgebra of $\mathbf{L}_n^{i+1}$ and $c_i \in L_n^{i+1}$ satisfies $a_i < c_i < b_i$. Now let $\mathbf{L}_{n+1}$ be the direct limit of the family $\{\mathbf{L}_n^i\}_{i\in\mathbb{N}}$ ordered by inclusion; that is, $L_{n+1} := \bigcup_{i\in\mathbb{N}} L_n^i$ and the operations of $\mathbf{L}_{n+1}$ are defined by referring to $\mathbf{L}_n^i$ for suitably large i. Clearly, $\mathbf{L}_{n+1}$ is countable and belongs to $\mathcal{V}$. Moreover, for any a, b in L_n satisfying $a < b$, there exists $c \in L_{n+1}$ such that $a < c < b$. Finally, let $\mathbf{M}$ be the direct limit of the family $\{\mathbf{L}_n\}_{n\in\mathbb{N}}$ ordered by inclusion; that is, $M := \bigcup_{n\in\mathbb{N}} L_n$ and the operations of $\mathbf{M}$ are defined by referring to $\mathbf{L}_n$ for suitably large n. By construction, $\mathbf{M}$ is a countable, dense, totally ordered member of $\mathcal{V}$ that contains $\mathbf{L} = \mathbf{L}_0$ as a subalgebra. □

We use this lemma to establish densifiability for varieties of semilinear residuated lattices defined by equations of a particular form. Recall that we call a formula or equation in which only operation symbols from a set S occur, an *S-formula* or *S-equation*, respectively; we also call $\alpha \leq \beta$ an *S-inequation*, for S-formulas α, β.

THEOREM 5.2.6. *Every variety defined by a set of $\{\wedge, \vee, \cdot, \mathrm{e}\}$-equations relative to the variety of semilinear, integral residuated lattices is densifiable.*

PROOF. Since the trivial variety is densifiable, let $\mathcal{V}$ be any non-trivial variety defined relative to the variety of semilinear, integral residuated lattices by a set of $\{\wedge, \vee, \cdot, \mathrm{e}\}$-equations. Using the distributivity of multiplication over binary joins and meets, and lattice distributivity, we may assume that these equations are of the form $\alpha_1 \wedge \cdots \wedge \alpha_n \leq \beta_1 \vee \cdots \vee \beta_m$, where $\alpha_1, \ldots, \alpha_n, \beta_1, \ldots, \beta_m$ are $\{\cdot, \mathrm{e}\}$-formulas. We can also assume that every variable occurring on the right in such an inequation occurs on the left. Suppose that this is not the case for some variable x. Then x does not occur in some β_j, since otherwise, substituting e for all the other variables in the inequation gives $\mathcal{V} \models \mathrm{e} \leq x$, which contradicts the fact that $\mathcal{V}$ is non-trivial. Now substitute x with β_j in the inequation to obtain $\alpha_1 \wedge \cdots \wedge \alpha_n \leq \beta_1' \vee \cdots \vee \beta_m'$. If the original inequation is satisfied by some semilinear, integral residuated lattice $\mathbf{L}$, then so is the new inequation. The converse also holds, since, by integrality, $\mathbf{L}$ satisfies $\beta_k' \leq \beta_j$, for each k such that x occurs in β_k. So we have obtained a new equational basis for $\mathcal{V}$, and the claim follows by repeating this argument.

To prove that $\mathcal{V}$ is densifiable, we check the condition in Lemma 5.2.5. Let $\mathbf{L}$ be any countable, totally ordered member of $\mathcal{V}$ and consider any $a, b \in L$ satisfying $a < b$. We show that $\mathbf{L}$ is a subalgebra of a totally ordered member $\mathbf{M}$ of $\mathcal{V}$ satisfying $a < c < b$, for some $c \in M$. We may as well assume therefore that, for all $d \in L$, either $d \leq a$ or $b \leq d$. We consider the set

$$M := \{\langle c, 1\rangle \mid c \in L\} \cup \{\langle b, 0\rangle\},$$

endowed with the order inherited from the lexicographic product of $\langle L, \leq\rangle$ and the two-element chain $\langle\{0,1\}, \leq\rangle$, and define

$$\langle u, r\rangle\langle v, s\rangle := \langle uv, 1\rangle \wedge \langle u, r\rangle \wedge \langle v, s\rangle.$$

Observe that

$$\langle u, r\rangle\langle \mathrm{e}, 1\rangle = \langle u\mathrm{e}, 1\rangle \wedge \langle u, r\rangle \wedge \langle \mathrm{e}, 1\rangle = \langle u, r\rangle = \langle \mathrm{e}u, 1\rangle \wedge \langle \mathrm{e}, 1\rangle \wedge \langle u, r\rangle = \langle \mathrm{e}, 1\rangle\langle u, r\rangle.$$

Also, if $\langle u, r\rangle \leq \langle v, s\rangle$, then $u \leq v$, so $uw \leq vw$ and $wu \leq wv$, and

$$\langle u, r\rangle\langle w, t\rangle = \langle uw, 1\rangle \wedge \langle u, r\rangle \wedge \langle w, t\rangle \leq \langle vw, 1\rangle \wedge \langle v, s\rangle \wedge \langle w, t\rangle = \langle v, s\rangle\langle w, t\rangle$$
$$\langle w, t\rangle\langle u, r\rangle = \langle wu, 1\rangle \wedge \langle w, t\rangle \wedge \langle u, r\rangle \leq \langle wv, 1\rangle \wedge \langle w, t\rangle \wedge \langle v, s\rangle = \langle w, t\rangle\langle v, s\rangle.$$

Since $\langle M, \leq\rangle$ is a chain, it follows also that

$$(\langle u,r\rangle \wedge \langle v,s\rangle)\langle w,t\rangle = \langle u,r\rangle\langle w,t\rangle \wedge \langle v,s\rangle\langle w,t\rangle.$$

Hence, for associativity,

$$\begin{aligned}(\langle u,r\rangle\langle v,s\rangle)\langle w,t\rangle &= (\langle uv,1\rangle \wedge \langle u,r\rangle \wedge \langle v,s\rangle)\langle w,t\rangle\\ &= \langle uv,1\rangle\langle w,t\rangle \wedge \langle u,r\rangle\langle w,t\rangle \wedge \langle v,s\rangle\langle w,t\rangle\\ &= \langle uvw,1\rangle \wedge \langle uv,1\rangle \wedge \langle uw,1\rangle \wedge \langle vw,1\rangle \wedge \langle u,r\rangle \wedge \langle v,s\rangle \wedge \langle w,t\rangle\\ &= \langle uvw,1\rangle \wedge \langle u,r\rangle \wedge \langle v,s\rangle \wedge \langle w,t\rangle\\ &= \langle u,r\rangle\langle vw,1\rangle \wedge \langle u,r\rangle\langle v,s\rangle \wedge \langle u,r\rangle\langle w,t\rangle\\ &= \langle u,r\rangle(\langle vw,1\rangle \wedge \langle v,s\rangle \wedge \langle w,t\rangle)\\ &= \langle u,r\rangle(\langle v,s\rangle\langle w,t\rangle).\end{aligned}$$

For the residuals, we observe that

$$\begin{aligned}\langle u,r\rangle\langle v,s\rangle \leq \langle w,1\rangle &\iff \langle uv,1\rangle \wedge \langle u,r\rangle \wedge \langle v,s\rangle \leq \langle w,1\rangle\\ &\iff uv \leq w\\ &\iff v \leq u\backslash w\\ &\iff \langle v,s\rangle \leq \langle u\backslash w,1\rangle,\end{aligned}$$

noting that in the second equivalence, if $\langle uv,1\rangle \wedge \langle u,r\rangle \wedge \langle v,s\rangle \leq \langle w,1\rangle$, then either $\langle uv,1\rangle \wedge \langle u,r\rangle \wedge \langle v,s\rangle = \langle uv,1\rangle$, or one of $uv = u \leq w$ and $uv = v \leq w$ holds. Hence

$$\langle u,r\rangle\backslash\langle w,1\rangle = \langle u\backslash w,1\rangle \quad \text{and, similarly,} \quad \langle w,1\rangle/\langle v,s\rangle = \langle w/v,1\rangle.$$

But also

$$\begin{aligned}\langle u,r\rangle\langle v,s\rangle \leq \langle b,0\rangle &\iff \langle uv,1\rangle \wedge \langle u,r\rangle \wedge \langle v,s\rangle \leq \langle b,0\rangle\\ &\iff \langle uv,1\rangle \leq \langle a,1\rangle \ \text{ or } \ \langle u,r\rangle \wedge \langle v,s\rangle \leq \langle b,0\rangle,\end{aligned}$$

so, writing $\Rightarrow$ for the Brouwerian implication of $\mathbf{M}$ viewed as a chain,

$$\begin{aligned}\langle u,r\rangle\backslash\langle b,0\rangle &= \langle u\backslash a,1\rangle \vee (\langle u,r\rangle \Rightarrow \langle b,0\rangle)\\ \langle b,0\rangle/\langle v,s\rangle &= \langle a/v,1\rangle \vee (\langle v,s\rangle \Rightarrow \langle b,0\rangle).\end{aligned}$$

Hence $\mathbf{M}$ is an integral, totally ordered residuated lattice. Observe also that

$$\langle u_1,r_1\rangle\langle u_2,r_2\rangle\cdots\langle u_n,r_n\rangle = \langle u_1u_2\cdots u_n,1\rangle \wedge \langle u_1,r_1\rangle \wedge \langle u_2,r_2\rangle \wedge \cdots \wedge \langle u_n,r_n\rangle.$$

We aim next to show that $\mathbf{M}$ satisfies all defining inequations of $\mathcal{V}$ of the form

$$\alpha_1 \wedge \cdots \wedge \alpha_n \leq \beta_1 \vee \cdots \vee \beta_m,$$

where $\alpha_1,\ldots,\alpha_n,\beta_1,\ldots,\beta_m$ are $\{\cdot,\mathrm{e}\}$-formulas containing variables $x_1,\ldots,x_k$ and each x_i occurs on the left. Let $u_i \in L$ and $r_i \in \{0,1\}$, for each $i \in \{1,\ldots,k\}$. Let $v_p := \alpha_p^{\mathbf{L}}(u_1,\ldots,u_k)$ and $w_q := \beta_q^{\mathbf{L}}(u_1,\ldots,u_k)$, for $p \in \{1,\ldots,n\}$, $q \in \{1,\ldots,m\}$. Without loss of generality, we may assume that $v_1 = \min\{v_1,\ldots,v_n\}$, $w_1 = \max\{w_1,\ldots,w_m\}$, and hence $v_1 \leq w_1$. Now let $v_p' := \alpha_p^{\mathbf{M}}(\langle u_1,r_1\rangle,\ldots,\langle u_k,r_k\rangle)$

and $w'_q := \beta_q^{\mathbf{M}}(\langle u_1, r_1\rangle, \dots, \langle u_k, r_k\rangle)$, for $p \in \{1, \dots, n\}$, $q \in \{1, \dots, m\}$. Then

$$\begin{aligned} v'_1 \wedge \cdots \wedge v'_n &= \textstyle\bigwedge_{p=1}^{n} \langle v_p, 1\rangle \wedge \langle u_1, r_1\rangle \wedge \cdots \wedge \langle u_k, r_k\rangle \\ &\leq \langle v_1, 1\rangle \wedge \langle u_1, r_1\rangle \wedge \cdots \wedge \langle u_k, r_k\rangle \\ &\leq \langle w_1, 1\rangle \wedge \langle u_1, r_1\rangle \wedge \cdots \wedge \langle u_k, r_k\rangle \\ &\leq \textstyle\bigvee_{q=1}^{m} \langle w_q, 1\rangle \wedge \langle u_1, r_1\rangle \wedge \cdots \wedge \langle u_k, r_k\rangle \\ &\leq w'_1 \vee \cdots \vee w'_m. \end{aligned}$$

We may therefore conclude that $\mathbf{M} \in \mathcal{V}$ and that $x \mapsto \langle x, 1\rangle$ is an embedding of $\mathbf{L}$ into $\mathbf{M}$ with $\langle a, 1\rangle < \langle b, 0\rangle < \langle b, 1\rangle$ as required. □

THEOREM 5.2.7. *Every variety defined by a set of $\{\cdot, \mathrm{e}\}$-inequations relative to the variety of cancellative, semilinear residuated lattices is densifiable.*

PROOF. Since the trivial variety is densifiable, let $\mathcal{V}$ be any non-trivial variety defined relative to the variety of cancellative, semilinear residuated lattices by a set of $\{\cdot, \mathrm{e}\}$-inequations. Let $\mathbf{K}$ be any non-trivial totally ordered member of $\mathcal{V}$, noting that its negative cone $\mathbf{K}^-$ is also a non-trivial totally ordered member of $\mathcal{V}$. To prove that $\mathcal{V}$ is densifiable, we again check the condition in Lemma 5.2.5.

Let $\mathbf{L}$ be any countable, totally ordered member of $\mathcal{V}$, and consider the lexicographic product $\langle M, \leq\rangle$ of $\langle L, \leq^{\mathbf{L}}\rangle$ and $\langle K^-, \leq^{\mathbf{K}^-}\rangle$. Equipping M with the operation

$$\langle a, x\rangle\langle b, y\rangle := \langle ab, xy\rangle$$

yields a cancellative monoid with neutral element $\langle \mathrm{e}, \mathrm{e}\rangle$. Consider $\langle a, x\rangle, \langle b, y\rangle \in M$ satisfying $\langle a, x\rangle \leq \langle b, y\rangle$, i.e., $a < b$ or ($a = b$ and $x \leq y$). Let $\langle c, z\rangle \in M$. Then $ac \leq bc$, and either $ac < bc$ or ($ac = bc$ and $xz \leq yz$). So $\langle a, x\rangle\langle c, z\rangle = \langle ac, xz\rangle \leq \langle bc, yz\rangle = \langle b, y\rangle\langle c, z\rangle$ and, similarly, $\langle c, z\rangle\langle a, x\rangle \leq \langle c, z\rangle\langle b, y\rangle$.

It is easily checked that, for any $\langle a, x\rangle, \langle b, y\rangle \in M$,

$$\langle a, x\rangle \backslash \langle b, y\rangle = \begin{cases} \langle a\backslash b, x\backslash y\rangle & \text{if } a(a\backslash b) = b; \\ \langle a\backslash b, \mathrm{e}\rangle & \text{if } a(a\backslash b) < b, \end{cases}$$

$$\langle a, x\rangle / \langle b, y\rangle = \begin{cases} \langle a/b, x/y\rangle & \text{if } (a/b)b = a; \\ \langle a/b, \mathrm{e}\rangle & \text{if } (a/b)b < a. \end{cases}$$

Hence $\mathbf{M}$ is a cancellative, totally ordered residuated lattice. Moreover, $\mathbf{M}$ satisfies any $\{\cdot, \mathrm{e}\}$-inequation satisfied by $\mathbf{L}$ and $\mathbf{K}^-$ and is therefore a member of $\mathcal{V}$. Finally, $a \mapsto \langle a, \mathrm{e}\rangle$ is an embedding of $\mathbf{L}$ into $\mathbf{M}$ and, whenever $a, b \in L$ satisfies $a < b$, also $\langle a, \mathrm{e}\rangle < \langle b, c\rangle < \langle b, \mathrm{e}\rangle$, for some $c \in K^- \backslash \{\mathrm{e}\}$. □

Similar densifiability results may be established for varieties of semilinear ℓ-groups and their negative cones (see the bibliographical remarks).

We turn our attention now to a further characterization of densifiability that is particularly useful for giving syntactic proofs of this property. Let $\mathrm{Var}(S)$ denote the set of variables occurring in any syntactic object (sequence, set, multiset, sequent, hypersequent, etc.) S defined over Fm.

THEOREM 5.2.8. *Let $\mathcal{V}$ be any variety of semilinear residuated lattices. Then $\mathcal{V}$ is densifiable if and only if, for any $\Sigma \cup \{\alpha, \beta, \gamma\} \subseteq \mathrm{Fm}$ and variable $x \notin \mathrm{Var}(\Sigma, \alpha, \beta, \gamma)$,*

$$\Sigma \vdash_{\mathcal{V}} (\alpha\backslash x) \vee (x\backslash \beta) \vee \gamma \implies \Sigma \vdash_{\mathcal{V}} (\alpha\backslash\beta) \vee \gamma.$$

Moreover, if $\mathcal{V}$ is any variety of commutative, semilinear residuated lattices, then $\mathcal{V}$ is densifiable if and only if, for any $\{\alpha, \beta, \gamma\} \subseteq \mathrm{Fm}$ and variable $x \notin \mathrm{Var}(\alpha, \beta, \gamma)$,

$$\vdash_{\mathcal{V}} (\alpha \to x) \vee (x \to \beta) \vee \gamma \implies \vdash_{\mathcal{V}} (\alpha \to \beta) \vee \gamma.$$

PROOF. Suppose first that $\mathcal{V}$ is densifiable. Consider $\Sigma \cup \{\alpha, \beta, \gamma\} \subseteq \mathrm{Fm}$ and $x \notin \mathrm{Var}(\Sigma, \alpha, \beta, \gamma)$. If $\Sigma \nvdash_{\mathcal{V}} (\alpha\backslash\beta) \vee \gamma$, then there exist a dense, totally ordered member $\mathbf{L}$ of $\mathcal{V}$ and homomorphism $\varphi\colon \mathbf{Fm} \to \mathbf{L}$ such that $\varphi(\alpha') \geq \mathrm{e}$, for each $\alpha' \in \Sigma$, $\varphi(\alpha) > \varphi(\beta)$, and $\varphi(\gamma) < \mathrm{e}$. Because $\mathbf{L}$ is dense, there exists $c \in L$ such that $\varphi(\alpha) > c > \varphi(\beta)$. We define $\psi\colon \mathbf{Fm} \to \mathbf{L}$ by setting $\psi(y) := c$ if $y = x$, and $\psi(y) := \varphi(y)$ otherwise. Then $\psi(\alpha') \geq \mathrm{e}$, for each $\alpha' \in \Sigma$, $\psi(\alpha) > \psi(x) > \psi(\beta)$, and $\psi(\gamma) < \mathrm{e}$. So $\Sigma \nvdash_{\mathcal{V}} (\alpha\backslash x) \vee (x\backslash\beta) \vee \gamma$.

For the converse, we again use Lemma 5.2.5. Let $\mathbf{L}$ be any countable, totally ordered member of $\mathcal{V}$ satisfying $a >^{\mathbf{L}} b$, for some $a, b \in L$; it suffices to prove that there exists a totally ordered member $\mathbf{M}$ of $\mathcal{V}$ such that $\mathbf{L}$ is a subalgebra of $\mathbf{M}$ and $a >^{\mathbf{M}} c >^{\mathbf{M}} b$, for some $c \in M$. Note first that because $\mathbf{L}$ is countable, we can assume that $\mathbf{L} = \mathbf{F}_{\mathcal{V}}(Y)/\mathrm{NC}[\overline{\Sigma}]$, for some countable set of variables Y and set of formulas $\Sigma \subseteq \mathrm{Fm}(Y)$ such that $x \notin Y$. For $\gamma \in \mathrm{Fm}(Y)$, let us write $[\gamma]$ for the equivalence class of $\bar{\gamma}$ in $\mathbf{L}$. Consider any $\alpha, \beta \in \mathrm{Fm}(Y)$ such that $a = [\alpha] >^{\mathbf{L}} [\beta] = b$, and define

$$\Delta := \{\alpha'\backslash\beta' \mid \alpha', \beta' \in \mathrm{Fm}(Y) \text{ and } [\alpha'] >^{\mathbf{L}} [\beta']\}.$$

Then $\Sigma \nvdash_{\mathcal{V}} (\alpha\backslash\beta) \vee \bigvee\Delta'$, for any finite $\Delta' \subseteq \Delta$, and hence also, by assumption,

$$(\star) \quad \Sigma \nvdash_{\mathcal{V}} (\alpha\backslash x) \vee (x\backslash\beta) \vee \bigvee\Delta', \text{ for any finite } \Delta' \subseteq \Delta.$$

We now enumerate all pairs of formulas $\langle\gamma_n, \delta_n\rangle$ from $\mathrm{Fm}(Y \cup \{x\})$. Let $\Sigma_0 := \Sigma$ and define Σ_{n+1} for $n \in \mathbb{N}$ such that $\Sigma_{n+1} := \Sigma_n \cup \{\gamma_n\backslash\delta_n\}$ or $\Sigma_{n+1} := \Sigma_n \cup \{\delta_n\backslash\gamma_n\}$ satisfies $(\star)$. If this were possible at step n, but not step $n+1$, we would have

$$\begin{aligned} \Sigma_n \cup \{\gamma_n\backslash\delta_n\} &\vdash_{\mathcal{V}} (\alpha\backslash x) \vee (x\backslash\beta) \vee \textstyle\bigvee\Delta_1, \text{ for some finite } \Delta_1 \subseteq \Delta \\ \Sigma_n \cup \{\delta_n\backslash\gamma_n\} &\vdash_{\mathcal{V}} (\alpha\backslash x) \vee (x\backslash\beta) \vee \textstyle\bigvee\Delta_2, \text{ for some finite } \Delta_2 \subseteq \Delta. \end{aligned}$$

But then, using the fact that $\mathcal{V}$ is generated as a quasivariety by its totally ordered members, $\Sigma_n \vdash_{\mathcal{V}} (\alpha\backslash x) \vee (x\backslash\beta) \vee \bigvee(\Delta_1 \cup \Delta_2)$, a contradiction. Let $\Sigma^* := \bigcup_{n\in\mathbb{N}} \Sigma_n$ and define $\mathbf{M} := \mathbf{F}_{\mathcal{V}}(Y \cup \{x\})/\mathrm{NC}[\overline{\Sigma^*}]$. Then $\mathbf{M}$ is a totally ordered member of $\mathcal{V}$ by construction. Also, writing now $[\gamma]$ for the equivalence class of $\bar{\gamma}$ in $\mathbf{M}$, we have $[\alpha] >_{\mathbf{M}} [x] >_{\mathbf{M}} [\beta]$. Moreover, $\mathbf{L}$ can be identified with a subalgebra of $\mathbf{M}$, since $[\gamma] >_{\mathbf{L}} [\delta]$ implies $\Sigma \nvdash_{\mathcal{V}} \gamma\backslash\delta$ and so also $\Sigma^* \nvdash_{\mathcal{V}} \gamma\backslash\delta$ and $[\gamma] >_{\mathbf{M}} [\delta]$.

Suppose finally that $\mathcal{V}$ is a variety of commutative, semilinear residuated lattices. The left-to-right direction follows immediately from the more general case above. For the other direction, suppose that, for any $\{\alpha, \beta, \gamma\} \subseteq \mathrm{Fm}$ and $x \notin \mathrm{Var}(\alpha, \beta, \gamma)$, whenever $\vdash_{\mathcal{V}} (\alpha \to x) \vee (x \to \beta) \vee \gamma$, also $\vdash_{\mathcal{V}} (\alpha \to \beta) \vee \gamma$. We prove that this implication holds also in the presence of a set of formulas Σ and hence that $\mathcal{V}$ is densifiable. Consider $\Sigma \cup \{\alpha, \beta, \gamma\} \subseteq \mathrm{Fm}$ and $x \notin \mathrm{Var}(\Sigma, \alpha, \beta, \gamma)$ such that

$$\Sigma \vdash_{\mathcal{V}} (\alpha \to x) \vee (x \to \beta) \vee \gamma.$$

By Proposition 4.4.10, for some $\{\delta_1, \ldots, \delta_m\} \subseteq \Sigma$ and $\delta = (\delta_1 \wedge \mathrm{e}) \cdots (\delta_m \wedge \mathrm{e})$,

$$\vdash_{\mathcal{V}} \delta \to ((\alpha \to x) \vee (x \to \beta) \vee \gamma)$$

and, using equations satisfied by all commutative, semilinear residuated lattices,

$$\vdash_{\mathcal{V}} ((\delta \cdot \alpha) \to x) \vee (x \to (\delta \to \beta)) \vee (\delta \to \gamma).$$

So then, by assumption,

$$\vdash_{\mathcal{V}} ((\delta \cdot \alpha) \to (\delta \to \beta)) \vee (\delta \to \gamma),$$

and, using equations satisfied by all commutative, semilinear residuated lattices,

$$\vdash_{\mathcal{V}} (\delta \cdot \delta) \to ((\alpha \to \beta) \vee \gamma).$$

But $\delta \cdot \delta$ is of the form $(\delta_1' \wedge \mathrm{e}) \cdots (\delta_k' \wedge \mathrm{e})$, for some $\{\delta_1', \ldots, \delta_k'\} \subseteq \Sigma$, so by Proposition 4.4.10 once more, $\Sigma \vdash_{\mathcal{V}} (\alpha \to \beta) \vee \gamma$. □

Theorem 5.2.8 provides a syntactic criterion for densifiability that can be verified (in some cases) by establishing a "density elimination" property for an appropriate hypersequent calculus. To illustrate, we consider the following notable case study.

PROPOSITION 5.2.9. *The variety of commutative, semilinear residuated lattices is densifiable.*

PROOF. Given hypersequents

$$\begin{aligned} \mathcal{G} &= ([\Gamma_i \Rightarrow x]_{i=1}^n \mid [\Pi_j, [x]^{\lambda_j} \Rightarrow \Delta_j]_{j=1}^m \mid [\Sigma_k, [x]^{\mu_k+1} \Rightarrow x]_{k=1}^l) \\ \mathcal{H}_x &= (\mathcal{H} \mid \Gamma \Rightarrow x \mid \Pi, x \Rightarrow \Delta), \end{aligned}$$

where $x \notin \mathrm{Var}(\Gamma_1, \ldots, \Gamma_n, \Pi_1, \ldots, \Pi_m, \Delta_1, \ldots, \Delta_m, \Sigma_1, \ldots, \Sigma_l, \mathcal{H}, \Gamma, \Pi, \Delta)$, let

$$(\mathcal{G}, \mathcal{H}_x)^{\mathcal{D}} := (\mathcal{H} \mid [\Gamma_i, \Pi \Rightarrow \Delta]_{i=1}^n \mid [\Pi_j, \Gamma^{\lambda_j} \Rightarrow \Delta_j]_{j=1}^m \mid [\Sigma_k, \Gamma^{\mu_k} \Rightarrow \mathrm{e}]_{k=1}^l).$$

It suffices to prove the following claim:

If $d_1 \vdash_{\mathsf{CSemFL}^\circ} \mathcal{G}$ and $d_2 \vdash_{\mathsf{CSemFL}^\circ} \mathcal{H}_x$, then $\vdash_{\mathsf{CSemFL}} (\mathcal{G}, \mathcal{H}_x)^{\mathcal{D}} \mid \Gamma, \Pi \Rightarrow \Delta$.

To see that this claim suffices, we check the condition provided by Theorem 5.2.8. Suppose that $\vdash_{\mathcal{CSemRL}} (\alpha \to x) \vee (x \to \beta) \vee \gamma$, for some $\alpha, \beta, \gamma \in \mathrm{Fm}$ and variable $x \notin \mathrm{Var}(\alpha, \beta, \gamma)$. Then $\vdash_{\mathsf{CSemFL}^\circ} \alpha \Rightarrow x \mid x \Rightarrow \beta \mid \; \Rightarrow \gamma$, by Theorems 2.5.3 and 2.5.4. Applying the claim with $\mathcal{G} = \mathcal{H}_x := (\alpha \Rightarrow x \mid x \Rightarrow \beta \mid \; \Rightarrow \gamma)$, it follows that $\vdash_{\mathsf{CSemFL}^\circ} \alpha \Rightarrow \beta \mid \alpha \Rightarrow \beta \mid \; \Rightarrow \gamma \mid \; \Rightarrow \gamma$. So $\vdash_{\mathcal{CSemRL}} (\alpha \to \beta) \vee \gamma$, by Theorem 2.5.3.

We prove the claim by induction on the height of d_1. If $\mathcal{G} = (\mathcal{G}' \mid x \Rightarrow x)$ or $\mathcal{G} = (\mathcal{G}' \mid \alpha \Rightarrow \alpha)$, for some other formula α, then the result follows by ($\Rightarrow$e) or (ID), respectively. Otherwise, we consider the last rule applied in d_1. The cases of (EC) and (EW) are immediate using the induction hypothesis. For the operation rules, there are many cases that follow a common pattern. Suppose for example that d_1 ends with

$$\dfrac{\dfrac{\vdots}{\mathcal{G}' \mid \Sigma_{1,1}, [x]^{\mu'} \Rightarrow \alpha} \quad \dfrac{\vdots}{\mathcal{G}' \mid \Sigma_{1,2}, \beta, [x]^{\mu_1+1-\mu'} \Rightarrow x}}{\mathcal{G}' \mid \Sigma_{1,1}, \Sigma_{1,2}, \alpha \to \beta, [x]^{\mu_1+1} \Rightarrow x,} \; (\to\Rightarrow)$$

where $\Sigma_1 = \Sigma_{1,1} \uplus \Sigma_{1,2} \uplus [\alpha \to \beta]$ and

$$\mathcal{G}' = ([\Gamma_i \Rightarrow x]_{i=1}^n \mid [\Pi_j, [x]^{\lambda_j} \Rightarrow \Delta_j]_{j=1}^m \mid [\Sigma_k, [x]^{\mu_k+1} \Rightarrow x]_{k=2}^l).$$

There are two subcases:

(a) If $\mu' < \mu_1 + 1$, then using the induction hypothesis twice,

$$\vdash_{\mathsf{CSemFL}} (\mathcal{G}', \mathcal{H}_x)^{\mathcal{D}} \mid \Sigma_{1,1}, \Gamma^{\mu'} \Rightarrow \alpha \mid \Gamma, \Pi \Rightarrow \Delta$$
$$\vdash_{\mathsf{CSemFL}} (\mathcal{G}', \mathcal{H}_x)^{\mathcal{D}} \mid \Sigma_{1,2}, \beta, \Gamma^{\mu_1-\mu'} \Rightarrow \mathrm{e} \mid \Gamma, \Pi \Rightarrow \Delta.$$

So by an application of $(\to\Rightarrow)$, as required

$$\vdash_{\mathsf{CSemFL}} (\mathcal{G}', \mathcal{H}_x)^{\mathcal{D}} \mid \Sigma_1, \Gamma^{\mu_1} \Rightarrow \mathrm{e} \mid \Gamma, \Pi \Rightarrow \Delta.$$

(b) If $\mu' = \mu_1 + 1$, then using the induction hypothesis twice,

$$\vdash_{\mathsf{CSemFL}} (\mathcal{G}', \mathcal{H}_x)^{\mathcal{D}} \mid \Sigma_{1,1}, \Gamma^{\mu_1+1} \Rightarrow \alpha \mid \Gamma, \Pi \Rightarrow \Delta$$
$$\vdash_{\mathsf{CSemFL}} (\mathcal{G}', \mathcal{H}_x)^{\mathcal{D}} \mid \Sigma_{1,2}, \beta, \Pi \Rightarrow \Delta \mid \Gamma, \Pi \Rightarrow \Delta.$$

So by an application of $(\to\Rightarrow)$,

$$\vdash_{\mathsf{CSemFL}} (\mathcal{G}', \mathcal{H}_x)^{\mathcal{D}} \mid \Sigma_1, \Gamma^{\mu_1+1}, \Pi \Rightarrow \Delta \mid \Gamma, \Pi \Rightarrow \Delta.$$

But clearly also, using $(\Rightarrow\mathrm{e})$,

$$\vdash_{\mathsf{CSemFL}} (\mathcal{G}', \mathcal{H}_x)^{\mathcal{D}} \mid \ \Rightarrow \mathrm{e} \mid \Gamma, \Pi \Rightarrow \Delta.$$

Hence by an application of (COM),

$$\vdash_{\mathsf{CSemFL}} (\mathcal{G}', \mathcal{H}_x)^{\mathcal{D}} \mid \Sigma_1, \Gamma^{\mu_1} \Rightarrow \mathrm{e} \mid \Gamma, \Pi \Rightarrow \Delta \mid \Gamma, \Pi \Rightarrow \Delta,$$

and the desired result follows by a further application of (EC).

Suppose that the last rule applied is (COM). We assume first that d_1 ends with

$$\dfrac{\dfrac{\vdots}{\mathcal{G}' \mid \Gamma_{1,1}, \Pi_{1,1}, [x]^{\lambda'+1} \Rightarrow x} \quad \dfrac{\vdots}{\mathcal{G}' \mid \Gamma_{1,2}, \Pi_{1,2}, [x]^{\lambda_1-\lambda'-1} \Rightarrow \Delta_1}}{\mathcal{G}' \mid \Gamma_1 \Rightarrow x \mid \Pi_1, [x]^{\lambda_1} \Rightarrow \Delta_1,} \ (\text{COM})$$

where $\Gamma_1 = \Gamma_{1,1} \uplus \Gamma_{1,2}$, $\Pi_1 = \Pi_{1,1} \uplus \Pi_{1,2}$, and

$$\mathcal{G}' = ([\Gamma_i \Rightarrow x]_{i=2}^n \mid [\Pi_j, [x]^{\lambda_j} \Rightarrow \Delta_j]_{j=2}^m \mid [\Sigma_k, [x]^{\mu_k+1} \Rightarrow x]_{k=1}^l).$$

By the induction hypothesis twice,

$$\vdash_{\mathsf{CSemFL}} \mathcal{H}' \mid \Gamma_{1,1}, \Pi_{1,1}, \Gamma^{\lambda'} \Rightarrow \mathrm{e} \quad \text{and} \quad \vdash_{\mathsf{CSemFL}} \mathcal{H}' \mid \Gamma_{1,2}, \Pi_{1,2}, \Gamma^{\lambda_1-\lambda'-1} \Rightarrow \Delta_1,$$

where $\mathcal{H}' = ((\mathcal{G}', \mathcal{H}_x)^{\mathcal{D}} \mid \Gamma, \Pi \Rightarrow \Delta)$. By an application of $(\mathrm{e}\Rightarrow)$,

$$\vdash_{\mathsf{CSemFL}} \mathcal{H}' \mid \Gamma_{1,2}, \Pi_{1,2}, \Gamma^{\lambda_1-\lambda'-1}, \mathrm{e} \Rightarrow \Delta_1,$$

and an application of (CUT) yields

$$\vdash_{\mathsf{CSemFL}} \mathcal{H}' \mid \Gamma_1, \Pi_1, \Gamma^{\lambda_1-1} \Rightarrow \Delta_1.$$

Now let $\alpha := \prod \Gamma_1$. By applying (EW), $(\mathrm{e}\Rightarrow)$, and $(\cdot\Rightarrow)$,

$$\vdash_{\mathsf{CSemFL}} \mathcal{H}' \mid \alpha, \Pi_1, \Gamma^{\lambda_1-1} \Rightarrow \Delta_1 \mid \Pi, \alpha \Rightarrow \Delta.$$

Moreover, by substituting all occurrences of x with α in the derivation d_2 and adding an application of (EW),

$$\vdash_{\mathsf{CSemFL}} \mathcal{H}' \mid \Gamma \Rightarrow \alpha \mid \Pi, \alpha \Rightarrow \Delta.$$

Hence an application of (CUT) yields

$$\vdash_{\mathsf{CSemFL}} \mathcal{H}' \mid \Pi_1, \Gamma^{\lambda_1} \Rightarrow \Delta_1 \mid \Pi, \alpha \Rightarrow \Delta.$$

But also, by applying (EW), $(\Rightarrow \mathrm{e})$, and $(\Rightarrow \cdot)$,

$$\vdash_{\mathsf{CSemFL}} \mathcal{H}' \mid \Pi_1, \Gamma^{\lambda_1} \Rightarrow \Delta_1 \mid \Gamma_1 \Rightarrow \alpha.$$

So an application of (CUT) gives the desired result

$$\vdash_{\mathsf{CSemFL}} \mathcal{H}' \mid \Pi_1, \Gamma^{\lambda_1} \Rightarrow \Delta_1 \mid \Gamma_1, \Pi \Rightarrow \Delta.$$

Now suppose that d_1 ends with:

$$\dfrac{\dfrac{\vdots}{\mathcal{G}' \mid \Pi_{1,1}, \Sigma_{1,1} \Rightarrow x} \quad \dfrac{\vdots}{\mathcal{G}' \mid \Pi_{1,2}, \Sigma_{1,2}, [x]^{\lambda_1+\mu_1+1} \Rightarrow \Delta_1}}{\mathcal{G}' \mid \Pi_1, [x]^{\lambda_1} \Rightarrow \Delta_1 \mid \Sigma_1, [x]^{\mu_1+1} \Rightarrow x,}\ (\textsc{com})$$

where $\Pi_1 = \Pi_{1,1} \uplus \Pi_{1,2}$, $\Sigma_1 = \Sigma_{1,1} \uplus \Sigma_{1,2}$, and

$$\mathcal{G}' = ([\Gamma_i \Rightarrow x]_{i=1}^n \mid [\Pi_j, [x]^{\lambda_j} \Rightarrow \Delta_j]_{j=2}^m \mid [\Sigma_k, [x]^{\mu_k+1} \Rightarrow x]_{k=2}^l).$$

By the induction hypothesis twice,

$$d_3 \vdash_{\mathsf{CSemFL}} \mathcal{H}' \mid \Pi_{1,1}, \Sigma_{1,1}, \Pi \Rightarrow \Delta \quad \text{and} \quad d_4 \vdash_{\mathsf{CSemFL}} \mathcal{H}' \mid \Pi_{1,2}, \Sigma_{1,2}, \Gamma^{\lambda_1+\mu_1+1} \Rightarrow \Delta_1,$$

where $\mathcal{H}' = ((\mathcal{G}', \mathcal{H}_x)^{\mathcal{D}} \mid \Gamma, \Pi \Rightarrow \Delta)$. So we can construct a derivation

$$\dfrac{\dfrac{\dfrac{\dfrac{\vdots\, d_3}{\mathcal{H}' \mid \Pi_{1,1}, \Sigma_{1,1}, \Pi \Rightarrow \Delta} \quad \dfrac{\vdots\, d_4}{\mathcal{H}' \mid \Pi_{1,2}, \Sigma_{1,2}, \Gamma^{\lambda_1+\mu_1+1} \Rightarrow \Delta_1}}{\mathcal{H}' \mid \Gamma, \Pi \Rightarrow \Delta \mid \Pi_1, \Sigma_1, \Gamma^{\lambda_1+\mu_1} \Rightarrow \Delta_1}\ (\textsc{com})}{\mathcal{H}' \mid \Pi_1, \Sigma_1, \Gamma^{\lambda_1+\mu_1} \Rightarrow \Delta_1}\ (\textsc{ec}) \quad \dfrac{}{\mathcal{H}' \mid\ \Rightarrow \mathrm{e}}\ (\Rightarrow \mathrm{e})}{\mathcal{H}' \mid \Pi_1, \Gamma^{\lambda_1} \Rightarrow \Delta_1 \mid \Sigma_1, \Gamma^{\mu_1} \Rightarrow \mathrm{e}.}\ (\textsc{com})$$

Suppose next that d_1 ends with

$$\dfrac{\dfrac{\vdots}{\mathcal{G}' \mid \Sigma_{1,1}, \Sigma_{2,1}, [x]^{\mu_1+\mu_2+2} \Rightarrow x} \quad \dfrac{\vdots}{\mathcal{G}' \mid \Sigma_{1,2}, \Sigma_{2,2} \Rightarrow x}}{\mathcal{G}' \mid \Sigma_1, [x]^{\mu_1+1} \Rightarrow x \mid \Sigma_2, [x]^{\mu_2+1} \Rightarrow x,}\ (\textsc{com})$$

where $\Sigma_1 = \Sigma_{1,1} \uplus \Sigma_{1,2}$, $\Sigma_2 = \Sigma_{2,1} \uplus \Sigma_{2,2}$, and

$$\mathcal{G}' = ([\Gamma_i \Rightarrow x]_{i=1}^n \mid [\Pi_j, [x]^{\lambda_j} \Rightarrow \Delta_j]_{j=1}^m \mid [\Sigma_k, [x]^{\mu_k+1} \Rightarrow x]_{k=3}^l).$$

By the induction hypothesis twice,

$$d_3 \vdash_{\mathsf{CSemFL}} \mathcal{H}' \mid \Sigma_{1,1}, \Sigma_{2,1}, \Gamma^{\mu_1+\mu_2+1} \Rightarrow \mathrm{e} \quad \text{and} \quad d_4 \vdash_{\mathsf{CSemFL}} \mathcal{H}' \mid \Sigma_{1,2}, \Sigma_{2,2}, \Pi \Rightarrow \Delta,$$

where $\mathcal{H}' = ((\mathcal{G}', \mathcal{H}_x)^{\mathcal{D}} \mid \Gamma, \Pi \Rightarrow \Delta)$. We first apply the rule $(\mathrm{e} \Rightarrow)$ to the last hypersequent of d_4, obtaining a derivation of $\mathcal{H}' \mid \Sigma_{1,2}, \Sigma_{2,2}, \Pi, \mathrm{e} \Rightarrow \Delta$. Then by (CUT) with the last hypersequent of d_3, we obtain a derivation

$$d_5 \vdash_{\mathsf{CSemFL}} \mathcal{H}' \mid \Sigma_1, \Sigma_2, \Gamma^{\mu_1+\mu_2+1}, \Pi \Rightarrow \Delta.$$

The required derivation is then

$$\dfrac{\dfrac{\dfrac{\dfrac{\vdots\, d_5}{\mathcal{H}' \mid \Sigma_1, \Sigma_2, \Gamma^{\mu_1+\mu_2+1}, \Pi \Rightarrow \Delta} \quad \dfrac{}{\mathcal{H}' \mid\ \Rightarrow \mathrm{e}}\,(\Rightarrow\mathrm{e})}{\mathcal{H}' \mid \Gamma, \Pi \Rightarrow \Delta \mid \Sigma_1, \Sigma_2, \Gamma^{\mu_1+\mu_2} \Rightarrow \mathrm{e}}\,(\textsc{com})}{\mathcal{H}' \mid \Sigma_1, \Sigma_2, \Gamma^{\mu_1+\mu_2} \Rightarrow \mathrm{e}}\,(\textsc{ec}) \qquad \dfrac{}{\mathcal{H}' \mid\ \Rightarrow \mathrm{e}}\,(\Rightarrow\mathrm{e})}{\mathcal{H}' \mid \Sigma_1, \Gamma^{\mu_1} \Rightarrow \mathrm{e} \mid \Sigma_2, \Gamma^{\mu_2} \Rightarrow \mathrm{e}.}\,(\textsc{com})$$

The remaining cases are all straightforward. □

5.3. Representations of distributive varieties

In this section, we investigate a class of residuated lattices that subsumes not only varieties of semilinear residuated lattices, but also ℓ-groups and their negative cones. Recall that in any residuated lattice, multiplication distributes over arbitrary joins, but may satisfy only the inequation $x(y \wedge z)w \leq xyw \wedge xzw$ for meets. Building on a more general result for ℓ-monoids (Theorem 5.3.3), we prove that a residuated lattice in which multiplication also distributes over binary meets can be represented via order-endomorphisms of a chain if and only if it is distributive (Corollary 5.3.6).

Let us call an algebra $\mathbf{M} = \langle M, \wedge, \vee, \cdot, \mathrm{e} \rangle$ of signature $\langle 2, 2, 2, 0 \rangle$ an *ℓ-monoid* (short for *lattice-ordered monoid*[2]) if $\langle M, \wedge, \vee \rangle$ is a lattice, $\langle M, \cdot, \mathrm{e} \rangle$ is a monoid, and $\mathbf{M}$ satisfies the equations

$$x(y \vee z)w \approx xyw \vee xzw \quad \text{and} \quad x(y \wedge z)w \approx xyw \wedge xzw.$$

We call an ℓ-monoid *distributive* if it has a distributive lattice reduct.

An important source of distributive ℓ-monoids is obtained by considering the set $\mathrm{End}(\mathbf{\Omega})$ of order-endomorphisms of a chain $\mathbf{\Omega} = \langle \Omega, \leq^{\mathbf{\Omega}} \rangle$. Clearly, $\mathrm{End}(\mathbf{\Omega})$ equipped with function composition and the identity map is a monoid, and a lattice-order on this monoid is obtained by defining for $\varphi, \psi \in \mathrm{End}(\mathbf{\Omega})$,

$$\varphi \leq \psi \;:\Longleftrightarrow\; \varphi(a) \leq^{\mathbf{\Omega}} \psi(a), \text{ for all } a \in \Omega.$$

The resulting structure is a distributive ℓ-monoid $\mathbf{End}(\mathbf{\Omega})$, where meets and joins are calculated pointwise.

The inverse-free reduct of any ℓ-group is a distributive ℓ-monoid. In particular, for any chain $\mathbf{\Omega}$, the ℓ-group $\mathbf{Aut}(\mathbf{\Omega})$ consists of the subalgebra of $\mathbf{End}(\mathbf{\Omega})$ containing all order-automorphisms of $\mathbf{\Omega}$ expanded with the unary function inverse operation.

Now consider any ℓ-monoid $\mathbf{L}$. For $\emptyset \neq H \subseteq L$ and $a \in L$, let

$$H /\!\!/ a := \{c \in L \mid ca \in H\}.$$

We obtain a binary relation Θ_H on L by defining, for $a, b \in L$,

$$a \,\Theta_H\, b \;:\Longleftrightarrow\; H /\!\!/ a = H /\!\!/ b.$$

Clearly, Θ_H is an equivalence relation on L. The following lemma collects some further useful properties of Θ_H for the case where H is a prime lattice-ideal of the lattice reduct of $\mathbf{L}$.

---

[2]Note that this terminology, which originated in the literature on ordered groups, conflicts with the definition of a pomonoid given in Chapter 1, since a pomonoid with a lattice-order need not be an ℓ-monoid.

LEMMA 5.3.1. *Let* $\mathbf{L}$ *be any* ℓ*-monoid with lattice reduct* $\mathbf{L}^*$ *and let* H *be any prime lattice-ideal of* $\mathbf{L}^*$.

(a) $\langle\{H/\!\!/a \mid a \in L\}, \subseteq\rangle$ *is a chain.*

(b) *The relation* Θ_H *is a lattice congruence of* $\mathbf{L}^*$ *that is compatible with left multiplication, i.e.,* $a \;\Theta_H\; b \implies ca \;\Theta_H\; cb$, *for all* $a, b, c \in L$.

(c) *The quotient lattice* $\mathbf{\Omega}_H := \mathbf{L}^*/\Theta_H$ *is a chain.*

(d) *The map* $\varphi_H\colon \mathbf{L} \to \mathbf{End}(\mathbf{\Omega}_H)$ *obtained by defining* $\varphi_H(a)([b]_{\Theta_H}) := [ab]_{\Theta_H}$, *for all* $a, b \in L$, *is a homomorphism.*

PROOF.

(a) Suppose that $H/\!\!/a \not\subseteq H/\!\!/b$, for some $a, b \in L$. Then there exists $c \in L$ such that $ca \in H$ and $cb \notin H$. Since $c(a \wedge b) \leq ca \in H$, also $c(a \wedge b) \in H$. To show that $H/\!\!/b \subseteq H/\!\!/a$, consider any $d \in H/\!\!/b$, i.e., $db \in H$. Then also $d(a \wedge b) \in H$ and

$$(c \vee d)a \wedge (c \vee d)b = (c \vee d)(a \wedge b) = c(a \wedge b) \vee d(a \wedge b) \in H.$$

But $(c \vee d)b \notin H$, since $(c \vee d)b \geq cb \notin H$, so, since H is a prime lattice-ideal, $(c \vee d)a \in H$. Hence also $da \in H$, i.e., $d \in Ha$. So $H/\!\!/b \subseteq H/\!\!/a$.

(b) We leave it to the reader to check that Θ_H is a lattice congruence of $\mathbf{L}^*$. For compatibility with left multiplication, consider $a, b, c \in L$ such that $a \;\Theta_H\; b$. Then $H/\!\!/ca = (H/\!\!/a)/\!\!/c = (H/\!\!/b)/\!\!/c = H/\!\!/cb$, that is, $ca \;\Theta_H\; cb$.

(c) Observe that, for any $a, b \in L$,

$$\begin{aligned}
[a]_{\Theta_H} \leq [b]_{\Theta_H} &\iff [a \vee b]_{\Theta_H} = [a]_{\Theta_H} \vee [b]_{\Theta_H} = [b]_{\Theta_H} \\
&\iff (c(a \vee b) \in H \iff cb \in H, \text{ for all } c \in L) \\
&\iff (ca \vee cb \in H \iff cb \in H, \text{ for all } c \in L) \\
&\iff (cb \in H \implies ca \in H, \text{ for all } c \in L) \\
&\iff H/\!\!/b \subseteq H/\!\!/a.
\end{aligned}$$

Hence $\mathbf{\Omega}_H := \mathbf{L}^*/\Theta_H$ is a chain, by (a).

(d) We prove first that φ_H is a well-defined map from L to $\mathrm{End}(\mathbf{\Omega}_H)$. Since Θ_H is compatible with left multiplication, by (b), clearly $[b]_{\Theta_H} = [c]_{\Theta_H}$ implies $[ab]_{\Theta_H} = [ac]_{\Theta_H}$, for any $a, b, c \in L$. Also, $\varphi_H(a)$ is an order-endomorphism of $\mathbf{\Omega}_H$, for each $a \in L$, since, for all $b, c \in L$,

$$\begin{aligned}
[b]_{\Theta_H} \leq [c]_{\Theta_H} &\implies H/\!\!/c \subseteq H/\!\!/b \\
&\implies H/\!\!/ac \subseteq H/\!\!/ab \\
&\implies \varphi_H(a)([b]_{\Theta_H}) = [ab]_{\Theta_H} \leq [ac]_{\Theta_H} = \varphi_H(a)([c]_{\Theta_H}).
\end{aligned}$$

The proof that φ_H is a homomorphism is left to the reader. □

Part (d) of the previous lemma admits the following converse.

LEMMA 5.3.2. *Let* $\mathbf{L}^*$ *be the lattice reduct of an* ℓ*-monoid* $\mathbf{L}$. *Then* $\mathbf{L}^*$ *has a prime lattice-ideal* $H \neq L$ *if and only if there exist a chain* $\mathbf{\Omega}$ *and non-trivial homomorphism* $\varphi\colon \mathbf{L} \to \mathbf{End}(\mathbf{\Omega})$.

PROOF. For the left-to-right direction, we show that for any prime lattice-ideal $H \neq L$ of $\mathbf{L}^*$, the homomorphism $\varphi_H : \mathbf{L} \to \mathbf{End}(\mathbf{\Omega}_H)$ provided by Lemma 5.3.1.(d) is non-trivial. Consider any $a \in H$ and $b \in L \backslash H$. Then $\mathrm{e} \in H /\!\!/ a$, but $\mathrm{e} \notin H /\!\!/ b$, so $\varphi_H(a)([\mathrm{e}]_{\Theta_H}) = [a]_{\Theta_H} \neq [b]_{\Theta_H} = \varphi_H(b)([\mathrm{e}]_{\Theta_H})$. Hence $\varphi_H(a) \neq \varphi_H(b)$.

For the converse, let $\varphi : \mathbf{L} \to \mathbf{End}(\mathbf{\Omega})$ be any non-trivial homomorphism, for some chain $\mathbf{\Omega}$. Then $\varphi(a) \neq \varphi(b)$, for some $a, b \in L$, and without loss of generality, $\varphi(a)(k) < \varphi(b)(k)$, for some $k \in \Omega$. Let

$$H_k := \{c \in L \mid \varphi(c)(k) \leq \varphi(a)(k)\}.$$

Then $a \in H_k$ and $b \notin H_k$. Moreover, $H_k \neq L$ is clearly closed downwards, closed under finite joins, and prime, and is therefore a prime lattice-ideal of $\mathbf{L}^*$. □

We are now able to give a precise characterization of the ℓ-monoids that admit a representation via order-endomorphisms of a chain.

THEOREM 5.3.3. *An ℓ-monoid $\mathbf{L}$ is distributive if and only if there exists a chain $\mathbf{\Omega}$ such that $\mathbf{L}$ embeds into $\mathbf{End}(\mathbf{\Omega})$.*

PROOF. The right-to-left direction follows immediately from the fact that $\mathbf{End}(\mathbf{\Omega})$ is a distributive ℓ-monoid, for any chain $\mathbf{\Omega}$. For the converse direction, let $\mathbf{L}$ be any distributive ℓ-monoid. We show first that $\mathbf{L}$ embeds into $\prod_{i \in I} \mathbf{End}(\mathbf{\Omega}_i)$, for some family of chains $\{\mathbf{\Omega}_i\}_{i \in I}$.

To this end, let I be the set of all prime lattice-ideals of the lattice reduct $\mathbf{L}^*$ of $\mathbf{L}$. It follows from Lemma 5.3.1 that $\mathbf{\Omega}_H := \mathbf{L}^*/\Theta_H$ is a chain, for each $H \in I$, and the map $\varphi_H : \mathbf{L} \to \mathbf{End}(\mathbf{\Omega}_H)$ defined by setting $\varphi_H(a)([b]_{\Theta_H}) := [ab]_{\Theta_H}$, for $a, b \in L$, is a homomorphism. Observe also that if $\varphi_H(a) = \varphi_H(b)$, then $[a]_{\Theta_H} = [a\mathrm{e}]_{\Theta_H} = \varphi_H(a)([\mathrm{e}]_{\Theta_H}) = \varphi_H(b)([\mathrm{e}]_{\Theta_H}) = [b\mathrm{e}]_{\Theta_H} = [b]_{\Theta_H}$, so $a \;\Theta_H\; b$. Hence $\ker(\varphi_H) \subseteq \Theta_H$.

Consider now the induced homomorphism

$$\varphi : \mathbf{L} \to \prod_{H \in I} \mathbf{End}(\mathbf{\Omega}_H); \quad a \mapsto \langle \varphi_H(a) \rangle_{H \in I}.$$

For any $a, b \in L$ with $a \neq b$, there exists, by the Prime ideal separation theorem for distributive lattices (see Section 1.1), a prime lattice-ideal H of $\mathbf{L}^*$ satisfying $a \in H$ and $b \notin H$, and hence $\langle a, b \rangle \notin \Theta_H$. So $\ker(\varphi) = \bigcap_{H \in I} \ker(\varphi_H) \subseteq \bigcap_{H \in I} \Theta_H = \Delta_L$ and φ is an embedding.

Finally, let us assume harmlessly that the chains $\mathbf{\Omega}_H$ $(H \in I)$ are disjoint and that I is totally ordered by $\preccurlyeq$. We order $\Omega := \bigcup_{H \in I} \Omega_H$ lexicographically by

$$a \leq^{\mathbf{\Omega}} b \;:\Longleftrightarrow\; (a \in \Omega_H,\, b \in \Omega_K, \text{ and } H \prec K) \text{ or } (a, b \in \Omega_H \text{ and } a \leq^{\mathbf{\Omega}_H} b).$$

Then it is easily checked that $\psi : \prod_{H \in I} \mathbf{End}(\mathbf{\Omega}_H) \to \mathbf{End}(\mathbf{\Omega})$ defined by

$$\psi(\varphi)(a) = \varphi(H)(a), \text{ for all } a \in \Omega_H \text{ and } H \in I,$$

is an embedding. □

For ℓ-groups, we obtain the following stronger result.

COROLLARY 5.3.4. *Every ℓ-group embeds into $\mathbf{Aut}(\mathbf{\Omega})$, for some chain $\mathbf{\Omega}$.*

PROOF. By Proposition 4.2.5, every ℓ-group $\mathbf{L}$ is distributive, and embeds as an ℓ-monoid into $\mathbf{End}(\mathbf{\Omega})$, for some chain $\mathbf{\Omega}$, by Theorem 5.3.3. This embedding must also preserve the inverse operation of $\mathbf{L}$, however, and hence map each element of $\mathbf{L}$ to a bijective member of $\mathbf{End}(\mathbf{\Omega})$. So $\mathbf{L}$ embeds as an ℓ-group into $\mathbf{Aut}(\mathbf{\Omega})$. □

Next, to extend these results to residuated lattices, we restrict our attention to the order-endomorphisms of a chain that are residuated. Observe first that the residuated maps on a chain $\mathbf{\Omega}$ form a subalgebra $\mathbf{Res}(\mathbf{\Omega})$ of $\mathbf{End}(\mathbf{\Omega})$ and that, for any $\psi_1, \psi_2 \in \mathrm{Res}(\mathbf{\Omega})$, writing φ^* for the residual of any $\varphi \in \mathrm{Res}(\mathbf{\Omega})$,

$$(\psi_1\psi_2)^* = \psi_1^*\psi_2^*, \quad (\psi_1 \vee \psi_2)^* = \psi_1^* \wedge \psi_2^*, \quad \text{and} \quad (\psi_1 \wedge \psi_2)^* = \psi_1^* \vee \psi_2^*.$$

Recall that the set of order-ideals of a chain $\mathbf{\Omega}$ forms a complete chain $\mathcal{L}(\mathbf{\Omega})$ under set-inclusion. Recall also from Example 1.2.10 that every order-endomorphism φ of a poset $\mathbf{P}$ induces a residuated map $\hat{\varphi}$ on $\mathcal{L}(\mathbf{P})$ and that the correspondence $\varphi \mapsto \hat{\varphi}$ is bijective. Hence we obtain the following result.

LEMMA 5.3.5. *For any chain $\mathbf{\Omega}$, the ℓ-monoid $\mathbf{End}(\mathbf{\Omega})$ embeds into $\mathbf{Res}(\mathcal{L}(\mathbf{\Omega}))$.*

Combining this lemma with Theorem 5.3.3 yields a more refined representation result for distributive ℓ-monoids.

COROLLARY 5.3.6. *An ℓ-monoid $\mathbf{L}$ is distributive if and only if it embeds into $\mathbf{Res}(\mathbf{\Omega})$ as an ℓ-monoid, for some chain $\mathbf{\Omega}$. In particular, the residual-free reduct of a residuated lattice is a distributive ℓ-monoid if and only if it embeds into $\mathbf{Res}(\mathbf{\Omega})$, for some chain $\mathbf{\Omega}$.*

5.4. Generation and decidability results

By Theorem 5.3.3 and Corollary 5.3.4, the classes of distributive ℓ-monoids and ℓ-groups are generated as varieties by the endomorphism ℓ-monoids of a chain and automorphism ℓ-groups of a chain, respectively. In this section, we refine these generation results, establish a correspondence between the equational theories of the two varieties, and show that both have a decidable equational theory.

Let us begin with some useful notation and terminology. Recall first that a binary relation $\preccurlyeq$ on a set P that is reflexive and transitive is called a *preorder* on P, and induces an equivalence relation on P defined by setting, for $a, b \in P$,

$$a \sim_{\preccurlyeq} b \;:\Longleftrightarrow\; a \preccurlyeq b \text{ and } b \preccurlyeq a.$$

We also write $a \prec b$ to denote that $a \preccurlyeq b$ and $b \not\preccurlyeq a$, and call $\preccurlyeq$ *total* if $a \preccurlyeq b$ or $b \preccurlyeq a$, for all $a, b \in P$.

Let us denote the formula algebras for monoids, groups, distributive ℓ-monoids, and ℓ-groups over a set X by $\mathbf{Fm}_m(X)$, $\mathbf{Fm}_g(X)$, $\mathbf{Fm}_d(X)$, and $\mathbf{Fm}_\ell(X)$, respectively, and corresponding free algebras by $\mathbf{F}_m(X)$, $\mathbf{F}_g(X)$, $\mathbf{F}_d(X)$, and $\mathbf{F}_\ell(X)$, assuming for convenience that $\mathrm{F}_m(X) \subseteq \mathrm{Fm}_m(X)$, $\mathrm{F}_g(X) \subseteq \mathrm{Fm}_g(X)$, $\mathrm{F}_d(X) \subseteq \mathrm{Fm}_d(X)$, and $\mathrm{F}_\ell(X) \subseteq \mathrm{Fm}_\ell(X)$.

We call a preorder $\preccurlyeq$ on a set of monoid formulas $T \subseteq \mathrm{F}_m(X)$ *left-X-invariant* if whenever $x \in X$, $\alpha \preccurlyeq \beta$, and $x\alpha, x\beta \in T$, also $x\alpha \preccurlyeq x\beta$, and *strictly left-$X$-invariant* if $\preccurlyeq$ is left X-invariant and whenever $x \in X$, $\alpha \prec \beta$, and $x\alpha, x\beta \in T$, also

$x\alpha \prec x\beta$. For any $S \subseteq \mathrm{F}_m(X)^2$, we define

$$\mathrm{sf}(S) := \{\gamma \in \mathrm{F}_m(X) \mid \langle \alpha\gamma, \beta\rangle \in S \text{ or } \langle \alpha, \beta\gamma\rangle \in S, \text{ for some } \alpha, \beta \in \mathrm{F}_m(X)\}.$$

LEMMA 5.4.1. *Let S be any finite subset of $\mathrm{F}_m(X)^2$ and let $\preccurlyeq$ be a total left-X-invariant preorder on $\mathrm{sf}(S)$ satisfying $\alpha \prec \beta$, for each $\langle \alpha, \beta\rangle \in S$.*

(a) *There exist a chain $\langle \Omega, \leq\rangle$ with $|\Omega| \leq |\mathrm{sf}(S)|$, $p \in \Omega$, and a homomorphism $\varphi\colon \mathbf{Fm}_d(X) \to \mathbf{End}(\langle \Omega, \leq\rangle)$ such that $\varphi(\alpha)(p) < \varphi(\beta)(p)$, for all $\langle \alpha, \beta\rangle \in S$.*

(b) *If $\preccurlyeq$ is strictly left-X-invariant, then there exist $q \in \mathbb{Q}$ and a homomorphism $\psi\colon \mathbf{Fm}_\ell(X) \to \mathbf{Aut}(\langle \mathbb{Q}, \leq\rangle)$ such that $\psi(\alpha)(q) < \psi(\beta)(q)$, for all $\langle \alpha, \beta\rangle \in S$.*

PROOF.

(a) Let $[\alpha]$ denote the $\sim_{\preccurlyeq}$-equivalence class of $\alpha \in \mathrm{sf}(S)$ and $\Omega := \{[\alpha] \mid \alpha \in \mathrm{sf}(S)\}$, noting that $|\Omega| \leq |\mathrm{sf}(S)|$. If $[\alpha] = [\alpha']$, $[\beta] = [\beta']$, and $\alpha \preccurlyeq \beta$, then $\alpha' \preccurlyeq \beta'$, so we can define for $[\alpha], [\beta] \in \Omega$,

$$[\alpha] \leq [\beta] \ :\Longleftrightarrow\ \alpha \preccurlyeq \beta.$$

Clearly, $\leq$ is a total order on Ω and $[\alpha] < [\beta]$, for each $\langle \alpha, \beta\rangle \in S$. Moreover, if $[\alpha], [\beta] \in \Omega$, $x \in X$, and $x\alpha, x\beta \in \mathrm{sf}(S)$, then, by the left-$X$-invariance of $\preccurlyeq$,

$$[\alpha] \leq [\beta] \implies [x\alpha] \leq [x\beta].$$

In particular, if $[\alpha] = [\beta] \in \Omega$, $x \in X$, and $x\alpha, x\beta \in \mathrm{sf}(S)$, then $[x\alpha] = [x\beta]$. Hence, for each $x \in X$, we obtain a partial map $\tilde{\chi}_x\colon \Omega \to \Omega$ by defining $\tilde{\chi}_x([\alpha]) := [x\alpha]$, whenever $x\alpha \in \mathrm{sf}(S)$. Moreover, each of these partial maps $\tilde{\chi}_x$ extends to an order-endomorphism $\chi_x\colon \Omega \to \Omega$ of $\langle \Omega, \leq\rangle$. Now let $\varphi\colon \mathbf{Fm}_d(X) \to \mathbf{End}(\langle \Omega, \leq\rangle)$ be the homomorphism extending the assignment $x \mapsto \chi_x$. Then $\varphi(\alpha)([\mathrm{e}]) = [\alpha]$, for every $\alpha \in \mathrm{sf}(S)$, and hence $\varphi(\alpha)([\mathrm{e}]) < \varphi(\beta)([\mathrm{e}])$, for all $\langle \alpha, \beta\rangle \in S$, as required.

For (b), suppose now that $\preccurlyeq$ is strictly left-X-invariant. Then the partial maps $\tilde{\chi}_x\colon \Omega \to \Omega$ defined in (a) are injective. Hence, identifying $\langle \Omega, \leq\rangle$ with a subchain of $\langle \mathbb{Q}, \leq\rangle$, each $\tilde{\chi}_x$ extends to an order-automorphism $\chi'_x\colon \mathbb{Q} \to \mathbb{Q}$ of $\langle \mathbb{Q}, \leq\rangle$. Finally, the homomorphism $\psi\colon \mathbf{Fm}_\ell(X) \to \mathbf{Aut}(\langle \mathbb{Q}, \leq\rangle)$ extending the assignment $x \mapsto \chi'_x$ satisfies $\psi(\alpha)([\mathrm{e}]) < \psi(\beta)([\mathrm{e}])$, for all $\langle \alpha, \beta\rangle \in S$, as required. □

To check the satisfaction of equations in $\mathcal{DLM}$, it suffices, using the distributivity of multiplication over binary meets and joins, and lattice distributivity, to consider equations of the form $\bigwedge_{i=1}^{n} \beta_i \leq \bigvee_{j=1}^{m} \alpha_j$, where $\beta_1, \dots, \beta_n, \alpha_1, \dots, \alpha_m$ are $\{\cdot, \mathrm{e}\}$-formulas. The following result shows that it suffices, in fact, to check the satisfaction of these equations in ℓ-monoids of order-endomorphisms of finite chains.

PROPOSITION 5.4.2. *Let ε be any equation of the form $\bigwedge_{i=1}^{n} \beta_i \leq \bigvee_{j=1}^{m} \alpha_j$, for some $\beta_1, \dots, \beta_n, \alpha_1, \dots, \alpha_m \in \mathrm{F}_m(X)$ and let $S := \{\langle \alpha_j, \beta_i\rangle \mid i \in \{1, \dots, n\}, j \in \{1, \dots, m\}\}$. Then ε is satisfied by all distributive ℓ-monoids if and only if it is satisfied by $\mathbf{End}(\langle \Omega, \leq\rangle)$, for every chain $\langle \Omega, \leq\rangle$ satisfying $|\Omega| \leq |\mathrm{sf}(S)|$.*

PROOF. Suppose that $\mathcal{DLM} \not\models \varepsilon$. Then, by Theorem 5.3.3, there exist a chain $\langle \Omega, \leq\rangle$, a homomorphism $\varphi\colon \mathbf{Fm}_d(X) \to \mathbf{End}(\langle \Omega, \leq\rangle)$, and some $p \in \Omega$ such that $\bigwedge_{i=1}^{n} \varphi(\beta_i)(p) > \bigvee_{j=1}^{m} \varphi(\alpha_j)(p)$. So $\varphi(\alpha)(p) < \varphi(\beta)(p)$, for each $\langle \alpha, \beta\rangle \in S$, and we define, for all $\alpha, \beta \in \mathrm{sf}(S)$,

$$\alpha \preccurlyeq \beta \ :\Longleftrightarrow\ \varphi(\alpha)(p) \leq \varphi(\beta)(p).$$

Clearly, $\preccurlyeq$ is a total preorder satisfying $\alpha \prec \beta$, for each $\langle \alpha, \beta \rangle \in S$. Also, since φ is a homomorphism, $\preccurlyeq$ is left-X-invariant on $\mathrm{sf}(S)$. Hence, by Lemma 5.4.1.(a), there exist a chain $\langle \Omega^*, \leq \rangle$ with $|\Omega^*| \leq |\mathrm{sf}(S)|$, $p^* \in \Omega^*$, and a homomorphism $\varphi^* \colon \mathbf{Fm}_d(X) \to \mathbf{End}(\langle \Omega^*, \leq \rangle)$ such that $\varphi^*(\alpha)(p^*) < \varphi^*(\beta)(p^*)$, for all $\langle \alpha, \beta \rangle \in S$. But then $\varphi^*(\bigwedge_{i=1}^n \beta_i)(p^*) > \varphi^*(\bigvee_{j=1}^m \alpha_j)(p^*)$, so $\mathbf{End}(\langle \Omega^*, \leq \rangle) \not\models \varepsilon$ as required. □

It follows from Proposition 5.4.2 that, given any $\{\wedge, \vee, \cdot, \mathrm{e}\}$-equation ε, a natural number k can be computed such that $\mathcal{DLM} \models \varepsilon$ if and only if ε is satisfied by the finite ℓ-monoid of order-endomorphisms of a chain with k elements. Since this latter property can be decided in finite time, we obtain the following result.

COROLLARY 5.4.3. *The equational theory of distributive ℓ-monoids is decidable.*

Moreover, the HSP theorem (Theorem B.12) yields a more refined generation result for $\mathcal{DLM}$.

COROLLARY 5.4.4. *The class of distributive ℓ-monoids is generated as a variety by the class of ℓ-monoids* $\mathbf{End}(\mathbf{\Omega})$ *such that* $\mathbf{\Omega}$ *is a finite chain.*

Corollary 5.4.4 states that $\mathcal{DLM}$ has the *finite model property*, one of the central notions studied in Chapter 9. Indeed, as will be explained carefully in that chapter, Corollary 5.4.3 is a direct consequence of Corollary 5.4.4 and the fact that $\mathcal{DLM}$ is finitely based.

Let us turn our attention now to ℓ-groups.

PROPOSITION 5.4.5. *The class of ℓ-groups is generated as a variety by* $\mathbf{Aut}(\langle \mathbb{Q}, \leq \rangle)$.

PROOF. It suffices, using the HSP theorem (Theorem B.12) and the distributivity properties of ℓ-groups, to show that $\mathcal{LG} \not\models \varepsilon$ implies $\mathbf{Aut}(\langle \mathbb{Q}, \leq \rangle) \not\models \varepsilon$, for every equation ε of the form $\mathrm{e} \leq \alpha_1 \vee \cdots \vee \alpha_n$, where $\alpha_1, \ldots, \alpha_n \in \mathrm{F}_g(X)$. Suppose that $\mathcal{LG} \not\models \varepsilon$. By Corollary 5.3.4, there exist a chain $\langle \Omega, \preccurlyeq \rangle$, $a \in \Omega$, and a homomorphism $\varphi \colon \mathbf{Fm}_\ell(X) \to \mathbf{Aut}(\langle \Omega, \preccurlyeq \rangle)$ such that $\varphi(\alpha_i)(q) \prec q$, for each $i \in \{1, \ldots, n\}$.

Let $\mathrm{sf}(\varepsilon)$ be the finite set consisting of all formulas $\beta \in \mathrm{F}_g(X)$ that end α_i, for some $i \in \{1, \ldots, n\}$, that is, such that $\alpha_i = \gamma\beta$, for some $\gamma \in \mathrm{F}_g(X)$, where the product of γ and β is taken in $\mathbf{Fm}_g(X)$. Now let $T := \{\varphi(\beta)(q) \mid \beta \in \mathrm{sf}(\varepsilon)\}$. Since T is finite, we may assume that $\langle T, \preccurlyeq \cap T^2 \rangle$ is a subposet of $\langle \mathbb{Q}, \leq \rangle$.

Consider any $x \in X$. Since $\varphi(x)$ is an order-automorphism of $\langle \Omega, \preccurlyeq \rangle$, we obtain a partial map $\tilde{\chi}_x \colon \mathbb{Q} \to \mathbb{Q}$ by fixing $\tilde{\chi}_x(\varphi(\beta)(q)) := \varphi(x\beta)(q)$, whenever $\beta, x\beta \in \mathrm{sf}(\varepsilon)$. Also, $\tilde{\chi}_x$ extends easily (e.g., linearly) to an order-automorphism χ_x of $\langle \mathbb{Q}, \leq \rangle$.

Let $\psi \colon \mathbf{Fm}_\ell(X) \to \mathbf{Aut}(\langle \mathbb{Q}, \leq \rangle)$ be the homomorphism extending the assignment $x \mapsto \chi_x$. Then $\psi(\alpha)(q) = \varphi(\alpha)(q)$, for every $\alpha \in \mathrm{sf}(\varepsilon)$, and hence $\psi(\alpha_i)(q) = \varphi(\alpha_i)(q) \prec q$, for each $i \in \{1, \ldots, n\}$. So $\mathbf{Aut}(\langle \mathbb{Q}, \leq \rangle) \not\models \varepsilon$. □

We prove now that the equational theory of $\mathcal{DLM}$ coincides with the equational theory of the class of ℓ-monoid reducts of ℓ-groups, or, equivalently, by the HSP theorem (Theorem B.12), that $\mathcal{DLM}$ is generated as a variety by this class.

LEMMA 5.4.6. *For any total left-X-invariant preorder $\preccurlyeq$ on $P \subseteq \mathrm{F}_m(X)$, there exists a total strictly left-X-invariant preorder $\trianglelefteq$ on P such that $\prec \subseteq \lhd$.*

PROOF. We define the following relations on P:

$$\begin{aligned} x_1\cdots x_k \lhd y_1\cdots y_l &:\Longleftrightarrow \text{ some } j\in\mathbb{N} \text{ satisfies } x_i\cdots x_k \sim_{\preccurlyeq} y_i\cdots y_l, \text{ for all } i<j, \\ &\qquad \text{and } (x_j\cdots x_k \prec y_j\cdots y_l \text{ or } j=k+2\le l+1) \\ x_k\cdots x_1 \equiv y_l\cdots y_1 &:\Longleftrightarrow k=l \text{ and } x_i\cdots x_k \sim_{\preccurlyeq} y_i\cdots y_l, \text{ for each } i\le k \\ \alpha \trianglelefteq \beta &:\Longleftrightarrow \alpha\lhd\beta \text{ or } \alpha\equiv\beta. \end{aligned}$$

Observe that setting $j=1$ in the definition of $\lhd$ yields $\prec\,\subseteq\,\lhd$. Also, $\alpha\lhd\beta$ implies $\alpha\not\equiv\beta$. It is straightforward to check that $\lhd$ is transitive and irreflexive, using the fact that $\preccurlyeq$ is a preorder. For the transitivity of $\trianglelefteq$, there are several cases. Clearly, if $\alpha\lhd\beta$ and $\beta\lhd\gamma$, then $\alpha\lhd\gamma$, by the transitivity of $\lhd$. If $\alpha\lhd\beta$ and $\beta\equiv\gamma$, then $\alpha\lhd\gamma$, using the definition of $\lhd$ and $\equiv$ and the transitivity of $\sim_{\preccurlyeq}$ and $\prec$. Similarly, if $\alpha\equiv\beta$ and $\beta\lhd\gamma$, then $\alpha\lhd\gamma$. Finally, if $\alpha\equiv\beta$ and $\beta\equiv\gamma$, then $\alpha\equiv\gamma$, by the transitivity of $\sim_{\preccurlyeq}$. Moreover, $\trianglelefteq$ is reflexive, since $\alpha\equiv\alpha$, for any $\alpha\in P$, so $\trianglelefteq$ is a preorder. Since $\preccurlyeq$ is total, $\alpha\not\lhd\beta$ and $\beta\not\lhd\alpha$ implies $\alpha\equiv\beta$; so $\trianglelefteq$ is total. Note also that $\alpha\lhd\beta$ if and only if $\alpha\trianglelefteq\beta$ and $\beta\ntrianglelefteq\alpha$ as suggested by the notation.

To prove that $\trianglelefteq$ is strictly left-X-invariant on P, consider $x\in X$ and $\alpha,\beta\in P$ such that $\alpha\trianglelefteq\beta$ and $x\alpha,x\beta\in P$. Suppose first that $\alpha\equiv\beta$, so α and β have the same length and $\alpha\sim_{\preccurlyeq}\beta$. Then $x\alpha$ and $x\beta$ have the same length and, since $\preccurlyeq$ is left-X-invariant, $x\alpha\sim_{\preccurlyeq}x\beta$. So $x\alpha\equiv x\beta$ and $x\alpha\trianglelefteq x\beta$. Now suppose that $\alpha\lhd\beta$. If $x\alpha\prec x\beta$, then $x\alpha\lhd x\beta$. Also, if $x\alpha\sim_{\preccurlyeq}x\beta$, then, since $\alpha\lhd\beta$, the definition of $\lhd$ gives $x\alpha\lhd x\beta$. Finally, suppose towards a contradiction that $x\alpha\not\preccurlyeq x\beta$. Since $\preccurlyeq$ is left-X-invariant, $\alpha\not\preccurlyeq\beta$. But then, since $\preccurlyeq$ is total, $\beta\prec\alpha$ and so $\beta\lhd\alpha$, contradicting $\alpha\lhd\beta$. □

PROPOSITION 5.4.7. *A $\{\wedge,\vee,\cdot,\mathrm{e}\}$-equation is satisfied by all distributive ℓ-monoids if and only if it is satisfied by all ℓ-groups.*

PROOF. The left-to-right direction follows from the fact that the $\{\wedge,\vee,\cdot,\mathrm{e}\}$-reduct of any ℓ-group is a distributive ℓ-monoid. For the right-to-left direction suppose, contrapositively, without loss of generality, that $\mathcal{DLM}\not\models\bigwedge_{i=1}^n\beta_i\le\bigvee_{j=1}^m\alpha_j$, where $\alpha_j,\beta_i\in\mathrm{F_m}(X)$ for $i\in\{1,\ldots,n\}$, $j\in\{1,\ldots,m\}$. By Theorem 5.3.3, there exist a chain $\langle\Omega,\le\rangle$, $p\in\Omega$, and a homomorphism $\varphi\colon\mathbf{Fm}_d(X)\to\mathbf{End}(\langle\Omega,\le\rangle)$ such that $\bigwedge_{i=1}^n\varphi(\beta_i)(p)>\bigvee_{j=1}^m\varphi(\alpha_j)(p)$. Let $S:=\{\langle\alpha_j,\beta_i\rangle\mid i\in\{1,\ldots,n\},\,j\in\{1,\ldots,m\}\}$. Then $\varphi(\alpha)(p)<\varphi(\beta)(p)$, for each $\langle\alpha,\beta\rangle\in S$, and we define, for $\alpha,\beta\in\mathrm{sf}(S)$,

$$\alpha\preccurlyeq\beta \;:\Longleftrightarrow\; \varphi(\alpha)(p)\le\varphi(\beta)(p).$$

Clearly, $\preccurlyeq$ is a total preorder on $\mathrm{sf}(S)$ satisfying $\alpha\prec\beta$, for each $\langle\alpha,\beta\rangle\in S$. Moreover, since φ is a homomorphism, $\preccurlyeq$ is left-X-invariant on $\mathrm{sf}(S)$.

By Lemma 5.4.6, there exists a total strictly left-X-invariant preorder $\trianglelefteq$ on $\mathrm{sf}(S)$ such that $\prec\,\subseteq\,\lhd$. In particular, $\alpha\lhd\beta$, for each $\langle\alpha,\beta\rangle\in S$. But then, by Lemma 5.4.1.(b), there exist a homomorphism $\psi\colon\mathbf{Fm}_\ell(X)\to\mathbf{Aut}(\langle\mathbb{Q},\le\rangle)$ and $q\in\mathbb{Q}$ such that $\psi(\alpha_j)(q)<\psi(\beta_i)(q)$, for all $i\in\{1,\ldots,n\}$, $j\in\{1,\ldots,m\}$. Hence $\mathbf{Aut}(\langle\mathbb{Q},\le\rangle)\not\models\bigwedge_{i=1}^n\beta_i\le\bigvee_{j=1}^m\alpha_j$. □

We show next that checking the satisfaction of any equation in $\mathcal{LG}$ can be reduced — via a process of iteratively eliminating inverses — to checking the satisfaction of finitely many equations in $\mathcal{DLM}$.

LEMMA 5.4.8. *Let* $\alpha, \beta, \gamma, \gamma_l, \gamma_r, \delta \in \mathrm{Fm}_\ell(X)$ *and* $y \notin X$.

(a) $\mathcal{LG} \models \mathrm{e} \leq \gamma \vee \alpha\beta \iff \mathcal{LG} \models \mathrm{e} \leq \gamma \vee \alpha y^{-1} \vee y\beta$.

(b) $\mathcal{LG} \models \gamma_l \leq \gamma_r \vee \alpha\delta^{-1}\beta \iff \mathcal{LG} \models \gamma_l y\delta \leq \gamma_r y\delta \vee \alpha \vee \gamma_l y\beta y\delta$.

PROOF.

(a) The left-to-right direction follows from the fact that every ℓ-group satisfies the quasiequation $\mathrm{e} \leq xy \vee z \implies \mathrm{e} \leq x \vee y \vee z$. For the converse, suppose that $\mathcal{LG} \not\models \mathrm{e} \leq \gamma \vee \alpha\beta$. Then $\mathbf{Aut}(\langle\mathbb{Q}, \leq\rangle) \not\models \mathrm{e} \leq \gamma \vee \alpha\beta$, by Proposition 5.4.5. Hence there exist a homomorphism $\varphi\colon \mathbf{Fm}_\ell(X) \to \mathbf{Aut}(\langle\mathbb{Q}, \leq\rangle)$ and $q \in \mathbb{Q}$ such that $\varphi(\gamma)(q) < q$ and $\varphi(\alpha\beta)(q) < q$. Let $p_1 < q < p_2$, for some $p_1, p_2 \in \mathbb{Q}$. Since $\varphi(\beta)(q) < \varphi(\alpha^{-1})(q)$ and $p_1 < p_2$, there exists an order-automorphism $\widehat{\varphi}(y)$ of $\langle\mathbb{Q}, \leq\rangle$ mapping $\varphi(\beta)(q)$ to p_1 and $\varphi(\alpha^{-1})(q)$ to p_2. Now let $\widehat{\varphi}(x) := \varphi(x)$, for each $x \in X$, to obtain a homomorphism $\widehat{\varphi}\colon \mathbf{Fm}_\ell(X \cup \{y\}) \to \mathbf{Aut}(\langle\mathbb{Q}, \leq\rangle)$ satisfying $q > \widehat{\varphi}(\gamma)(q)$, $q > \widehat{\varphi}(\alpha y^{-1})(q)$, and $q > \widehat{\varphi}(y\beta)(q)$. Hence $\mathcal{LG} \not\models \mathrm{e} \leq \gamma \vee \alpha y^{-1} \vee y\beta$.

(b) We apply (a) to obtain

$$\begin{aligned}\mathcal{LG} \models \gamma_l \leq \gamma_r \vee \alpha\delta^{-1}\beta &\iff \mathcal{LG} \models \mathrm{e} \leq \gamma_l{}^{-1}\gamma_r \vee \gamma_l{}^{-1}\alpha\delta^{-1}\beta\\ &\iff \mathcal{LG} \models \mathrm{e} \leq \gamma_l{}^{-1}\gamma_r \vee \gamma_l{}^{-1}\alpha\delta^{-1}y^{-1} \vee y\beta\\ &\iff \mathcal{LG} \models \gamma_l y\delta \leq \gamma_r y\delta \vee \alpha \vee \gamma_l y\beta y\delta.\end{aligned}$$

□

PROPOSITION 5.4.9. *Let* ε *be any* $\{\wedge, \vee, \cdot, ^{-1}, \mathrm{e}\}$*-equation with variables in a set* X. *Then a finite set* Y *and finite set of* $\{\wedge, \vee, \cdot, \mathrm{e}\}$*-equations* Σ *with variables in* $X \cup Y$ *can be effectively constructed such that* ε *is satisfied by all* ℓ*-groups if and only if the equations in* Σ *are satisfied by all distributive* ℓ*-monoids.*

PROOF. By Proposition 5.4.7, it suffices to define an algorithm that, given any $\alpha_0 \in \mathrm{Fm}_m(X)$ and $\alpha_1, \ldots, \alpha_n \in \mathrm{Fm}_g(X)$, constructs $\beta_0, \beta_1, \ldots, \beta_m \in \mathrm{Fm}_m(X \cup Y)$, for some finite set Y, such that

$$\mathcal{LG} \models \alpha_0 \leq \alpha_1 \vee \cdots \vee \alpha_n \iff \mathcal{LG} \models \beta_0 \leq \beta_1 \vee \cdots \vee \beta_m.$$

Without loss of generality, let $\alpha_1 = \gamma x^{-1}\delta$. By Lemma 5.4.8, for any $y \notin X$,

$$\mathcal{LG} \models \alpha_0 \leq \alpha_1 \vee \cdots \vee \alpha_n \iff \mathcal{LG} \models \alpha_0 yx \leq \gamma \vee \alpha_0 y\delta yx \vee \alpha_2 yx \vee \cdots \vee \alpha_n yx.$$

The equation $\alpha_0 yx \leq \gamma \vee \alpha_0 y\delta yx \vee \alpha_2 yx \vee \cdots \vee \alpha_n yx$ contains fewer inverses than $\alpha_0 \leq \alpha_1 \vee \cdots \vee \alpha_n$, so iterating this procedure produces a $\{\wedge, \vee, \cdot, \mathrm{e}\}$-equation after finitely many steps. □

Combining this last proposition with Corollary 5.4.3, yields the following result.

COROLLARY 5.4.10. *The equational theory of* ℓ*-groups is decidable.*

We conclude this section by considering the variety $\mathcal{AbLG}$ of Abelian ℓ-groups.

PROPOSITION 5.4.11. *The class of Abelian* ℓ*-groups is generated as a quasivariety by the ordered group* $\mathbf{Z} = \langle\mathbb{Z}, \min, \max, +, -, 0\rangle$.

PROOF. It suffices to prove that $\{\alpha\} \vdash_{\mathbf{Z}} \beta$ implies $\{\alpha\} \vdash_{\mathcal{AbLG}} \beta$, for any $\alpha, \beta \in \mathrm{Fm}$, proceeding by induction on $|\mathrm{Var}(\beta)|$. Suppose first that $\mathrm{Var}(\beta) = \emptyset$. Then, clearly, $\vdash_{\mathcal{AbLG}} \beta$, and hence $\{\alpha\} \vdash_{\mathcal{AbLG}} \beta$.

For the inductive step, suppose that $x \in \mathrm{Var}(\beta)$ and $\{\alpha\} \vdash_{\mathbf{Z}} \beta$. Using distributivity properties of $\mathcal{AbLG}$, we may assume that $\alpha = \alpha_1 \vee \cdots \vee \alpha_m$ and $\beta = \beta_1 \wedge \cdots \wedge \beta_n$, where each α_i is a meet of $\{\cdot, ^{-1}, \mathrm{e}\}$-formulas and each β_j is a join of $\{\cdot, ^{-1}, \mathrm{e}\}$-formulas. Fix $i \in \{1, \dots, m\}$ and $j \in \{1, \dots, n\}$. Then $\{\alpha_i\} \vdash_{\mathbf{Z}} \beta_j$ and it suffices to prove that $\{\alpha_i\} \vdash_{\mathcal{AbLG}} \beta_j$. If $x \notin \mathrm{Var}(\beta_j)$, then $\{\alpha_i\} \vdash_{\mathcal{AbLG}} \beta_j$, by the induction hypothesis. Suppose then that $x \in \mathrm{Var}(\beta_j)$. Note now that, for any class $\mathcal{K}$ of Abelian ℓ-groups, $\gamma, \gamma_1, \gamma_2, \delta, \delta_1, \delta_2 \in \mathrm{Fm}$, and $k \in \mathbb{N}^{>0}$,

$$\{\delta\} \vdash_{\mathcal{K}} \gamma_1 \vee \gamma_2 \iff \{\delta\} \vdash_{\mathcal{K}} \gamma_1 \vee \gamma_2^k \quad \text{and} \quad \{\delta_1 \wedge \delta_2\} \vdash_{\mathcal{K}} \gamma \iff \{\delta_1 \wedge \delta_2^k\} \vdash_{\mathcal{K}} \gamma.$$

Hence we may assume, without loss of generality, that β_j is γ' or $\gamma_0 \vee \gamma'$, where γ' is the join of at least one of $\gamma_1 x^k$ and $\gamma_2 x^{-k}$, for some $k \in \mathbb{N}^{>0}$ and $\gamma_0, \gamma_1, \gamma_2 \in \mathrm{Fm}$ with $x \notin \mathrm{Var}(\gamma_0, \gamma_1, \gamma_2)$. Similarly, we may assume that α_i is the meet of δ_0 and zero or more of $\delta_1 x^k$ and $\delta_2 x^{-k}$, for some $\delta_0, \delta_1, \delta_2 \in \mathrm{Fm}$ with $x \notin \mathrm{Var}(\delta_0, \delta_1, \delta_2)$.

Note that some combinations are not possible. If $\beta_j = \gamma_1 x^k$ and α_i does not contain $\delta_1 x^k$, then an evaluation mapping x to -1 and all other variables to 0 yields a contradiction to $\{\alpha_i\} \vdash_{\mathbf{Z}} \beta_j$. Similarly, if $\beta_j = \gamma_2 x^{-k}$ and α_i does not contain $\delta_2 x^{-k}$, then an evaluation mapping x to 1 and all other variables to 0 yields a contradiction to $\{\alpha_i\} \vdash_{\mathbf{Z}} \beta_j$. We consider just the case where $\beta_j = \gamma_0 \vee \gamma_1 x^k \vee \gamma_2 x^{-k}$ and $\alpha_i = \delta_0 \wedge \delta_1 x^k \wedge \delta_2 x^{-k}$, the other remaining cases being very similar. Let

$$\alpha_i' := \delta_0 \wedge \delta_1 \delta_2 \quad \text{and} \quad \beta_j' := \gamma_0 \vee \gamma_1 \gamma_2 \vee \gamma_1 \delta_1^{-1} \vee \gamma_2 \delta_2^{-1}.$$

Reasoning in $\mathbf{Z}$, it follows that $\{\alpha_i'\} \vdash_{\mathbf{Z}} \beta_j'$ and hence, by the induction hypothesis, $\{\alpha_i'\} \vdash_{\mathcal{AbLG}} \beta_j'$. But then, using the fact that $\{uv \vee w\} \vdash_{\mathcal{AbLG}} u \vee v \vee w$, for any $u, v, w \in \mathrm{Fm}$, it follows that $\{\alpha_i\} \vdash_{\mathcal{AbLG}} \beta_j$ as required. □

The proof of Proposition 5.4.11 yields a method for checking the satisfaction of an arbitrary quasi-equation in $\mathcal{AbLG}$: to check $\{\alpha\} \vdash_{\mathcal{AbLG}} \beta$, rewrite α and β using distributivity properties of Abelian ℓ-groups, so that it can be assumed that α and β have a certain syntactic form, as described in the proof, then iteratively eliminate the variables occurring in β by constructing new formulas.

A variety $\mathcal{V}$ is said to have a *decidable quasi-equational theory* — equivalently, *decidable word problem*; see Chapter 9 — if there exists an effective procedure for deciding $\Sigma \vdash_{\mathcal{V}} \varepsilon$, for any finite $\Sigma \cup \{\varepsilon\} \subseteq \mathrm{Eq}$.

COROLLARY 5.4.12. *The quasi-equational theory of Abelian ℓ-groups is decidable.*

5.5. Bibliographical remarks

The characterizations of semilinearity provided by Theorems 5.1.2 and 5.1.7 are based on results from **[44]**, and generalize analogous theorems in the literature for ℓ-groups **[9]**, pseudo-MV-algebras **[114]**, pseudo-BL-algebras **[226]**, GBL-algebras (DRℓ-monoids) **[225]**, and integral residuated lattices **[361]**. Corollary 5.1.3 was first proved in **[180]**.

In **[229]**, Kühr obtained a purely implicational characterization of semilinearity for integral residuated lattices (conjectured in **[361]**) via the equation

$$((x\backslash y)\backslash u)\backslash(((w/(w/(((y\backslash x)\backslash z)\backslash z)))\backslash u)\backslash u) \approx \mathrm{e}.$$

This result provided the inspiration in **[49]** for a characterization of semilinearity for residuated lattices that does not involve multiplication. For any variables u and x, let $\lambda_u^*(x) := ((x\backslash u)\backslash u) \wedge \mathrm{e}$ and $\rho_u^*(x) := (u/(u/x)) \wedge \mathrm{e}$. Then the following statements are equivalent:

(1) $\mathcal{V}$ is a variety of semilinear residuated lattices.

(2) $\mathcal{V}$ is a variety of residuated lattices satisfying one of the following equations:

$$\lambda_u^*((x \vee y)\backslash x) \vee \rho_v^*((x \vee y)\backslash y) \approx \mathrm{e}$$
$$\lambda_u^*(x/(x \vee y)) \vee \rho_v^*(y/(x \vee y)) \approx \mathrm{e}.$$

(3) $\mathcal{V}$ is a variety of prelinear residuated lattices satisfying

$$x \vee y \approx \mathrm{e} \implies \lambda_u^*(x) \vee \rho_v^*(y) \approx \mathrm{e}.$$

This characterization can be greatly simplified for varieties satisfying the equation $(x \vee y) \wedge \mathrm{e} \approx (x \wedge \mathrm{e}) \vee (y \wedge \mathrm{e})$ **[232]**. For further characterizations of semilinearity in different signatures, see also **[76, 79]**.

The study of densifiable varieties began in earnest with the publication of Hájek's monograph **[174]** and the subsequent proof that his basic fuzzy logic BL is complete with respect to the class of bounded residuated lattices whose monoidal operation is a continuous t-norm **[74]**. This "standard completeness" result implies (but is not equivalent to) the fact that the variety of BL-algebras is densifiable. The embedding method used in the proof of Theorem 5.2.6 was first used to establish standard completeness for monoidal t-norm logic **[202]** — implying that the variety of commutative, semilinear, integral bounded residuated lattices is densifiable — and subsequently to obtain similar results for many other logics and varieties; see, e.g., **[122, 191, 262, 273]**. The proof of Theorem 5.2.7, a generalization of standard completeness proofs given in **[122, 192]**, may be found in **[262]**, along with related results for varieties of semilinear ℓ-groups and their negative cones.

Theorem 5.2.4 was first proved in **[375]**; the shorter proof given here is taken from **[191]**. A version of Theorem 5.2.8 was first formulated in **[255]** (see also **[72, 78, 259]**) and used to prove Proposition 5.2.9 and other densifiability and standard completeness results. The method of "density elimination by substitutions" used in the proof given here was introduced in **[72]**, while algebraic interpretations of the method may be found in **[17, 142]**.

The fundamental embedding theorem for ℓ-groups, stated here as Corollary 5.3.4, was established by Holland in **[187]**. The account given in Section 5.3, culminating in Theorem 5.3.3 and Corollary 5.3.6, is based on **[306]** (see also **[141]**), which built in turn on earlier results of **[8, 254]**. Proposition 5.4.5, Corollary 5.4.10, and Proposition 5.4.11 were first proved in **[188]**, **[189]**, and **[369]**, respectively.

Propositions 5.4.7 and 5.4.9, which together provide a bridge between the equational theories of ℓ-groups and distributive ℓ-monoids, are taken from **[82]**. Note, however, that this close relationship does not extend to the commutative setting; there exist $\{\wedge, \vee, \cdot, \mathrm{e}\}$-equations such as

$$x_1x_2x_3 \wedge x_4x_5x_6 \wedge x_7x_8x_9 \leq x_1x_4x_7 \vee x_2x_5x_8 \vee x_3x_6x_9$$

that are satisfied by all Abelian ℓ-groups, but not by all commutative, distributive ℓ-monoids **[326]**. Similarly, there are $\{\wedge, \vee, \cdot, \mathrm{e}\}$-equations that are satisfied by all totally ordered groups, but not by all totally ordered monoids **[82]**.

CHAPTER 6

Cancellativity

In this chapter, we focus on *cancellativity*, a natural property satisfied by some of the most important structures in mathematics, including the monoid reduct of any ℓ-group or its negative cone. Cancellativity plays a pivotal role in the structure theory of algebras for a broad family of substructural logics, including Łukasiewicz logic (MV-algebras) and Hájek's basic fuzzy logic (BL-algebras). In Chapter 7, we develop this structure theory in the general setting of divisible residuated lattices.

An algebra with a monoid reduct $\mathbf{M}$ is called *cancellative* if $cad = cbd$ implies $a = b$, for all $a, b, c, d \in M$. In Section 6.1 we show that cancellative residuated lattices form a variety extending well beyond ℓ-groups and their negative cones; in fact, *every* lattice is a subreduct of some cancellative residuated lattice.

In Section 6.2, we present the construction of the group of left quotients of a right reversible, cancellative monoid, a standard technique from algebra that generalizes the construction of the field of rational numbers from the ring of integers. We then use this construction in Section 6.3 to establish an equivalence between a category of cancellative residuated lattices with a right reversible monoid reduct and a category of ℓ-groups augmented with a fractional co-nucleus.

6.1. Cancellative residuated lattices

We first provide an equational characterization of cancellative residuated lattices.

LEMMA 6.1.1. *A residuated lattice* $\mathbf{L}$ *is cancellative if and only if it satisfies the equations* $xy/y \approx x$ *and* $y\backslash yx \approx x$.

PROOF. First, consider any $a, b \in L$. Clearly, $(ab/b)b \leq ab$. Also $a \leq ab/b$, so $ab \leq (ab/b)b$. That is, $(ab/b)b = ab$, and, symmetrically, $b(b\backslash ba) = ba$. Hence, if $\mathbf{L}$ is cancellative, $ab/b = b\backslash ba = a$. Conversely, if $\mathbf{L}$ satisfies $xy/y \approx x$ and $y\backslash yx \approx x$, then $cad = cbd$ implies $a = c\backslash(cad/d) = c\backslash(cbd/d) = b$, for any $a, b, c, d \in L$. □

Cancellative residuated lattices therefore form a variety $\mathcal{CanRL}$ that contains all ℓ-groups and their negative cones. The next result collects some further useful properties of these algebras.

LEMMA 6.1.2. *Let* $\mathbf{L}$ *be any residuated lattice.*

(a) *If* $\mathbf{L}$ *satisfies* $x\backslash x \approx \mathrm{e}$, *then it satisfies* $x/x \approx \mathrm{e}$ *and is* e*-cyclic.*
(b) *If* $\mathbf{L}$ *satisfies* $x\backslash x \approx \mathrm{e}$ *and has a greatest element, then it is integral.*
(c) *If* $\mathbf{L}$ *is cancellative and has a least element, then it is trivial.*
(d) *If* $\mathbf{L}$ *is cancellative and non-trivial, then it is infinite.*

PROOF.

(a) Suppose that $\mathbf{L}$ satisfies $x\backslash x \approx \mathrm{e}$. Then $\mathbf{L}$ also satisfies $x/x \approx \mathrm{e}$, since every residuated lattice satisfies $x/x \approx (x/x)\backslash(x/x)$. Now consider any $a \in L$. Clearly, $a(a\backslash \mathrm{e}) \leq \mathrm{e}$, so $a(a\backslash \mathrm{e})a \leq a$ and $(a\backslash \mathrm{e})a \leq a\backslash a = \mathrm{e}$. Hence $a\backslash \mathrm{e} \leq \mathrm{e}/a$, and, symmetrically, since $\mathbf{L}$ satisfies $x/x \approx \mathrm{e}$, also $\mathrm{e}/a \leq a\backslash \mathrm{e}$. So $\mathbf{L}$ is e-cyclic.

(b) If $\mathbf{L}$ satisfies $x\backslash x \approx \mathrm{e}$ and has a greatest element $\top$, then $a \leq \top = \top\backslash\top = \mathrm{e}$, for any $a \in L$, so $\mathbf{L}$ is integral.

(c) If $\mathbf{L}$ is cancellative and has a least element $\bot$, then it has a greatest element $\top = \bot\backslash\bot = \bot\backslash(\bot \cdot \bot) = \bot$, so $\mathbf{L}$ is trivial.

(d) Follows from (c), since every finite residuated lattice has a least element. □

A totally ordered, commutative, integral, cancellative residuated lattice that is not the negative cone of any ℓ-group was described in Example 1.4.14. The next result implies that, in fact, $\mathcal{CanRL}$ satisfies no non-trivial lattice equation, i.e., any $\{\wedge, \vee\}$-equation that is satisfied by all cancellative residuated lattices is satisfied by all lattices.

THEOREM 6.1.3. *Every lattice is a subreduct of a simple, cancellative, integral residuated lattice.*

PROOF. Let $\mathbf{L} = \langle L, \wedge, \vee\rangle$ be any lattice. Since every lattice embeds into a bounded lattice, we may assume that $\mathbf{L}$ has a greatest element 1. Now let $\langle L^*, \circ, \varepsilon\rangle$ be the free monoid over L. We define a simple, integral, cancellative residuated lattice $\mathbf{L}^*$ with monoid reduct $\langle L^*, \circ, \varepsilon\rangle$ such that $\mathbf{L}$ is a subreduct of $\mathbf{L}^*$. Let $l(w)$ denote the length of $w \in L^*$ and define $L_n := \{w \in L^* \mid l(w) = n\}$, for each $n \in \mathbb{N}$. We obtain a lattice-order $\preccurlyeq$ on L^* with meet and join operations $\curlywedge$ and $\curlyvee$ by defining

$$u \preccurlyeq v \;:\Longleftrightarrow\; l(u) > l(v) \text{ or } (u = u_1 \circ \cdots \circ u_n \in L_n,\ v = v_1 \circ \cdots \circ v_n \in L_n, \text{ and } u_i \leq v_i, \text{ for each } i \in \{1, \ldots, n\}).$$

Then $\varepsilon \in L_0$ is the greatest element of $\langle L^*, \preccurlyeq\rangle$ and each L_n forms a sublattice of $\langle L^*, \curlywedge, \curlyvee\rangle$ that is isomorphic to the direct power $\mathbf{L}^n$ (as depicted in Figure 6.1). If $u \preccurlyeq v$, then, clearly, $u\backslash v = \varepsilon$. If $u \not\preccurlyeq v$ with $u = u_1 \circ \cdots \circ u_m \in L_m$ and $v = v_1 \circ \cdots \circ v_n \in L_n$, then $m \leq n$ and

$$u\backslash v = \begin{cases} v_{m+1} \circ \cdots \circ v_n & \text{if } m < n \text{ and } u_i \leq v_i, \text{ for each } i \in \{1, \ldots, n\}; \\ \underbrace{1 \circ \cdots \circ 1}_{n-m+1} & \text{otherwise.} \end{cases}$$

The operation / has a symmetric characterization. Hence $\mathbf{L}^* := \langle L^*, \curlywedge, \curlyvee, \circ, \backslash, /, \varepsilon\rangle$ is a cancellative, integral residuated lattice, and $\mathbf{L}$ is a sublattice of $\langle L^*, \curlywedge, \curlyvee\rangle$. Moreover, $1 \in L_1$ generates the (normal) convex subuniverse L^*, so $\mathbf{L}^*$ is simple, by Theorem 4.3.7. □

The construction of $\mathbf{L}^*$ in the proof of Theorem 6.1.3 resembles the ordinal sum construction to be presented in Chapter 7. Indeed, ordinal sums of posets are obtained by stacking summands on top of each other, and the preceding proof describes one particular instance of this construction. Let us note also that, since

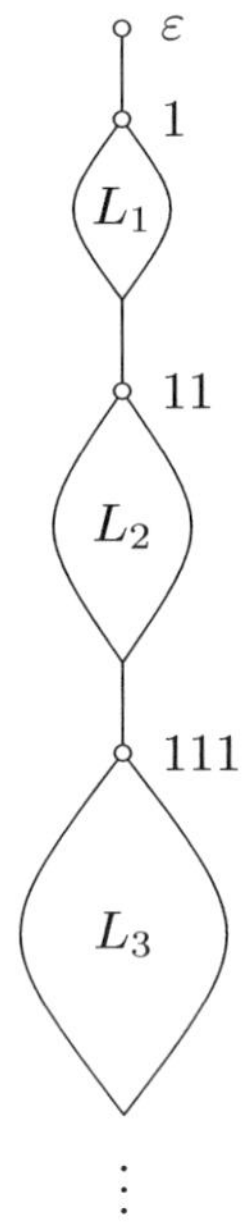

FIGURE 6.1. A cancellative expansion of a lattice $\mathbf{L}$

varieties of lattices are closed under taking ordinal sums, the lattice reduct of $\mathbf{L}^*$ will always belong to the variety generated by $\mathbf{L}$.

More generally, for any variety $\mathcal{V}$ of lattices, $\hat{\mathcal{V}} := \mathbb{HSP}(\{\mathbf{L}^* \mid \mathbf{L} \in \mathcal{V}\})$ is a variety of cancellative, integral residuated lattices, and the map $\mathcal{V} \mapsto \hat{\mathcal{V}}$ is an order-embedding of the lattice of varieties of lattices into the lattice of varieties of cancellative, integral residuated lattices. However, it is not known if a similar result holds for commutative, cancellative, integral residuated lattices, or, indeed, if there is any non-trivial lattice equation satisfied by all such algebras.

When $\mathbf{L}$ is the two-element chain $\langle\{0,1\}, \wedge, \vee\rangle$, the preceding construction produces a distributive, cancellative, integral residuated lattice $\mathbf{L}^*$ that is not semilinear or even prelinear: just note that $x = 01$ and $y = 10$ gives $x\backslash y = y\backslash x = 1$ and $\mathrm{e} \not\leq (x\backslash y) \vee (y\backslash x)$. The following example shows that there are also distributive, commutative, cancellative, integral residuated lattices that are not semilinear.

EXAMPLE 6.1.4. The set $B := (\mathbb{Z}^{<0})^2 \cup \{\langle 0,0\rangle\}$ is closed under the product, meet, and join operations of $(\mathbf{Z}^-)^2$, recalling that $\mathbf{Z}^-$ is the negative cone of the ordered group of integers $\mathbf{Z}$. Moreover, the product operation on $\langle B, \leq\rangle$ has a residual

$$\langle a,b\rangle \to \langle c,d\rangle := \begin{cases} \langle 0,0\rangle & \text{if } a \leq c \text{ and } b \leq d; \\ \langle \min(c-a,-1), \min(d-b,-1)\rangle & \text{otherwise.} \end{cases}$$

It is easily checked that $\mathbf{B} = \langle B, \wedge, \vee, \cdot, \to, \langle 0,0\rangle\rangle$ is a distributive, commutative, cancellative, integral residuated lattice. However, $\mathbf{B}$ is not semilinear, since, e.g.,

$$(\langle -1,-2\rangle \to \langle -2,-1\rangle) \vee (\langle -2,-1\rangle \to \langle -1,-2\rangle) = \langle -1,-1\rangle \neq \langle 0,0\rangle.$$

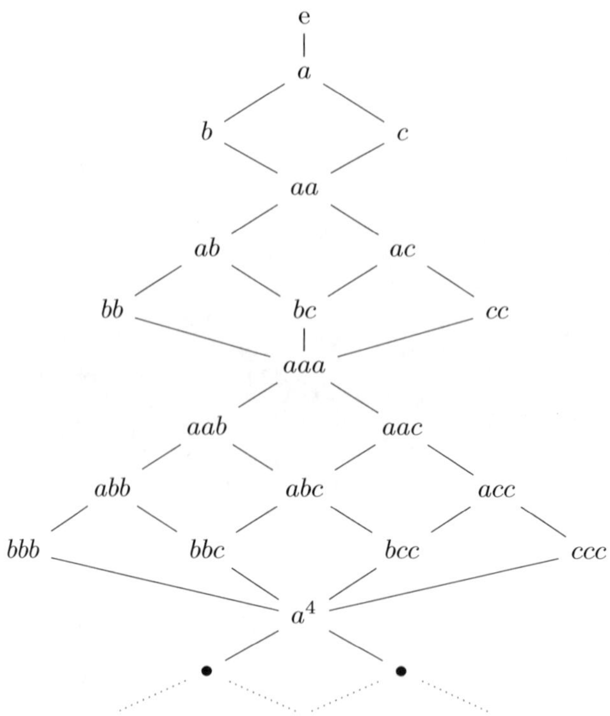

FIGURE 6.2. A lattice-order on the free commutative monoid over $\{a, b, c\}$

The following example describes a commutative, cancellative, integral residuated lattices that fails to be distributive.

EXAMPLE 6.1.5. Let $\mathbf{F}_3 = \langle F_3, \cdot, \mathrm{e}\rangle$ be the free commutative monoid over the set $\{a, b, c\}$, consisting of sequences $a^i b^j c^k$ with $i, j, k \in \mathbb{N}$, where e denotes the empty sequence. For $u = a^i b^j c^k \in F_3$, let $l(u) := i+j+k$, $l_a(u) := i$, $l_b(u) := j$, $l_c(u) := k$. Note that $\mathbf{F}_3$ is cancellative. For $u, v \in F_3$, define (as depicted in Figure 6.2)

$$u \leq v \;:\Longleftrightarrow\; l(u) > l(v) \text{ or } (l(u) = l(v),\ l_b(u) \geq l_b(v),\ \text{and } l_c(u) \geq l_c(v)).$$

Each block of sequences of the same length is a finite join-subsemilattice of a product of two chains, so $\langle F_3, \leq\rangle$ is a lattice in which every non-empty join of elements exists and is attained by a finite subjoin. In particular, for any $u, v \in F_3$, the joins of the non-empty sets $\{w \in F_3 \mid uw \leq v\}$ and $\{w \in F_3 \mid wu \leq v\}$ are the residuals of the monoid operation of $\mathbf{F}_3$, with respect to $\leq$. To show this, it suffices to check that the monoid operation of $\mathbf{F}_3$ preserves finite joins. If $u, v \in F_3$ satisfy $l(u) = l(v)$, then $l_b(u \vee v) = \min(l_b(u), l_b(v))$, and, for any $w \in F_3$,

$$\begin{aligned} l_b((u \vee v)w) &= \min(l_b(u), l_b(v)) + l_b(w) \\ &= \min(l_b(u) + l_b(w), l_b(v) + l_b(w)) \\ &= l_b(uw \vee vw), \end{aligned}$$

and, similarly, $l_c((u \vee v)w) = l_c(uw \vee vw)$. We therefore obtain a commutative, cancellative, integral residuated lattice that is not distributive, since $bb \vee (ab \wedge cc) = bb \vee aaa = bb$, but $(bb \vee ab) \wedge (bb \vee cc) = ab \wedge aa = ab$.

The situation changes, however, if multiplication distributes over binary meets.

PROPOSITION 6.1.6. *Every commutative, cancellative residuated lattice satisfying* $x(y \wedge z) \approx xy \wedge xz$ *is distributive.*

PROOF. Consider any commutative, cancellative residuated lattice $\mathbf{L}$ satisfying $x(y \wedge z) \approx xy \wedge xz$. For any $a, b \in L$ such that $a \leq b$, the map $\varphi\colon L \to L;\ c \mapsto a \to cb$ satisfies $\varphi(a) = a \to ab = b$ and $\varphi(c) = a \to cb \geq b \to cb = c$, for all $c \in L$, and is a meet-endomorphism, i.e., for any $c, d \in L$,

$$\varphi(c \wedge d) = a \to (c \wedge d)b = a \to (cb \wedge db) = (a \to cb) \wedge (a \to db) = \varphi(c) \wedge \varphi(d).$$

Hence, by Proposition 1.1.10, the lattice reduct of $\mathbf{L}$ is distributive. □

We conclude this section by providing a description of totally ordered, cancellative residuated lattices.

PROPOSITION 6.1.7. *Let* $\mathbf{L}$ *be any totally ordered, cancellative residuated lattice. Then* $\mathbf{L}$ *is a homomorphic image of a totally ordered residuated lattice whose monoid reduct is a free monoid.*

PROOF. Let $\langle L^*, \circ, \varepsilon \rangle$ be the free monoid over L, and let $\trianglelefteq$ be any dual well-order on L. We define a total order $\preccurlyeq$ on L^* by setting, for any $u, v \in L^*$,

$$u \prec v \ :\Longleftrightarrow\ l(v) < l(u) \ \text{ or } \ (l(v) = l(u),\ u = wau',\ v = wbv', \text{ and } b \lhd a).$$

Next, let $\varphi\colon L^* \to L$ be the monoid homomorphism mapping $a_1 \circ \cdots \circ a_n$ to $a_1 \cdots a_n$, for any $a_1, \ldots, a_n \in L$, and define, for any $u, v \in L^*$,

$$u \leq v \ :\Longleftrightarrow\ \varphi(u) < \varphi(v) \ \text{ or } \ (\varphi(u) = \varphi(v) \text{ and } u \preccurlyeq v).$$

Clearly, $\leq$ is a total order and multiplication in L^* is order-preserving. We show now that multiplication is also residuated with respect to this order and that φ is a homomorphism from the resulting residuated lattice $\mathbf{L}^*$ onto $\mathbf{L}$.

Let $u, v \in L^*$. For any $w \in L^*$, if $w \circ v \leq u$, then $\varphi(w \circ v) = \varphi(w)\varphi(v) \leq \varphi(u)$, and hence $\varphi(w) \leq \varphi(u)/\varphi(v)$. So, if u/v exists in L^*, then $\varphi(u/v) \leq \varphi(u)/\varphi(v)$. (Similarly, if $v\backslash u$ exists in L^*, then $\varphi(v\backslash u) \leq \varphi(v)\backslash\varphi(u)$.) Now consider the set $S := \{w \in L^* \mid \varphi(w) = \varphi(u)/\varphi(v)\}$, fix $n := \max(l(u) - l(v), 0)$, and let

$$w' := (\varphi(u)/\varphi(v)) \circ \underbrace{e \circ \ldots \circ e}_{n} \in S.$$

Observe that $\varphi(w' \circ v) = (\varphi(u)/\varphi(v))\varphi(v) \leq \varphi(u)$ and $l(u) < 1 + n + l(v) = l(w' \circ v)$. So $w' \circ v \leq u$ and the set $S' := \{w \in S \mid w \circ v \leq u\}$ is non-empty. Note also that S is dually well-ordered, so there exists $w_0 := \bigvee S'$ satisfying

$$w_0 \circ v = \bigvee \{w \circ v \mid w \in S'\} \leq u.$$

Hence if $w \leq w_0$, then $w \circ v \leq u$. Moreover, $w \circ v \leq u$ implies $\varphi(w) \leq \varphi(u)/\varphi(v)$, which implies $w \leq w_0$. So $w_0 = u/v$. Similarly, it can be shown that $v\backslash u$ is the join of the set $\{w \in L^* \mid \varphi(w) = \varphi(v)\backslash\varphi(u) \text{ and } v \circ w \leq u\}$. It follows also that $\varphi(u/v) = \varphi(u)/\varphi(v)$ and $\varphi(v\backslash u) = \varphi(v)\backslash\varphi(u)$. So $\mathbf{L}^*$ is a totally ordered residuated lattice and φ is a residuated lattice homomorphism from $\mathbf{L}^*$ onto $\mathbf{L}$. □

COROLLARY 6.1.8. *The variety of cancellative, semilinear residuated lattices is generated by the class of totally ordered residuated lattices whose monoid reduct is a free monoid.*

6.2. Lattice-ordered groups of left quotients

In this section, we first present the well-known construction of the group of left quotients of a right reversible, cancellative monoid, then apply this construction to the monoid reduct of a cancellative residuated lattices that is right reversible and define a lattice-order on the resulting group of left quotients.

A group $\mathbf{G}$ is said to be a *group of left quotients* of a submonoid $\mathbf{M}$ of $\mathbf{G}$ if every element of G can be expressed in the form $a^{-1}b$, for some $a, b \in M$.

LEMMA 6.2.1. *Let* $\mathbf{G}$ *be a group of left quotients of a monoid* $\mathbf{M}$. *Then* $a, b, c, d \in M$ *satisfy* $a^{-1}b = c^{-1}d$ *in* $\mathbf{G}$ *if and only if* $xa = yc$ *and* $xb = yd$, *for some* $x, y \in M$.

PROOF. Let $a, b, c, d \in M$. Suppose first that $a^{-1}b = c^{-1}d$ in $\mathbf{G}$. By assumption, $ac^{-1} = x^{-1}y$, for some $x, y \in M$. Hence $xa = yc$ and $xb = yca^{-1}b = ycc^{-1}d = yd$. Now suppose that $xa = yc$ and $xb = yd$, for some $x, y \in M$. It follows that $a^{-1}b = (xa)^{-1}xb = (yc)^{-1}yd = c^{-1}d$ in $\mathbf{G}$. □

A monoid $\mathbf{M}$ is said to be *right reversible* if $Ma \cap Mb \neq \emptyset$, for all $a, b \in M$; that is, for all $a, b \in M$, there exist $x, y \in M$ such that $xa = yb$. The next result shows that cancellativity and right reversibility are necessary and sufficient conditions for a monoid to have a group of left quotients.

THEOREM 6.2.2. *A monoid* $\mathbf{M} = \langle M, \cdot, \mathrm{e}\rangle$ *has a group* $\mathbf{G}(\mathbf{M})$ *of left quotients if and only if it is cancellative and right reversible. Moreover, if* $\mathbf{G}(\mathbf{M})$ *exists, then it is unique up to isomorphism.*

PROOF. Let us suppose first that $\mathbf{G}(\mathbf{M})$ is a group of left quotients of $\mathbf{M}$. Clearly, $\mathbf{M}$ is cancellative. Moreover, given any $a, b \in M$, there exist $x, y \in M$ such that $ab^{-1} = x^{-1}y$ and hence $xa = yb$; that is, $\mathbf{M}$ is right reversible.

Conversely, suppose that $\mathbf{M}$ is cancellative and right reversible. We define the following relation on M^2:

$$\Theta := \{\langle\langle a, b\rangle, \langle c, d\rangle\rangle \in (M^2)^2 \mid xa = yc \text{ and } xb = yd, \text{ for some } x, y \in M\}.$$

Let us check that Θ is an equivalence relation. Reflexivity and symmetry are clear, so we focus on transitivity and consider $\langle\langle a, b\rangle, \langle c, d\rangle\rangle, \langle\langle c, d\rangle, \langle f, g\rangle\rangle \in \Theta$; that is, for some $x_1, x_2, y_1, y_2 \in M$,

$$x_1a = y_1c, \quad x_1b = y_1d, \quad x_2c = y_2f, \quad \text{and} \quad x_2d = y_2g.$$

By right reversibility, there exist $u_1, u_2 \in M$ such that $u_1y_1 = u_2x_2$. Hence, letting $u = u_1x_1$ and $v = u_2y_2$,

$$ua = u_1x_1a = u_1y_1c = u_2x_2c = u_2y_2f = vf$$
$$ub = u_1x_1b = u_1y_1d = u_2x_2d = u_2y_2g = vg.$$

That is, $\langle\langle a, b\rangle, \langle f, g\rangle\rangle \in \Theta$ as required.

Let us denote the Θ-equivalence class of $\langle a, b\rangle \in M^2$ by $[a, b]$, and the set of Θ-equivalence classes by M^2/Θ. Observe that $[xa, xb] = [a, b]$, for all $x, a, b \in M$. Next, note that given any $[a, b], [c, d] \in M^2/\Theta$, there exist $x, y, u \in M$ such that $[a, b] = [x, u]$ and $[c, d] = [u, y]$; that is, by right reversibility, there exist $v, w \in M$

such that $vb = wc$, where also $[a,b] = [va, vb]$ and $[c,d] = [wc, wd]$. Hence we may define a binary operation on M^2/Θ by setting

$$[a,b] \star [b,c] := [a,c].$$

To check that $\star$ is well-defined, consider $[a',b'] = [a,b]$ and $[b',c'] = [b,c]$. There exist $u,u',v,v' \in M$ such that $ua' = va$, $ub' = vb$, $u'b' = v'b$, and $u'c' = v'c$. By right reversibility, $xu' = yu$, for some $x,y \in M$. Also $xv'b = xu'b' = yub' = yvb$ and, by cancellativity, $xv' = yv$. But then $(yu)a' = (yv)a$ and $(yu)c' = xu'c' = xv'c = (yv)c$, so $[a',c'] = [a,c]$.

Observe also that $\star$ is associative; that is, for any $a,b,c,d \in M$,

$$([a,b] \star [b,c]) \star [c,d] = [a,c] \star [c,d] = [a,d] = [a,b] \star [b,d] = [a,b] \star ([b,c] \star [c,d]).$$

Note next that, for any $a \in M$, there exist, by right reversibility, $x,y \in M$ such that $xa = y\mathrm{e} = y$, so $[a,a] = [\mathrm{e},\mathrm{e}]$. Hence the element $[\mathrm{e},\mathrm{e}]$ is a neutral element for $\star$; just observe that, for any $a,b \in M$,

$$[a,a] \star [a,b] = [a,b] = [a,b] \star [b,b].$$

Moreover, each $[a,b] \in M^2/\Theta$ has an inverse $[a,b]^{-1} := [b,a]$, since

$$[a,b] \star [b,a] = [a,a] = [b,b] = [b,a] \star [a,b].$$

So $\langle M^2/\Theta, \star, ^{-1}, [\mathrm{e},\mathrm{e}]\rangle$ is a group.

Now let φ be the map sending each $a \in M$ to $[\mathrm{e},a] \in M^2/\Theta$. Then, for any $a,b \in M$, it follows that $[\mathrm{e},a] \star [\mathrm{e},b] = [\mathrm{e},a] \star [a,ab] = [\mathrm{e},ab]$, and, if $[\mathrm{e},a] = [\mathrm{e},b]$, then $xa = xb$, for some $x \in M$, and so, by cancellativity, $a = b$. Hence φ is a monoid embedding. Moreover, for all $a,b \in M$,

$$\varphi(a)^{-1} \star \varphi(b) = [\mathrm{e},a]^{-1} \star [\mathrm{e},b] = [a,\mathrm{e}] \star [\mathrm{e},b] = [a,b].$$

Taking a suitable isomorphic copy $\mathbf{G(M)} = \langle G(\mathbf{M}), \star, ^{-1}, \mathrm{e}\rangle$ of $\langle M^2/\Theta, \star, ^{-1}, [\mathrm{e},\mathrm{e}]\rangle$, we obtain a group of left quotients of $\mathbf{M}$. Let us show finally that $\mathbf{G(M)}$ is the unique such group up to isomorphism. Suppose that $\mathbf{G(M)}' = \langle G(\mathbf{M})', \star', ^{-1}, \mathrm{e}\rangle$ is also a group of left quotients of $\mathbf{M}$ and let ψ be a map sending $a^{-1} \star b \in G(\mathbf{M})$ to $a^{-1} \star' b \in G(\mathbf{M})'$, for $a,b \in M$. To see that this map is well-defined, consider $a,b,c,d \in M$ and suppose that $a^{-1} \star b = c^{-1} \star d$. There exist $x,y \in M$ such that $xa = yc$ and $xb = yd$, so $yc \star' a^{-1} \star' b = xa \star' a^{-1} \star' b = xb = yd$ and hence $c \star' a^{-1} \star' b = d$, that is, $a^{-1} \star' b = c^{-1} \star' d$. Injectivity follows by a symmetrical argument, and since ψ also preserves multiplication, it is a group isomorphism. $\square$

We now apply the construction described in the proof of Theorem 6.2.2 to monoid reducts of cancellative residuated lattices that are right reversible, with the aim of showing that the corresponding groups of left quotients can be lattice-ordered. Let us call a cancellative residuated lattice with a right reversible monoid reduct an *Ore residuated lattice*. These algebras form a proper subclass $\mathcal{ORL}$ of $\mathcal{CanRL}$. For example, a free monoid generated by a set with more than one element is the monoid reduct of a cancellative residuated lattice, as shown in the proof of Theorem 6.1.3, but is not right reversible. Nevertheless, $\mathcal{ORL}$ contains important subvarieties of $\mathcal{CanRL}$, including the variety of commutative, cancellative residuated lattices; in fact, every monoid that does not generate the variety of all monoids is right reversible (see the bibliographical remarks for references).

Given any Ore residuated lattice $\mathbf{L}$, let $\mathbf{G}(\mathbf{L}) = \langle G(\mathbf{L}), \star, ^{-1}, \mathrm{e}\rangle$ denote the group of left quotients of the monoid reduct of $\mathbf{L}$. The following lemma shows that there exists a lattice-order on $\mathbf{G}(\mathbf{L})$ that extends the order of $\mathbf{L}$.

LEMMA 6.2.3. *Let $\mathbf{L}$ be any Ore residuated lattice. Then the binary relation on $G(\mathbf{L})$ defined by setting, for $a, b, c, d \in L$,*

$$a^{-1}b \preccurlyeq c^{-1}d \;:\Longleftrightarrow\; xa = yc \text{ and } xb \le yd, \text{ for some} x, y \in L,$$

is the unique lattice-order on the group $\mathbf{G}(\mathbf{L})$ extending $\le$. Moreover, finite joins in $\mathbf{L}$ coincide with the corresponding joins in $\mathbf{G}(\mathbf{L})$, with respect to $\preccurlyeq$, and, for any $a, b, c, d, x, y \in L$ satisfying $xa = yc$,

$$(a^{-1}b) \vee (c^{-1}d) = (xa)^{-1}(xb \vee yd).$$

PROOF. We begin by showing that

$$S := \{a^{-1}b \mid a, b \in L \text{ and } b \le a\} \subseteq G(\mathbf{L})$$

is the negative cone of a partial order on $\mathbf{G}(\mathbf{L})$ (see Example 1.4.1), that is,

(i) $\{x \in S \mid x^{-1} \in S\} = \{\mathrm{e}\}$;

(ii) $xy \in S$, for all $x, y \in S$;

(iii) $yxy^{-1} \in S$, for all $x \in S$ and $y \in G(\mathbf{L})$.

Consider any $c, d \in L$ such that $c^{-1}d \in S$. Then $c^{-1}d = a^{-1}b$, for some $a, b \in L$ such that $b \le a$. Since $\mathbf{L}$ is an Ore residuated lattice, $xc = ya$, for some $x, y \in L$. So $xd = xcc^{-1}d = yaa^{-1}b = yb \le ya = xc$, and, by cancellativity, $d \le c$.

Condition (i) is now clear, since if $a^{-1}b \in S$ and $b^{-1}a = (a^{-1}b)^{-1} \in S$, then $b \le a$ and $a \le b$, so $a = b$ and $a^{-1}b = \mathrm{e}$. For condition (ii), consider any $a, b, c, d \in L$ such that $a^{-1}b, c^{-1}d \in S$, and hence $b \le a$ and $d \le c$. Since $\mathbf{L}$ is an Ore residuated lattice, $yc = xb$, for some $x, y \in L$. So $(a^{-1}b)(c^{-1}d) = a^{-1}x^{-1}yd = (xa)^{-1}(yd)$, where $yd \le yc = xb \le xa$, and hence $(a^{-1}b)(c^{-1}d) \in S$. For condition (iii), it suffices to show that, for any $a^{-1}b \in S$ and $c \in L$, both $c^{-1}a^{-1}bc \in S$ and $ca^{-1}bc^{-1} \in S$. That $c^{-1}a^{-1}bc \in S$ is easily checked. Now let $x, y, z, w \in L$ be such that $xc = ya$ and $wc = zyb$, yielding $ca^{-1} = x^{-1}y$ and $ybc^{-1} = z^{-1}w$. Then $ca^{-1}bc^{-1} = x^{-1}ybc^{-1} = x^{-1}z^{-1}w = (zx)^{-1}w$. Finally, $wc = zyb \le zya = zxc$, since $b \le a$, and hence, by cancellativity, $w \le zx$. So $ca^{-1}bc^{-1} \in S$.

We may now check that, for any $a, b, c, d \in L$,

$$a^{-1}b \preccurlyeq c^{-1}d \iff d^{-1}ca^{-1}b \in S.$$

For the left-to-right direction, suppose that $xa = yc$ and $xb \le yd$, for some $x, y \in L$. Then $d^{-1}ca^{-1}b = d^{-1}c(xa)^{-1}xb = d^{-1}c(yc)^{-1}xb = d^{-1}y^{-1}xb = (yd)^{-1}xb \in S$. Conversely, suppose that $d^{-1}ca^{-1}b \in S$. There exist $x, y \in L$ such that $xa = yc$, so, since again $(yd)^{-1}xb = d^{-1}ca^{-1}b \in S$, it follows that $xb \le yd$, and $a^{-1}b \preccurlyeq c^{-1}d$.

Hence $\preccurlyeq$ is a partial order on $\mathbf{G}(\mathbf{L})$. Moreover, it is easily checked that $\preccurlyeq$ is the only partial order on $\mathbf{G}(\mathbf{L})$ that extends $\le$, so it remains to prove that $\preccurlyeq$ is a lattice-order. We show first that if $a, b \in G(\mathbf{L})$ have a join $a \vee^{\mathbf{G}(\mathbf{L})} b$ with respect to $\preccurlyeq$, then $a \vee^{\mathbf{G}(\mathbf{L})} b = a \vee^{\mathbf{L}} b$. Obviously, $a \vee^{\mathbf{L}} b$ is an upper bound of a and b in $\mathbf{G}(\mathbf{L})$. If $c^{-1}d$ is another upper bound of a and b, with $c, d \in L$, then $ca \le d$ and $cb \le d$. Hence $c(a \vee^{\mathbf{L}} b) = ca \vee^{\mathbf{L}} cb \le d$, and $a \vee^{\mathbf{L}} b \le c^{-1}d$. To prove that joins are defined for any pair of elements in $G(\mathbf{L})$, it suffices to show that $a^{-1}b \vee^{\mathbf{G}(\mathbf{L})} \mathrm{e}$ exists, for all $a, b \in L$, since multiplication distributes over existing joins in every partially ordered group. Consider any $a, b \in L$, recalling that $b \vee^{\mathbf{G}(\mathbf{L})} a$ exists. The map $\varphi_{a^{-1}} \colon G(\mathbf{L}) \to G(\mathbf{L});\ x \mapsto a^{-1}x$ is an order-automorphism of $\langle G(\mathbf{L}), \preccurlyeq\rangle$ and hence preserves all existing joins. So $\varphi_{a^{-1}}(b \vee^{\mathbf{G}(\mathbf{L})} a) = \varphi_{a^{-1}}(b) \vee^{\mathbf{G}(\mathbf{L})} \varphi_{a^{-1}}(a) = a^{-1}b \vee^{\mathbf{G}(\mathbf{L})} \mathrm{e}$ exists as required.

In light of the previous step, let us denote the join operation in $\mathbf{G}(\mathbf{L})$ by $\vee$. Consider $a^{-1}b, c^{-1}d \in G(\mathbf{L})$, with $a, b, c, d \in L$, and, using the right reversibility of the monoid reduct of $\mathbf{L}$, let $x, y \in L$ be such that $x^{-1}y = ac^{-1}$, that is, $xa = yc$. Then, using the fact that multiplication distributes over joins,

$$(a^{-1}b) \vee (c^{-1}d) = a^{-1}(b \vee ac^{-1}d) = a^{-1}(b \vee x^{-1}yd) = (xa)^{-1}(xb \vee yd). \qquad \square$$

Let us also denote by $\mathbf{G}(\mathbf{L})$ the unique ℓ-group of left quotients of an Ore residuated lattice $\mathbf{L}$ as constructed in Lemma 6.2.3. The following result is then a direct consequence of this lemma.

COROLLARY 6.2.4. *Any integral Ore residuated lattice* $\mathbf{L}$ *embeds as a monoid into the negative cone of* $\mathbf{G}(\mathbf{L})$.

6.3. A categorical equivalence

Let us call a co-nucleus δ on an ℓ-group $\mathbf{G}$ *fractional* if the group reduct of $\mathbf{G}$ is the group of left quotients of the monoid reduct of $\mathbf{G}_\delta$. In this case, we call the expansion of the ℓ-group $\mathbf{G}$ with the unary operation δ, denoted by $\langle \mathbf{G}, \delta\rangle$, an *$\ell$-group with a fractional co-nucleus.*

Our goal in this section will be to establish an equivalence between

1. $\mathcal{ORL}$, the category whose objects are Ore residuated lattices, and morphisms are (ordinary) homomorphisms;
2. $\mathcal{LG}_{cn}$, the category whose objects are ℓ-groups with a fractional co-nucleus, and morphisms are (ordinary) homomorphisms.

The equivalence will follow as the culmination of several preliminary results. First, let $\mathbf{L}$ be any Ore residuated lattice and suppose that $a^{-1}b = c^{-1}d$ in $\mathbf{G}(\mathbf{L})$, for some $a, b, c, d \in L$. By Lemma 6.2.1, there exist $x, y \in L$ such that $xa = yc$ and $xb = yd$, so, by cancellativity, $a\backslash b = a\backslash(x\backslash xb) = xa\backslash xb = yc\backslash yd = c\backslash(y\backslash yd) = c\backslash d$. We may therefore unambiguously define $\delta_{\mathbf{L}}(a^{-1}b) := a\backslash b$, for any $a, b \in L$.

LEMMA 6.3.1. *Let* $\mathbf{L}$ *be any Ore residuated lattice. Then* $\delta_{\mathbf{L}}$ *is a fractional co-nucleus on* $\mathbf{G}(\mathbf{L})$.

PROOF. Let $a, b \in L$. Then $a\backslash b = \max\{z \in L \mid az \le b\}$. But also, $az \le b$ in $\mathbf{L}$ if and only if $z \le a^{-1}b$ in $\mathbf{G}(\mathbf{L})$, for any $z \in L$. So $a\backslash b = \max\{z \in L \mid z \le a^{-1}b\}$. Hence $\mathbf{L}$ is a co-closure system of $\mathbf{G}(\mathbf{L})$ and, by Proposition 1.3.10, the associated co-closure operator $\delta_{\mathbf{L}}$ is a fractional co-nucleus on $\mathbf{G}(\mathbf{L})$; that is, $\mathbf{L} = \mathbf{G}(\mathbf{L})_{\delta_{\mathbf{L}}}$. $\square$

We can therefore associate with any Ore residuated lattice $\mathbf{L}$ in $\mathcal{ORL}$, an ℓ-group with a fractional co-nucleus $\Omega(\mathbf{L}) := \langle \mathbf{G}(\mathbf{L}), \delta_{\mathbf{L}}\rangle$ in $\mathcal{LG}_{cn}$. Next, we take care of the morphisms. Consider a homomorphism $\varphi\colon \mathbf{L} \to \mathbf{K}$ between Ore residuated lattices $\mathbf{L}$ and $\mathbf{K}$, and $a, b, c, d \in L$ satisfying $a^{-1}b = c^{-1}d$ in $\mathbf{G}(\mathbf{L})$. By Lemma 6.2.1, there exist $x, y \in L$ such that $xa = yc$ and $xb = yd$. So $\varphi(x)\varphi(a) = \varphi(y)\varphi(c)$ and $\varphi(x)\varphi(b) = \varphi(y)\varphi(d)$, and also $\varphi(a)^{-1}\varphi(b) = \varphi(c)^{-1}\varphi(d)$. Hence we can unambiguously define the map

$$\Omega(\varphi)\colon \langle \mathbf{G}(\mathbf{L}), \delta_{\mathbf{L}}\rangle \to \langle \mathbf{G}(\mathbf{K}), \delta_{\mathbf{K}}\rangle; \quad a^{-1}b \mapsto \varphi(a)^{-1}\varphi(b).$$

LEMMA 6.3.2. *If $\varphi\colon \mathbf{L} \to \mathbf{K}$ is a homomorphism between Ore residuated lattices $\mathbf{L}$ and $\mathbf{K}$, then $\Omega(\varphi)$ is the unique homomorphism from $\langle \mathbf{G}(\mathbf{L}), \delta_{\mathbf{L}}\rangle$ to $\langle \mathbf{G}(\mathbf{K}), \delta_{\mathbf{K}}\rangle$ that extends φ.*

PROOF. Clearly, any homomorphism from $\langle \mathbf{G}(\mathbf{L}), \delta_{\mathbf{L}}\rangle$ to $\langle \mathbf{G}(\mathbf{K}), \delta_{\mathbf{K}}\rangle$ extending φ must coincide with $\Omega(\varphi)$, so it suffices to prove that $\Omega(\varphi)$ is a homomorphism. First we show that $\Omega(\varphi)$ is an ℓ-group homomorphism. Clearly, $\Omega(\varphi)$ preserves the group operations and, since $x \wedge y \approx (x^{-1} \vee y^{-1})^{-1}$ is satisfied by all ℓ-groups, it remains to prove that $\Omega(\varphi)$ preserves finite joins. Let $a, b, c, d \in L$. By Lemma 6.2.3, for any $x, y \in L$ such that $xa = yc$,

$$(a^{-1}b) \vee (c^{-1}d) = (xa)^{-1}(xb \vee yd).$$

Since φ is a homomorphism and $\Omega(\varphi)$ preserves the group operations,

$$\Omega(\varphi)((a^{-1}b) \vee (c^{-1}d)) = (\varphi(x)\varphi(a))^{-1}(\varphi(x)\varphi(b) \vee \varphi(y)\varphi(d)).$$

A further application of Lemma 6.2.3 yields as required

$$\Omega(\varphi)((a^{-1}b) \vee (c^{-1}d)) = (\varphi(a)^{-1}\varphi(b)) \vee (\varphi(c)^{-1}\varphi(d)) = \Omega(\varphi)(a^{-1}b) \vee \Omega(\varphi)(c^{-1}d).$$

Finally, we check that $\Omega(\varphi)$ commutes with the co-nuclei. For any $a, b \in L$,

$$\Omega(\varphi)(\delta_{\mathbf{L}}(a^{-1}b)) = \varphi(a\backslash b) = \varphi(a)\backslash\varphi(b) = \delta_{\mathbf{K}}(\varphi(a)^{-1}\varphi(b)) = \delta_{\mathbf{K}}(\Omega(\varphi)(a^{-1}b)).$$

Hence $\Omega(\varphi)$ is a homomorphism from $\langle \mathbf{G}(\mathbf{L}), \delta_{\mathbf{L}}\rangle$ to $\langle \mathbf{G}(\mathbf{K}), \delta_{\mathbf{K}}\rangle$ that extends φ. □

So far, we have shown that Ω is a functor from $\mathcal{ORL}$ to $\mathcal{LG}_{cn}$. To obtain an equivalence, we define now a functor Ξ from $\mathcal{LG}_{cn}$ to $\mathcal{ORL}$. Note first that, for any fractional co-nucleus δ on an ℓ-group $\mathbf{G}$, we obtain, by virtue of Lemma 1.5.11 and the assumption on δ, an Ore residuated lattice $\mathbf{G}_\delta$. Moreover, for any morphism $\varphi\colon \langle \mathbf{G}, \delta_1\rangle \to \langle \mathbf{H}, \delta_2\rangle$ in $\mathcal{LG}_{cn}$, the restriction of φ to G_{δ_1} is a homomorphism from $\mathbf{G}_{\delta_1}$ to $\mathbf{H}_{\delta_2}$. Hence we may define the functor Ξ as follows:

(a) For every object $\langle \mathbf{G}, \delta\rangle$ of $\mathcal{LG}_{cn}$, let $\Xi(\langle \mathbf{G}, \delta\rangle) := \mathbf{G}_\delta$.
(b) For every morphism $\varphi\colon \langle \mathbf{G}, \delta_1\rangle \to \langle \mathbf{H}, \delta_2\rangle$ in $\mathcal{LG}_{cn}$, let $\Xi(\varphi)\colon \mathbf{G}_{\delta_1} \to \mathbf{H}_{\delta_2}$ be the restriction of φ to G_{δ_1}.

To establish the equivalence between $\mathcal{ORL}$ and $\mathcal{LG}_{cn}$, it suffices, by general results from category theory, to prove the following lemma.

LEMMA 6.3.3.

(a) *Every object $\langle \mathbf{G}, \delta\rangle$ in $\mathcal{LG}_{cn}$ is isomorphic to $\Omega(\Xi(\langle \mathbf{G}, \delta\rangle))$.*
(b) *The functor Ω is faithful and full, i.e., the map sending a morphism $\varphi\colon \mathbf{L} \to \mathbf{K}$ in $\mathcal{ORL}$ to a morphism $\Omega(\varphi)\colon \langle \mathbf{G}(\mathbf{L}), \delta_{\mathbf{L}}\rangle \to \langle \mathbf{G}(\mathbf{K}), \delta_{\mathbf{K}}\rangle$ in $\mathcal{LG}_{cn}$ is injective and surjective.*

PROOF.

(a) Consider any $\langle \mathbf{G}, \delta \rangle$ in $\mathcal{LG}_{cn}$ and let $\mathbf{L} := \mathbf{G}_\delta$. We need to prove that $\langle \mathbf{G}, \delta \rangle$ is isomorphic to $\langle \mathbf{G}(\mathbf{L}), \delta_\mathbf{L} \rangle$. Note first that both $\mathbf{G}$ and $\mathbf{G}(\mathbf{L})$ are ℓ-groups of left quotients of $\mathbf{L}$. Hence, by Lemma 6.3.2, there exists an ℓ-group isomorphism $\varphi\colon \mathbf{G} \to \mathbf{G}(\mathbf{L})$ that fixes the elements of $\mathbf{L}$. It remains therefore to prove that $\varphi\delta = \delta_\mathbf{L}\varphi$. Let $\cdot_1$ and $\cdot_2$ denote multiplication in $\mathbf{G}$ and $\mathbf{G}(\mathbf{L})$, respectively, and let ${}^{-1}$ denote the inverse operation for both algebras. Consider any element $a^{-1} \cdot_1 b$ of $\mathbf{G}$ with $a, b \in L$. Denoting the left residual of $\mathbf{L}$ by $\backslash_\mathbf{L}$,

$$\varphi(\delta(a^{-1} \cdot_1 b)) = \varphi(a\backslash_\mathbf{L} b) = a\backslash_\mathbf{L} b = \delta_\mathbf{L}(a^{-1} \cdot_2 b) = \delta_\mathbf{L}(\varphi(a^{-1} \cdot_1 b)).$$

So $\varphi\delta = \delta_\mathbf{L}\varphi$ as required.

(b) For distinct morphisms $\varphi_1\colon \mathbf{L} \to \mathbf{K}$ and $\varphi_2\colon \mathbf{L} \to \mathbf{K}$ in $\mathcal{ORL}$, the morphisms $\Omega(\varphi_1)$ and $\Omega(\varphi_2)$ in $\mathcal{LG}_{cn}$ extend φ_1 and φ_2, respectively, and are hence also distinct. So Ω is faithful. To check that Ω is full, consider any morphism $\varphi\colon \langle \mathbf{G}, \delta_1 \rangle \to \langle \mathbf{H}, \delta_2 \rangle$ in $\mathcal{LG}_{cn}$. By (a), we may safely assume that $\langle \mathbf{G}, \delta_1 \rangle = \Omega(\mathbf{L})$ and $\langle \mathbf{H}, \delta_2 \rangle = \Omega(\mathbf{K})$, for appropriate objects $\mathbf{L}$ and $\mathbf{K}$ in $\mathcal{ORL}$. So $\Xi(\varphi)$ is a homomorphism from $\mathbf{L}$ to $\mathbf{K}$, and both φ and $\Omega(\Xi(\varphi))$ are homomorphisms from $\Omega(\mathbf{L})$ to $\Omega(\mathbf{K})$ that extend $\Xi(\varphi)$. Hence $\varphi = \Omega(\Xi(\varphi))$, by the uniqueness part of Lemma 6.3.2. □

Our main result now follows directly.

THEOREM 6.3.4. *The pair of functors $\Omega\colon \mathcal{ORL} \to \mathcal{LG}_{cn}$ and $\Xi\colon \mathcal{LG}_{cn} \to \mathcal{ORL}$ constitutes an equivalence between the categories $\mathcal{ORL}$ and $\mathcal{LG}_{cn}$.*

We conclude this section by considering some further notable equivalences obtained as specializations of Theorem 6.3.4. Let $\mathcal{CC}\mathrm{an}\mathcal{RL}$ be the category of commutative, cancellative residuated lattices with homomorphisms, and let $\mathcal{A}b\mathcal{LG}_{cn}$ be the full subcategory of $\mathcal{LG}_{cn}$ containing objects $\langle \mathbf{G}, \delta \rangle$ such that $\mathbf{G}$ is an Abelian ℓ-group.

COROLLARY 6.3.5. *The restrictions of Ω and Ξ constitute an equivalence between the categories $\mathcal{CC}an\mathcal{RL}$ and $\mathcal{A}b\mathcal{LG}_{cn}$.*

The next result collects together equivalences for classes of cancellative residuated lattices satisfying equations considered in Chapters 4 and 5.

PROPOSITION 6.3.6. *The restrictions of Ω and Ξ yield equivalences between*

(i) *the full subcategory of $\mathcal{ORL}$ whose objects satisfy $x(y \wedge z) \approx xy \wedge xz$ and the full subcategory of $\mathcal{LG}_{cn}$ whose objects satisfy $\delta(x \wedge y) \approx \delta(x) \wedge \delta(y)$;*
(ii) *the subcategory of $\mathcal{CC}an\mathcal{RL}$ whose objects satisfy $x(y \wedge z) \approx xy \wedge xz$ and the subcategory of $\mathcal{A}b\mathcal{LG}_{cn}$ whose objects satisfy $\delta(x \wedge y) \approx \delta(x) \wedge \delta(y)$;*
(iii) *the full subcategory of $\mathcal{ORL}$ whose objects satisfy $x\backslash(y \vee z) \approx (x\backslash y) \vee (x\backslash z)$ and the full subcategory of $\mathcal{LG}_{cn}$ whose objects satisfy $\delta(x \vee y) \approx \delta(x) \vee \delta(y)$;*
(iv) *the full subcategory of $\mathcal{CC}an\mathcal{RL}$ whose objects satisfy $x\backslash(y \vee z) \approx (x\backslash y) \vee (x\backslash z)$ and the full subcategory of $\mathcal{A}b\mathcal{LG}_{cn}$ whose objects satisfy $\delta(x \vee y) \approx \delta(x) \vee \delta(y)$.*

PROOF.

(i) It suffices to prove that $\langle \mathbf{G}, \delta \rangle$ in $\mathcal{LG}_{cn}$ satisfies $\delta(x \wedge y) \approx \delta(x) \wedge \delta(y)$ if and only if $\mathbf{G}_\delta$ satisfies $x(y \wedge z) \approx xy \wedge xz$. Suppose first that $\langle \mathbf{G}, \delta \rangle$ satisfies $\delta(x \wedge y) \approx$

$\delta(x) \wedge \delta(y)$. Then the residual-free reduct of $\mathbf{G}_\delta$ is a subalgebra of the residual-free reduct of $\mathbf{G}$ and, since $x(y \wedge z) \approx xy \wedge xz$ is satisfied by every ℓ-group, it is satisfied by $\mathbf{G}$ and hence also $\mathbf{G}_\delta$.

Suppose now that $\mathbf{G}_\delta$ satisfies $x(y \wedge z) \approx xy \wedge xz$. We show first that the meet operation $\wedge$ of $\mathbf{G}_\delta$ is the restriction of the meet operation $\wedge^{\mathbf{G}}$ of $\mathbf{G}$. Consider $a, b \in G_\delta$. Clearly, $a \wedge b$ is a lower bound of a and b in $\mathbf{G}$. Consider an arbitrary element $c^{-1}d$ of G with $c, d \in G_\delta$. If $c^{-1}d$ is a lower bound of a and b, then $d \leq ca$ and $d \leq cb$ in $\mathbf{G}_\delta$. By assumption, $d \leq c(a \wedge b)$ in $\mathbf{G}_\delta$, and so $c^{-1}d \leq a \wedge b$ in $\mathbf{G}$. So $a \wedge b$ is the greatest lower bound of a and b in $\mathbf{G}$. It follows that the meet of arbitrary elements $c_1^{-1}d_1$ and $c_2^{-1}d_2$ of G with $c_1, c_2, d_1, d_2 \in G_\delta$ is

$$c_1^{-1}d_1 \wedge^{\mathbf{G}} c_2^{-1}d_2 = (ac_1)^{-1}(ad_1 \wedge bd_2),$$

where a, b are any elements of G_δ satisfying $ac_1 = bc_2$. Hence, using the fact that the left division operation in $\mathbf{G}_\delta$ is given by $c \backslash d = \delta(c^{-1}d)$, for $a, b \in G_\delta$,

$$\begin{aligned}
\delta(c_1^{-1}d_1 \wedge^{\mathbf{G}} c_2^{-1}d_2) &= \delta((ac_1)^{-1}(ad_1 \wedge bd_2)) \\
&= ac_1 \backslash (ad_1 \wedge bd_2) \\
&= (ac_1 \backslash ad_1) \wedge (ac_1 \backslash bd_2) \\
&= (ac_1 \backslash ad_1) \wedge (bc_2 \backslash bd_2) \\
&= (c_1 \backslash d_1) \wedge (c_2 \backslash d_2) \\
&= \delta(c_1^{-1}d_1) \wedge^{\mathbf{G}} \delta(c_2^{-1}d_2).
\end{aligned}$$

That is, $\langle \mathbf{G}, \delta \rangle$ satisfies $\delta(x \wedge y) \approx \delta(x) \wedge \delta(y)$ as required.

(ii) follows directly from (i).

(iii) It suffices to prove that $\langle \mathbf{G}, \delta \rangle$ in $\mathcal{LG}_{cn}$ satisfies $\delta(x \vee y) \approx \delta(x) \vee \delta(y)$ if and only if $\mathbf{G}_\delta$ satisfies $x \backslash (y \vee z) \approx (x \backslash y) \vee (x \backslash z)$. Suppose first that $\langle \mathbf{G}, \delta \rangle$ satisfies $\delta(x \vee y) \approx \delta(x) \vee \delta(y)$. Then, for any $a, b, c, d \in G_\delta$,

$$a \backslash (b \vee c) = \delta(a^{-1}(b \vee c)) = \delta((a^{-1}b) \vee (a^{-1}c)) = \delta(a^{-1}b) \vee \delta(a^{-1}c)) = (a \backslash b) \vee (a \backslash c).$$

That is, $\mathbf{G}_\delta$ satisfies $x \backslash (y \vee z) \approx (x \backslash y) \vee (x \backslash z)$ as required.

Suppose now that $\mathbf{G}_\delta$ satisfies $x \backslash (y \vee z) \approx (x \backslash y) \vee (x \backslash z)$ and let $c_1^{-1}d_1$ and $c_2^{-1}d_2$ be elements of G with $c_1, c_2, d_1, d_2 \in G_\delta$. Their join in $\mathbf{G}$ is given by

$$c_1^{-1}d_1 \vee c_2^{-1}d_2 = (ac_1)^{-1}(ad_1 \vee bd_2),$$

where a, b are any elements of G_δ satisfying $ac_1 = bc_2$. Hence we obtain

$$\begin{aligned}
\delta(c_1^{-1}d_1 \vee c_2^{-1}d_2) &= \delta((ac_1)^{-1}(ad_1 \vee bd_2)) \\
&= ac_1 \backslash (ad_1 \vee bd_2) \\
&= (ac_1 \backslash ad_1) \vee (ac_1 \backslash bd_2) \\
&= (ac_1 \backslash ad_1) \vee (bc_2 \backslash bd_2) \\
&= (c_1 \backslash d_1) \vee (c_2 \backslash d_2) \\
&= \delta(c_1^{-1}d_1) \vee \delta(c_2^{-1}d_2).
\end{aligned}$$

That is, $\langle \mathbf{G}, \delta \rangle$ satisfies $\delta(x \vee y) \approx \delta(x) \vee \delta(y)$ as required.

(iv) follows directly from (iii). □

6.4. Bibliographical remarks

The class of residuated lattices with a cancellative monoid reduct was first shown to form a variety and studied in detail in [**16**], and the results and examples of Section 6.1 are all taken from this paper, with the exception of Example 6.1.4, which may be found in [**38**]. Let us remark also in passing that it was shown in [**139**] that every cancellative residuated lattice is either an ℓ-group or contains the negative cone of the ordered group of integers as a subalgebra, and hence that the only cancellative atoms in the subvariety lattice of residuated lattices are the varieties of Abelian ℓ-groups and negative cones of Abelian ℓ-groups.

Theorem 6.2.2 originated in the work on rings by Ore [**301**] and Dubreil [**106**]. A readable historical account of these developments is given in [**190**, Chapter 5], and an elegant proof of the result can be found in [**81**, Section 1.10]. The proof we present here follows Ore's ordered pair approach, familiar to any algebraist (at least in the commutative setting) from the construction of groups of fractions. Cancellativity and right reversibility are jointly sufficient, but by no means jointly necessary, conditions for embeddability of a monoid into a group. Necessary and sufficient conditions were first provided by Mal'cev in [**246**], while the fact that every monoid that does not generate the variety of all monoids is right reversible was proved by Neumann and Taylor in [**286**].

The categorical equivalence between Ore residuated lattices and ℓ-groups with a fractional co-nucleus, and its restrictions to notable subcategories, were established in [**272**]. In [**377**], these results were placed in a common framework with the well-known correspondence between Heyting algebras and Boolean algebras with a co-nucleus, and the categorical equivalence between integral GMV-algebras and negative cones of ℓ-groups with a strong order unit, to be presented in Chapter 7.

To date, no analytic proof system has been obtained for the variety of cancellative residuated lattices or indeed any of its subvarieties satisfying one or more of the integrality, commutativity, and semilinearity conditions. Moreover, it is not known if any of these varieties have a decidable equational theory, except in the case of commutative, cancellative, semilinear, integral residuated lattices. Decidability for this latter variety was established, via a description of its totally ordered members, in [**193**] (see also [**194**]); a similar description was obtained for varieties of nilpotent, cancellative, integral residuated lattices in [**83**].

CHAPTER 7

Divisibility

In this chapter, we turn our attention to *divisibility*, a familiar algebraic property that, like cancellativity, is satisfied by the pomonoid reduct of any ℓ-group and its negative cone. The class of divisible (bounded) residuated lattices is both general enough to encompass algebras for a wide range of non-classical logics — including intuitionistic propositional logic (Heyting algebras) and Hájek's basic fuzzy logic (BL-algebras) — and specific enough to admit elegant representation theorems.

A pomonoid $\langle M, \cdot, \mathrm{e}, \leq \rangle$ is said to be *divisible* if, for all $a, b \in M$ with $a \leq b$, there exist $c, d \in M$ satisfying $bc = db = a$. In Sections 7.1–7.3, we study the structure of residuated lattices with a divisible pomonoid reduct, referred to as *GBL-algebras*. We show that these algebras have a distributive lattice reduct and form a variety whose cancellative integral members are precisely the negative cones of ℓ-groups. Two structure theorems are presented. First, we prove that every GBL-algebra decomposes as the inner direct product of an ℓ-group and an integral GBL-algebra. We then prove that every totally ordered, integral GBL-algebra decomposes as an ordinal sum of totally ordered, integral *GMV-algebras*, the latter forming a variety of GBL-algebras that encompasses all ℓ-groups and their negative cones.

In Sections 7.4 and 7.5, we delve deeper into the structure of integral GMV-algebras. Using an equational description of $\{\backslash, /, \mathrm{e}\}$-subreducts of negative cones of ℓ-groups as *cone algebras*, we obtain a representation of integral GMV-algebras as nucleus-systems of negative cones of ℓ-groups. This representation allows us to establish an equivalence between a category of integral GMV-algebras and a category of negative cones of ℓ-groups augmented with a dense nucleus. Finally, in Section 7.6, we show that a broad class of GBL-algebras, encompassing all GMV-algebras, admits a generalization of Otto Hölder's well-known theorem for totally ordered groups.

7.1. GBL-algebras and GMV-algebras

Recall from Chapter 1 that a residuated lattice $\mathbf{L} = \langle L, \wedge, \vee, \cdot, \backslash, /, \mathrm{e} \rangle$ is called a *GBL-algebra* (short for *generalized basic logic algebra*) if it satisfies the equations

$$y(y\backslash(x \wedge y)) \approx x \wedge y \quad \text{and} \quad ((x \wedge y)/y)y \approx x \wedge y.$$

For integral residuated lattices, these defining equations may be replaced (see Proposition 7.1.4) by the simplified versions

$$y(y\backslash x) \approx x \wedge y \quad \text{and} \quad (x/y)y \approx x \wedge y.$$

The variety $\mathcal{GBL}$ of GBL-algebras includes all Brouwerian algebras, ℓ-groups, and negative cones of ℓ-groups, but excludes, for example, the cancellative residuated lattices defined in Example 1.4.14 and the proof of Theorem 6.1.3.

Let us now establish some elementary properties of GBL-algebras. We call an element a of a residuated lattice $\mathbf{L}$ *positive* if $\mathrm{e} \leq a$, *idempotent* if $aa = a$, *integral* if $a\backslash\mathrm{e} = \mathrm{e} = \mathrm{e}/a$, and *invertible* if there exists a (necessarily unique) element $a^{-1} \in L$ satisfying $a^{-1}a = \mathrm{e} = aa^{-1}$.

LEMMA 7.1.1. *Let* $\mathbf{L}$ *be any residuated lattice.*

(a) *If* $a \in L$ *is invertible, then* $a\backslash b = a^{-1}b$ *and* $b/a = ba^{-1}$, *for any* $b \in L$.
(b) $a \in L$ *is invertible if and only if* $a(a\backslash\mathrm{e}) = \mathrm{e} = (\mathrm{e}/a)a$, *i.e.,* $a^{-1} = a\backslash\mathrm{e} = \mathrm{e}/a$.

PROOF.

(a) Suppose that $a \in L$ is invertible. Then, for any $b, c \in L$,

$$c \leq a^{-1}b \iff ac \leq b \iff c \leq a\backslash b.$$

Hence $a\backslash b = a^{-1}b$ and, symmetrically, $b/a = ba^{-1}$.

(b) The non-trivial direction follows directly from (a), taking $b = \mathrm{e}$. □

LEMMA 7.1.2. *Let* $\mathbf{L}$ *be any GBL-algebra.*

(a) *Every positive element of* $\mathbf{L}$ *is invertible.*
(b) $\mathbf{L}$ *satisfies the equations* $x\backslash x \approx \mathrm{e}$ *and* $x/x = \mathrm{e}$, *and is* e*-cyclic.*
(c) *If* $\mathbf{L}$ *has a greatest element, then it is integral.*

PROOF.

(a) Just observe that, for any positive element a of $\mathbf{L}$,

$$a(a\backslash\mathrm{e}) = a(a\backslash(\mathrm{e} \wedge a)) = \mathrm{e} \wedge a = \mathrm{e} = \mathrm{e} \wedge a = ((\mathrm{e} \wedge a)/a)a = (\mathrm{e}/a)a.$$

(b) Let $a \in L$. Then $a\backslash a$ is positive and, by (a), invertible. Hence, using also Proposition 1.4.4.(viii),

$$\mathrm{e} = (a\backslash a)((a\backslash a)\backslash\mathrm{e}) = (a\backslash a)^2((a\backslash a)\backslash\mathrm{e}) = (a\backslash a)\mathrm{e} = a\backslash a.$$

So $\mathbf{L}$ satisfies $x\backslash x \approx \mathrm{e}$ and, by Lemma 6.1.2.(a), it satisfies $x/x \approx \mathrm{e}$, and is e-cyclic.

(c) Follows immediately from (b) and Lemma 6.1.2.(b). □

By Lemma 7.1.2.(c), every bounded residuated lattice with a GBL-algebra reduct — referred to as a *pseudo-BL-algebra* — is integral. A commutative, semilinear pseudo-BL-algebra is called a *BL-algebra*. The following lemma shows that every finite, semilinear GBL-algebra is not only integral, but also commutative, and hence that every finite, semilinear pseudo-BL-algebra is a BL-algebra. In fact, it can be shown that all finite GBL-algebras are commutative, but the proof is more involved and will not be given here (see the bibliographical remarks).

LEMMA 7.1.3. *Every finite, semilinear GBL-algebra is integral and commutative.*

PROOF. Since every finite GBL-algebra is integral, by Lemma 7.1.2.(c), it suffices to show that any non-trivial, finite, totally ordered GBL-algebra $\mathbf{L}$ is commutative. Let $a \in L$ be the unique coatom of $\mathbf{L}$. We claim that $\mathbf{L}$ is generated as a monoid by a, and is therefore commutative. Suppose towards a contradiction that there exist $b \in L$ and $n \in \mathbb{N}$ satisfying $a^{n+1} < b < a^n$. Then $a^n a = a^{n+1} < b$ and $a^n(a^n\backslash b) = b$, so $a^n\backslash b > a$. But a is a coatom, so $a^n\backslash b = \mathrm{e}$ and $a^n \leq b$, a contradiction. □

The next result clarifies the relationship between GBL-algebras and divisibility.

PROPOSITION 7.1.4. *Let* $\mathbf{L}$ *be any residuated lattice. The following statements are equivalent:*

(1) $\mathbf{L}$ *is a GBL-algebra.*
(2) $\mathbf{L}$ *satisfies the quasi-equation* $x \leq y \implies x \approx y(y\backslash x)$ *and its mirror image.*
(3) $\mathbf{L}$ *satisfies the equations* $y(y\backslash x \wedge \mathrm{e}) \approx x \wedge y$ *and* $(x/y \wedge \mathrm{e})y \approx x \wedge y$.
(4) *The pomonoid* $\langle L, \cdot, \mathrm{e}, \leq\rangle$ *is divisible.*

PROOF.

(1) ⇔ (2) Obvious.

(1) ⇒ (3) If $\mathbf{L}$ is a GBL-algebra, then, for any $a, b \in L$, using Lemma 7.1.2.(b),

$$b(b\backslash a \wedge \mathrm{e}) = b(b\backslash a \wedge b\backslash b) = b(b\backslash(a \wedge b)) = a \wedge b.$$

For the mirror image, we argue analogously.

(3) ⇒ (4) If (3) holds, then, for any $a, b \in L$ with $a \leq b$,

$$a = a \wedge b = b(b\backslash a \wedge \mathrm{e}) = (a/b \wedge \mathrm{e})b.$$

(4) ⇒ (2) Suppose that $\langle L, \cdot, \mathrm{e}, \leq\rangle$ is divisible, and consider any $a, b \in L$ with $a \leq b$. By divisibility, $a = bc$, for some $c \in L$. So $c \leq b\backslash a$ and $b(b\backslash a) \leq a = bc \leq b(b\backslash a)$, that is, $b(b\backslash a) = a$. The proof for the mirror image is again analogous. □

It is well-known that ℓ-groups and BL-algebras are distributive; the following result shows that this is the case, in fact, for all GBL-algebras.

PROPOSITION 7.1.5. *Every GBL-algebra is distributive.*

PROOF. It suffices to prove that the equation $x \wedge (y \vee z) \leq (x \wedge y) \vee (x \wedge z)$ is satisfied by every GBL-algebra $\mathbf{L}$. For any $a, b, c \in L$, using Propositions 7.1.4 and 1.4.2,

$$\begin{aligned} a \wedge (b \vee c) &= (b \vee c)((b \vee c)\backslash a \wedge \mathrm{e}) \\ &= b((b \vee c)\backslash a \wedge \mathrm{e}) \vee c((b \vee c)\backslash a \wedge \mathrm{e}) \\ &\leq b(b\backslash a \wedge \mathrm{e}) \vee c(c\backslash a \wedge \mathrm{e}) \\ &= (a \wedge b) \vee (a \wedge c). \end{aligned}$$

An alternative proof is obtained by applying Proposition 1.1.10 using the join-homomorphism defined by $x \mapsto x(b\backslash(a \wedge b))$, for any $a, b \in L$ with $a \leq b$. □

Below, we extend a familiar decomposition property for elements of ℓ-groups to the elements of any GBL-algebra.

PROPOSITION 7.1.6. *Let* $\mathbf{L}$ *be any GBL-algebra. Then every element* $a \in L$ *is the product of its positive part* $a \vee \mathrm{e}$ *and negative part* $a \wedge \mathrm{e}$*; that is,* $\mathbf{L}$ *satisfies the equations* $x \approx (x \vee \mathrm{e})(x \wedge \mathrm{e})$ *and* $x \approx (x \wedge \mathrm{e})(x \vee \mathrm{e})$.

PROOF. Consider any element $a \in L$. Then $(\mathrm{e}/a \wedge \mathrm{e})a = a \wedge \mathrm{e}$, by Proposition 7.1.4. Also $a \vee \mathrm{e}$ is invertible and $(a \vee \mathrm{e})^{-1} = (a \vee \mathrm{e})\backslash\mathrm{e} = a\backslash\mathrm{e} \wedge \mathrm{e}$, by Lemma 7.1.2.(a). Hence $(a \vee \mathrm{e})^{-1}a = (a\backslash\mathrm{e} \wedge \mathrm{e})a = (\mathrm{e}/a \wedge \mathrm{e})a = a \wedge \mathrm{e}$, by Lemma 7.1.2.(b), and so $a = (a \vee \mathrm{e})(a \wedge \mathrm{e})$. An analogous proof shows that $a = (a \wedge \mathrm{e})(a \vee \mathrm{e})$. □

We turn our attention now to an important subvariety of $\mathcal{GBL}$ that contains every ℓ-group and its negative cone. A residuated lattice is called a *GMV-algebra* (short for *generalized MV-algebra*) if it satisfies the equations

$$x/((x \vee y)\backslash x) \approx x \vee y \quad \text{and} \quad (x/(x \vee y))\backslash x \approx x \vee y.$$

LEMMA 7.1.7. *Every GMV-algebra is a GBL-algebra.*

PROOF. Let $\mathbf{L}$ be any GMV-algebra. By Proposition 7.1.4, it suffices to prove that $\mathbf{L}$ satisfies the quasi-equation $x \leq y \Longrightarrow x \approx y(y\backslash x)$ and its mirror image. Consider any $a, b \in L$ with $a \leq b$, and let $c := b(b\backslash a)$. Then Lemma 1.4.5.(ii) yields $b\backslash c \leq a\backslash c$ and, using the defining equations for GMV-algebras and Proposition 1.4.4.(v),

$$a\backslash c = ((b\backslash c)/(a\backslash c))\backslash(b\backslash c) = (b\backslash(c/(a\backslash c)))\backslash(b\backslash c).$$

Since $c \leq a$, also $c/(a\backslash c) = a$ and $a\backslash c = (b\backslash a)\backslash(b\backslash c) = b(b\backslash a)\backslash c = c\backslash c$. Hence $a = c/(a\backslash c) = c/(c\backslash c) = c = b(b\backslash a)$. The proof that $a = (a/b)b$ is analogous. □

Let us denote the variety of GMV-algebras by $\mathcal{GMV}$. As in the case of $\mathcal{GBL}$, this variety admits useful alternative characterizations.

PROPOSITION 7.1.8. *Let* $\mathbf{L}$ *be any residuated lattice. The following statements are equivalent:*

(1) $\mathbf{L}$ *is a GMV-algebra.*
(2) $\mathbf{L}$ *satisfies the quasi-equation* $x \leq y \Longrightarrow x/(y\backslash x) \approx y$ *and its mirror image.*
(3) $\mathbf{L}$ *satisfies the equations* $x/(y\backslash x \wedge \mathrm{e}) \approx x \vee y$ *and* $(x/y \wedge \mathrm{e})\backslash x \approx x \vee y$.

PROOF.

(1) ⇔ (2) Obvious.

(1) ⇒ (3) If $\mathbf{L}$ is a GMV-algebra, then, for any $a, b \in L$, using Lemma 7.1.2.(b),

$$a/(b\backslash a \wedge \mathrm{e}) = a/(b\backslash a \wedge a\backslash a) = a/((a \vee b)\backslash a) = a \vee b$$
$$(a/b \wedge \mathrm{e})\backslash a = (a/b \wedge a/a)\backslash a = (a/(a \vee b))\backslash a = a \vee b.$$

(3) ⇒ (1) Assuming (3), for any $a, b \in L$, using Proposition 1.4.2.(b),

$$a/((a \vee b)\backslash a) = a/(b\backslash a \wedge a\backslash a) \leq a/(b\backslash a \wedge \mathrm{e}) = a \vee b \leq a/((a \vee b)\backslash a)$$
$$(a/(a \vee b))\backslash a = (a/b \wedge a/a)\backslash a \leq (a/b \wedge \mathrm{e})\backslash a = a \vee b \leq (a/(a \vee b))\backslash a.$$ □

It follows from condition (3) of this last proposition that the variety $\mathcal{IGMV}$ of integral GMV-algebras can also be defined relative to $\mathcal{RL}$ by the equations

$$x/(y\backslash x) \approx x \vee y \quad \text{and} \quad (x/y)\backslash x \approx x \vee y.$$

BL-algebras and pseudo-BL-algebras satisfying these latter equations are known as *MV-algebras* (see Example 1.4.12) and *pseudo-MV-algebras*, respectively.

It will be shown below (Corollary 7.2.5) that every GMV-algebra is prelinear. The following lemma provides a first step towards a proof of this result.

LEMMA 7.1.9. *Every integral GMV-algebra is prelinear.*

PROOF. By Propositions 7.1.5 and 4.2.4, it suffices to observe that every integral GMV-algebra $\mathbf{L}$ satisfies $(y \wedge z)\backslash x \approx (y\backslash x) \vee (z\backslash x)$; that is, for any $a, b, c \in L$,

$$\begin{aligned}(b\backslash a) \vee (c\backslash a) &= ((b\backslash a)/(c\backslash a))\backslash(b\backslash a)\\ &= (b\backslash(a/(c\backslash a)))\backslash(b\backslash a)\\ &= (b\backslash(a \vee c))\backslash(b\backslash a)\\ &= b(b\backslash(a \vee c))\backslash a\\ &= (b \wedge (a \vee c))\backslash a\\ &= ((b \wedge a) \vee (b \wedge c))\backslash a\\ &= ((b \wedge a)\backslash a) \wedge ((b \wedge c)\backslash a)\\ &= \mathrm{e} \wedge ((b \wedge c)\backslash a)\\ &= (b \wedge c)\backslash a.\end{aligned}$$

□

It is easily checked that ℓ-groups and their negative cones are GMV-algebras and integral GMV-algebras, respectively. Moreover, as we now prove, the negative cones of ℓ-groups are precisely the integral GBL-algebras that are also cancellative.

THEOREM 7.1.10. *Let* $\mathbf{L}$ *be any residuated lattice. The following statements are equivalent:*

(1) $\mathbf{L}$ *is the negative cone of an* ℓ*-group.*
(2) $\mathbf{L}$ *is a cancellative, integral GMV-algebra.*
(3) $\mathbf{L}$ *is a cancellative, integral GBL-algebra.*

PROOF.

(1) $\Rightarrow$ (2) Let $\mathbf{L}$ be the negative cone of an ℓ-group. Clearly, $\mathbf{L}$ is cancellative and integral. Moreover, for any $a, b \in L$ with $a \leq b$, denoting by x^{-1} the inverse of an element x of the ℓ-group,

$$a/(b\backslash a) = a(b^{-1}a \wedge \mathrm{e})^{-1} \wedge \mathrm{e} = a(a^{-1}b \vee \mathrm{e}) = b \vee a = b.$$

So $\mathbf{L}$ satisfies the quasi-equation $x \leq y \implies x/(y\backslash x) \approx y$ and, by symmetry, also its mirror image. Hence $\mathbf{L}$ is a GMV-algebra, by Proposition 7.1.8.

(2) $\Rightarrow$ (3) Follows immediately by Lemma 7.1.7.

(3) $\Rightarrow$ (1) Let $\mathbf{L}$ be a cancellative, integral GBL-algebra. Then $(b/a)a = a \wedge b = (a/b)b$, for any $a, b \in L$. Hence $\mathbf{L}$ is an Ore residuated lattice. By Theorem 6.2.2 and Lemma 6.2.3, the monoid $\langle L, \cdot, \mathrm{e}\rangle$ embeds into the monoid reduct of the ℓ-group of left quotients $\mathbf{G}(\mathbf{L})$, whose elements have the form $a^{-1}b$, for some $a, b \in L$. Moreover, since $\mathbf{L}$ is integral, every element of L is in the negative cone of $\mathbf{G}(\mathbf{L})$. Note also that $\leq^{\mathbf{L}} = \leq^{\mathbf{G}(\mathbf{L})} \cap L^2$, by Lemma 6.2.3, so we may drop all superscripts.

Next, we check that, for all $a, b \in L$,

$$a \leq b \iff b^{-1}a = b\backslash a \text{ and } ab^{-1} = a/b.$$

If $a \leq b$, then $bb^{-1}a = a = a \wedge b = b(b\backslash a)$, and cancellation yields $b^{-1}a = b\backslash a$. Similarly, $ab^{-1} = a/b$. Conversely, if $ab^{-1} = a/b$, then $ab^{-1} \in L$ is a negative element of $\mathbf{G}(\mathbf{L})$, giving $a \leq b$. Hence $G(\mathbf{L})^- = L$. Finally, $b\backslash a = (b^{-1}a) \wedge \mathrm{e}$ and

$a/b = (ab^{-1}) \wedge \mathrm{e}$, for all $a, b \in L$, and the remaining operations of $\mathbf{L}$ are restrictions of the corresponding operations in $\mathbf{G}(\mathbf{L})$, so $\mathbf{L}$ is the negative cone of $\mathbf{G}(\mathbf{L})$. □

The next two corollaries present useful applications of Theorem 7.1.10.

COROLLARY 7.1.11. *Let* $\mathbf{L}$ *be the negative cone of an* ℓ*-group. For any* $a, b, c \in L$,

$$ab = c \iff (b = a\backslash c \text{ and } c \leq a) \iff (a = c/b \text{ and } c \leq b).$$

PROOF. We just prove the first equivalence. First, if $ab = c$, then, by integrality, $c = ab \leq a$, and, since $\mathbf{L}$ is cancellative, $b = a\backslash ab = a\backslash c$. Conversely, if $b = a\backslash c$ and $c \leq a$, then, since $\mathbf{L}$ is divisible, $ab = a(a\backslash c) = a \wedge c = c$. □

COROLLARY 7.1.12. *Any map between negative cones of* ℓ*-groups that preserves multiplication and binary meets also preserves binary joins and residuals.*

PROOF. Suppose that $\varphi\colon L_1 \to L_2$ is a map between negative cones of ℓ-groups $\mathbf{L}_1$ and $\mathbf{L}_2$ that preserves multiplication and binary meets, and consider any $a, b \in L_1$. By assumption, $\varphi(a \wedge b) = \varphi(a(a\backslash b)) = \varphi(a)\varphi(a\backslash b)$, and, using Corollary 7.1.11,

$$\varphi(a\backslash b) = \varphi(a)\backslash\varphi(a \wedge b) = \varphi(a)\backslash(\varphi(a) \wedge \varphi(b)) = \mathrm{e} \wedge \varphi(a)\backslash\varphi(b) = \varphi(a)\backslash\varphi(b).$$

Preservation of the other residual follows analogously and preservation of binary joins follows using the fact that $\mathbf{L}_1$ is a GMV-algebra and hence $a \vee b = b/(a\backslash b)$. □

EXAMPLE 7.1.13. Let $\mathbb{N}^X$ be the set of maps from a set X to $\mathbb{N}$ and consider the algebra $\langle \mathbb{N}^X, \wedge, \vee, \cdot, \to, \mathrm{e}\rangle$, where e is the zero map, and, for all $\varphi, \psi \in \mathbb{N}^X$, $x \in X$,

$$(\varphi \wedge \psi)(x) := \max(\varphi(x), \psi(x)) \qquad (\varphi \vee \psi)(x) := \min(\varphi(x), \psi(x))$$
$$(\varphi \cdot \psi)(x) := \varphi(x) + \psi(x) \qquad (\varphi \to \psi)(x) := \max(\psi(x) - \varphi(x), 0).$$

It is easily checked that this algebra is a commutative, cancellative, integral GBL-algebra, and therefore, by Theorem 7.1.10, the negative cone of an Abelian ℓ-group. Indeed, it is the negative cone of an isomorphic copy of the X-th Cartesian power of the ordered group of the integers.

7.2. Direct decomposition

In this section, we prove that every GBL-algebra decomposes as an inner direct product of an ℓ-group and an integral GBL-algebra (Theorem 7.2.3).

A residuated lattice $\mathbf{L}$ is said to be the *inner direct product*, denoted by $\mathbf{K}_1 \otimes \mathbf{K}_2$, of two subalgebras $\mathbf{K}_1$ and $\mathbf{K}_2$ of $\mathbf{L}$ if the map $\langle a, b\rangle \mapsto ab$ is an isomorphism from $\mathbf{K}_1 \times \mathbf{K}_2$ onto $\mathbf{L}$. We also let $\mathrm{Inv}(\mathbf{L})$ and $\mathrm{Int}(\mathbf{L})$ denote, respectively, the sets of invertible and integral elements of a residuated lattice $\mathbf{L}$.

LEMMA 7.2.1. *Let* $\mathbf{L}$ *be any GBL-algebra.*

(a) $\mathrm{Inv}(\mathbf{L})$ *is a subuniverse of* $\mathbf{L}$.
(b) $\mathrm{Int}(\mathbf{L})$ *is a subuniverse of* $\mathbf{L}$.
(c) $\mathrm{Inv}(\mathbf{L}) \cap \mathrm{Int}(\mathbf{L}) = \{\mathrm{e}\}$.

PROOF.

(a) Clearly, $\mathrm{Inv}(\mathbf{L})$ forms a group with product operation $\cdot$, neutral element e, and inverse operation $a^{-1} = a\backslash \mathrm{e} = \mathrm{e}/a$. Now consider any $a, b \in \mathrm{Inv}(\mathbf{L})$. Then $a\backslash b = a^{-1}b \in \mathrm{Inv}(\mathbf{L})$ and $b/a = ba^{-1} \in \mathrm{Inv}(\mathbf{L})$, by Lemma 7.1.1.(a). Moreover, $a \vee b = ab^{-1}b \vee b = (ab^{-1} \vee \mathrm{e})b \in \mathrm{Inv}(\mathbf{L})$, by Lemma 7.1.2.(a), and hence also $a \wedge b = (a^{-1} \vee b^{-1})^{-1} \in \mathrm{Inv}(\mathbf{L})$.

(b) Recall that $\mathbf{L}$ is e-cyclic, by Lemma 7.1.2.(b), so $a \in \mathrm{Int}(\mathbf{L})$ if and only if $a\backslash \mathrm{e} = \mathrm{e}$. Consider any $a, b \in \mathrm{Int}(\mathbf{L})$, i.e., $a\backslash \mathrm{e} = \mathrm{e}$ and $b\backslash \mathrm{e} = \mathrm{e}$. Then $ab \in \mathrm{Int}(\mathbf{L})$, since $ab\backslash \mathrm{e} = b\backslash (a\backslash \mathrm{e}) = b\backslash \mathrm{e} = \mathrm{e}$, and $a\backslash b \in \mathrm{Int}(\mathbf{L})$, since $\mathrm{e} = (a\backslash \mathrm{e})\backslash \mathrm{e} \leq (a\backslash b)\backslash \mathrm{e} \leq b\backslash \mathrm{e} = \mathrm{e}$. Similarly, $b/a \in \mathrm{Int}(\mathbf{L})$. Also, $a \vee b \in \mathrm{Int}(\mathbf{L})$, since $(a \vee b)\backslash \mathrm{e} = a\backslash \mathrm{e} \wedge b\backslash \mathrm{e} = \mathrm{e} \wedge \mathrm{e} = \mathrm{e}$. Finally, since $b, b\backslash a \in \mathrm{Int}(\mathbf{L})$, also $b(b\backslash a) \in \mathrm{Int}(\mathbf{L})$ and $b\backslash a \wedge \mathrm{e} = b\backslash a$. Hence $(a \wedge b)\backslash \mathrm{e} = b(b\backslash a \wedge \mathrm{e})\backslash \mathrm{e} = b(b\backslash a)\backslash \mathrm{e} = \mathrm{e}$, using Proposition 7.1.4. So $a \wedge b \in \mathrm{Int}(\mathbf{L})$.

(c) If $a \in \mathrm{Inv}(\mathbf{L}) \cap \mathrm{Int}(\mathbf{L})$, then $a^{-1} = a\backslash \mathrm{e} = \mathrm{e}$, so $a = \mathrm{e}$. □

Following this lemma, we may denote the subalgebras of invertible elements and integral elements of a GBL-algebra $\mathbf{L}$ by $\mathbf{Inv}(\mathbf{L})$ and $\mathbf{Int}(\mathbf{L})$, respectively. Clearly, $\mathbf{Inv}(\mathbf{L})$ is an ℓ-group and $\mathbf{Int}(\mathbf{L})$ is an integral GBL-algebra. The next lemma describes the interaction of elements from these two subalgebras.

LEMMA 7.2.2. *Let* $\mathbf{L}$ *be any GBL-algebra.*

(a) *If* $a, b \in L$ *and* $a \vee b = \mathrm{e}$, *then* $ab = a \wedge b$.
(b) *If* $a \in \mathrm{Inv}(\mathbf{L})^-$ *and* $b \in \mathrm{Int}(\mathbf{L})$, *then* $a \vee b = \mathrm{e}$ *and* $ab = ba = a \wedge b$.
(c) *If* $a \in \mathrm{Inv}(\mathbf{L})$ *and* $b \in \mathrm{Int}(\mathbf{L})$, *then* $ab = ba$.
(d) *For all* $a_1, a_2 \in \mathrm{Inv}(\mathbf{L})$ *and* $b_1, b_2 \in \mathrm{Int}(\mathbf{L})$,
$$a_1 b_1 \leq a_2 b_2 \iff a_1 \leq a_2 \text{ and } b_1 \leq b_2.$$

PROOF.

(a) If $a, b \in L$ and $a \vee b = \mathrm{e}$, then, using Lemma 7.1.2.(b),
$$a = a/\mathrm{e} = a/(a \vee b) = (a/a) \wedge (a/b) = \mathrm{e} \wedge (a/b) = (b/b) \wedge (a/b) = (b \wedge a)/b,$$
and hence, using divisibility, $ab = ((a \wedge b)/b)b = a \wedge b$.

(b) Consider any $a \in \mathrm{Inv}(\mathbf{L})^-$ and $b \in \mathrm{Int}(\mathbf{L})$. Observe first that
$$(a \vee b)\backslash \mathrm{e} = a^{-1} \wedge (b\backslash \mathrm{e}) = a^{-1} \wedge \mathrm{e} = \mathrm{e}.$$
That is, $a \vee b \in \mathrm{Int}(\mathbf{L})$. But also
$$a \vee b = (a \vee b)a^{-1}a = (\mathrm{e} \vee ba^{-1})a,$$
and therefore $a \vee b \in \mathrm{Inv}(\mathbf{L})$, by Lemmas 7.1.2.(a) and 7.2.1.(a). Hence $a \vee b = \mathrm{e}$, by Lemma 7.2.1.(c), and $ab = ba = a \wedge b$, by (a).

(c) Consider any $b \in \mathrm{Int}(\mathbf{L})$. If $a \in \mathrm{Inv}(\mathbf{L})^-$, then $ab = ba$, by (b). If $a^{-1} \in \mathrm{Inv}(\mathbf{L})^-$, then $a^{-1}b = ba^{-1}$, by (b), and $ab = aba^{-1}a = aa^{-1}ba = ba$. Finally, for any $a \in \mathrm{Inv}(\mathbf{L})$, using Proposition 7.1.6,
$$ab = (a \vee \mathrm{e})(a \wedge \mathrm{e})b = b(a \vee \mathrm{e})(a \wedge \mathrm{e}) = ba.$$

(d) For the non-trivial direction, suppose that $a_1 b_1 \leq a_2 b_2$, for some $a_1, a_2 \in \mathrm{Inv}(\mathbf{L})$ and $b_1, b_2 \in \mathrm{Int}(\mathbf{L})$. Then $a_2^{-1} a_1 b_1 \leq a_2^{-1} a_2 b_2 = b_2$ and hence $a_2^{-1} a_1 \leq b_2/b_1 \leq \mathrm{e}$.

So $a_1 \le a_2$ and $\mathrm{e} \le (b_2/b_1)(a_2^{-1}a_1)^{-1}$. Now let $d := (b_2/b_1)((a_2\backslash\mathrm{e})a_1\backslash\mathrm{e})$. Then d is invertible, by Lemma 7.1.2.(a), and

$$d\backslash\mathrm{e} = ((b_2/b_1)((a_2\backslash\mathrm{e})a_1\backslash\mathrm{e}))\backslash\mathrm{e} = ((a_2\backslash\mathrm{e})a_1\backslash\mathrm{e})\backslash((b_2/b_1)\backslash\mathrm{e}) = ((a_2\backslash\mathrm{e})a_1\backslash\mathrm{e})\backslash\mathrm{e}.$$

So $\mathrm{e} = d(d\backslash\mathrm{e}) = b_2/b_1$ and $b_1 \le b_2$. □

We are now ready to establish the main result of this section.

THEOREM 7.2.3. *Let* $\mathbf{L}$ *be any GBL-algebra. Then* $\mathbf{L} = \mathbf{Inv}(\mathbf{L}) \otimes \mathbf{Int}(\mathbf{L})$.

PROOF. $\mathbf{Inv}(\mathbf{L})$ and $\mathbf{Int}(\mathbf{L})$ are subalgebras of $\mathbf{L}$, by Lemma 7.2.1, so it remains to show that the map $\varphi\colon \mathrm{Inv}(\mathbf{L}) \times \mathrm{Int}(\mathbf{L}) \to L$; $\langle a, b\rangle \mapsto ab$ is an isomorphism from $\mathbf{Inv}(\mathbf{L}) \times \mathbf{Int}(\mathbf{L})$ onto $\mathbf{L}$. Note first that φ is an injective lattice homomorphism, by Lemma 7.2.2.(d), and $\varphi(\langle \mathrm{e}, \mathrm{e}\rangle) = \mathrm{e}$. Now consider any $a_1, a_2 \in \mathrm{Inv}(\mathbf{L})$ and $b_1, b_2 \in \mathrm{Int}(\mathbf{L})$. Then $a_1a_2b_1b_2 = a_1b_1a_2b_2$, by Lemma 7.2.2.(c), so φ preserves multiplication. Also, for any $a_3 \in \mathrm{Inv}(\mathbf{L}), b_3 \in \mathrm{Int}(\mathbf{L})$, by Lemma 7.2.2.(c)-(d),

$$\begin{aligned} a_3b_3 \le a_1b_1/a_2b_2 &\iff a_3b_3a_2b_2 \le a_1b_1 \\ &\iff a_3a_2b_3b_2 \le a_1b_1 \\ &\iff a_3a_2 \le a_1 \text{ and } b_3b_2 \le b_1 \\ &\iff a_3 \le a_1/a_2 \text{ and } b_3 \le b_1/b_2 \\ &\iff a_3b_3 \le (a_1/a_2)(b_1/b_2). \end{aligned}$$

So $a_1b_1/a_2b_2 = (a_1/a_2)(b_1/b_2)$, and, symmetrically, $a_1b_1\backslash a_2b_2 = (a_1\backslash a_2)(b_1\backslash b_2)$. That is, φ preserves the residual operations.

Finally, to show that φ is surjective, consider any $c \in L$, and let

$$a = (c \vee \mathrm{e})(((c \wedge \mathrm{e})\backslash\mathrm{e})\backslash\mathrm{e}) \quad \text{and} \quad b = ((c \wedge \mathrm{e})\backslash\mathrm{e})(c \wedge \mathrm{e}),$$

observing that, using Proposition 7.1.6,

$$c = (c \vee \mathrm{e})(c \wedge \mathrm{e}) = (c \vee \mathrm{e})(((c \wedge \mathrm{e})\backslash\mathrm{e})\backslash\mathrm{e})((c \wedge \mathrm{e})\backslash\mathrm{e})(c \wedge \mathrm{e}) = ab.$$

By Lemma 7.1.2.(a), both $c \vee \mathrm{e}$ and $(c \wedge \mathrm{e})\backslash\mathrm{e}$ are invertible, so also a is invertible. Moreover, using Proposition 1.4.4.(iv) and Lemma 7.1.2.(b),

$$\begin{aligned} b\backslash\mathrm{e} &= ((c \wedge \mathrm{e})\backslash\mathrm{e})(c \wedge \mathrm{e})\backslash\mathrm{e} \\ &= \mathrm{e}/(\mathrm{e}/(c \wedge \mathrm{e}))(c \wedge \mathrm{e}) \\ &= (\mathrm{e}/(c \wedge \mathrm{e}))/(\mathrm{e}/(c \wedge \mathrm{e})) \\ &= \mathrm{e}, \end{aligned}$$

so b is integral, as required. □

COROLLARY 7.2.4. *$\mathcal{GBL}$ is the class of all residuated lattices that are isomorphic to the direct product of an ℓ-group and an integral GBL-algebra.*

This result allows us to transfer certain properties of classes of ℓ-groups and integral GBL-algebras to a corresponding class of GBL-algebras. In particular, ℓ-groups and integral GMV-algebras are prelinear (Lemma 7.1.9), so the same is true of their direct products, and hence of all GMV-algebras.

COROLLARY 7.2.5. *Every GMV-algebra is prelinear.*

COROLLARY 7.2.6. *Every commutative GMV-algebra is semilinear.*

7.3. Ordinal decomposition

In this section, we obtain a description of totally ordered, integral GBL-algebras as "ordinal sums" of totally ordered, integral GMV-algebras with a least element and negative cones of totally ordered groups (Theorems 7.3.4 and 7.3.7).

Let $\langle I, \leq \rangle$ be any non-empty chain and suppose that $\mathbf{L}_i$ is a totally ordered, integral GBL-algebra, for each $i \in I$. Let us assume also that $L_i \cap L_j = \{\mathrm{e}\}$, for all $i, j \in I$ with $i \neq j$. Then the *ordinal sum* of the family $\{\mathbf{L}_i\}_{i \in I}$ is the algebra

$$\mathbf{L} = \bigoplus_{i \in I} \mathbf{L}_i := \langle \bigcup_{i \in I} L_i, \wedge, \vee, \cdot, \backslash, /, \mathrm{e} \rangle,$$

where the lattice operations are induced by the total order

$$x \leq^{\mathbf{L}} y :\Longleftrightarrow (x \in L_i \backslash \{\mathrm{e}\},\ y \in L_j \backslash \{\mathrm{e}\}, \text{ and } i < j) \text{ or } (x, y \in L_i \text{ and } x \leq^{\mathbf{L}_i} y),$$

and the monoid operation defined by

$$xy := \begin{cases} x \cdot^{\mathbf{L}_i} y & \text{if } x, y \in L_i; \\ x \wedge y & \text{if } x \in L_i,\ y \in L_j, \text{ and } i \neq j \end{cases}$$

is residuated, with left residual (the right residual behaves dually)

$$x \backslash y = \begin{cases} x \backslash^{\mathbf{L}_i} y & \text{if } x, y \in L_i; \\ \mathrm{e} & \text{if } x \in L_i \backslash \{\mathrm{e}\},\ y \in L_j \backslash \{\mathrm{e}\}, \text{ and } i < j; \\ y & \text{if } x \in L_i \backslash \{\mathrm{e}\},\ y \in L_j \backslash \{\mathrm{e}\}, \text{ and } j < i. \end{cases}$$

It is easily checked that $\mathbf{L}$ is a totally ordered, integral GBL-algebra.

We call a totally ordered, integral GBL-algebra $\mathbf{L}$ *ordinally irreducible* if, whenever $\mathbf{L}$ is an ordinal sum of a family of subalgebras $\{\mathbf{L}_i\}_{i \in I}$ of $\mathbf{L}$, there exists $j \in I$ such that $\mathbf{L} \cong \mathbf{L}_j$ and $L_i = \{\mathrm{e}\}$, for all $i \in I \backslash \{j\}$. We show below that every totally ordered, integral GBL-algebra can be uniquely represented as the ordinal sum of a family of ordinally irreducible, totally ordered, integral GBL-algebras. We then show that a totally ordered, integral GBL-algebra is ordinally irreducible if and only if it is either a totally ordered, integral GMV-algebras with a least element or the negative cone of a totally ordered group.

We begin with the following useful lemma.

LEMMA 7.3.1. *Let $\mathbf{L}$ be any totally ordered, integral GBL-algebra.*

(a) *No elements $a, b, c \in L$ satisfy $a < b < c$, $ca = a$, $ba < a$, and $bc < b$.*
(b) *No elements $a, b, c \in L$ satisfy $a < b < c$, $ac = a$, $ab < a$, and $cb < b$.*

PROOF. We prove (a), the proof of (b) being very similar. Suppose that there exist elements $a, b, c \in L$ such that $a < b < c$, $ca = a$, $ba < a$, and $bc < b$. Note first that $ba/a < c$, since otherwise $ba < a = ca \leq (ba/a)a \leq ba$, a contradiction. Hence

$$ba/a = c \wedge ba/a = ((ba/a)/c)c = (ba/ca)c = (ba/a)c.$$

But then we obtain the following contradiction:

$$b = ba/a \wedge b = (b/(ba/a))(ba/a) = (b/(ba/a))(ba/a)c = (ba/a \wedge b)c = bc < b.$$

So no elements $a, b, c \in L$ satisfy $a < b < c$, $ca = a$, $ba < a$, and $bc < b$. $\square$

Now let $\mathbf{L}$ be any totally ordered, integral GBL-algebra. We call an ordered pair $\langle D, U\rangle$ of subsets of L a *cut* (in $\mathbf{L}$) if (i) $D \cap U = \emptyset$; (ii) $D \cup U = L$; (iii) $\uparrow U = U \neq \emptyset$. It is easily checked that in this case also (iv) $\downarrow D = D$; (v) $\mathrm{e} \in U \backslash D$; (vi) $b < c$, for all $b \in D, c \in U$. We call $\langle D, U\rangle$ an *α-cut* (in $\mathbf{L}$) if, additionally,

(α1) U is closed under multiplication;

(α2) $b \in D, c \in U \implies bc = cb = b \wedge c = b$.

The following lemma describes a useful way of defining α-cuts.

LEMMA 7.3.2. *For any totally ordered, integral GBL-algebra* $\mathbf{L}$ *and* $a \in L$, *let*

$$U_a := \{z \in L \mid za = a\} \quad \text{and} \quad D_a := L \backslash U_a.$$

Then $\langle D_a, U_a\rangle$ *is an* α*-cut in* $\mathbf{L}$ *and*

$$a \text{ is idempotent} \iff a \in U_a \iff U_a = \uparrow a.$$

PROOF. For the second claim, observe that a is idempotent, by definition, if and only if $a \in U_a$, and clearly, if $U_a = \uparrow a$, then $a \in U_a$. For the remaining implication, suppose that $a \in U_a$. If $b \in U_a$, then $a = ba \leq b$, so $b \in \uparrow a$; conversely, if $b \in \uparrow a$, then $a = aa \leq ba \leq a$, so $b \in U_a$. That is, $U_a = \uparrow a$.

Let us now prove that $\langle D_a, U_a\rangle$ is an α-cut in $\mathbf{L}$. Notice first that $\langle D_a, U_a\rangle$ is a cut: clearly, $D_a \cap U_a = \emptyset$, $D_a \cup U_a = L$, and $\mathrm{e} \in U_a$; moreover, if $z \in U_a$ and $z \leq y$, then $a = za \leq ya \leq a$, so $y \in U_a$. Observe also that U_a is closed under multiplication: if $b, c \in U_a$, then $bca = ba = a$, so $bc \in U_a$. It remains to show that $bc = cb = b$, for any $b \in D_a$ and $c \in U_a$. Since $a = ca \leq c$ and $b < c$, there are two cases:

(a) If $b \leq a \leq c$, then $cb = c(b \wedge a) = ca(a\backslash b) = a(a\backslash b) = b \wedge a = b$, and, similarly, $bc = b$.

(b) Suppose now that $a < b < c$. Since $ca = a$ and $ba < a$, Lemma 7.3.1.(a) yields $bc = b$. Next, we show that $a/c = a$. Clearly, $a \leq a/c$, so suppose towards a contradiction that $a < a/c$. Note that $a \notin U_a$, since $a < b$, and hence, since $a = c \wedge a = (a/c)c$, also $a/c \notin U_a$. So $a/c < c$. But then $a < a/c < c$, $ca = a$, and $(a/c)c = c \wedge a = a < a/c$, and Lemma 7.3.1.(a) yields $(a/c)a = a$, i.e., $a/c \in U_a$, a contradiction. Hence $a/c = a$ as claimed, and $ac = (a/c)c = c \wedge a = a$. Similarly, it can be shown that $ab < a$. So Lemma 7.3.1.(b) yields $cb = b$, as required. □

Every α-cut of a totally ordered, integral GBL-algebra yields a decomposition into two subalgebras that intersect only at the neutral element.

LEMMA 7.3.3. *Let* $\langle D, U\rangle$ *be an* α*-cut of a totally ordered, integral GBL-algebra* $\mathbf{L}$. *Then* $D \cup \{\mathrm{e}\}$ *and* U *are subuniverses of* $\mathbf{L}$ *and*

$$a \in D,\, a \neq a^2,\, b \in U \implies b\backslash a = a = a/b.$$

PROOF. We check first that $D \cup \{\mathrm{e}\}$ is closed under the residual operations. Let $a, b \in D \cup \{\mathrm{e}\}$. If $b \leq a$, then $b\backslash a = \mathrm{e} \in D \cup \{\mathrm{e}\}$. If $a < b$ and $b\backslash a \notin D \cup \{\mathrm{e}\}$, then $b\backslash a \in U$, and $a = a \wedge b = b(b\backslash a) = b$, by ($\alpha$2), a contradiction. So $b\backslash a \in D \cup \{\mathrm{e}\}$. The case for the right residual is analogous. The closure of $D \cup \{\mathrm{e}\}$ under multiplication and U under residuals follows from the fact that $D = \downarrow D$ and $U = \uparrow U$, and $\mathbf{L}$ is integral. Hence both $D \cup \{\mathrm{e}\}$ and U are subuniverses of $\mathbf{L}$.

Now suppose that $a \in D$, $a \neq a^2$, and $b \in U$. Then $a < b$. Note also that $b \in U_a$, since $a = ba$. Moreover, $b\backslash a \notin U_a$, since otherwise $a = b \wedge a = b(b\backslash a) \in U_a$ and $a^2 = a$, a contradiction. Hence $b\backslash a \in D_a$ and $a \leq b\backslash a < b$. Finally, if $a < b\backslash a$, then, since $ab = a$ and $b(b\backslash a) = a < b$, an application of Lemma 7.3.1.(b) yields $a(b\backslash a) = a$, contradicting the fact that $b\backslash a \in D_a$. So $a = b\backslash a$. The case for the right residual is analogous. □

Observe next that α-cuts in a totally ordered, integral GBL-algebra $\mathbf{L}$ can be totally ordered by stipulating

$$\langle D_1, U_1\rangle \leq \langle D_2, U_2\rangle \;:\Longleftrightarrow\; D_1 \subseteq D_2.$$

If $\langle D, U\rangle$ has a successor in this ordering, we call it a β-cut (in $\mathbf{L}$).

THEOREM 7.3.4. *Every totally ordered, integral GBL-algebra* $\mathbf{L}$ *is isomorphic to an ordinal sum of a unique family* $\{\mathbf{L}_\lambda\}_{\lambda\in\Lambda}$ *of ordinally irreducible subalgebras of* $\mathbf{L}$, *for some chain* $\langle \Lambda, \leq\rangle$.

PROOF. Let $\mathbf{L}$ be any totally ordered, integral GBL-algebra. If $\mathbf{L}$ is trivial, it is ordinally irreducible, so we may assume that this is not the case. Let $\langle \Lambda, \leq\rangle$ be the chain consisting of all β-cuts in $\mathbf{L}$. For each $\lambda = \langle D_\lambda, U_\lambda\rangle \in \Lambda$ with successor $\mu = \langle D_\mu, U_\mu\rangle$ (which may not belong to Λ), let $L_\lambda := U_\lambda \cap (D_\mu \cup \{\mathrm{e}\})$. Clearly, each L_λ is a subuniverse of $\mathbf{L}$ and $L_\lambda\backslash\{\mathrm{e}\}$ is convex. We let $\mathbf{L}_\lambda$ denote the subalgebra of $\mathbf{L}$ with universe L_λ. (Intuitively, the reader may think of $L_\lambda\backslash\{\mathrm{e}\}$ as the set of upper elements that cross over to the cut's lower set upon passing from λ to μ.) It remains to show that

(i) each $\mathbf{L}_\lambda$ is ordinally irreducible;

(ii) $\mathbf{L} = \bigoplus_{\lambda\in\Lambda} \mathbf{L}_\lambda$;

(iii) this decomposition is unique.

(i) Suppose towards a contradiction that $\mathbf{L}_\lambda = \mathbf{L}_1 \oplus \mathbf{L}_2$, where $\mathbf{L}_1$ and $\mathbf{L}_2$ are both non-trivial subalgebras of $\mathbf{L}_\lambda$. We show that the cut $\langle D_\lambda \cup (L_1\backslash\{\mathrm{e}\}), U_\mu \cup L_2\rangle$ is an α-cut, and hence that

$$\langle D_\lambda, U_\lambda\rangle < \langle D_\lambda \cup (L_1\backslash\{\mathrm{e}\}), U_\mu \cup L_2\rangle < \langle D_\mu, U_\mu\rangle,$$

contradicting the assumption that μ is the successor of λ. Let us just check (α2). Consider $b \in D_\lambda \cup (L_1\backslash\{\mathrm{e}\})$, $c \in U_\mu \cup L_2$. If $c = \mathrm{e}$, then $bc = cb = b$, so suppose that $c < \mathrm{e}$. If $b \in D_\lambda$, then $c \in U_\mu \cup L_2 \subseteq U_\lambda$ implies that $bc = cb = b$. If $b \in L_1 \subseteq D_\mu$ and $c \in U_\mu$, then again $bc = cb = b$. Finally, if $b \in L_1$ and $c \in L_2$, then, by the definition of an ordinal sum, $bc = cb = b$.

(ii) The family $\{\mathbf{L}_\lambda\}_{\lambda\in\Lambda}$ satisfies the requirements for an ordinal sum, so it suffices to prove that each $a \in L$ belongs to L_λ, for some $\lambda \in \Lambda$. Since e belongs to every L_λ and, by assumption, $\mathbf{L}$ is non-trivial, we may assume that $a \neq \mathrm{e}$. Define

$$D_0^a := \bigcup\{D \mid \langle D, U\rangle \text{ is an } \alpha\text{-cut and } a \notin D\}$$
$$D_1^a := \bigcap\{D \mid \langle D, U\rangle \text{ is an } \alpha\text{-cut and } a \in D\}.$$

Then $\langle D_0^a, L\backslash D_0^a\rangle$ and $\langle D_1^a, L\backslash D_1^a\rangle$ are α-cuts and $\langle D_1^a, L\backslash D_1^a\rangle$ is the successor of $\langle D_0^a, L\backslash D_0^a\rangle$. So $\langle D_0^a, L\backslash D_0^a\rangle = \lambda$, for some $\lambda \in \Lambda$, and $a \in L_\lambda$.

(iii) Suppose that $\mathbf{L}$ is an ordinal sum of a family $\{\mathbf{M}_i\}_{i\in I}$ of ordinally irreducible, integral GBL-algebras, for some chain $\langle I, \leq\rangle$. For $j \in I$, let

$$D_j := \bigcup_{i<j} M_i,\ U_j := \bigcup_{i\geq j} M_i \quad \text{and} \quad D'_j := \bigcup_{i\leq j} M_i,\ U'_j := \bigcup_{i>j} M_i.$$

Then $\langle D_j\backslash\{\mathrm{e}\}, U_j\rangle$ and $\langle D'_j\backslash\{\mathrm{e}\}, U'_j\rangle$ are α-cuts, where the latter is the successor of the former. So $\langle D_j\backslash\{\mathrm{e}\}, U_j\rangle$ is a β-cut in $\mathbf{L}$. This means that, for some $\lambda \in \Lambda$,

$$M_j = (U_j \cap D'_j) \cup \{\mathrm{e}\} = L_\lambda.$$

Hence the two representations coincide. □

Let us now investigate the "building blocks" of the ordinal sum representation given in Theorem 7.3.4: the totally ordered, integral GBL-algebras that are ordinally irreducible. First, we prove two auxiliary lemmas.

LEMMA 7.3.5. *The only idempotent elements of an ordinally irreducible, totally ordered, integral GBL-algebra* $\mathbf{L}$ *are* e *and, if it has one, the least element of* $\mathbf{L}$.

PROOF. Suppose that $a \in L\backslash\{\mathrm{e}\}$ is an idempotent element of $\mathbf{L}$, and consider the cut $\langle D_a, U_a\rangle$ defined in Lemma 7.3.2. Since U_a contains a and e, it is non-trivial. Hence, because $\mathbf{L}$ is ordinally irreducible, $D_a = \emptyset$, and $L = U_a = {\uparrow}a$. That is, a is the least element of $\mathbf{L}$. □

LEMMA 7.3.6. *Every ordinally irreducible, totally ordered, integral GBL-algebra* $\mathbf{L}$ *satisfies the following conditions, for all* $a, b \in L$:

(U1) *if* $a\backslash b = b$, *then either* $a = \mathrm{e}$ *or* $b = \mathrm{e}$;

(U2) *if* $b/a = b$, *then either* $a = \mathrm{e}$ *or* $b = \mathrm{e}$.

If $\mathbf{L}$ *has no least element, it satisfies also the following conditions, for all* $a, b \in L$:

(U3) *if* $ab = a$, *then* $b = \mathrm{e}$;

(U4) *if* $ba = a$, *then* $b = \mathrm{e}$.

PROOF. We just prove (U4) as an example. Consider any $a, b \in L$ such that $ba = a$. If $a = \mathrm{e}$, then $b = \mathrm{e}$. Otherwise, $a < \mathrm{e}$ and we consider the α-cut $\langle D_a, U_a\rangle$ defined in Lemma 7.3.2. Then $a \in D_a$, since a cannot be an idempotent element of $\mathbf{L}$, by Lemma 7.3.5. Since $\mathbf{L}$ is ordinally irreducible, U_a is trivial. But $b \in U_a$, so $b = \mathrm{e}$ as required. □

THEOREM 7.3.7. *Every ordinally irreducible, totally ordered, integral GBL-algebra is either a totally ordered, integral GMV-algebra with a least element or the negative cone of a totally ordered group.*

PROOF. First, we show that every ordinally irreducible, totally ordered, integral GBL-algebra $\mathbf{L}$ is a GMV-algebra. Consider any $a, b \in L$ such that $a \leq b$. We prove that, for any $c \in {\uparrow}b$, if $b\backslash a = c\backslash a$ or $a/b = a/c$, then $b = c$. The fact that $\mathbf{L}$ is a GMV-algebra then follows by noting that $c = (a/b)\backslash a$ satisfies $b\backslash a = c\backslash a$, and $c = (a\backslash b)/a$ satisfies $a/b = a/c$, and applying Proposition 7.1.8. Suppose then that $c \in {\uparrow}b$ satisfies $b\backslash a = c\backslash a$, and observe that $c\backslash a = b\backslash a = c(c\backslash b)\backslash a = (c\backslash b)\backslash(c\backslash a)$. So, using (U1), either $c\backslash a = \mathrm{e}$, giving $c \leq a$ and hence $a = b = c$, or $c\backslash b = \mathrm{e}$, giving $c \leq b$ and hence $b = c$. Similarly, $b = c$ if $c \in {\uparrow}b$ satisfies $a/b = a/c$.

Finally, suppose that $\mathbf{L}$ is a totally ordered, integral GMV-algebra that does not have a least element. Then, by Theorem 7.1.10, to show that $\mathbf{L}$ is the negative cone of a totally ordered group, it suffices to prove that it is cancellative. Suppose that $ca = cb$ and, without loss of generality, $a \leq b$, so $a = b(b\backslash a)$. Then $cb = ca = cb(b\backslash a)$, and $b\backslash a = \mathrm{e}$, by (U3). So also $b \leq a$, i.e., $a = b$. The other cancellation condition is symmetrical. $\square$

7.4. Cone algebras and negative cones

In this section and the next, we focus our attention on integral GMV-algebras, recalling that every GMV-algebra decomposes as the product of an integral member of this class and an ℓ-group. In particular, we establish an equivalence between a category of integral GMV-algebras and a category of negative cones of ℓ-groups augmented with a dense nucleus (Theorem 7.5.10). The key step in obtaining this equivalence will be the construction of an ℓ-group from the $\{\backslash, /, \mathrm{e}\}$-reduct of an integral GMV-algebra. With this goal in mind, we first provide an equational description of the $\{\backslash, /, \mathrm{e}\}$-subreducts of negative cones of ℓ-groups (Theorem 7.4.5).

A *cone algebra* is an algebra $\mathbf{C} = \langle C, \backslash, /, \mathrm{e}\rangle$ of signature $\langle 2, 2, 0\rangle$ that satisfies the following equations and their mirror images:

$$\begin{array}{ll}
\text{(C1)} & (x\backslash y)\backslash(x\backslash z) \approx (y\backslash x)\backslash(y\backslash z) \\
\text{(C2)} & \mathrm{e}\backslash x \approx x \\
\text{(C3)} & x\backslash(y/z) \approx (x\backslash y)/z \\
\text{(C4)} & x\backslash x \approx \mathrm{e} \\
\text{(C5)} & x/(y\backslash x) \approx (y/x)\backslash y.
\end{array}$$

LEMMA 7.4.1. *The $\{\backslash, /, \mathrm{e}\}$-reduct of any integral GMV-algebra is a cone algebra.*

PROOF. The only condition to be checked is (C1). Just observe that, for any elements a, b, c of an integral GMV-algebra, using Proposition 1.4.4 and divisibility,

$$(a\backslash b)\backslash(a\backslash c) = a(a\backslash b)\backslash c = (a \wedge b)\backslash c = b(b\backslash a)\backslash c = (b\backslash a)\backslash(b\backslash c). \qquad \square$$

We will prove in Theorem 7.4.5 that every cone algebra embeds into the $\{\backslash, /, \mathrm{e}\}$-reduct of the negative cone of an ℓ-group, and hence that cone algebras are precisely the $\{\backslash, /, \mathrm{e}\}$-subreducts of negative cones of ℓ-groups. First, we establish some useful properties of these algebras.

LEMMA 7.4.2. *Let $\mathbf{C}$ be any cone algebra and define, for all $a, b \in C$,*

$$a \leq b \;:\Longleftrightarrow\; a\backslash b = \mathrm{e}.$$

Then $\langle C, \leq\rangle$ is a semilattice with greatest element e and binary joins given by $a \vee b = a/(b\backslash a) = (b/a)\backslash b$, for any $a, b \in C$, such that

(i) *$\backslash$ and $/$ are order-preserving in the numerator argument, and order-reversing in the denominator argument;*

(ii) *the following equalities and their mirror images hold for all $a, b, c \in C$:*

$$(a \vee b)\backslash a = b\backslash a, \quad (a\backslash b) \vee (b\backslash a) = \mathrm{e}, \quad \text{and} \quad a\backslash(b \vee c) = (a\backslash b) \vee (a\backslash c).$$

PROOF. Note first that, for any $a \in C$, by (C2), (C1), and (C4),

$$a\backslash e = (e\backslash a)\backslash(e\backslash e) = (a\backslash e)\backslash(a\backslash e) = e.$$

Moreover, if $a\backslash b = e$, then, by (C2), (C5), (C3), and (C4),

$$b/a = (b/e)/a = (b/(a\backslash b))/a = ((a/b)\backslash a)/a = (a/b)\backslash(a/a) = (a/b)\backslash e = e.$$

The converse is symmetrical. Reflexivity and antisymmetry for $\leq$ then follow from (C2), (C4), and (C5). For transitivity, if $a\backslash b = b\backslash c = e$, then, by (C1) and (C2),

$$a\backslash c = e\backslash(a\backslash c) = (a\backslash b)\backslash(a\backslash c) = (b\backslash a)\backslash(b\backslash c) = (b\backslash a)\backslash e = e.$$

Hence $\langle C, \leq\rangle$ is a poset with greatest element e.

We show next that $\backslash$ (and therefore, symmetrically, $/$) is order-preserving and order-reversing in the numerator and denominator arguments, respectively. Observe first that $e = b\backslash e = b\backslash(a/a) = (b\backslash a)/a$, and hence, by (C2), (C5), (C3), and (C4),

$$a\backslash(b\backslash a) = a\backslash(((b\backslash a)/a)\backslash(b\backslash a)) = a\backslash(a/((b\backslash a)\backslash a)) = (a\backslash a)/((b\backslash a)\backslash a) = e.$$

Suppose that $a \leq b$, i.e., $a\backslash b = e$. Then, using (C1) and the previous observation,

$$(c\backslash a)\backslash(c\backslash b) = (a\backslash c)\backslash(a\backslash b) = (a\backslash c)\backslash e = e$$
$$(b\backslash c)\backslash(a\backslash c) = (b\backslash c)\backslash((a\backslash b)\backslash(a\backslash c)) = (b\backslash c)\backslash((b\backslash a)\backslash(b\backslash c)) = e.$$

That is, $c\backslash a \leq c\backslash b$ and $b\backslash c \leq a\backslash c$.

We show now that $\langle C, \leq\rangle$ is a semilattice with binary joins given by $a \vee b := a/(b\backslash a)$, for any $a, b \in C$. Observe first that $a, b \leq a \vee b$, since, using (C3), (C4), and (C2),

$$a\backslash(a/(b\backslash a)) = (a\backslash a)/(b\backslash a) = e/(b\backslash a) = e \quad \text{and} \quad b\backslash(a/(b\backslash a)) = (b\backslash a)/(b\backslash a) = e.$$

Moreover, if $a, b \leq c$, then $a \vee b \leq c$, since, using (i), (C5), and (C2),

$$a/(b\backslash a) \leq a/(c\backslash a) = c/(a\backslash c) = c/e = c.$$

Finally, for (ii), suppose that $a, b, c \in C$. Then, using (i),

$$(a \vee b)\backslash a = (a/(b\backslash a))\backslash a = a \vee (b\backslash a) = b\backslash a,$$

and by (C1), (C2), and (C4),

$$\begin{aligned}(a\backslash b) \vee (b\backslash a) &= ((a \vee b)\backslash b) \vee ((a \vee b)\backslash a)\\ &= ((a \vee b)\backslash a)/(((a \vee b)\backslash b)\backslash((a \vee b)\backslash a))\\ &= ((a \vee b)\backslash a)/((b\backslash(a \vee b))\backslash(b\backslash a))\\ &= ((a \vee b)\backslash a)/(b\backslash a)\\ &= (b\backslash a)/(b\backslash a)\\ &= e.\end{aligned}$$

For the last equation, using (C1) and (i),

$$c\backslash b = (b \vee c)\backslash b \leq (a\backslash(b \vee c))\backslash(a\backslash b) \quad \text{and} \quad b\backslash c = (b \vee c)\backslash c \leq (a\backslash(b \vee c))\backslash(a\backslash c),$$

and so, using the previous equation,

$$\begin{aligned}e &= (c\backslash b) \vee (b\backslash c)\\ &\leq ((a\backslash(b \vee c))\backslash(a\backslash b)) \vee (((a\backslash(b \vee c))\backslash(a\backslash c)))\\ &\leq (a\backslash(b \vee c))\backslash((a\backslash b) \vee (a\backslash c)),\end{aligned}$$

yielding $a\backslash(b \vee c) = (a\backslash b) \vee (a\backslash c)$. The mirror images are proved symmetrically. ☐

LEMMA 7.4.3. *Let* **C** *be any cone algebra and let* $a, b, c, d \in C$. *If* $a \vee b = \mathrm{e}$, *then the following equalities and their mirror images hold:*

$$a\backslash b = b, \quad c\backslash b = (a\backslash c)\backslash b = (c/a)\backslash b, \quad \textit{and} \quad (c\backslash d)\backslash b = ((c/a)\backslash d)\backslash b.$$

PROOF. Assume $a \vee b = \mathrm{e}$. First, using Lemma 7.4.2.(ii),

$$a\backslash b = (a \vee b)\backslash b = \mathrm{e}\backslash b = b.$$

Next, using $a \vee b \vee c = \mathrm{e}$ and the previous equality, (C1), and Lemma 7.4.2.(ii),

$$(a\backslash c)\backslash b = ((a \vee c)\backslash c)\backslash((a \vee c)\backslash b) = (c\backslash(a \vee c))\backslash(c\backslash b) = \mathrm{e}\backslash(c\backslash b) = c\backslash b,$$

and similar reasoning yields $(c/a)\backslash b = c\backslash b$. Finally, using (C2), Lemma 7.4.2.(ii), (C1), and noting once again that $a \vee b \vee c = \mathrm{e}$,

$$\begin{aligned}(c\backslash d)\backslash b &= (\mathrm{e}\backslash(c\backslash d))\backslash b \\ &= ((c\backslash(c/a))\backslash(c\backslash d))\backslash b \\ &= (((c/a)\backslash c)\backslash((c/a)\backslash d))\backslash b \\ &= ((c/a)\backslash d)\backslash b.\end{aligned}$$

The mirror images are proved symmetrically. □

LEMMA 7.4.4. *Let* **C** *be any cone algebra, and let* $a, b, c, d \in C$.

(a) $(a/b)\backslash c = ((a/(c\backslash a))/(b/(c\backslash a)))\backslash c$.

(b) $(d/a)/(c/b) = ((d/(b\backslash a))/(c/(a\backslash b)))/((b/c)/((a\backslash b)/c))$.

PROOF.

(a) Using Lemma 7.4.2.(ii), (C3), and (C1),

$$(a/b)\backslash c = ((a/b) \vee c)\backslash c = ((a/b)/((c\backslash a)/b))\backslash c = ((a/(c\backslash a))/(b/(c\backslash a)))\backslash c.$$

(b) Just apply (C1) (three times) and (C3). □

We are now ready to prove the main theorem of this section.

THEOREM 7.4.5. *Every cone algebra* **C** *embeds into the* $\{\backslash, /, \mathrm{e}\}$*-reduct of a negative cone* $\hat{\mathbf{C}}$ *of an* ℓ*-group, where the monoid reduct of* $\hat{\mathbf{C}}$ *is generated by* C *as a monoid.*

PROOF. The desired negative cone $\hat{\mathbf{C}}$ is obtained by defining the direct limit of a family $\{\mathbf{C}_n\}_{n\in\mathbb{N}}$ of cone algebras ordered by inclusion, where products in $\hat{\mathbf{C}}$ are constructed stepwise such that each $\mathbf{C}_{n+1}$ contains the products of members of $\mathbf{C}_n$.

Let **C** be any cone algebra. For $\langle a, b\rangle, \langle c, d\rangle$ in C^2, define

$$\begin{aligned}\langle a, b\rangle \mathbin{\backslash\!\!\backslash} \langle c, d\rangle &:= \langle b\backslash(a\backslash c), ((a\backslash c)\backslash b)\backslash((c\backslash a)\backslash d)\rangle \\ \langle d, c\rangle \mathbin{/\!\!/} \langle b, a\rangle &:= \langle (d/(a/c))/(b/(c/a)), (c/a)/b\rangle.\end{aligned}$$

The rationale for this definition is the fact that all negative cones of ℓ-groups satisfy the following equation and its mirror image:

$$\text{(Id)} \quad xy\backslash zw \approx (y\backslash(x\backslash z)) \cdot (((x\backslash z)\backslash y)\backslash((z\backslash x)\backslash w)).$$

That is, we use the Cartesian product operation to represent the product in **C**.

We now show that the following algebra of signature $\langle 2,2,0\rangle$ satisfies all the defining axioms of cone algebras except, possibly, (C5):

$$\mathbf{D} := \langle C^2, \mathbin{\backslash\!\!\backslash}, /\!\!/, \langle \mathrm{e},\mathrm{e}\rangle\rangle.$$

We first make the crucial observation that, for all $a,b,c,d \in C$,

(E1) $\langle a,b\rangle \mathbin{\backslash\!\!\backslash} \langle c,d\rangle = \langle b,\mathrm{e}\rangle \mathbin{\backslash\!\!\backslash} (\langle a,\mathrm{e}\rangle \mathbin{\backslash\!\!\backslash} \langle c,d\rangle)$

(E2) $\langle d,c\rangle /\!\!/ \langle b,a\rangle = (\langle d,c\rangle /\!\!/ \langle a,\mathrm{e}\rangle) /\!\!/ \langle b,\mathrm{e}\rangle$.

We then check (C3), leaving the other axioms as exercises for the reader. For any $a,b,c,d,u,v \in C$,

$$\begin{aligned}
(\langle a,\mathrm{e}\rangle \mathbin{\backslash\!\!\backslash} \langle u,v\rangle) /\!\!/ \langle c,\mathrm{e}\rangle &= \langle a\backslash u, (u\backslash a)\backslash v\rangle /\!\!/ \langle c,\mathrm{e}\rangle \\
&= \langle (a\backslash u)/(c/((u\backslash a)\backslash v)), ((u\backslash a)\backslash v)/c\rangle \\
&= \langle (a\backslash u)/(c/v), (u\backslash a)\backslash(v/c)\rangle \\
&= \langle a\backslash(u/(c/v)), ((u/(c/v))\backslash a)\backslash(v/c)\rangle \\
&= \langle a,\mathrm{e}\rangle \mathbin{\backslash\!\!\backslash} \langle u/(c/v), v/c\rangle \\
&= \langle a,\mathrm{e}\rangle \mathbin{\backslash\!\!\backslash} (\langle u,v\rangle /\!\!/ \langle c,\mathrm{e}\rangle).
\end{aligned}$$

In the third and fourth row, the use of the third equality in Lemma 7.4.3 is justified since $(c/v) \vee (v/c) = \mathrm{e} = (a\backslash u) \vee (u\backslash a)$. Repeatedly using this fact, (E1), and (E2),

$$\begin{aligned}
(\langle a,b\rangle \mathbin{\backslash\!\!\backslash} \langle u,v\rangle) /\!\!/ \langle c,d\rangle &= ((\langle b,\mathrm{e}\rangle \mathbin{\backslash\!\!\backslash} (\langle a,\mathrm{e}\rangle \mathbin{\backslash\!\!\backslash} \langle u,v\rangle)) /\!\!/ \langle d,\mathrm{e}\rangle) /\!\!/ \langle c,\mathrm{e}\rangle \\
&= (\langle b,\mathrm{e}\rangle \mathbin{\backslash\!\!\backslash} ((\langle a,\mathrm{e}\rangle \mathbin{\backslash\!\!\backslash} \langle u,v\rangle) /\!\!/ \langle d,\mathrm{e}\rangle)) /\!\!/ \langle c,\mathrm{e}\rangle \\
&= (\langle b,\mathrm{e}\rangle \mathbin{\backslash\!\!\backslash} (\langle a,\mathrm{e}\rangle \mathbin{\backslash\!\!\backslash} (\langle u,v\rangle /\!\!/ \langle d,\mathrm{e}\rangle))) /\!\!/ \langle c,\mathrm{e}\rangle \\
&= \langle b,\mathrm{e}\rangle \mathbin{\backslash\!\!\backslash} ((\langle a,\mathrm{e}\rangle \mathbin{\backslash\!\!\backslash} (\langle u,v\rangle /\!\!/ \langle d,\mathrm{e}\rangle)) /\!\!/ \langle c,\mathrm{e}\rangle) \\
&= \langle b,\mathrm{e}\rangle \mathbin{\backslash\!\!\backslash} (\langle a,\mathrm{e}\rangle \mathbin{\backslash\!\!\backslash} ((\langle u,v\rangle /\!\!/ \langle d,\mathrm{e}\rangle) /\!\!/ \langle c,\mathrm{e}\rangle)) \\
&= \langle a,b\rangle \mathbin{\backslash\!\!\backslash} (\langle u,v\rangle /\!\!/ \langle c,d\rangle).
\end{aligned}$$

Next we prove that the following relation is a congruence of $\mathbf{D}$:

$$\Theta := \{\langle\langle a,b\rangle, \langle c,d\rangle\rangle \in (C^2)^2 \mid \langle a,b\rangle \mathbin{\backslash\!\!\backslash} \langle c,d\rangle = \langle \mathrm{e},\mathrm{e}\rangle = \langle c,d\rangle \mathbin{\backslash\!\!\backslash} \langle a,b\rangle\}.$$

Clearly, Θ is reflexive and symmetric. Also, transitivity follows using (C1) and (C2), as in the proof of Lemma 7.4.2.(i). Next, we verify that $\mathbin{\backslash\!\!\backslash}$ is preserved in both arguments. Let $a,b,c,d,u,v \in C$ and suppose that $\langle c,d\rangle\ \Theta\ \langle u,v\rangle$. Then, using (C2) and (C1),

$$\begin{aligned}
\langle c,d\rangle \mathbin{\backslash\!\!\backslash} \langle a,b\rangle &= (\langle c,d\rangle \mathbin{\backslash\!\!\backslash} \langle u,v\rangle) \mathbin{\backslash\!\!\backslash} (\langle c,d\rangle \mathbin{\backslash\!\!\backslash} \langle a,b\rangle) \\
&= (\langle u,v\rangle \mathbin{\backslash\!\!\backslash} \langle c,d\rangle) \mathbin{\backslash\!\!\backslash} (\langle u,v\rangle \mathbin{\backslash\!\!\backslash} \langle a,b\rangle) \\
&= \langle u,v\rangle \mathbin{\backslash\!\!\backslash} \langle a,b\rangle,
\end{aligned}$$

so $\langle c,d\rangle \mathbin{\backslash\!\!\backslash} \langle a,b\rangle\ \Theta\ \langle u,v\rangle \mathbin{\backslash\!\!\backslash} \langle a,b\rangle$. For the denominator argument, it suffices, without loss of generality, to prove that $(\langle a,b\rangle \mathbin{\backslash\!\!\backslash} \langle c,d\rangle) \mathbin{\backslash\!\!\backslash} (\langle a,b\rangle \mathbin{\backslash\!\!\backslash} \langle u,v\rangle) = \langle \mathrm{e},\mathrm{e}\rangle$. Observing that the proof of $a\backslash\mathrm{e} = \mathrm{e}$ in Lemma 7.4.2 does not depend on (C5), we obtain

$$(\langle a,b\rangle \mathbin{\backslash\!\!\backslash} \langle c,d\rangle) \mathbin{\backslash\!\!\backslash} (\langle a,b\rangle \mathbin{\backslash\!\!\backslash} \langle u,v\rangle) = (\langle c,d\rangle \mathbin{\backslash\!\!\backslash} \langle a,b\rangle) \mathbin{\backslash\!\!\backslash} (\langle c,d\rangle \mathbin{\backslash\!\!\backslash} \langle u,v\rangle) = \langle \mathrm{e},\mathrm{e}\rangle.$$

To confirm that $/\!\!/$ is also preserved in both arguments, it suffices to show that, for all $a,b,c,d \in C$,

$$\langle a,b\rangle \mathbin{\backslash\!\!\backslash} \langle c,d\rangle = \langle \mathrm{e},\mathrm{e}\rangle \iff \langle c,d\rangle /\!\!/ \langle a,b\rangle = \langle \mathrm{e},\mathrm{e}\rangle.$$

Suppose first that $\langle a, b\rangle \backslash\!\backslash \langle c, d\rangle = \langle \mathrm{e}, \mathrm{e}\rangle$. The definition of $\backslash\!\backslash$ yields $b \leq a\backslash c$ and $(a\backslash c)\backslash b \leq (c\backslash a)\backslash d$. Using also the properties established in Lemma 7.4.2 and the fact that $(a\backslash c) \vee ((c\backslash a) \vee d) = \mathrm{e}$,

$$\begin{aligned}
a\backslash c &= (a\backslash c)/((d/b)\backslash(d \vee (c\backslash a))) && \text{Lemma 7.4.3}\\
&= (a\backslash c)/(((d/b)\backslash d)/((c\backslash a)\backslash d)) && \text{(C3)}\\
&\geq (a\backslash c)/(((b/d)\backslash b)/((a\backslash c)\backslash b)) && \text{(C5)}, (a\backslash c)\backslash b \leq (c\backslash a)\backslash d\\
&= (a\backslash c)/((b/d)\backslash(b/((a\backslash c)\backslash b))) && \text{(C3)}\\
&= (a\backslash c)/((b/d)\backslash(a\backslash c)) && b \leq a\backslash c\\
&= (b/d)/((a\backslash c)\backslash(b/d)) && \text{(C5)}\\
&= (b/d)/(((a\backslash c)\backslash b)/d) && \text{(C3)}\\
&\geq (b/d)/(((c\backslash a)\backslash d)/d) && (a\backslash c)\backslash b \leq (c\backslash a)\backslash d\\
&= (b/d)/((c\backslash a)\backslash(d/d)) = b/d && \text{(C3), (C2).}
\end{aligned}$$

On the other hand, observe that, since $b \leq (a\backslash c)\backslash b \leq (c\backslash a)\backslash d$,

$$(c\backslash a)\backslash(d/b) = ((c\backslash a)\backslash d)/b = \mathrm{e},$$

so $a \leq c\backslash a \leq d/b$, and this last inequality, together with the inequality $b/d \leq a\backslash c$ established above, yields

$$a/(d/b) \leq a/(c\backslash a) = c/(a\backslash c) \leq c/(b/d).$$

So $\langle c, d\rangle /\!\!/ \langle a, b\rangle = \langle \mathrm{e}, \mathrm{e}\rangle$. Hence Θ is a congruence and the quotient $s(\mathbf{C}) := \mathbf{D}/\Theta$ is again an algebra of signature $\langle 2, 2, 0\rangle$ satisfying (C1)–(C4). In this case, however, (C5) also holds. Given $a, b, c, d \in C$, let $x := d\backslash(c\backslash a)$ and $y := ((c\backslash a)\backslash d)\backslash((a\backslash c)\backslash b)$. An application of the defining conditions for $\backslash\!\backslash$ and $/\!\!/$ in $\mathbf{D}$ yields

$$\langle a, b\rangle /\!\!/ (\langle c, d\rangle \backslash\!\backslash \langle a, b\rangle) = \langle (a/(y/b))/(x/(b/y)), (b/y)/x\rangle.$$

Since $((c\backslash a)\backslash d) \vee x = x \vee (a\backslash c) \vee b = \mathrm{e}$,

$$\begin{aligned}
(a/(y/b))/(x/(b/y)) &= a/(x/(b/y)) && \text{(C2), Lemma 7.4.2}\\
&= a/(x/(b/((a\backslash c)\backslash b))) && \text{Lemma 7.4.3}\\
&= a/x && \text{Lemma 7.4.3.}
\end{aligned}$$

But then, by Lemma 7.4.3 again, since $x \vee (b/y) = \mathrm{e}$, also $(b/y)/x = b/y$. That is,

$$\langle a, b\rangle /\!\!/ (\langle c, d\rangle \backslash\!\backslash \langle a, b\rangle) = \langle a/x, b/y\rangle.$$

Similarly, letting $u := b\backslash(a\backslash c)$ and $v := ((a\backslash c)\backslash b)\backslash((c\backslash a)\backslash d)$,

$$\langle c, d\rangle /\!\!/ (\langle a, b\rangle \backslash\!\backslash \langle c, d\rangle) = \langle c/u, d/v\rangle.$$

We obtain now

$$\begin{aligned}
(a/x)\backslash(c/u) &= ((a/x)\backslash c)/u && \text{(C3)}\\
&= ((a/(c\backslash a))\backslash c)/u && \text{Lemmas 7.4.4.(a) and 7.4.2}\\
&= ((c/(a\backslash c))\backslash c)/u && \text{(C5)}\\
&= (a\backslash c)/u && \text{Lemma 7.4.2}\\
&= b/((a\backslash c)\backslash b) && \text{(C5)}\\
&= (b/((a\backslash c)\backslash b))/(d/((c\backslash a)\backslash d)) && \text{Lemma 7.4.3}\\
&= (b/y)/(d/v) && \text{Lemmas 7.4.4.(b) and 7.4.2.}
\end{aligned}$$

This yields $(\langle c,d\rangle /\!\!/ (\langle a,b\rangle \backslash\!\!\backslash \langle c,d\rangle)) \backslash\!\!\backslash (\langle a,b\rangle /\!\!/ (\langle c,d\rangle \backslash\!\!\backslash \langle a,b\rangle)) = \langle \mathrm{e},\mathrm{e}\rangle$ and, similarly, $(\langle a,b\rangle /\!\!/ (\langle c,d\rangle \backslash\!\!\backslash \langle a,b\rangle)) \backslash\!\!\backslash (\langle c,d\rangle /\!\!/ (\langle a,b\rangle \backslash\!\!\backslash \langle c,d\rangle)) = \langle \mathrm{e},\mathrm{e}\rangle$, as required.

Moreover, the map $\varphi\colon C \to C^2/\Theta;\ a \mapsto [\langle a,\mathrm{e}\rangle]_\Theta$ is an embedding of $\mathbf{C}$ into $s(\mathbf{C})$. For injectivity, observe that $\varphi(a) = \varphi(b)$ implies $\mathrm{e} = a\backslash b = b\backslash a$ and hence $a = b$. For preservation of the operations, we just check $\backslash\!\!\backslash$, leaving other cases to the reader:

$$\varphi(a\backslash b) = [\langle a\backslash b,\mathrm{e}\rangle]_\Theta = [\langle \mathrm{e}\backslash(a\backslash b), ((a\backslash b)\backslash \mathrm{e})\backslash((b\backslash a)\backslash \mathrm{e})\rangle]_\Theta = \varphi(a)\backslash\!\!\backslash\varphi(b).$$

To obtain our target negative cone, we repeat this construction, defining $\mathbf{C}_0 := \mathbf{C}$ and $\mathbf{C}_{n+1} := s(\mathbf{C}_n)$, for each $n \in \mathbb{N}$, for convenience viewing $\mathbf{C}_n$ as a subalgebra of $\mathbf{C}_{n+1}$. It then follows that, for each $n \in \mathbb{N}$, letting $ab := [\langle a,b\rangle]_\Theta$, for $a,b \in C_n$, the equation (Id) is satisfied in $\mathbf{C}_{n+1}$ by the elements of C_n.

Now let $\overline{\mathbf{C}}$ be the direct limit of the family $\{\mathbf{C}_n\}_{n\in\mathbb{N}}$ ordered by inclusion; that is, $\overline{C} := \bigcup_{n\in\mathbb{N}} C_n$ and the operations of $\overline{\mathbf{C}}$ are defined by referring to $\mathbf{C}_n$ for suitably large n. Since each $\mathbf{C}_n$ is a subalgebra of $\mathbf{C}_{n+1}$, this is a cone algebra that still contains $\mathbf{C}$ as a subalgebra. Moreover, given any $a,b \in \overline{C}$, we may assume that $a,b \in C_n$, for some $n \in \mathbb{N}$, and define

$$a \cdot b := [\langle a,b\rangle]_\Theta.$$

That this operation is well-defined follows from the construction of $\overline{\mathbf{C}}$.

We now define the algebra of signature $\langle 2,2,2,2,2,0\rangle$

$$\widehat{\mathbf{C}} := \langle \overline{C}, \wedge, \vee, \cdot, \backslash^{\overline{\mathbf{C}}}, /^{\overline{\mathbf{C}}}, \mathrm{e}^{\overline{\mathbf{C}}}\rangle,$$

where $a\vee b := a/^{\overline{\mathbf{C}}}(b\backslash^{\overline{\mathbf{C}}}a)$ and $a\wedge b := a(a\backslash^{\overline{\mathbf{C}}}b)$. Clearly, $\overline{\mathbf{C}}$ is the $\{\backslash, /, \mathrm{e}\}$-reduct of $\widehat{\mathbf{C}}$ and it remains to show that $\widehat{\mathbf{C}}$ is a negative cone. Indeed, in light of Theorem 7.1.10, it suffices to show that $\widehat{\mathbf{C}}$ is a cancellative, integral GBL-algebra.

For notational convenience, we omit superscripts for the rest of this proof. Observe first that $\langle \overline{C}, \vee\rangle$ is a semilattice with greatest element e, by Lemma 7.4.2. To show that $\widehat{\mathbf{C}}$ is a residuated lattice, we need to prove that multiplication is associative, order-preserving, and residuated, and that $\wedge$ is the meet operation. In what follows, we make repeated use of the fact that (Id) and its mirror image are now satisfied throughout $\overline{C}$.

Consider first any $a,c \in \overline{C}$ with $a \le c$, i.e., $a\backslash c = \mathrm{e}$. Then, for any $b \in \overline{C}$,

$$\begin{aligned} ab\backslash cb &= (b\backslash \mathrm{e})((\mathrm{e}\backslash b)\backslash((c\backslash a)\backslash b)) && a \le c \\ &= \mathrm{e}(b\backslash((c\backslash a)\backslash b)) && \text{Lemma 7.4.2, (C2)} \\ &= \mathrm{e} && \text{Lemma 7.4.2.} \end{aligned}$$

That is, $ab \le cb$, and, similarly, $ba \le ca$. Next, consider any $a,b,c \in \overline{C}$. Note that $ab\backslash c = ab\backslash ce = b\backslash(a\backslash c)$, by (Id) and Lemma 7.4.2. Hence, in particular, $ab \le c$ if and only if $b \le a\backslash c$, and analogously for the other residuation conditions. Now, using what we have just proved, as well as (Id), observe that, for any $a,b,c,d \in \overline{C}$,

$$a(bc) \le d \iff bc \le a\backslash d \iff c \le b\backslash(a\backslash d) = ab\backslash d \iff (ab)c \le d,$$

so multiplication is associative. Hence $\langle \overline{C}, \cdot, \backslash, /, \mathrm{e}, \le\rangle$ is an integral residuated pomonoid. Now consider any $a,b \in \overline{C}$. Clearly, $a(a\backslash b) \le a,b$. Also, if $c \le a,b$, for some $c \in \overline{C}$, then, using (Id) and (C1),

$$c\backslash a(a\backslash b) = (c\backslash a) \cdot ((a\backslash c)\backslash(a\backslash b)) = (c\backslash a) \cdot ((c\backslash a)\backslash(c\backslash b)) = \mathrm{e}.$$

Hence $a \wedge b := a(a\backslash b)$ is the meet operation of the lattice $\langle C, \leq \rangle$, and $\widehat{\mathbf{C}}$ is an integral GBL-algebra. Finally, an application of (Id) and its mirror image, show that $\widehat{\mathbf{C}}$ satisfies the equations $xy/y \approx x$ and $y\backslash yx \approx x$, as required. □

7.5. A categorical equivalence

Recall that a nucleus γ on a negative cone $\mathbf{G}^-$ of an ℓ-group $\mathbf{G}$ (referred to in this section, more briefly, as a *negative cone*) is called *dense* if its image G^-_γ generates $\mathbf{G}^-$ as a monoid. In this case, we call the expansion of the algebra $\mathbf{G}^-$ with the unary operation γ, denoted by $\langle \mathbf{G}^-, \gamma \rangle$, a *negative cone with a dense nucleus.*

Our goal in this section will be to establish an equivalence between

1. $\mathcal{IGMV}$, the category whose objects are integral GMV-algebras, and morphisms are (ordinary) homomorphisms;
2. $\mathcal{LG}^-_*$, the category whose objects are negative cones with a dense nucleus, and morphisms are (ordinary) homomorphisms.

Our first significant step towards this goal will be to prove that every integral GMV-algebra is the image of a dense nucleus on a negative cone. The next lemma collects some useful further properties of nuclei on GMV-algebras.

LEMMA 7.5.1. *Let γ be a nucleus on a GMV-algebra $\mathbf{L} = \langle L, \wedge, \vee, \cdot, \backslash, /, \mathrm{e} \rangle$.*

(a) *$a \vee_\gamma b = a \vee b$, for all $a, b \in L_\gamma$, and $\gamma(\mathrm{e}) = \mathrm{e}$.*

(b) *$\mathbf{L}_\gamma$ is a GMV-algebra.*

(c) *L_γ is a lattice-filter of $\mathbf{L}$.*

(d) *If $\mathbf{L}$ is integral and $a \in L$, then $\gamma(b) = \gamma_{\gamma(a)}(b) = \gamma(a) \vee b$, for all $b \in {\uparrow}a$.*

PROOF.

(a) For any $a, b \in L_\gamma$, the defining equations for GMV-algebras and Lemma 1.5.2 yield $a \vee b = a/((a \vee b)\backslash a) \in L_\gamma$ and hence $a \vee_\gamma b = a \vee b$. Also, $\gamma(\mathrm{e})\gamma(\mathrm{e}) \leq \gamma(\mathrm{e})$, so $\mathrm{e} \leq \gamma(\mathrm{e}) \leq \gamma(\mathrm{e})\backslash\gamma(\mathrm{e}) = \mathrm{e}$, by Lemma 7.1.2.(b), and $\gamma(\mathrm{e}) = \mathrm{e}$.

(b) Recalling Lemma 1.5.4, observe that the defining equations for GMV-algebras involve only the join and residual operations, which in $\mathbf{L}_\gamma$ are the restrictions of the corresponding operations in $\mathbf{L}$. So $\mathbf{L}_\gamma$ is a GMV-algebra.

(c) $\langle L_\gamma, \wedge, \vee \rangle$ is a sublattice of the lattice reduct of $\mathbf{L}$, so L_γ is closed under meets. Moreover, if $a \in L_\gamma$ and $b \in L$ satisfy $a \leq b$, the defining equations for GMV-algebras and Lemma 1.5.2 yield $b = a \vee b = a/((a \vee b)\backslash a) \in L_\gamma$.

(d) Suppose that $\mathbf{L}$ is integral and consider any $a \in L$ and $b \in {\uparrow}a$. Then, using the properties of nuclei and Lemma 1.5.2,

$$\gamma(b) \leq \gamma(a)/(\gamma(b)\backslash\gamma(a)) = \gamma(a)/(b\backslash\gamma(a)) = \gamma(a) \vee b \leq \gamma(b).$$

So $\gamma(b) = \gamma(a) \vee b = \gamma_{\gamma(a)}(b)$. □

THEOREM 7.5.2. *A residuated lattice* $\mathbf{L}$ *is an integral GMV-algebra if and only if* $\mathbf{L} = \mathbf{G}^-_\gamma$, *for a negative cone* $\mathbf{G}^-$ *and dense nucleus* γ *on* $\mathbf{G}^-$.

PROOF. The right-to-left direction follows from Lemma 7.5.1. For the converse, let $\mathbf{L} = \langle L, \wedge, \vee, \circ, \backslash, /, \mathrm{e}\rangle$ be an integral GMV-algebra. By Lemma 7.4.1, its $\{\backslash, /, \mathrm{e}\}$-reduct $\langle L, \backslash, /, \mathrm{e}\rangle$ is a cone algebra that, by Theorem 7.4.5, embeds into the $\{\backslash, /, \mathrm{e}\}$-reduct of a negative cone $\mathbf{G}^-$, where G^- is generated by L as a monoid. The residuals of $\mathbf{L}$, neutral element e, and term-definable join operation $\vee$ may be viewed as restrictions of the corresponding operations of $\mathbf{G}^-$, so we use the same symbols $\backslash$, $/$, e, and $\vee$. However, since the product in $\mathbf{L}$ does not coincide, in general, with the product in $\mathbf{G}^-$, we use the symbol "$\cdot$" to denote the latter.

Claim 1. If $a_1, \ldots, a_n, b \in L$, then $b \vee (a_1 \cdots a_n) = b \vee (a_1 \circ \cdots \circ a_n)$.

Proof. Using the defining equations for GMV-algebras and Proposition 1.4.4,

$$\begin{aligned} b \vee (a_1 \cdots a_n) &= b/(a_1 \cdots a_n \backslash b) \\ &= b/(a_n \backslash (\ldots (a_2 \backslash (a_1 \backslash b)) \ldots)) \\ &= b/(a_1 \circ \cdots \circ a_n \backslash b) \\ &= b \vee (a_1 \circ \cdots \circ a_n). \end{aligned}$$ □

Claim 2. If $a_1, \ldots, a_n, b_1, \ldots, b_m \in L$ and $a_1 \cdots a_n = b_1 \cdots b_m$, then $a_1 \circ \cdots \circ a_n = b_1 \circ \cdots \circ b_m$.

Proof. By Claim 1,

$$\begin{aligned} (a_1 \circ \cdots \circ a_n) \vee (b_1 \circ \cdots \circ b_m) &= (a_1 \circ \cdots \circ a_n) \vee (b_1 \cdots b_m) \\ &= (a_1 \circ \cdots \circ a_n) \vee (a_1 \cdots a_n) \\ &= a_1 \circ \cdots \circ a_n. \end{aligned}$$

So $b_1 \circ \cdots \circ b_m \le a_1 \circ \cdots \circ a_n$ and, by symmetry, $a_1 \circ \cdots \circ a_n \le b_1 \circ \cdots \circ b_m$. □

By Claim 2, we obtain a well-defined map by fixing $\gamma(a) := a_1 \circ \cdots \circ a_n$, for all $a = a_1 \cdots a_n \in G^-$ with $a_1, \ldots, a_n \in L$. We now show that γ is a nucleus on $\mathbf{G}^-$ and that $\mathbf{G}^-_\gamma = \mathbf{L}$.

Note first that $\gamma(a) \in L$, for all $a \in G^-$, so by setting $b = \gamma(a)$ in the statement of Claim 1, we get $\gamma(a) \vee a = \gamma(a)$. That is, $a \le \gamma(a)$, for all $a \in G^-$. Moreover, if $a \le b$, then, using Claim 1,

$$\gamma(a) \le \gamma(b) \vee \gamma(a) = \gamma(b) \vee a \le \gamma(b) \vee b = \gamma(b).$$

Clearly, $\gamma(\gamma(a)) = \gamma(a)$, so γ is a closure operator. Moreover, given $a = a_1 \cdots a_n$ and $b = b_1 \cdots b_m$, for some $a_1, \ldots, a_n, b_1, \ldots, b_m \in L$,

$$\gamma(a)\gamma(b) \le \gamma(\gamma(a)\gamma(b)) = \gamma(a) \circ \gamma(b) = (a_1 \circ \cdots \circ a_n) \circ (b_1 \circ \cdots \circ b_m) = \gamma(ab).$$

The definition of γ gives $G^-_\gamma = L$, so γ is a dense nucleus. We have already observed that all the operations except multiplication and meet agree on $\mathbf{G}^-_\gamma$ and $\mathbf{L}$. Also, $a \cdot_\gamma b = \gamma(ab) = a \circ b$, for all $a, b \in G^-$. Finally, meets coincide since integral GMV-algebras satisfy the equation $x \wedge y \approx (x/y)y$. So $\mathbf{G}^-_\gamma = \mathbf{L}$. □

In order to extend the relationship established in Theorem 7.5.2 to an equivalence between the categories $\mathcal{IGMV}$ and $\mathcal{LG}^-_*$, we represent the elements of a negative cone with a dense nucleus as "canonical" products of members of the image of that nucleus. Let γ be a nucleus on a negative cone $\mathbf{G}^-$, recalling that the monoid multiplication $\circ_\gamma$ of $\mathbf{G}^-_\gamma$ is defined by $a \circ_\gamma b := \gamma(ab)$, for $a, b \in G^-_\gamma$. A sequence $\langle a_1, \dots, a_n \rangle$ of elements of G^-_γ is called a *γ-decomposition* of $a \in G^-$ if $a = a_1 \cdots a_n$. It is called *canonical* if, in addition, $a_i \circ_\gamma a_{i+1} = a_i$, for each $i \in \{1, \dots, n-1\}$.

LEMMA 7.5.3. *Let γ be a dense nucleus on a negative cone $\mathbf{G}^-$. Then every element of G^- has a canonical γ-decomposition. Moreover, if $\langle a_1, \dots, a_n \rangle$ and $\langle a'_1, \dots, a'_m \rangle$ are canonical γ-decompositions of the same element and $m \geq n$, then $a_i = a'_i$, for each $i \in \{1, \dots, n\}$ and $a'_i = \mathrm{e}$, for each $i \in \{n+1, \dots, m\}$.*

PROOF. We begin by constructing a canonical γ-decomposition of an arbitrary element a of G^-. Let $a_1 := \gamma(a)$ and $a_{i+1} := \gamma(a_1 \cdots a_i \backslash a)$, for each $i \in \mathbb{N}^{>0}$. We prove that there exists $n \in \mathbb{N}$ such that $a = a_1 \cdots a_n$ and $a_i \circ_\gamma a_{i+1} = a_i$, for each $i \in \{1, \dots, n-1\}$.

First, we show by induction that $a \leq a_1 \cdots a_k$, for each $k \in \mathbb{N}^{>0}$. For $k = 1$, we have $a \leq \gamma(a) = a_1$. If $a \leq a_1 \cdots a_k$, then

$$\begin{aligned} a &= (a_1 \cdots a_k) \wedge a \\ &= (a_1 \cdots a_k) \cdot (a_1 \cdots a_k \backslash a) \\ &\leq (a_1 \cdots a_k) \cdot \gamma(a_1 \cdots a_k \backslash a) \\ &= (a_1 \cdots a_k) \cdot a_{k+1}. \end{aligned}$$

Next, let $u := \gamma(b)$, for any $b \leq a$. By Lemma 7.5.1.(d), the maps γ and γ_u agree on $\uparrow a$. The arguments of γ in the definition of the elements a_i, and in the equality $\gamma(a_i \cdot a_{i+1}) = a_i$, are all in $\uparrow a$, so we can replace γ by γ_u. Hence a γ-decomposition of an element a is canonical if and only if it is a canonical γ_u-decomposition.

Applying Corollary 7.1.11, we obtain, for all $i \in \mathbb{N}^{>0}$,

$$a_i(a_i \backslash (a_1 \cdots a_{i-1} \backslash a)) = (a_1 \cdots a_{i-1}) \backslash a,$$

where $a_1 \cdots a_{i-1} = \mathrm{e}$ for $i = 1$. It follows that, for all $i \in \mathbb{N}^{>0}$,

$$\begin{aligned} a_i \circ_\gamma a_{i+1} &= \gamma_u(a_i a_{i+1}) \\ &= u \vee a_i a_{i+1} \\ &= u \vee a_i \gamma_u(a_1 \cdots a_i \backslash a) \\ &= u \vee a_i(u \vee (a_1 \cdots a_i \backslash a)) \\ &= u \vee a_i u \vee a_i(a_1 \cdots a_i \backslash a) \\ &= u \vee a_i u \vee a_i(a_i \backslash (a_1 \cdots a_{i-1} \backslash a)) \\ &= u \vee a_i u \vee (a_1 \cdots a_{i-1} \backslash a) \\ &= u \vee (a_1 \cdots a_{i-1} \backslash a) \\ &= \gamma_u(a_1 \cdots a_{i-1} \backslash a) \\ &= a_i. \end{aligned}$$

We next show by induction on k that $a_1 \cdots a_k \backslash a = u^k \backslash a$, for all $k \in \mathbb{N}^{>0}$. Observe first that

$$a_1 \backslash a = \gamma(a) \backslash a = \gamma_u(a) \backslash a = (a \vee u) \backslash a = (a \backslash a) \wedge (u \backslash a) = u \backslash a.$$

For the inductive step, suppose that $a_1 \cdots a_k \backslash a = u^k \backslash a$. Using Proposition 1.4.4,

$$\begin{aligned} a_1 \cdots a_k a_{k+1} \backslash a &= a_{k+1} \backslash (a_1 \cdots a_k \backslash a) \\ &= \gamma(a_1 \cdots a_k \backslash a) \backslash (a_1 \cdots a_k \backslash a) \\ &= ((a_1 \cdots a_k \backslash a) \vee u) \backslash (a_1 \cdots a_k \backslash a) \\ &= (a_1 \cdots a_k \backslash a) \backslash (a_1 \cdots a_k \backslash a) \wedge (u \backslash (a_1 \cdots a_k \backslash a)) \\ &= \mathrm{e} \wedge (u \backslash (a_1 \cdots a_k \backslash a)) \\ &= u \backslash (u^k \backslash a) \\ &= u^{k+1} \backslash a. \end{aligned}$$

Since G^- is the universe of the monoid generated by G^-_γ, there exist $n \in \mathbb{N}$ and elements $c_1, \ldots, c_n \in G^-_\gamma$ such that $a = c_1 \cdots c_n$. Hence, for each $i \in \mathbb{N}^{>0}$,

$$u = \gamma(b) \leq \gamma(a) = \gamma(c_1 \cdots c_n) = c_1 \circ_\gamma \cdots \circ_\gamma c_n \leq c_i.$$

It follows that $u^n \leq c_1 \cdots c_n = a$. Hence $\mathrm{e} \leq u^n \backslash a = a_1 \cdots a_n \backslash a$, and $a_1 \cdots a_n \leq a$. Since the reverse inequality was established above, $a = a_1 \cdots a_n$.

For uniqueness, let $\langle a_1, \ldots, a_n \rangle$ and $\langle a'_1, \ldots, a'_m \rangle$ be canonical γ-decompositions of an element a in G^- with $m \geq n$. Then

$$a_1 = a_1 \circ_\gamma \cdots \circ_\gamma a_n = a'_1 \circ_\gamma \cdots \circ_\gamma a'_m = a'_1,$$

and hence also, by Corollary 7.1.11,

$$a_2 \cdots a_n = a_1 \backslash a_1 a_2 \cdots a_n = a'_1 \backslash a'_1 a'_2 \cdots a'_m = a'_2 \cdots a'_m.$$

Inductively, $a_i = a'_i$, for each $i \in \{1, \ldots, n\}$. Another application of Corollary 7.1.11 yields $\mathrm{e} = a'_{n+1} \cdots a'_m$, so $a'_i = \mathrm{e}$, for each $i \in \{n+1, \ldots, m\}$. □

It follows from the preceding lemma that, for any dense nucleus γ on a negative cone $\mathbf{G}^-$, each $a \in G^-$ has at least one canonical γ-decomposition, and all canonical γ-decompositions of a can be obtained from the shortest one by appending some extra e's at the end of the sequence. We may therefore harmlessly assume that canonical γ-decompositions are unique and, when considering canonical γ-decompositions of finitely many elements, assume that they all have the same length.

Corollary 7.5.4. *Let γ be a dense nucleus on a negative cone $\mathbf{G}^-$. If $\langle a_1, \ldots, a_n \rangle$ and $\langle b_1, \ldots, b_n \rangle$ are canonical γ-decompositions of $a, b \in G^-$, respectively, then $a \leq b$ implies $a_i \leq b_i$, for each $i \in \{1, \ldots, n\}$.*

Proof. In view of Lemma 7.5.3, we may assume that $a_i = \gamma(a_1 \cdots a_{i-1} \backslash a)$ and $b_i = \gamma(b_1 \cdots b_{i-1} \backslash b)$, for each $i \in \{1, \ldots, n\}$. Suppose that $a \leq b$ and let $u := \gamma(a \wedge b) = \gamma(a)$. From the proof of Lemma 7.5.3, we obtain $a_1 \cdots a_k \backslash a = u^k \backslash a$ and $b_1 \cdots b_k \backslash b = u^k \backslash b$, for each $k \in \{1, \ldots, n\}$. Hence, for each $i \in \{1, \ldots, n\}$,

$$a_i = \gamma(a_1 \cdots a_{i-1} \backslash a) = \gamma(u^{i-1} \backslash a) \leq \gamma(u^{i-1} \backslash b) = \gamma(b_1 \cdots b_{i-1} \backslash b) = b_i. \qquad \square$$

We now check that canonical γ-decompositions respect meets and joins.

LEMMA 7.5.5. *Let γ be a dense nucleus on a negative cone $\mathbf{G}^-$. If $\langle a_1, \dots, a_n\rangle$ and $\langle b_1, \dots, b_n\rangle$ are canonical γ-decompositions of $a, b \in G^-$, respectively, then*

$$a \wedge b = \prod_{i=1}^{n}(a_i \wedge b_i) \quad \text{and} \quad a \vee b = \prod_{i=1}^{n}(a_i \vee b_i).$$

PROOF. Let $\langle c_1, \dots, c_n\rangle$ be a canonical γ-decomposition of $a \wedge b$. Since $a \wedge b \le a, b$, Corollary 7.5.4 yields $c_i \le a_i, b_i$ and hence $c_i \le a_i \wedge b_i$, for each $i \in \{1, \dots, n\}$. So

$$a \wedge b = \prod_{i=1}^{n} c_i \le \prod_{i=1}^{n}(a_i \wedge b_i) \le \prod_{i=1}^{n} a_i \wedge \prod_{i=1}^{n} b_i = a \wedge b.$$

The proof for joins is analogous. □

We now describe a property that characterizes finite sequences of elements that decompose the same element of a negative cone $\mathbf{G}^-$. Let us say that $\langle a_1, \dots, a_n\rangle$ and $\langle b_1, \dots, b_m\rangle$ have the *orthogonal refinement property* ORP if there exist elements c_{ij} of G^-, for $i \in \{1, \dots, n\}$ and $j \in \{1, \dots, m\}$, such that

$$a_j = \prod_{i=1}^{m} c_{ij}, \quad b_i = \prod_{j=1}^{n} c_{ij}, \quad \text{and} \quad \prod_{l=j+1}^{m} c_{il} \vee \prod_{k=i+1}^{n} c_{kj} = \mathrm{e}.$$

In a nutshell, the ORP states that in the matrix

$$\begin{array}{cccc} & a_1 & \cdots & a_n \\ b_1 & c_{11} & \cdots & c_{1n} \\ \vdots & \vdots & & \vdots \\ b_m & c_{m1} & \cdots & c_{mn}, \end{array}$$

each a_j is the product of the elements of the j-th column, each b_i is the product of the elements of the i-th row, and the product of the elements to the right of c_{ij} is orthogonal to the product of elements below it. Matrices of this form will be used to indicate that sequences $\langle a_1, \dots, a_n\rangle$ and $\langle b_1, \dots, b_m\rangle$ have the ORP.

LEMMA 7.5.6. *Let $\mathbf{G}^-$ be a negative cone. The following statements are equivalent for any $a_1, \dots, a_n, b_1, \dots, b_m \in G^-$:*

(1) $a_1 \cdots a_n = b_1 \cdots b_m$.
(2) $\langle a_1, \dots, a_n\rangle$ *and* $\langle b_1, \dots, b_m\rangle$ *have the* ORP.

PROOF.

(2) $\Rightarrow$ (1) We prove the implication by induction on the ordered pair $\langle m, n\rangle$ in the lexicographically ordered product of two copies of $\langle \mathbb{N}, \le\rangle$. For $m = n = 2$, recall that if $a \vee b = \mathrm{e}$, then $ab = ba$, by Lemma 7.2.2. So $a_1a_2 = c_{11}c_{21}c_{12}c_{22} = c_{11}c_{12}c_{21}c_{22} = b_1b_2$. Now, let $m \in \mathbb{N}^{>1}$ and $n \in \mathbb{N}^{>2}$, and assume that the lemma is true for $\langle m, k\rangle$, for each $k \in \{2, \dots, n-1\}$. We will show that it is true for $\langle m, n\rangle$.

It is easy to see that, if $c = c_1 \cdots c_m$, the ORP implies

$$\begin{array}{c} \\ c_1 \\ \vdots \\ c_m \end{array} \left(\begin{array}{ccc} a_2 & \cdots & a_n \\ c_{12} & \cdots & c_{1n} \\ \vdots & & \vdots \\ c_{m2} & \cdots & c_{mn} \end{array} \right) \quad \text{and} \quad \begin{array}{c} \\ b_1 \\ \vdots \\ b_m \end{array} \left(\begin{array}{cc} a_1 & c \\ c_{11} & c_1 \\ \vdots & \vdots \\ c_{m1} & c_m \end{array} \right).$$

So $a_1a_2\cdots a_n = a_1(c_1 \ldots c_m) = a_1c = b_1b_2\cdots b_m$. Moreover, the lemma holds for $\langle m,n\rangle$ if and only if it holds for $\langle n,m\rangle$, completing the inductive proof.

(1) $\Rightarrow$ (2) We again prove the implication by induction on the ordered pair $\langle m,n\rangle$ in the lexicographically ordered product of two copies of $\langle \mathbb{N}, \leq\rangle$. For the base case, let $m = n = 2$. Assume that $a_1a_2 = b_1b_2 = c$ and let

$$c_{11} := a_1 \vee b_1, \quad c_{12} := a_2/c_{22}, \quad c_{21} := c_{11}\backslash a_1, \quad \text{and} \quad c_{22} := a_2 \vee b_2.$$

By Proposition 1.4.4, Corollary 7.1.11, and Corollary 7.2.5,

$$\begin{aligned} c_{12} &= a_2/c_{22} \\ &= a_2/(a_2 \vee b_2) \\ &= (a_1\backslash c)/((a_1\backslash c) \vee (b_1\backslash c)) \\ &= (a_1\backslash c)/((a_1 \wedge b_1)\backslash c) \\ &= a_1\backslash(c/((a_1 \wedge b_1)\backslash c)) \\ &= a_1\backslash((a_1 \wedge b_1) \vee c) \\ &= a_1\backslash(a_1 \wedge b_1) \\ &= (a_1\backslash a_1) \wedge (a_1\backslash b_1) \\ &= \mathrm{e} \wedge (a_1\backslash b_1) \\ &= (a_1\backslash b_1) \wedge (b_1\backslash b_1) \\ &= (a_1 \vee b_1)\backslash b_1 \\ &= c_{11}\backslash b_1. \end{aligned}$$

Similarly, $c_{21} = b_2/c_{22}$. Hence

$$\begin{aligned} c_{11}c_{21} &= c_{11}(c_{11}\backslash a_1) = c_{11} \wedge a_1 = a_1, \\ c_{12}c_{22} &= (a_2/c_{22})c_{22} = a_2 \wedge c_{22} = a_2, \\ c_{11}c_{12} &= c_{11}(c_{11}\backslash b_1) = c_{11} \wedge b_1 = b_1, \\ c_{21}c_{22} &= (b_2/c_{22})c_{22} = b_2 \wedge c_{22} = b_2. \end{aligned}$$

Finally, $c_{21} \vee c_{12} = (c_{11}\backslash a_1) \vee (c_{11}\backslash b_1) = c_{11}\backslash(a_1 \vee b_1) = c_{11}\backslash c_{11} = \mathrm{e}$.

For the general case, let $m \in \mathbb{N}^{>1}$ and $n \in \mathbb{N}^{>2}$ and assume that the lemma is true for $\langle m,k\rangle$, for each $k \in \{2,\ldots,n-1\}$. We will show that it is true for $\langle m,n\rangle$. Assume that $a_1a_2\cdots a_n = b_1b_2\cdots b_m$ and let $a = a_2a_3\cdots a_n$. So $a_1a = b_1b_2\cdots b_m$.

By the induction hypothesis,

$$\begin{array}{c} \\ b_1 \\ \vdots \\ b_m \end{array} \begin{array}{c} a_1 \quad\; a \\ \left(\begin{array}{cc} c_{11} & c_{12} \\ \vdots & \vdots \\ c_{m1} & c_{m2} \end{array}\right) \end{array} \quad \text{and} \quad \begin{array}{c} \\ c_{12} \\ \vdots \\ c_{m2} \end{array} \begin{array}{c} a_2 \quad \dots \quad a_n \\ \left(\begin{array}{ccc} d_{12} & \dots & d_{1n} \\ \vdots & & \vdots \\ d_{m2} & \dots & d_{mn} \end{array}\right) \end{array},$$

for some c_{ij}, d_{kl} with appropriate indices. So we have

$$\begin{array}{c} \\ b_1 \\ \vdots \\ b_m \end{array} \begin{array}{c} a_1 \quad a_2 \quad \dots \quad a_n \\ \left(\begin{array}{cccc} c_{11} & d_{12} & \dots & d_{1n} \\ \vdots & \vdots & & \vdots \\ c_{m1} & d_{m2} & \dots & d_{mn} \end{array}\right) \end{array}.$$

Hence the claim holds for $\langle m, n\rangle$ as required. □

LEMMA 7.5.7. *Let γ be a dense nucleus on a negative cone $\mathbf{G}^-$ and let $a, a_1, \dots, a_n$ be any elements of G_γ^-. Then $a = a_1 \cdots a_n$ if and only if $a = a_1 \circ_\gamma \cdots \circ_\gamma a_n$ and $a_k = (a_k \circ_\gamma \cdots \circ_\gamma a_n)/(a_{k+1} \circ_\gamma \cdots \circ_\gamma a_n)$, for each $k \in \{1, \dots, n-1\}$.*

PROOF. We prove the claim by induction on n. The case $n = 1$ is immediate. For $n = 2$, if $a = a_1a_2$, then $\gamma(a) = \gamma(a_1a_2)$, so $a = a_1 \circ_\gamma a_2$. Moreover, by Corollary 7.1.11, $a_1 = a/a_2$, so $a_1 = (a_1 \circ_\gamma a_2)/a_2$. Conversely, if $a = a_1 \circ_\gamma a_2$ and $a_1 = (a_1 \circ_\gamma a_2)/a_2$, then $a = \gamma(a_1a_2) \leq \gamma(a_2) = a_2$ and $a_1 = a/a_2$, so $a = a_1a_2$, by Corollary 7.1.11.

For the inductive step, note first that if $a_1a_2 \cdots a_n \in G_\gamma^-$, then $a_2 \cdots a_n \in G_\gamma^-$, since, by Theorem 7.5.1, G_γ^- is a lattice-filter of $\mathbf{G}^-$. So $a = a_1(a_2 \cdots a_n)$ if and only if $a = a_1b$ and $b = a_2 \cdots a_n$, for some $b \in G_\gamma^-$. But the latter holds, by the induction hypothesis and the case $n = 2$, if and only if $a = a_1 \circ_\gamma b$, $a_1 = a/b$, $b = a_2 \circ_\gamma \cdots \circ_\gamma a_n$, and $a_k = (a_k \circ_\gamma \cdots \circ_\gamma a_n)/(a_{k+1} \circ_\gamma \cdots \circ_\gamma a_n)$, for each $k \in \{2, \dots, n-1\}$. Hence $a = a_1 \cdots a_n$ if and only if $a = a_1 \circ_\gamma \cdots \circ_\gamma a_n$ and $a_k = (a_k \circ_\gamma \cdots \circ_\gamma a_n)/(a_{k+1} \circ_\gamma \cdots \circ_\gamma a_n)$, for each $k \in \{1, \dots, n-1\}$, as required. □

We now turn our attention to morphisms between nucleus-systems. We need to show that morphisms on nucleus-images of negative cones mesh nicely with the nuclei themselves. In particular, given dense nuclei γ_1 and γ_2 on negative cones $\mathbf{H}^-$ and $\mathbf{G}^-$, respectively, if $\langle a_1, \dots, a_n\rangle$ and $\langle b_1, \dots, b_m\rangle$ (with $a_i, b_j \in H_{\gamma_1}^-$) decompose the same element in H^- and $\varphi\colon \mathbf{H}_{\gamma_1}^- \to \mathbf{G}_{\gamma_2}^-$ is a residuated lattice homomorphism, then $\langle \varphi(a_1), \dots, \varphi(a_n)\rangle$ and $\langle \varphi(b_1), \dots, \varphi(b_m)\rangle$ decompose the same element in G^-.

LEMMA 7.5.8. *Suppose that γ_1 and γ_2 are dense nuclei on negative cones $\mathbf{H}^-$ and $\mathbf{G}^-$, respectively, and $\varphi\colon \mathbf{H}_{\gamma_1}^- \to \mathbf{G}_{\gamma_2}^-$ is a residuated lattice homomorphism.*

(a) *For all $c_1, \dots, c_n \in H_{\gamma_1}^-$ satisfying $c_1 \cdots c_n \in H_{\gamma_1}^-$,*

$$\varphi(c_1 \cdots c_n) = \varphi(c_1) \cdots \varphi(c_n).$$

(b) *For all $a_1, \dots, a_n, b_1, \dots, b_m \in H_{\gamma_1}^-$ satisfying $a_1 \cdots a_n = b_1 \cdots b_m$,*

$$\varphi(a_1) \cdots \varphi(a_n) = \varphi(b_1) \cdots \varphi(b_m).$$

PROOF.

(a) Suppose that $c_1, \ldots, c_n \in H_{\gamma_1}^-$ and $c_1 \cdots c_n \in H_{\gamma_1}^-$. By Lemma 7.5.7,

$$c_1 \cdots c_n = c_1 \circ_{\gamma_1} \cdots \circ_{\gamma_1} c_n,$$

and, for each $k \in \{1, \ldots, n-1\}$,

$$c_k = (c_k \circ_{\gamma_1} \cdots \circ_{\gamma_1} c_n)/(c_{k+1} \circ_{\gamma_1} \cdots \circ_{\gamma_1} c_n).$$

Moreover, since $\varphi\colon \mathbf{H}_{\gamma_1}^- \to \mathbf{G}_{\gamma_2}^-$ is a residuated lattice homomorphism,

$$\varphi(c_1 \cdots c_n) = \varphi(c_1 \circ_{\gamma_1} \cdots \circ_{\gamma_1} c_n) = \varphi(c_1) \circ_{\gamma_2} \cdots \circ_{\gamma_2} \varphi(c_n),$$

and, for each $k \in \{1, \ldots, n-1\}$,

$$\varphi(c_k) = (\varphi(c_k) \circ_{\gamma_2} \cdots \circ_{\gamma_2} \varphi(c_n))/(\varphi(c_{k+1}) \circ_{\gamma_2} \cdots \circ_{\gamma_2} \varphi(c_n)).$$

Hence another application of Lemma 7.5.7 yields $\varphi(c_1 \cdots c_n) = \varphi(c_1) \cdots \varphi(c_n)$.

(b) Suppose that $a_1, \ldots, a_n, b_1, \ldots, b_m \in H_{\gamma_1}^-$ satisfy $a_1 \cdots a_n = b_1 \cdots b_m$. Then, by Lemma 7.5.6, there exist $c_{ij} \in H^-$, for $i \in \{1, \ldots, m\}$ and $j \in \{1, \ldots, n\}$, satisfying

$$a_j = \prod_{i=1}^{m} c_{ij}, \quad b_i = \prod_{j=1}^{n} c_{ij}, \quad \text{and} \quad \prod_{l=j+1}^{m} c_{il} \vee \prod_{k=i+1}^{n} c_{kj} = \mathrm{e}.$$

Note that, for any $d, d_1, \ldots, d_k \in H_{\gamma_1}^-$ satisfying $d = d_1 \cdots d_k$,

$$d \le \gamma_1(d_1) \cdots \gamma_1(d_k) \le \gamma_1(\gamma_1(d_1) \cdots \gamma_1(d_k)) = \gamma_1(d_1 \cdots d_k) = d,$$

that is, $d = \gamma_1(d_1) \cdots \gamma_1(d_k)$. Hence we may assume that $c_{ij} \in H_{\gamma_1}^-$, for any $i \in \{1, \ldots, m\}$ and $j \in \{1, \ldots, n\}$ (swapping c_{ij} for $\gamma_1(c_{ij})$ where necessary), and that the above products can be taken in either $\mathbf{H}^-$ or $\mathbf{H}_{\gamma_1}^-$.

Using (a) and the fact that φ preserves joins (recall that the join operation in $\mathbf{H}_{\gamma_1}^-$ is the restriction of the join operation in $\mathbf{H}^-$, by Theorem 7.5.1), we obtain, for any $i \in \{1, \ldots, m\}$ and $j \in \{1, \ldots, n\}$,

$$\varphi(a_j) = \prod_{i=1}^{m} \varphi(c_{ij}), \quad \varphi(b_i) = \prod_{j=1}^{n} \varphi(c_{ij}), \quad \text{and} \quad \prod_{l=j+1}^{m} \varphi(c_{il}) \vee \prod_{k=i+1}^{n} \varphi(c_{kj}) = \mathrm{e}.$$

A further application of Lemma 7.5.6 yields $\varphi(a_1) \cdots \varphi(a_n) = \varphi(b_1) \cdots \varphi(b_m)$. ☐

We now show that morphisms between dense nucleus-systems of negative cones can be uniquely lifted to morphisms between the entire negative cones, in such a way as to appropriately commute with the relevant nuclei.

LEMMA 7.5.9. *Suppose that γ_1 and γ_2 are dense nuclei on negative cones $\mathbf{H}^-$ and $\mathbf{G}^-$, respectively. If $\varphi\colon \mathbf{H}_{\gamma_1}^- \to \mathbf{G}_{\gamma_2}^-$ is a residuated lattice homomorphism, then there exists a unique residuated lattice homomorphism $\bar{\varphi}\colon \mathbf{H}^- \to \mathbf{G}^-$ that makes the following diagram commute:*

$$\begin{array}{ccc}
\mathbf{H}^- & \xrightarrow{\ \bar{\varphi}\ } & \mathbf{G}^- \\
\Big\downarrow{\scriptstyle \gamma_1} & & \Big\downarrow{\scriptstyle \gamma_2} \\
\mathbf{H}_{\gamma_1}^- & \xrightarrow{\ \varphi\ } & \mathbf{G}_{\gamma_2}^- .
\end{array}$$

Moreover, if φ is injective or surjective, then so is $\bar{\varphi}$.

PROOF. Since γ_1 is dense, each element of $\mathbf{H}^-$ is a product of elements of $\mathbf{H}^-_{\gamma_1}$. By Lemma 7.5.8, the map $\bar{\varphi}\colon H^- \to G^-$ defined by $\bar{\varphi}(a_1 \cdots a_n) := \varphi(a_1)\cdots\varphi(a_n)$, for $a_1,\ldots,a_n \in H^-_{\gamma_1}$, is well-defined and preserves multiplication.

If $a \in H^-$, then there exist $a_1,\ldots,a_n \in H^-_{\gamma_1}$ such that $a = a_1\cdots a_n$. Hence $\varphi(\gamma_1(a)) = \varphi(\gamma_1(a_1\cdots a_n)) = \varphi(a_1 \circ_{\gamma_1} \cdots \circ_{\gamma_1} a_n) = \varphi(a_1) \circ_{\gamma_2} \cdots \circ_{\gamma_2} \varphi(a_n) = \gamma_2(\varphi(a_1)\cdots\varphi(a_n)) = \gamma_2(\bar{\varphi}(a))$. So $\varphi\circ\gamma_1 = \gamma_2\circ\bar{\varphi}$ and the above diagram commutes.

Moreover, if $\langle a_1,\ldots,a_n\rangle$ is a canonical γ_1-decomposition for a, then $a = a_1\cdots a_n$ and $a_i \circ_{\gamma_1} a_{i+1} = a_i$. So $\bar{\varphi}(a) = \varphi(a_1)\cdots\varphi(a_n)$ and $\varphi(a_i) \circ_{\gamma_2} \varphi(a_{i+1}) = \varphi(a_i)$, i.e., $\langle\varphi(a_1),\ldots,\varphi(a_n)\rangle$ is a canonical γ_2-decomposition for $\bar{\varphi}(a)$.

By Corollary 7.1.12, to conclude that $\bar{\varphi}$ is a residuated lattice homomorphism, it suffices to show that $\bar{\varphi}$ also preserves meets. Let $\langle a_1,\ldots,a_n\rangle$ and $\langle b_1,\ldots,b_n\rangle$ be canonical γ_1-decompositions for a, b, respectively. Then

$$\bar{\varphi}(a \wedge b) = \bar{\varphi}(\prod_{i=1}^{n}(a_i \wedge b_i)) = \prod_{i=1}^{n}\varphi(a_i \wedge b_i) = \prod_{i=1}^{n}(\varphi(a_i)\wedge\varphi(b_i)) = \bar{\varphi}(a)\wedge\bar{\varphi}(b),$$

where, since φ preserves canonical decompositions, the last equality is given by Lemma 7.5.5.

For the last claim, suppose first that φ is surjective and let $b \in G^-$. Then there exist $b_1,\ldots,b_n \in G^-_{\gamma_2}$ with $b = b_1\cdots b_n$. Moreover, there exist $a_1,\ldots,a_n \in H^-_{\gamma_1}$, such that $\varphi(a_i) = b_i$, for each $i \in \{1,\ldots,n\}$. So $\bar{\varphi}(a_1\cdots a_n) = \varphi(a_1)\cdots\varphi(a_n) = b_1\cdots b_n = b$. Now suppose that φ is injective. If $\langle a_1,\ldots,a_n\rangle$ and $\langle b_1,\ldots,b_n\rangle$ are canonical γ_1-decompositions for a, b and $\bar{\varphi}(a) = \bar{\varphi}(b)$, namely $\varphi(a_1)\cdots\varphi(a_n) = \varphi(b_1)\cdots\varphi(b_n)$, then canonicity of the decomposition is preserved under $\bar{\varphi}$, as shown in the proof of Lemma 7.5.8, so $\varphi(a_i) = \varphi(b_i)$, for each $i \in \{1,\ldots,n\}$. By the injectivity of φ we obtain $a_i = b_i$, for each $i \in \{1,\ldots,n\}$, so $a = b$. □

That the restriction of any morphism $\varphi\colon \langle\mathbf{G}^-,\gamma_1\rangle \to \langle\mathbf{H}^-,\gamma_2\rangle$ in $\mathcal{LG}^-_*$ to $G^-_{\gamma_1}$ is a morphism from $\mathbf{G}^-_{\gamma_1}$ to $\mathbf{H}^-_{\gamma_2}$ follows using the fact that φ commutes with the nuclei γ_1 and γ_2. Hence we may define the functor Γ as follows:

(a) For every object $\langle\mathbf{G}^-,\gamma\rangle$ of $\mathcal{LG}^-_*$, let $\Gamma(\langle\mathbf{G}^-,\gamma\rangle) := \mathbf{G}^-_\gamma$.

(b) For every morphism $\varphi\colon \langle\mathbf{G}^-,\gamma_1\rangle \to \langle\mathbf{H}^-,\gamma_2\rangle$ in $\mathcal{LG}^-_*$, let $\Gamma(\varphi)\colon \mathbf{G}^-_{\gamma_1} \to \mathbf{H}^-_{\gamma_2}$ be the restriction of φ to $G^-_{\gamma_1}$.

Clearly, Γ preserves composition of arrows and the identity morphism. Moreover, by Theorem 7.5.2, every object in $\mathcal{IGMV}$ is the image under Γ of an object in $\mathcal{LG}^-_*$, and, by Lemma 7.5.9, every morphism in $\mathcal{IGMV}$ is the image under Γ of a morphism in $\mathcal{LG}^-_*$. So Γ is full. Finally, $\mathbf{G}^-_{\gamma_1} = \mathbf{H}^-_{\gamma_2}$ implies, using the density of γ_1 and γ_2 that $\mathbf{G}^- = \mathbf{H}^-$, and, for homomorphisms $\varphi,\psi\colon \langle\mathbf{G}^-,\gamma_1\rangle \to \langle\mathbf{H}^-,\gamma_2\rangle$, similarly, if $\varphi{\restriction}G^-_{\gamma_1} = \psi{\restriction}G^-_{\gamma_1}$, then $\varphi = \psi$. So Γ is faithful. Our main result now follows directly.

THEOREM 7.5.10. *The categories $\mathcal{IGMV}$ and $\mathcal{LG}^-_*$ are equivalent.*

More specialized equivalences between categories related to (but not necessarily subcategories of) $\mathcal{IGMV}$ and $\mathcal{LG}^-_*$ follow as corollaries of this result. Let us call an element u of a negative cone $\mathbf{G}^-$ a *strong order unit* of $\mathbf{G}^-$ if, for every $a \in G^-$, there exists $n \in \mathbb{N}^{>0}$ such that $u^n \le a$. In this case, we call the expansion of the algebra $\mathbf{G}^-$ with the constant u, denoted by $\langle \mathbf{G}^-, u\rangle$, a *negative cone with a strong order unit*.

Of particular relevance, historically and conceptually, is the equivalence between

1. $ps\mathcal{MV}$, the category whose objects are pseudo-MV-algebras, and morphisms are (ordinary) homomorphisms;
2. $\mathcal{ULG}^-$, the category whose objects are negative cones with a strong order unit, and morphisms are (ordinary) homomorphisms.

COROLLARY 7.5.11. *The categories $ps\mathcal{MV}$ and $\mathcal{ULG}^-$ are equivalent.*

PROOF. Let $\mathcal{ULG}^-_*$ be the full subcategory of $\mathcal{LG}^-_*$, whose objects are negative cones with a dense nucleus such that its nucleus-system has a least element. We show that this category essentially coincides with $\mathcal{ULG}^-$. Let γ be any nucleus on a negative cone $\mathbf{G}^-$. Observe that if $\mathbf{G}^-_\gamma$ has a least element u, then, for all $a \in G^-$,

$$\gamma(a) = \gamma(a) \vee u = u/(\gamma(a)\backslash u).$$

Hence it follows from Proposition 1.5.9 that $\gamma = \gamma_u$ and $G^-_\gamma = \{x \in G^- \mid u \le x\}$. Now, if γ is also dense, it can be deduced from the proof of Theorem 7.5.2 that u is a strong order unit of $\mathbf{G}^-$. Conversely, if u is a strong order unit of $\mathbf{G}^-$, then γ_u is a dense nucleus on $\mathbf{G}^-$ such that $\mathbf{G}^-_{\gamma_u}$ has a least element u. Hence $\mathcal{ULG}^-_*$ and $\mathcal{ULG}^-$ are, essentially, one and the same category which, by Theorem 7.5.10, is equivalent to $ps\mathcal{MV}$. □

7.6. Strongly simple GBL-algebras

Around the turn of the last century, Hölder proved that every totally ordered group with no proper non-trivial convex ℓ-subgroup embeds into the additive ordered group $\mathbf{R}$ of real numbers. Such representations extend well beyond ordered groups, however. Let us call a residuated lattice $\mathbf{L}$ *strongly simple* if $\mathcal{C}(\mathbf{L}) = \{L, \{\mathrm{e}\}\}$. In this section, we provide representations first for strongly simple, totally ordered groups (giving Hölder's theorem) and totally ordered pseudo-MV-algebras, then use Theorem 7.3.7 to extend these representations to the class of all strongly simple, totally ordered GBL-algebras (Theorem 7.6.3).

PROPOSITION 7.6.1. *Every strongly simple, totally ordered group is isomorphic to a subalgebra of* $\mathbf{R}$.

PROOF. Let $\mathbf{L} = \langle L, \wedge, \vee, \cdot, \backslash, /, \mathrm{e}\rangle$ be any strongly simple, totally ordered group. If $\mathbf{L}$ is trivial, it obviously embeds into $\mathbf{R}$, so fix $a \in L$ such that $a < \mathrm{e}$. Since $\mathrm{C}[a] = L$, there exists, for each $b \in L$, some $k_b \in \mathbb{N}$ satisfying $a^{k_b} \le |b|$. Now let $a^m := (a\backslash \mathrm{e})^{-m}$, for each $m \in \mathbb{Z}^{<0}$, and define

$$L(b) := \{\tfrac{m}{n} \mid m \in \mathbb{Z},\, n \in \mathbb{N}^{>0}, \text{ and } b^n \le a^m\}$$
$$U(b) := \{\tfrac{m}{n} \mid m \in \mathbb{Z},\, n \in \mathbb{N}^{>0}, \text{ and } b^n > a^m\}.$$

It follows easily that

(i) if $b = \mathrm{e}$, then $L(b) = \mathbb{Q}^{\le 0}$ and $U(b) = \mathbb{Q}^{>0}$;
(ii) if $b \neq \mathrm{e}$ and $b \le a$, then $1 \in L(b)$ and $U(b)$ is non-empty;

(iii) if $b \neq \mathrm{e}$ and $b > a$, then $1 \in U(b)$ and $L(b)$ is non-empty.

Moreover, any $k_b = \frac{k_b}{1}$ satisfying $a^{k_b} \leq |b|$ is an upper bound for $L(b)$, and every element of $L(b)$ is a lower bound of $U(b)$.

Next, we show that $L(b)$ is an order-ideal of $\langle \mathbb{Q}, \leq \rangle$. Consider any $m \in \mathbb{Z}$ and $n \in \mathbb{N}^{>0}$ such that $b^n \leq a^m$, so $\frac{m}{n} \in L(b)$. Given any $r \in \mathbb{Q}$ such that $r \leq \frac{m}{n}$, we may assume without loss of generality that $r = \frac{p}{n}$, for some $p \in \mathbb{Z}$. So $np \leq nm$. Observe that if $b^n > a^p$, then $a^{nm} \leq a^{np} < b^{n^2} \leq a^{nm}$, a contradiction. So $r = \frac{p}{n} \in L(b)$. A similar argument shows that $U(b)$ is an order-filter of $\langle \mathbb{Q}, \leq \rangle$. Hence $\langle L(b), R(b) \rangle$ is a Dedekind cut of $\mathbb{Q}$ with separating element $b^* := \sup L(b) \in \mathbb{R}$.

We claim that the map $\varphi \colon L \to \mathbb{R};\ b \mapsto b^*$ provides the required embedding. Clearly, $\varphi(\mathrm{e}) = 0$ and φ is order-preserving. Also $\varphi(bc) = \varphi(b) + \varphi(c)$. To check that $\varphi(b) + \varphi(c) \leq \varphi(bc)$, it suffices to show that $\frac{m}{n} \in U(b)$ and $\frac{p}{n} \in U(c)$ implies $\frac{m+p}{n} \in U(bc)$. Suppose that $b^n > a^m, c^n > a^p$ and, without loss of generality, $bc \leq cb$. Then $b^n c^n > a^{m+p}$ and $b^n c^n \leq (bc)^n$, so $(bc)^n \geq b^n c^n > a^{m+p}$. The converse, $\varphi(bc) \leq \varphi(b) + \varphi(c)$, is established similarly. Finally, if $b^* = c^*$, then, for any m, n we have that $b^n > a^m$ implies $c^n > a^m$, so $b \leq c$. A parallel argument establishes that $c \leq b$, so φ is injective. □

Now let $\mathbf{[0,1]} = \langle [0,1], \wedge, \vee, \cdot, \to, 1, 0 \rangle$ be the standard MV-algebra on the real unit interval $[0,1]$, where $a \cdot b := \max(0, a+b-1)$ and $a \to b := \min(1, 1-a+b)$, for all $a, b \in [0,1]$.

PROPOSITION 7.6.2. *Every strongly simple, totally ordered pseudo-MV-algebra is isomorphic to a subalgebra of* $\mathbf{[0,1]}$.

PROOF. Let $\mathbf{L}$ be a strongly simple, totally ordered pseudo-MV-algebra. Then, by Corollary 7.5.11, $\mathbf{L}$ is isomorphic to $\mathbf{G}^-_{\gamma_u}$, for some totally ordered group $\mathbf{G}$ and strong order unit u of $\mathbf{G}^-$. But Lemma 4.1.9 implies that $\mathbf{G}$ is also strongly simple, so it is isomorphic to a subalgebra of $\mathbf{R}$, by Proposition 7.6.1. Hence $\mathbf{G}^-_{\gamma_u}$ is isomorphic to a subalgebra of $\langle [u,0], \wedge, \vee, \cdot, \to, 0, u \rangle$, where $a \cdot b = \max(a+b, u)$, and hence also to a subalgebra of $\mathbf{[0,1]}$. □

We now extend these representation results to all strongly simple, totally ordered GBL-algebras.

THEOREM 7.6.3. *A totally ordered GBL-algebra is strongly simple if and only if it is isomorphic to a subalgebra of* $\mathbf{R}$, $\mathbf{R}^-$, *or the residuated lattice reduct of* $\mathbf{[0,1]}$.

PROOF. It is easily checked that $\mathcal{C}(\mathbf{L}) = \{L, \{\mathrm{e}\}\}$, for any subalgebra $\mathbf{L}$ of $\mathbf{R}$, $\mathbf{R}^-$, or the residuated lattice reduct of $\mathbf{[0,1]}$, so these algebras are all strongly simple.

Conversely, let $\mathbf{L}$ be any strongly simple, totally ordered GBL-algebra. Since $\mathbf{L}$ is totally ordered, it follows from Corollary 7.2.4 that $\mathbf{L}$ is either a strongly simple, totally ordered group, or a strongly simple, totally ordered, integral GBL-algebra. For the first case, Theorem 7.6.1 yields the required embedding into $\mathbf{R}$. For the second case, we observe that $\mathbf{L}$ is ordinally irreducible, since otherwise $\mathbf{L}$ would admit a non-trivial α-cut $\langle D, U \rangle$ and hence have a non-trivial convex subuniverse U. By Theorem 7.3.7, there are two possibilities:

(a) If $\mathbf{L}$ is a totally ordered, integral GMV-algebra with a least element, then, by Proposition 7.6.2, it is isomorphic to a subalgebra of the residuated lattice reduct of $[\mathbf{0}, \mathbf{1}]$.

(b) Suppose, on the other hand, that $\mathbf{L}$ is the negative cone of some totally ordered group $\mathbf{G}$. Since the only convex subuniverses of $\mathbf{L}$ are L and $\{e\}$ and the convex subalgebras of an ℓ-group are uniquely determined by their negative cones, it follows that $\mathbf{G}$ is itself a strongly simple ℓ-group. Hence, resorting once more to Theorem 7.6.1, there exists an embedding of $\mathbf{G}$ into $\mathbf{R}$ that restricts to an embedding of $\mathbf{L}$ into $\mathbf{R}^-$. □

Note that by Corollary 4.2.8, every strongly simple, prelinear, e-cyclic residuated lattice is totally ordered, so Theorem 7.6.3 applies in particular to all strongly simple, prelinear GBL-algebras.

7.7. Bibliographical remarks

GBL-algebras were introduced, with the name of *DRℓ-semigroups*, by Swamy [**352**]. They were rediscovered independently in the setting of residuated lattices in [**206**] and studied in greater detail in [**16**, **152**, **203**–**205**]. An early axiomatization of ℓ-groups in the signature with the multiplication and positive and negative part operations was given in [**237**]. The fact that any finite GBL-algebra is commutative, which implies our Lemma 7.1.3, was first proved in [**203**, Theorem 4], and broadened to cover all n-potent GBL-algebras in [**204**, Theorem 4.5]. Theorem 7.1.10 appeared as Theorem 6.2 in [**16**] (see also [**30**, Theorem 12]), although the proof given here takes advantage of results presented in [**272**, Section 6]. Corollary 7.1.12 is an immediate consequence of Theorem 1.4.5 of [**27**].

MV-algebras were introduced by Chang as algebraic semantics for Łukasiewicz infinite-valued logic [**66**]; for a wealth of further material on these structures, we refer to [**73**, **281**]. BL-algebras were introduced by Hájek in [**175**] as a common generalization of MV-algebras, Gödel algebras, and other structures considered in mathematical fuzzy logic, and studied in depth in [**3**, **55**, **174**]. Both classes were subsequently generalized to the non-commutative setting; pseudo-MV-algebras were introduced in [**156**] and, independently, as non-commutative MV-algebras in [**318**], and pseudo-BL-algebras were introduced in [**104**, Part 1].

Theorem 7.2.3 is Theorem 5.2 of [**152**]; an alternative proof, presented as part of a systematic treatment of direct and semidirect decompositions of residuated lattices, can be found in [**212**].

The ordinal decomposition of totally ordered, integral GBL-algebras presented in Theorem 7.3.4 generalizes several representation theorems found in different areas of algebra. In particular, we follow here quite closely Fuchs' analysis of ordinal sums of (not necessarily residuated) po-semigroups in [**136**]. The ordinal sum representation of totally ordered BL-algebras over the real unit interval may be found in [**278**] and was extended to all totally ordered BL-algebras in [**3**] (see also [**55**]) and to totally ordered pseudo BL-algebras in [**116**], using intermediate lemmas from [**104**] and [**117**]. Let us remark also that a generalization of the ordinal sum construction called a "poset sum" was introduced in [**204**, **205**] to give embedding theorems for GBL-algebras that need not be integral or totally ordered.

All the results in Section 7.4 are essentially due to Bosbach. Our presentation is heavily indebted to his papers [**47**, **48**], although occasional simplifications have been introduced with some help from [**331**].

The main source for the results presented in Section 7.5 is [**152**]. Lemmas 7.5.5 and 7.5.7 are variants of results available in [**73**], while Lemma 7.5.6 is Theorem 1 on page 68 of [**136**]. Theorem 6.9 of [**152**] establishes a more general equivalence between the category whose objects are (not necessarily integral) GMV-algebras and the category of ℓ-groups with a *core* (essentially, the composition of a certain co-closure operator on the ℓ-group and a nucleus on the image of this operator). Variants of this construction can be applied, more generally, to groupoids endowed with a single residual and a lattice-order with respect to which integrality and divisibility are both satisfied — such algebras include a term-equivalent version of orthomodular lattices. In particular, an equivalence is established in [**332**] between orthomodular lattices and negative cones of *right ℓ-groups* (groups endowed with a lattice-order that is preserved by right multiplication) with a strong order unit.

Let us also comment on a further consequence of Corollary 7.5.11. Example 1.4.9 implicitly defines a functor Ξ from the concrete category of ℓ-groups to the concrete category of negative cones of ℓ-groups; that is, for any ℓ-groups $\mathbf{G}$ and $\mathbf{K}$ and homomorphism $\varphi\colon \mathbf{G} \to \mathbf{K}$, let $\Xi(\mathbf{G}) := \mathbf{G}^-$ and let $\Xi(\varphi)$ be the restriction of φ to $\mathbf{G}^-$. In [**16**], it is shown that this functor induces an isomorphism between the mentioned categories. Combining this result with Corollary 7.5.11, we obtain the equivalence between the categories of pseudo-MV-algebras and unital ℓ-groups established in [**115**]. This result generalizes to the non-commutative setting the well-known equivalence between the categories of MV-algebras and Abelian unital ℓ-groups established in [**279**].

Hölder's theorem, first proved in [**186**], is a standard result in the structure theory of ℓ-groups (see, e.g., [**9**, Theorem 2.3]). Our Theorem 7.6.3 can be found in [**233**] and has antecedents in [**115**, **118**, **228**]. This theorem is also presented in a more general setting in [**50**], where it is shown that every strongly simple, totally ordered, integral residuated lattice with no coatom is a commutative, integral GMV-algebra and therefore either a subalgebra of $\mathbf{R}^-$ or the residuated lattice reduct of $[\mathbf{0}, \mathbf{1}]$. Note, however, that although in a strongly simple, totally ordered, integral residuated lattice $\mathbf{L}$ that is not divisible and has a coatom a and no least element, the powers of a form a totally ordered subalgebra of $\mathbf{L}$ that is order-isomorphic to the negative integers, we do not know much more in general about its structure.

CHAPTER 8

Bridges between algebra and logic

This chapter focuses on fascinating relationships — or "bridges" — existing between algebraic properties of a variety and syntactic properties of equational consequences in that variety. The scaffolding for these bridges is provided by Proposition 3.2.7, which relates consequences of sets of equations to congruences of free algebras.

In Sections 8.1–8.3, we build such a bridge between the amalgamation property, which combines algebras while preserving a common subalgebra, and the deductive interpolation property, which restricts consequences to a common set of variables. We also build a bridge between the congruence extension property and possession of a deduction theorem for equational consequence known as the extension property.

Amalgamation is a relatively rare property satisfied by the varieties of lattices, Heyting algebras, and MV-algebras, for example, but not the varieties of ℓ-groups, GBL-algebras, or semilinear residuated lattices. In Section 8.4, we develop and apply syntactic and algebraic methods for identifying varieties of residuated lattices that do or do not have this property.

This chapter builds on a wealth of accumulated results and methods from universal algebra and algebraic logic. The goal here is to explain the main ideas and establish the most important theorems without overburdening the reader with too many concepts or an overly general setting. Key contributions from the literature are acknowledged, and pointers given to further reading, in the bibliographical remarks.

8.1. The amalgamation property

Let $\mathcal{K}$ be a class of $\mathcal{L}$-algebras. A *V-formation* in $\mathcal{K}$ is a 5-tuple $\langle \mathbf{A}, \mathbf{B}, \mathbf{C}, \varphi_B, \varphi_C \rangle$, where $\mathbf{A}, \mathbf{B}, \mathbf{C} \in \mathcal{K}$ and φ_B and φ_C are embeddings of $\mathbf{A}$ into $\mathbf{B}$ and $\mathbf{C}$, respectively. A 3-tuple $\langle \mathbf{D}, \psi_B, \psi_C \rangle$ is an *amalgam* of this V-formation in a class $\mathcal{K}'$ of $\mathcal{L}$-algebras if $\mathbf{D} \in \mathcal{K}'$ and ψ_B, ψ_C are embeddings of $\mathbf{B}$ and $\mathbf{C}$, respectively, into $\mathbf{D}$ such that $\psi_B\varphi_B = \psi_C\varphi_C$. Pictorially, we obtain the commutative diagram

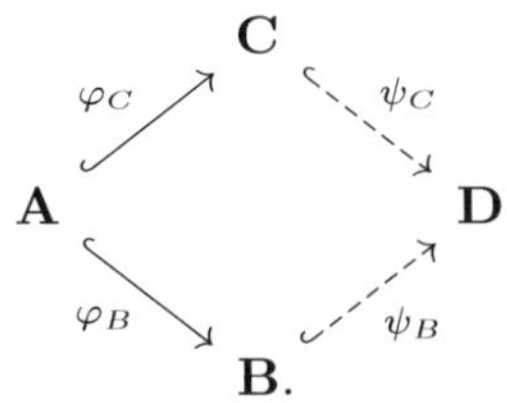

The class $\mathcal{K}$ has the *amalgamation property* (for short, AP) if each V-formation in $\mathcal{K}$ has an amalgam in $\mathcal{K}$. The main result of this section, Theorem 8.1.5, relates the AP for a variety $\mathcal{V}$ to a property of equational consequence in $\mathcal{V}$.

EXAMPLE 8.1.1. The variety of lattices has the AP. Consider any V-formation consisting of lattices $\mathbf{B}$ and $\mathbf{C}$ with a common sublattice $\mathbf{A}$ such that $A = B \cap C$. Let $R :=\leq^{\mathbf{B}} \cup \leq^{\mathbf{C}}$ and define, for any $x, y \in B \cup C$,

$$x \preceq y :\Longleftrightarrow Rxy \text{ or } (Rxz \text{ and } Rzy, \text{ for some } z \in B \cap C).$$

It is easily confirmed that $\preceq$ is a partial order on $B \cup C$; indeed, it is the smallest partial order on $B \cup C$ extending $\leq^{\mathbf{B}}$ and $\leq^{\mathbf{C}}$. Now let $\mathbf{D}$ be the Dedekind-MacNeille completion (see Section 9.3) of the poset $\langle B \cup C, \preceq \rangle$, i.e., the set of subsets X of $B \cup C$ satisfying $(X^u)^l = X$, ordered by set-inclusion, where Y^l and Y^u denote the sets of lower bounds and upper bounds of $Y \subseteq B \cup C$, respectively. Then the maps ψ_B and ψ_C sending an element x to the order-ideal $\downarrow x$ in $\langle B \cup C, \preceq \rangle$ are embeddings of $\mathbf{B}$ and $\mathbf{C}$, respectively, into $\mathbf{D}$, and satisfy $\psi_B(x) = \psi_C(x)$, for each $x \in A$. Hence $\langle \mathbf{D}, \psi_B, \psi_C \rangle$ is an amalgam of the given V-formation.

EXAMPLE 8.1.2. Let $\mathcal{V}$ be any variety generated by a totally ordered Brouwerian algebra with $n \geq 3$ elements. Then there exist algebras $\mathbf{A}, \mathbf{B}, \mathbf{C}$ in $\mathcal{V}$ such that $A = \{0, a_1, \dots, a_{n-3}, 1\}$, $B = \{0, b, a_1, \dots, a_{n-3}, 1\}$, $C = \{0, a_1, \dots, a_{n-3}, c, 1\}$, $0 < b < a_1 < \cdots < a_{n-3} < 1$, and $0 < a_1 < \cdots < a_{n-3} < c < 1$. If the V-formation consisting of $\mathbf{B}$ and $\mathbf{C}$ with common subalgebra $\mathbf{A}$ had an amalgam $\langle \mathbf{D}, \psi_B, \psi_C \rangle$ in $\mathcal{V}$, then $\mathcal{V}$ would contain the $n+1$-element, totally ordered Brouwerian algebra generated by $\psi_B[B] \cup \psi_C[C]$, contradicting the fact that $\mathcal{V}$ satisfies the equation $\mathrm{e} \leq x_1 \vee (x_1 \to x_2) \vee (x_2 \to x_3) \vee \cdots \vee (x_{n-1} \to x_n)$. So $\mathcal{V}$ does not have the AP.

The following lemma provides a useful necessary and sufficient condition for the existence of amalgams in a variety.

LEMMA 8.1.3. *Let $\langle \mathbf{A}, \mathbf{B}, \mathbf{C}, \varphi_B, \varphi_C \rangle$ be a V-formation in a variety $\mathcal{V}$. The following statements are equivalent:*

(1) *$\langle \mathbf{A}, \mathbf{B}, \mathbf{C}, \varphi_B, \varphi_C \rangle$ has an amalgam in $\mathcal{V}$.*
(2) *For any distinct $x, y \in B$ (respectively, $x, y \in C$), there exist $\mathbf{D}_{xy} \in \mathcal{V}$ and homomorphisms $\psi_B^{xy} \colon \mathbf{B} \to \mathbf{D}_{xy}$ and $\psi_C^{xy} \colon \mathbf{C} \to \mathbf{D}_{xy}$ such that $\psi_B^{xy} \varphi_B = \psi_C^{xy} \varphi_C$ and $\psi_B^{xy}(x) \neq \psi_B^{xy}(y)$ (respectively, $\psi_C^{xy}(x) \neq \psi_C^{xy}(y)$).*

PROOF.

(1) $\Rightarrow$ (2) Let each $\langle \mathbf{D}_{xy}, \psi_B^{xy}, \psi_C^{xy} \rangle$ be any amalgam of $\langle \mathbf{A}, \mathbf{B}, \mathbf{C}, \varphi_B, \varphi_C \rangle$ in $\mathcal{V}$.

(2) $\Rightarrow$ (1) Let $\mathbf{D}$ be the direct product of the algebras $\mathbf{D}_{xy}$, for all distinct x, y in B, C, as given by (2). The homomorphisms $\psi_B^{xy} \colon \mathbf{B} \to \mathbf{D}_{xy}$ and $\psi_C^{xy} \colon \mathbf{C} \to \mathbf{D}_{xy}$ induce homomorphisms $\psi_B \colon \mathbf{B} \to \mathbf{D}$ and $\psi_C \colon \mathbf{C} \to \mathbf{D}$, and the assumptions on ψ_B^{xy} and ψ_C^{xy} ensure that ψ_B and ψ_C are injective and satisfy $\psi_B \varphi_B = \psi_C \varphi_C$. $\square$

Next, we give a characterization of the AP for a variety in terms of the congruences of its free algebras.

LEMMA 8.1.4. *A variety $\mathcal{V}$ has the amalgamation property if and only if, for any non-disjoint sets Y, Z and $\Theta_Y \in \mathrm{Con}(\mathbf{F}(Y))$, $\Theta_Z \in \mathrm{Con}(\mathbf{F}(Z))$ satisfying*

$$\Theta_Y \cap \mathrm{F}(Y \cap Z)^2 = \Theta_Z \cap \mathrm{F}(Y \cap Z)^2,$$

there exists $\Theta \in \mathrm{Con}(\mathbf{F}(Y \cup Z))$ such that $\Theta_Y = \Theta \cap \mathrm{F}(Y)^2$ and $\Theta_Z = \Theta \cap \mathrm{F}(Z)^2$.

PROOF.

($\Rightarrow$) Suppose first that $\mathcal{V}$ has the AP and consider any non-disjoint sets Y, Z and $\Theta_Y \in \mathrm{Con}(\mathbf{F}(Y))$, $\Theta_Z \in \mathrm{Con}(\mathbf{F}(Z))$ satisfying

$$\Theta_0 := \Theta_Y \cap \mathrm{F}(Y \cap Z)^2 = \Theta_Z \cap \mathrm{F}(Y \cap Z)^2.$$

Let $\mathbf{A} := \mathbf{F}(Y \cap Z)/\Theta_0$, $\mathbf{B} := \mathbf{F}(Y)/\Theta_Y$, and $\mathbf{C} := \mathbf{F}(Z)/\Theta_Z$. Then the following maps are embeddings:

$$\varphi_B \colon \mathbf{A} \to \mathbf{B};\ [a]_{\Theta_0} \mapsto [a]_{\Theta_Y} \quad \text{and} \quad \varphi_C \colon \mathbf{A} \to \mathbf{C};\ [a]_{\Theta_0} \mapsto [a]_{\Theta_Z}.$$

Since $\mathcal{V}$ has the AP, the V-formation $\langle \mathbf{A}, \mathbf{B}, \mathbf{C}, \varphi_B, \varphi_C \rangle$ has an amalgam $\langle \mathbf{D}, \psi_B, \psi_C \rangle$ in $\mathcal{V}$. Moreover, we may assume without loss of generality that $\mathbf{D}$ is generated by $\psi_B[B] \cup \psi_C[C]$. Define the map $\psi \colon Y \cup Z \to D$ by

$$\psi(x) := \begin{cases} \psi_B([x]_{\Theta_Y}) & \text{if } x \in Y; \\ \psi_C([x]_{\Theta_Z}) & \text{if } x \in Z. \end{cases}$$

Note that ψ is well-defined, since if $x \in Y \cap Z$, then

$$\psi_B([x]_{\Theta_Y}) = \psi_B(\varphi_B([x]_{\Theta_0})) = \psi_C(\varphi_C([x]_{\Theta_0})) = \psi_C([x]_{\Theta_Z}).$$

Now let $\tilde{\psi} \colon \mathbf{F}(Y \cup Z) \to \mathbf{D}$ be the unique homomorphism extending ψ. Clearly, $\tilde{\psi}$ is surjective, since $\mathbf{D}$ is generated by $\psi_B[B] \cup \psi_C[C]$, and we obtain the following commutative diagram:

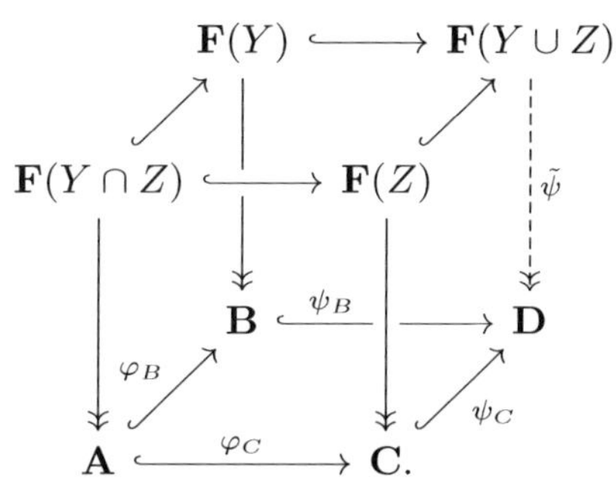

Let $\Theta := \ker(\tilde{\psi})$. We claim that $\Theta \cap \mathrm{F}(Y)^2 = \Theta_Y$ and $\Theta \cap \mathrm{F}(Z)^2 = \Theta_Z$, proving just the non-trivial inclusion of the first equality. Let $\langle u, v \rangle \in \Theta \cap \mathrm{F}(Y)^2$. Then, using the uniqueness of $\tilde{\psi}$,

$$\psi_B([u]_{\Theta_Y}) = \tilde{\psi}(u) = \tilde{\psi}(v) = \psi_B([v]_{\Theta_Y}).$$

Since ψ_B is injective, $[u]_{\Theta_Y} = [v]_{\Theta_Y}$ and $\langle u, v \rangle \in \Theta_Y$.

($\Leftarrow$) Suppose next that $\mathcal{V}$ satisfies the condition described in the lemma and consider any V-formation $\langle \mathbf{A}, \mathbf{B}, \mathbf{C}, \varphi_B, \varphi_C \rangle$ in $\mathcal{V}$, assuming without loss of generality that φ_B and φ_C are inclusion maps. The sets A, B, and C will play the roles of $Y \cap Z$, Y, and Z, respectively, in the condition. Consider the surjective homomorphisms extending the identity maps on A, B, and C, respectively,

$$\pi_A \colon \mathbf{F}(A) \to \mathbf{A}, \quad \pi_B \colon \mathbf{F}(B) \to \mathbf{B}, \quad \text{and} \quad \pi_C \colon \mathbf{F}(C) \to \mathbf{C}.$$

Let $\Theta_A := \ker(\pi_A)$, $\Theta_B := \ker(\pi_B)$, and $\Theta_C := \ker(\pi_C)$. The restrictions of π_B and π_C to A are both π_A, so $\Theta_A = \Theta_B \cap \mathrm{F}(A)^2 = \Theta_C \cap \mathrm{F}(A)^2$. Hence, by assumption, there exists $\Theta \in \mathrm{Con}(\mathbf{F}(B \cup C))$ such that $\Theta \cap \mathrm{F}(B)^2 = \Theta_B$ and $\Theta \cap \mathrm{F}(C)^2 = \Theta_C$.

Let $\mathbf{D} := \mathbf{F}(B \cup C)/\Theta$, let $\pi_\Theta\colon \mathbf{F}(B \cup C) \to \mathbf{D};\ u \mapsto [u]_\Theta$ be the canonical map, and let $\chi_B\colon \mathbf{F}(B) \to \mathbf{F}(B \cup C)$ and $\chi_C\colon \mathbf{F}(C) \to \mathbf{F}(B \cup C)$ be inclusion maps. Then $\ker(\pi_\Theta\chi_B) = \ker(\pi_B)$ and $\ker(\pi_\Theta\chi_C) = \ker(\pi_C)$. Hence, by the General homomorphism theorem (Theorem B.2), there exist embeddings $\psi_B\colon \mathbf{B} \to \mathbf{D}$ and $\psi_C\colon \mathbf{C} \to \mathbf{D}$ such that $\psi_B\pi_B = \pi_\Theta\chi_B$ and $\psi_C\pi_C = \pi_C\chi_C$, as illustrated by the following commutative diagram:

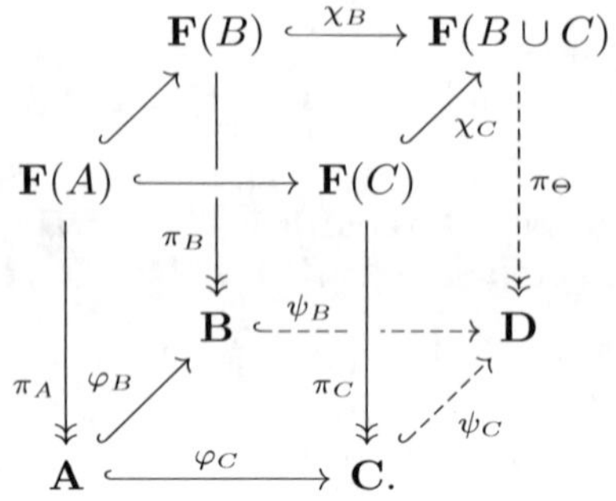

So $\psi_B\varphi_B = \psi_C\varphi_C$ and $\langle \mathbf{D}, \psi_B, \psi_C\rangle$ is an amalgam of $\langle \mathbf{A}, \mathbf{B}, \mathbf{C}, \varphi_B, \varphi_C\rangle$ in $\mathcal{V}$. $\square$

The property of congruences of free algebras described in Lemma 8.1.4 can also be formulated in terms of consequences. Recall that $\mathrm{Var}(S)$ denotes the set of variables occurring in some syntactic object S defined over a set of formulas (in this setting, an equation or a set of equations).

A variety $\mathcal{V}$ has the *Robinson property* (for short, RP) if, for any set X, whenever

(i) $\Sigma \cup \Pi \cup \{\varepsilon\} \subseteq \mathrm{Eq}(X)$, $\mathrm{Var}(\Sigma) \cap \mathrm{Var}(\Pi) \neq \emptyset$, and $\mathrm{Var}(\varepsilon) \cap \mathrm{Var}(\Pi) \subseteq \mathrm{Var}(\Sigma)$;

(ii) $\Sigma \vdash_\mathcal{V} \delta \iff \Pi \vdash_\mathcal{V} \delta$, for all $\delta \in \mathrm{Eq}(X)$ with $\mathrm{Var}(\delta) \subseteq \mathrm{Var}(\Sigma) \cap \mathrm{Var}(\Pi)$;

(iii) $\Sigma \cup \Pi \vdash_\mathcal{V} \varepsilon$,

then $\Sigma \vdash_\mathcal{V} \varepsilon$.

The RP is expressed here for sets X of arbitrary cardinality. Suppose, however, that the property is known to hold for finite sets, and that conditions (i)–(iii) are satisfied for some set X and $\Sigma \cup \Pi \cup \{\varepsilon\} \subseteq \mathrm{Eq}(X)$. Since $\vdash_\mathcal{V}$ is finitary, there exist finite sets $\Sigma' \subseteq \Sigma$ and $\Pi' \subseteq \Pi$ such that $\mathrm{Var}(\Sigma') \cap \mathrm{Var}(\Pi') \neq \emptyset$, $\mathrm{Var}(\varepsilon) \cap \mathrm{Var}(\Pi') \subseteq \mathrm{Var}(\Sigma')$, and $\Sigma' \cup \Pi' \vdash_\mathcal{V} \varepsilon$. Consider the finite set $Y := \mathrm{Var}(\Sigma' \cup \Pi' \cup \{\varepsilon\}) \subseteq X$, and let

$$\Sigma^* := \{\delta \in \mathrm{Eq}(Y) \mid \mathrm{Var}(\delta) \subseteq \mathrm{Var}(\Sigma') \text{ and } \Sigma \vdash_\mathcal{V} \delta\}$$
$$\Pi^* := \{\delta \in \mathrm{Eq}(Y) \mid \mathrm{Var}(\delta) \subseteq \mathrm{Var}(\Pi') \text{ and } \Pi \vdash_\mathcal{V} \delta\}.$$

Clearly, $\Sigma' \subseteq \Sigma^*$, $\Pi' \subseteq \Pi^*$, $\mathrm{Var}(\Sigma^*) = \mathrm{Var}(\Sigma')$, and $\mathrm{Var}(\Pi^*) = \mathrm{Var}(\Pi')$. Hence $\mathrm{Var}(\Sigma^*) \cap \mathrm{Var}(\Pi^*) \neq \emptyset$, $\mathrm{Var}(\varepsilon) \cap \mathrm{Var}(\Pi^*) \subseteq \mathrm{Var}(\Sigma^*)$, and $\Sigma^* \cup \Pi^* \vdash_\mathcal{V} \varepsilon$. Suppose now that $\Sigma^* \vdash_\mathcal{V} \delta$, for some $\delta \in \mathrm{Eq}(Y)$ with $\mathrm{Var}(\delta) \subseteq \mathrm{Var}(\Sigma^*) \cap \mathrm{Var}(\Pi^*)$. Then $\Sigma \vdash_\mathcal{V} \delta$. So $\Pi \vdash_\mathcal{V} \delta$, by (ii), and, by definition, $\Pi^* \vdash_\mathcal{V} \delta$. Using the same reasoning, $\Sigma^* \vdash_\mathcal{V} \delta$ if and only if $\Pi^* \vdash_\mathcal{V} \delta$, for all $\delta \in \mathrm{Eq}(Y)$ satisfying $\mathrm{Var}(\delta) \subseteq \mathrm{Var}(\Sigma^*) \cap \mathrm{Var}(\Pi^*)$. So $\Sigma^* \vdash_\mathcal{V} \varepsilon$, by the RP for the finite set Y, and therefore also $\Sigma \vdash_\mathcal{V} \varepsilon$. Hence we may assume without further comment that the RP can be restricted to finite sets.

We now prove the main result of this section.

THEOREM 8.1.5. *A variety $\mathcal{V}$ has the amalgamation property if and only if it has the Robinson property.*

PROOF.

($\Rightarrow$) Suppose that $\mathcal{V}$ has the AP and that conditions (i)–(iii) of the RP are satisfied for some set X and $\Sigma \cup \Pi \cup \{\varepsilon\} \subseteq \mathrm{Eq}(X)$. We may also assume, by replacing Σ with $\Sigma \cup \{x \approx x \mid x \in \mathrm{Var}(\varepsilon)\backslash\mathrm{Var}(\Pi)\}$ if necessary, that $\mathrm{Var}(\varepsilon) \subseteq \mathrm{Var}(\Sigma)$. Define

$$Y := \mathrm{Var}(\Sigma), \quad Z := \mathrm{Var}(\Pi), \quad \Theta_Y := \mathrm{Cg}^{\mathbf{F}(Y)}(\overline{\Sigma}), \quad \text{and} \quad \Theta_Z := \mathrm{Cg}^{\mathbf{F}(Z)}(\overline{\Pi}).$$

It follows from (ii), using Proposition 3.2.7, that $\Theta_Y \cap \mathrm{F}(Y \cap Z)^2 = \Theta_Z \cap \mathrm{F}(Y \cap Z)^2$. So, by Lemma 8.1.4, there exists $\Theta \in \mathrm{Con}(\mathbf{F}(Y \cup Z))$ such that $\Theta_Y = \Theta \cap \mathrm{F}(Y)^2$ and $\Theta_Z = \Theta \cap \mathrm{F}(Z)^2$. By Proposition 3.2.7 again and (iii), $\bar{\varepsilon} \in \mathrm{Cg}^{\mathbf{F}(Y \cup Z)}(\Theta_Y \cup \Theta_Z) \subseteq \Theta$. Hence $\bar{\varepsilon} \in \Theta \cap \mathrm{F}(Y)^2 = \Theta_Y = \mathrm{Cg}^{\mathbf{F}(Y)}(\overline{\Sigma})$ and applying Proposition 3.2.7 once more yields $\Sigma \vdash_{\mathcal{V}} \varepsilon$.

($\Leftarrow$) Suppose now that $\mathcal{V}$ has the RP. We establish the equivalent condition for the AP presented in Lemma 8.1.4. Consider any non-disjoint sets Y, Z and $\Theta_Y \in \mathrm{Con}(\mathbf{F}(Y))$, $\Theta_Z \in \mathrm{Con}(\mathbf{F}(Z))$ satisfying $\Theta_Y \cap \mathrm{F}(Y \cap Z)^2 = \Theta_Z \cap \mathrm{F}(Y \cap Z)^2$. Choose any $\Sigma \subseteq \mathrm{Eq}(Y)$ and $\Pi \subseteq \mathrm{Eq}(Z)$ such that $\Theta_Y = \overline{\Sigma} = \mathrm{Cg}^{\mathbf{F}(Y)}(\overline{\Sigma})$ and $\Theta_Z = \overline{\Pi} = \mathrm{Cg}^{\mathbf{F}(Z)}(\overline{\Pi})$, and define $\Theta := \mathrm{Cg}^{\mathbf{F}(Y \cup Z)}(\Theta_Y \cup \Theta_Z)$.

Let $\bar{\varepsilon} \in \Theta \cap \mathrm{F}(Y)^2$. Then $\mathrm{Var}(\varepsilon) \subseteq \mathrm{Var}(\Sigma)$ and Proposition 3.2.7 yields $\Sigma \cup \Pi \vdash_{\mathcal{V}} \varepsilon$. The conditions (i)–(iii) of the RP hold, so $\Sigma \vdash_{\mathcal{V}} \varepsilon$ and, using Proposition 3.2.7 again, $\bar{\varepsilon} \in \Theta_Y$. Hence $\Theta_Y = \Theta \cap \mathrm{F}(Y)^2$, and, symmetrically, also $\Theta_Z = \Theta \cap \mathrm{F}(Z)^2$. □

Observe, moreover, that if every V-formation of finitely generated members of $\mathcal{V}$ has an amalgam in $\mathcal{V}$, then the proof of Theorem 8.1.5 establishes the RP for finite sets, and hence that $\mathcal{V}$ has the RP. From this, we deduce the following result.

PROPOSITION 8.1.6. *A variety $\mathcal{V}$ has the amalgamation property if and only if every V-formation of finitely generated members of $\mathcal{V}$ has an amalgam in $\mathcal{V}$. In particular, a locally finite variety $\mathcal{V}$ has the amalgamation property if and only if every V-formation of finite members of $\mathcal{V}$ has an amalgam in $\mathcal{V}$.*

We now establish a further useful sufficient condition for a variety to have the AP.

PROPOSITION 8.1.7. *Let $\mathcal{S}$ be a subclass of a variety $\mathcal{V}$ satisfying*

(i) *every subdirectly irreducible member of $\mathcal{V}$ belongs to $\mathcal{S}$;*
(ii) *$\mathcal{S}$ is closed under isomorphisms and subalgebras;*
(iii) *for any $\mathbf{B} \in \mathcal{V}$ and subalgebra $\mathbf{A}$ of $\mathbf{B}$, if $\Theta \in \mathrm{Con}(\mathbf{A})$ and $\mathbf{A}/\Theta \in \mathcal{S}$, then there exists $\Phi \in \mathrm{Con}(\mathbf{B})$ such that $\Phi \cap A^2 = \Theta$ and $\mathbf{B}/\Phi \in \mathcal{S}$;*
(iv) *every V-formation of finitely generated members of $\mathcal{S}$ has an amalgam in $\mathcal{V}$.*

Then $\mathcal{V}$ has the amalgamation property.

PROOF. By Proposition 8.1.6, to show that $\mathcal{V}$ has the AP, it suffices to prove that any V-formation $\langle \mathbf{A}, \mathbf{B}, \mathbf{C}, \varphi_B, \varphi_C \rangle$ of finitely generated members of $\mathcal{V}$ has an amalgam in $\mathcal{V}$. We check condition (2) of Lemma 8.1.3. Consider any distinct $x, y \in B$ and let Ψ be a congruence of $\mathbf{B}$ that is maximal with respect to $\langle x, y \rangle \notin \Psi$. Then $\mathbf{B}/\Psi$ is a subdirectly irreducible member of $\mathcal{V}$ and belongs to $\mathcal{S}$, by (i). Define $\Theta := \Psi \cap A^2$. The map φ'_B sending $[a]_\Theta \in \mathbf{A}/\Theta$ to $[\varphi_B(a)]_\Psi$ is an embedding of $\mathbf{A}/\Theta$ into $\mathbf{B}/\Psi$, so $\mathbf{A}/\Theta \in \mathcal{S}$, by (ii). Hence, by (iii), there exists $\Phi \in \mathrm{Con}(\mathbf{C})$ such that $\Phi \cap A^2 = \Theta$ and $\mathbf{C}/\Phi \in \mathcal{S}$. Moreover, the map φ'_C sending any $[a]_\Theta \in \mathbf{A}/\Theta$ to $[\varphi_C(a)]_\Phi$ is an embedding of $\mathbf{A}/\Theta$ into $\mathbf{C}/\Phi$.

Since $\mathbf{A}$, $\mathbf{B}$, and $\mathbf{C}$ are finitely generated, $\langle \mathbf{A}/\Theta, \mathbf{B}/\Psi, \mathbf{C}/\Phi, \varphi'_B, \varphi'_C\rangle$ is a V-formation of finitely generated members of $\mathcal{S}$ and has an amalgam $\langle \mathbf{D}_{xy}, \chi_B, \chi_C\rangle$ in $\mathcal{V}$, by (iv). Consider the homomorphisms

$$\psi_B^{xy}\colon \mathbf{B} \to \mathbf{D}_{xy};\ \ b \mapsto \chi_B([b]_\Psi) \quad \text{and} \quad \psi_C^{xy}\colon \mathbf{C} \to \mathbf{D}_{xy};\ \ c \mapsto \chi_C([c]_\Phi).$$

Then $\psi_B^{xy}(x) \neq \psi_B^{xy}(y)$ (since χ_B is injective and $[x]_\Psi \neq [y]_\Psi$) and, for all $a \in A$,

$$\begin{aligned}
\psi_B^{xy}(\varphi_B(a)) &= \chi_B([\varphi_B(a)]_\Psi)\\
&= \chi_B(\varphi'_B([a]_\Theta))\\
&= \chi_C(\varphi'_C([a]_\Theta))\\
&= \chi_C([\varphi_C(a)]_\Phi)\\
&= \psi_C^{xy}(\varphi_C(a)). \qquad \square
\end{aligned}$$

In certain circumstances, the AP for varieties $\mathcal{V}_1$ and $\mathcal{V}_2$ of $\mathcal{L}$-algebras transfers to their join $\mathcal{V}_1 \vee \mathcal{V}_2$ in the lattice of varieties of $\mathcal{L}$-algebras.

Two varieties $\mathcal{V}_1$ and $\mathcal{V}_2$ of $\mathcal{L}$-algebras are called

(i) *disjoint* if their intersection consists of trivial $\mathcal{L}$-algebras;

(ii) *independent* if there exists $\alpha \in \mathrm{Fm}(\{x, y\})$ (called a *witness formula*) such that $\mathcal{V}_1 \models \alpha \approx x$ and $\mathcal{V}_2 \models \alpha \approx y$.

EXAMPLE 8.1.8. The varieties of integral GBL-algebras and ℓ-groups are disjoint, since any integral ℓ-group is trivial, and also independent, as witnessed by the formula $\alpha := x(x\backslash \mathrm{e})((y\backslash \mathrm{e})\backslash \mathrm{e})$.

We now establish some relationships between these properties.

PROPOSITION 8.1.9. *Let $\mathcal{V}_1$ and $\mathcal{V}_2$ be varieties of $\mathcal{L}$-algebras.*

(a) *If $\mathcal{V}_1$ and $\mathcal{V}_2$ are independent, then they are disjoint and every member of $\mathcal{V}_1 \vee \mathcal{V}_2$ is isomorphic to an algebra $\mathbf{A}_1 \times \mathbf{A}_2$, for some $\mathbf{A}_1 \in \mathcal{V}_1$ and $\mathbf{A}_2 \in \mathcal{V}_2$.*

(b) *If $\mathcal{V}_1 \vee \mathcal{V}_2$ is congruence-permutable and $\mathcal{V}_1$ and $\mathcal{V}_2$ are disjoint, then $\mathcal{V}_1$ and $\mathcal{V}_2$ are independent.*

PROOF.

(a) Suppose that $\mathcal{V}_1$ and $\mathcal{V}_2$ are independent with witness formula $\alpha \in \mathrm{Fm}(\{x, y\})$. Clearly, $\mathcal{V}_1$ and $\mathcal{V}_2$ are disjoint, since $\mathcal{V}_1 \models \alpha(x, y) \approx x$ and $\mathcal{V}_2 \models \alpha(x, y) \approx y$. For the second part, let $\mathcal{K}$ be the class of algebras isomorphic to $\mathbf{A}_1 \times \mathbf{A}_2$, for some $\mathbf{A}_1 \in \mathcal{V}_1$ and $\mathbf{A}_2 \in \mathcal{V}_2$. Then $\mathcal{V}_1, \mathcal{V}_2 \subseteq \mathcal{K} \subseteq \mathcal{V}_1 \vee \mathcal{V}_2$ and $\mathcal{K}$ is closed under taking products, so if $\mathcal{K}$ is also closed under taking subalgebras and homomorphic images, $\mathcal{K} = \mathcal{V}_1 \vee \mathcal{V}_2$. Suppose first that $\mathbf{B}$ is a subalgebra of $\mathbf{A}_1 \times \mathbf{A}_2$ with $\mathbf{A}_1 \in \mathcal{V}_1$ and $\mathbf{A}_2 \in \mathcal{V}_2$. Let $B_1 := \{b_1 \in A_1 \mid \langle b_1, c_2\rangle \in B$, for some $c_2 \in A_2\}$ and $B_2 := \{b_2 \in A_2 \mid \langle c_1, b_2\rangle \in B$, for some $c_1 \in A_1\}$. Then $B \subseteq B_1 \times B_2$ and B_1, B_2 are non-empty subuniverses of $\mathbf{A}_1, \mathbf{A}_2$, respectively, so $\mathbf{B}_1 \in \mathcal{V}_1$ and $\mathbf{B}_2 \in \mathcal{V}_2$. Moreover, for each $\langle b_1, b_2\rangle \in B_1 \times B_2$, there exist $\langle b_1, c_2\rangle, \langle c_1, b_2\rangle \in B$ such that

$$\langle b_1, b_2\rangle = \langle \alpha^{\mathbf{A}_1}(b_1, c_1), \alpha^{\mathbf{A}_2}(c_2, b_2)\rangle = \alpha^{\mathbf{B}}(\langle b_1, c_2\rangle, \langle c_1, b_2\rangle) \in B.$$

So $\mathbf{B} = \mathbf{B}_1 \times \mathbf{B}_2 \in \mathcal{K}$. Now suppose that Θ is a congruence of $\mathbf{A}_1 \times \mathbf{A}_2$ with $\mathbf{A}_1 \in \mathcal{V}_1$ and $\mathbf{A}_2 \in \mathcal{V}_2$. Since Θ can be viewed as a subalgebra of an algebra with universe $A_1^2 \times A_2^2$, the same argument as above yields congruences Θ_1 of $\mathbf{A}_1$ and Θ_2 of $\mathbf{A}_2$ such that $\langle\langle a_1, a_2\rangle, \langle b_1, b_2\rangle\rangle \in \Theta$ if and only if $\langle a_1, b_1\rangle \in \Theta_1$ and $\langle a_2, b_2\rangle \in \Theta_2$. It follows that $\mathcal{K}$ is also closed under taking homomorphic images.

(b) Suppose that $\mathcal{V}_1 \vee \mathcal{V}_2$ is congruence-permutable and $\mathcal{V}_1$ and $\mathcal{V}_2$ are disjoint. Let $\Theta_1 := \Theta_{\mathcal{V}_1}(\{x, y\})$ and $\Theta_2 := \Theta_{\mathcal{V}_2}(\{x, y\})$. Since any two congruences of $\mathbf{Fm}(\{x, y\})$ that contain $\Theta_{\mathcal{V}_1 \vee \mathcal{V}_2}(\{x, y\})$ permute, $\Theta_1 \vee \Theta_2 = \Theta_1 \circ \Theta_2$. But also $\mathbf{Fm}(\{x, y\})/(\Theta_1 \vee \Theta_2) = \mathbf{Fm}(\{x, y\})/(\Theta_1 \circ \Theta_2)$ is trivial, as it belongs to both $\mathcal{V}_1$ and $\mathcal{V}_2$, which were assumed to be disjoint. Hence $x\,(\Theta_1 \circ \Theta_2)\,y$ and there exists $\alpha \in \mathrm{Fm}(\{x, y\})$ such that $x\,\Theta_1\,\alpha$ and $\alpha\,\Theta_2\,y$. Proposition 3.2.7 then yields $\mathcal{V}_1 \models \alpha(x, y) \approx x$ and $\mathcal{V}_2 \models \alpha(x, y) \approx y$ as required. □

Let us show now that the AP transfers from any join $\mathcal{V}_1 \vee \mathcal{V}_2$ of independent varieties $\mathcal{V}_1$ and $\mathcal{V}_2$ to both $\mathcal{V}_1$ and $\mathcal{V}_2$, and vice versa.

PROPOSITION 8.1.10. *Let $\mathcal{V}_1$ and $\mathcal{V}_2$ be independent varieties of $\mathcal{L}$-algebras. Then $\mathcal{V}_1 \vee \mathcal{V}_2$ has the amalgamation property if and only if both $\mathcal{V}_1$ and $\mathcal{V}_2$ have the amalgamation property.*

PROOF. Suppose first that $\mathcal{V}_1$ and $\mathcal{V}_2$ both have the AP and let $\langle \mathbf{A}, \mathbf{B}, \mathbf{C}, \varphi_B, \varphi_C\rangle$ be a V-formation of members of $\mathcal{V}_1 \vee \mathcal{V}_2$. By Proposition 8.1.9, we may assume without loss of generality that $\mathbf{A} = \mathbf{A}_1 \times \mathbf{A}_2$, $\mathbf{B} = \mathbf{B}_1 \times \mathbf{B}_2$, and $\mathbf{C} = \mathbf{C}_1 \times \mathbf{C}_2$ with $\mathbf{A}_1, \mathbf{B}_1, \mathbf{C}_1 \in \mathcal{V}_1$ and $\mathbf{A}_2, \mathbf{B}_2, \mathbf{C}_2 \in \mathcal{V}_2$. Moreover, reasoning as in the proof of Proposition 8.1.9, we may assume that $\mathbf{A}_i$ is a subalgebra of $\mathbf{B}_i$ and $\mathbf{C}_i$, for $i \in \{1, 2\}$. By assumption, the V-formation $\langle \mathbf{A}_i, \mathbf{B}_i, \mathbf{C}_i\rangle$ (suppressing identity maps) has an amalgam $\langle \mathbf{D}_i, \psi_i, \chi_i\rangle$, for $i \in \{1, 2\}$. Let $\mathbf{D} := \mathbf{D}_1 \times \mathbf{D}_2$. We define the maps

$$\psi\colon B \to D;\ \langle b_1, b_2\rangle \mapsto \langle \psi_1(b_1), \psi_2(b_2)\rangle,$$
$$\chi\colon C \to D;\ \langle c_1, c_2\rangle \mapsto \langle \chi_1(c_1), \chi_2(c_2)\rangle.$$

Then ψ is a homomorphism, and symmetrically also χ is a homomorphism, since, for any $f \in \mathcal{L}_n$ ($n \in \mathbb{N}$) and $\langle x_1, y_1\rangle, \dots, \langle x_n, y_n\rangle \in B$,

$$\begin{aligned}
\psi(f^{\mathbf{B}}(\langle x_1, y_1\rangle, \dots, \langle x_n, y_n\rangle)) &= \psi(\langle f^{\mathbf{B}_1}(x_1, \dots, x_n), f^{\mathbf{B}_2}(y_1, \dots, y_n)\rangle) \\
&= \langle \psi_1(f^{\mathbf{B}_1}(x_1, \dots, x_n)), \psi_2(f^{\mathbf{B}_2}(y_1, \dots, y_n))\rangle \\
&= \langle f^{\mathbf{D}_1}(\psi_1(x_1), \dots, \psi_1(x_n)), f^{\mathbf{D}_2}(\psi_2(y_1), \dots, \psi_2(y_n))\rangle \\
&= f^{\mathbf{D}}(\langle \psi_1(x_1), \psi_2(y_1)\rangle, \dots, \langle \psi_1(x_n), \psi_2(y_n)\rangle) \\
&= f^{\mathbf{D}}(\psi(\langle x_1, y_1\rangle), \dots, \psi(\langle x_n, y_n\rangle)).
\end{aligned}$$

Since ψ_1, ψ_2, χ_1, and χ_2 are injective, ψ and χ are also injective. Finally, to see that the resulting diagram is commutative, observe that, for any $\langle a_1, a_2\rangle \in A$,

$$\psi(\langle a_1, a_2\rangle) = \langle \psi_1(a_1), \psi_2(a_2)\rangle = \langle \chi_1(a_1), \chi_2(a_2)\rangle = \chi(\langle a_1, a_2\rangle).$$

Suppose now for the converse that $\mathcal{V}_1 \vee \mathcal{V}_2$ has the AP and consider a V-formation $\langle \mathbf{A}, \mathbf{B}, \mathbf{C}, \varphi_B, \varphi_C\rangle$ of members of $\mathcal{V}_1$ with amalgam $\langle \mathbf{D}, \psi_B, \psi_C\rangle$ in $\mathcal{V}_1 \vee \mathcal{V}_2$. By Proposition 8.1.9, we may assume that $\mathbf{D} = \mathbf{D}_1 \times \mathbf{D}_2$, where $\mathbf{D}_i \in \mathcal{V}_i$, for $i \in \{1, 2\}$.

Let $\alpha \in \mathrm{Fm}(\{x,y\})$ be the witness formula for the independence of $\mathcal{V}_1$ and $\mathcal{V}_2$. Observe that, for any $x, y \in B$ with $\psi_B(x) = \langle a_1, a_2\rangle$ and $\psi_B(y) = \langle b_1, b_2\rangle$,

$$\begin{aligned}\langle a_1, a_2\rangle &= \psi_B(x)\\ &= \psi_B(\alpha^{\mathbf{B}}(x,y))\\ &= \alpha^{\mathbf{D}}(\psi_B(x), \psi_B(y))\\ &= \alpha^{\mathbf{D}}(\langle a_1, a_2\rangle, \langle b_1, b_2\rangle)\\ &= \langle \alpha^{\mathbf{D}_1}(a_1, b_1), \alpha^{\mathbf{D}_2}(a_2, b_2)\rangle\\ &= \langle a_1, b_2\rangle,\end{aligned}$$

so $a_2 = b_2$. Hence there exist $u \in D_2$ and an embedding $\psi'_B \colon \mathbf{B} \to \mathbf{D}_1$ such that $\psi_B(x) = \langle \psi'_B(x), u\rangle$, for each $x \in B$. Reasoning symmetrically, there exist $v \in D_2$ and an embedding $\psi'_C \colon \mathbf{C} \to \mathbf{D}_1$ such that $\psi_C(x) = \langle \psi'_C(x), v\rangle$, for each $x \in C$. Moreover, for any $x \in A$,

$$\langle \psi'_B(\varphi_B(x)), u\rangle = \psi_B(\varphi_B(x)) = \psi_C(\varphi_C(x)) = \langle \psi'_C(\varphi_C(x)), v\rangle.$$

Hence $\psi'_B\varphi_B = \psi'_C\varphi_C$ and $u = v$. Therefore $\langle \mathbf{D}_1, \psi'_B, \psi'_C\rangle$ is an amalgam of the V-formation $\langle \mathbf{A}, \mathbf{B}, \mathbf{C}, \varphi_B, \varphi_C\rangle$ in $\mathcal{V}_1$. □

8.2. The congruence extension property

An algebra $\mathbf{A}$ has the *congruence extension property* (for short, CEP) if, for any subalgebra $\mathbf{B}$ of $\mathbf{A}$ and $\Theta \in \mathrm{Con}(\mathbf{B})$, there exists $\Phi \in \mathrm{Con}(\mathbf{A})$ such that $\Phi \cap B^2 = \Theta$. A class of algebras has the CEP if each of its members has the CEP. In this section, we relate the CEP for a variety $\mathcal{V}$ to a syntactic property that can be understood as a (local) deduction theorem for equational consequence in $\mathcal{V}$ (Theorem 8.2.5).

EXAMPLE 8.2.1. The variety of Abelian groups has the CEP. The congruences of an Abelian group $\mathbf{G}$ are in one-to-one correspondence with its subgroups, and every subgroup of a subgroup of $\mathbf{G}$ is also a subgroup of $\mathbf{G}$. However, the variety of groups does not have the CEP, since it is not the case in general that every normal subgroup of a subgroup of a group $\mathbf{G}$ is a normal subgroup of $\mathbf{G}$.

EXAMPLE 8.2.2. The variety of lattices does not have the CEP. Consider the simple lattice $\mathbf{M}_5$ (depicted in Figure 1.1) and the sublattice $\mathbf{B}$ obtained by removing its middle element c. Let Θ be the congruence of $\mathbf{B}$ with congruence classes $\{a, \bot\}$ and $\{b, d\}$. Then $\Delta_{M_5} \cap B^2 = \Delta_B \neq \Theta$ and $(M_5)^2 \cap B^2 = B^2 \neq \Theta$. So there does not exist any $\Phi \in \mathrm{Con}(\mathbf{M}_5)$ such that $\Phi \cap B^2 = \Theta$, and $\mathbf{M}_5$ does not have the CEP.

We begin with two straightforward observations about extending congruences.

LEMMA 8.2.3. *Let* $\mathbf{A}$ *be any algebra.*

(a) *Let* $\mathbf{B}$ *be any subalgebra of* $\mathbf{A}$ *and let* $R \subseteq B^2$. *If there exists* $\Phi \in \mathrm{Con}(\mathbf{A})$ *such that* $\mathrm{Cg}^{\mathbf{B}}(R) = \Phi \cap B^2$, *then* $\mathrm{Cg}^{\mathbf{B}}(R) = \mathrm{Cg}^{\mathbf{A}}(R) \cap B^2$.

(b) *Suppose that for any subalgebra* $\mathbf{B}$ *of* $\mathbf{A}$ *and finitely generated congruence* Θ *of* $\mathbf{B}$, *there exists* $\Phi \in \mathrm{Con}(\mathbf{A})$ *such that* $\Theta = \Phi \cap B^2$. *Then* $\mathbf{A}$ *has the congruence extension property.*

PROOF.

(a) Suppose that $\mathrm{Cg}^{\mathbf{B}}(R) = \Phi \cap B^2$, for some $\Phi \in \mathrm{Con}(\mathbf{A})$. Clearly, $\mathrm{Cg}^{\mathbf{B}}(R) \subseteq \mathrm{Cg}^{\mathbf{A}}(R) \cap B^2$. But also $R \subseteq \Phi \in \mathrm{Con}(\mathbf{A})$, so $\mathrm{Cg}^{\mathbf{A}}(R) \cap B^2 \subseteq \Phi \cap B^2 = \mathrm{Cg}^{\mathbf{B}}(R)$.

(b) Let $\mathbf{B}$ be any subalgebra of $\mathbf{A}$. Then, using the assumption and part (a), $\mathrm{Cg}^{\mathbf{B}}(R) = \mathrm{Cg}^{\mathbf{A}}(R) \cap B^2$, for any finite subset R of B^2. Now, given any $\Theta \in \mathrm{Con}(\mathbf{B})$, let $S := \{\mathrm{Cg}^{\mathbf{A}}(R) \mid R \subseteq \Theta \text{ and } R \text{ is finite}\}$. Then $\Phi := \bigcup S \in \mathrm{Con}(\mathbf{A})$, since $\langle S, \subseteq \rangle$ is an up-directed subposet of the algebraic lattice $\mathrm{Con}(\mathbf{A})$. Hence

$$\begin{aligned} \Theta &= \bigcup \{\mathrm{Cg}^{\mathbf{B}}(R) \mid R \subseteq \Theta \text{ and } R \text{ is finite}\} \\ &= \bigcup \{\mathrm{Cg}^{\mathbf{A}}(R) \cap B^2 \mid R \subseteq \Theta \text{ and } R \text{ is finite}\} \\ &= \Phi \cap B^2. \end{aligned}$$

□

Like the AP, the CEP for a variety can be described via congruences of free algebras.

LEMMA 8.2.4. *A variety $\mathcal{V}$ has the congruence extension property if and only if, for any sets Y, Z satisfying $\emptyset \neq Z \subseteq Y$, $\Theta \in \mathrm{Con}(\mathbf{F}(Y))$, and $R \subseteq \mathrm{F}(Z)^2$,*

$$(\mathrm{Cg}^{\mathbf{F}(Y)}(R) \vee \Theta) \cap \mathrm{F}(Z)^2 = \mathrm{Cg}^{\mathbf{F}(Z)}(R) \vee (\Theta \cap \mathrm{F}(Z)^2).$$

PROOF.

($\Rightarrow$) Suppose that $\mathcal{V}$ has the CEP and consider any sets Y, Z satisfying $\emptyset \neq Z \subseteq Y$, $\Theta \in \mathrm{Con}(\mathbf{F}(Y))$, and $R \subseteq \mathrm{F}(Z)^2$. Let $\pi_\Theta \colon \mathbf{F}(Y) \to \mathbf{F}(Y)/\Theta;\ u \mapsto [u]_\Theta$ be the canonical map and let φ be the restriction of π_Θ to $\mathbf{F}(Z)$, defining $\mathbf{A} := \pi_\Theta(\mathbf{F}(Y)) = \mathbf{F}(Y)/\Theta$ and $\mathbf{B} := \varphi(\mathbf{F}(Z))$. Note that $\ker(\pi_\Theta) = \Theta$ and $\ker(\varphi) = \Theta \cap \mathrm{F}(Z)^2$. Using the Correspondence theorem (Theorem B.4.(d)) for the first and last equalities and the CEP for the third, we obtain

$$\begin{aligned} (\mathrm{Cg}^{\mathbf{F}(Y)}(R) \vee \Theta) \cap \mathrm{F}(Z)^2 &= \pi_\Theta^{-1}[\mathrm{Cg}^{\mathbf{A}}(\pi_\Theta[R])] \cap \mathrm{F}(Z)^2 \\ &= \pi_\Theta^{-1}[\mathrm{Cg}^{\mathbf{A}}(\varphi[R]) \cap B^2] \\ &= \pi_\Theta^{-1}[\mathrm{Cg}^{\mathbf{B}}(\varphi[R])] \\ &= \varphi^{-1}[\mathrm{Cg}^{\mathbf{B}}(\varphi[R])] \\ &= \mathrm{Cg}^{\mathbf{F}(Z)}(R) \vee (\Theta \cap \mathrm{F}(Z)^2). \end{aligned}$$

($\Leftarrow$) Suppose that the condition is satisfied. To prove that $\mathcal{V}$ has the CEP, it suffices to establish $\mathrm{Cg}^{\mathbf{A}}(R) \cap B^2 = \mathrm{Cg}^{\mathbf{B}}(R)$, for any $\mathbf{A} \in \mathcal{V}$, subalgebra $\mathbf{B}$ of $\mathbf{A}$, and $R \subseteq B^2$. Let $\varphi_A \colon \mathbf{F}(A) \to \mathbf{A}$ and $\varphi_B \colon \mathbf{F}(B) \to \mathbf{B}$ be the surjective homomorphisms extending the identity maps on A and B, respectively, and let $\Theta := \ker(\varphi_A)$, noting that $\ker(\varphi_B) = \Theta \cap \mathrm{F}(B)^2$. Using the Correspondence theorem (Theorem B.4.(d)) for the first and third equalities and the assumption for the second, we obtain

$$\begin{aligned} \mathrm{Cg}^{\mathbf{A}}(R) \cap B^2 &\subseteq \varphi_A^{-1}[\mathrm{Cg}^{\mathbf{A}}(\varphi_A[R])] \cap \mathrm{F}(B)^2 \\ &= (\mathrm{Cg}^{\mathbf{F}(A)}(R) \vee \Theta) \cap \mathrm{F}(B)^2 \\ &= \mathrm{Cg}^{\mathbf{F}(B)}(R) \vee (\Theta \cap \mathrm{F}(B)^2) \\ &= \varphi_B^{-1}[\mathrm{Cg}^{\mathbf{B}}(\varphi_B[R])] \\ &= \mathrm{Cg}^{\mathbf{B}}(R). \end{aligned}$$

Hence, since the converse inclusion is trivial, $\mathrm{Cg}^{\mathbf{A}}(R) \cap B^2 = \mathrm{Cg}^{\mathbf{B}}(R)$. □

A variety $\mathcal{V}$ has the *extension property* (for short, EP) if, for any set X, whenever

(i) $\Sigma \cup \Pi \cup \{\varepsilon\} \subseteq \mathrm{Eq}(X)$;
(ii) $\Sigma \cup \Pi \vdash_{\mathcal{V}} \varepsilon$,

there exists $\Delta \subseteq \mathrm{Eq}(X)$ such that

(iii) $\Sigma \vdash_{\mathcal{V}} \Delta$;
(iv) $\Delta \cup \Pi \vdash_{\mathcal{V}} \varepsilon$;
(v) $\mathrm{Var}(\Delta) \subseteq \mathrm{Var}(\Pi \cup \{\varepsilon\})$.

Similarly to the RP, it is easily shown that the EP (for all sets) is implied by the EP restricted to the case where X is finite. We may also add the restriction $\mathrm{Var}(\Pi \cup \{\varepsilon\}) \neq \emptyset$, dealing with the case where $\mathrm{Var}(\Pi \cup \{\varepsilon\}) = \emptyset$ by adding $x \approx x$ to Π, for some new variable x, and then substituting x with any constant in the resulting Δ. The main theorem of this section may now be stated as follows.

THEOREM 8.2.5. *A variety has the congruence extension property if and only if it has the extension property.*

PROOF. We use the equivalent characterization of the CEP supplied by Lemma 8.2.4. Suppose first that $\mathcal{V}$ has the EP and consider any sets Y, Z satisfying $\emptyset \neq Z \subseteq Y$, $\Theta \in \mathrm{Con}(\mathbf{F}(Y))$, and $R \subseteq \mathrm{F}(Z)^2$. Let $\langle u, v\rangle \in (\mathrm{Cg}^{\mathbf{F}(Y)}(R) \vee \Theta) \cap \mathrm{F}(Z)^2$. Choose $\Sigma \subseteq \mathrm{Eq}(Y)$ and $\Pi \cup \{\varepsilon\} \subseteq \mathrm{Eq}(Z)$ such that $\overline{\Sigma} = \Theta$, $\overline{\Pi} = R$, and $\bar{\varepsilon} = \langle u, v\rangle$. Since $\langle u, v\rangle \in \Theta \vee \mathrm{Cg}^{\mathbf{F}(Y)}(R)$, an application of Proposition 3.2.7 yields $\Sigma \cup \Pi \vdash_{\mathcal{V}} \varepsilon$. Hence, by assumption, there exists $\Delta \subseteq \mathrm{Eq}(Z)$ such that $\Sigma \vdash_{\mathcal{V}} \Delta$ and $\Delta \cup \Pi \vdash_{\mathcal{V}} \varepsilon$. Applying again Proposition 3.2.7, we obtain $\langle u, v\rangle \in \mathrm{Cg}^{\mathbf{F}(Z)}(R) \vee (\Theta \cap \mathrm{F}(Z)^2)$. So

$$(\mathrm{Cg}^{\mathbf{F}(Y)}(R) \vee \Theta) \cap \mathrm{F}(Z)^2 \subseteq \mathrm{Cg}^{\mathbf{F}(Z)}(R) \vee (\Theta \cap \mathrm{F}(Z)^2),$$

and, since the reverse inclusion is trivial, $\mathcal{V}$ has the CEP.

For the converse, suppose that $\mathcal{V}$ has the CEP and consider $\Sigma \cup \Pi \cup \{\varepsilon\} \subseteq \mathrm{Eq}(Y)$ and $\Sigma \cup \Pi \vdash_{\mathcal{V}} \varepsilon$, assuming without loss of generality that $Z := \mathrm{Var}(\Pi \cup \{\varepsilon\})$ is non-empty. Let $\Theta := \mathrm{Cg}^{\mathbf{F}(Y)}(\overline{\Sigma})$. Since $\Sigma \cup \Pi \vdash_{\mathcal{V}} \varepsilon$, using Proposition 3.2.7 and the fact that $\mathcal{V}$ has the CEP,

$$\bar{\varepsilon} \in (\mathrm{Cg}^{\mathbf{F}(Y)}(\overline{\Pi}) \vee \Theta) \cap \mathrm{F}(Z)^2 = \mathrm{Cg}^{\mathbf{F}(Z)}(\overline{\Pi}) \vee (\Theta \cap \mathrm{F}(Z)^2).$$

Choose $\Delta \subseteq \mathrm{Eq}(Z)$ such that $\overline{\Delta} = \Theta \cap \mathrm{F}(Z)^2$. Then, by Proposition 3.2.7 again, $\Sigma \vdash_{\mathcal{V}} \Delta$ and $\Delta \cup \Pi \vdash_{\mathcal{V}} \varepsilon$ as required. □

In the presence of the CEP, we obtain also a further useful criterion for a variety to have the AP.

PROPOSITION 8.2.6. *Let $\mathcal{V}$ be a variety with the congruence extension property, and suppose that the class $\mathcal{S}$ of finitely subdirectly irreducible members of $\mathcal{V}$ is closed under subalgebras and that every V-formation of finitely generated members of $\mathcal{S}$ has an amalgam in $\mathcal{V}$. Then $\mathcal{V}$ has the amalgamation property.*

PROOF. Since every subdirectly irreducible member of $\mathcal{V}$ is finitely subdirectly irreducible and $\mathcal{S}$ is closed under isomorphisms and subalgebras, it suffices to check condition (iii) of Proposition 8.1.7. Let $\mathbf{A}$ be any subalgebra of $\mathbf{B} \in \mathcal{V}$ and suppose that $\mathbf{A}/\Theta \in \mathcal{S}$, for some $\Theta \in \mathrm{Con}(\mathbf{A})$. Our goal is to prove that $\Phi \cap A^2 = \Theta$ and $\mathbf{B}/\Phi \in \mathcal{S}$, for some $\Phi \in \mathrm{Con}(\mathbf{B})$. Note first that $\mathrm{Cg}^{\mathbf{B}}(\Theta) \cap A^2 = \Theta$, by the CEP. Hence $T := \{\Psi \in \mathrm{Con}(\mathbf{B}) \mid \Psi \cap A^2 = \Theta\} \neq \emptyset$. Moreover, every chain in $\langle T, \subseteq\rangle$ has an upper bound in T, so, by Zorn's lemma, $\langle T, \subseteq\rangle$ has a maximal element Φ. Clearly, $\Phi \cap A^2 = \Theta$, so it remains to check that $\mathbf{B}/\Phi \in \mathcal{S}$.

Since an algebra $\mathbf{C}$ is finitely subdirectly irreducible if and only if Δ_C is meet-irreducible in $\mathrm{Con}(\mathbf{C})$ (Proposition B.6), it follows, by the Correspondence theorem (Theorem B.4.(c)), that Θ is meet-irreducible in $\mathrm{Con}(\mathbf{A})$. Suppose that $\Phi = \Phi_1 \cap \Phi_2$ with $\Phi_1, \Phi_2 \in \mathrm{Con}(\mathbf{B})$. Then $(\Phi_1 \cap A^2) \cap (\Phi_2 \cap A^2) = \Phi_1 \cap \Phi_2 \cap A^2 = \Phi \cap A^2 = \Theta$ and, since Θ is meet-irreducible in $\mathrm{Con}(\mathbf{A})$, either $\Phi_1 \cap A^2 = \Theta$ or $\Phi_2 \cap A^2 = \Theta$. Hence $\Phi_1 \in T$ or $\Phi_2 \in T$, and, by maximality, $\Phi_1 = \Phi$ or $\Phi_2 = \Phi$. So Φ is meet-irreducible in $\mathrm{Con}(\mathbf{B})$ and $\mathbf{B}/\Phi$ is finitely subdirectly irreducible, that is, $\mathbf{B}/\Phi \in \mathcal{S}$. □

8.3. Interpolation properties

In the last two sections, we built bridges from well-known algebraic properties of varieties to properties of their consequence relations. In this section, we proceed in the opposite direction. We relate well-known syntactic "interpolation" properties to amalgamation and other notable algebraic properties (Theorems 8.3.1 and 8.3.3).

A variety $\mathcal{V}$ has the *deductive interpolation property* (for short, DIP) if, for any set X, whenever

(i) $\Sigma \cup \{\varepsilon\} \subseteq \mathrm{Eq}(X)$ and $\mathrm{Var}(\Sigma) \cap \mathrm{Var}(\varepsilon) \neq \emptyset$;

(ii) $\Sigma \vdash_{\mathcal{V}} \varepsilon$,

there exists $\Delta \subseteq \mathrm{Eq}(X)$ such that

(iii) $\Sigma \vdash_{\mathcal{V}} \Delta$;

(iv) $\Delta \vdash_{\mathcal{V}} \varepsilon$;

(v) $\mathrm{Var}(\Delta) \subseteq \mathrm{Var}(\Sigma) \cap \mathrm{Var}(\varepsilon)$.

It follows easily that $\mathcal{V}$ has the DIP if this property holds for all finite sets X and finite $\Sigma \cup \{\varepsilon\} \subseteq \mathrm{Eq}(X)$.

The relationship of the DIP to the AP is described by the following bridge theorem.

THEOREM 8.3.1. *Let $\mathcal{V}$ be any variety.*

(a) *If $\mathcal{V}$ has the amalgamation property, then it has the deductive interpolation property.*

(b) *If $\mathcal{V}$ has the deductive interpolation property and the extension property, then it has the amalgamation property.*

PROOF.

(a) Suppose that $\mathcal{V}$ has the AP and hence, by Theorem 8.1.5, the RP. Consider any set X and $\Sigma \cup \{\varepsilon\} \subseteq \mathrm{Eq}(X)$ such that $\mathrm{Var}(\Sigma) \cap \mathrm{Var}(\varepsilon) \neq \emptyset$ and $\Sigma \vdash_{\mathcal{V}} \varepsilon$. Let

$$\Delta := \{\delta \in \mathrm{Eq}(X) \mid \mathrm{Var}(\delta) \subseteq \mathrm{Var}(\Sigma) \cap \mathrm{Var}(\varepsilon) \text{ and } \Sigma \vdash_{\mathcal{V}} \delta\}.$$

Clearly, $\Sigma \vdash_{\mathcal{V}} \Delta$ and $\mathrm{Var}(\Delta) = \mathrm{Var}(\Sigma) \cap \mathrm{Var}(\varepsilon)$, using the fact that $x \approx x \in \Delta$, for each $x \in \mathrm{Var}(\Sigma) \cap \mathrm{Var}(\varepsilon)$. Moreover, (i)–(iii) in the definition of the RP are satisfied (with Δ taking the role of Σ and Σ taking the role of Π), so $\Delta \vdash_{\mathcal{V}} \varepsilon$.

(b) Suppose that $\mathcal{V}$ has the DIP and the EP. By Theorem 8.1.5, it suffices to show that $\mathcal{V}$ has the RP, so suppose that, for some set X,

(i) $\Sigma \cup \Pi \cup \{\varepsilon\} \subseteq \mathrm{Eq}(X)$, $\mathrm{Var}(\Sigma) \cap \mathrm{Var}(\Pi) \neq \emptyset$, and $\mathrm{Var}(\varepsilon) \cap \mathrm{Var}(\Pi) \subseteq \mathrm{Var}(\Sigma)$;
(ii) $\Sigma \vdash_{\mathcal{V}} \delta \iff \Pi \vdash_{\mathcal{V}} \delta$, for all $\delta \in \mathrm{Eq}(X)$ with $\mathrm{Var}(\delta) \subseteq \mathrm{Var}(\Sigma) \cap \mathrm{Var}(\Pi)$;
(iii) $\Sigma \cup \Pi \vdash_{\mathcal{V}} \varepsilon$.

By the EP, there exists $\Delta' \subseteq \mathrm{Eq}(X)$ with $\mathrm{Var}(\Delta') \subseteq \mathrm{Var}(\Sigma) \cup \mathrm{Var}(\varepsilon)$ such that $\Sigma \cup \Delta' \vdash_{\mathcal{V}} \varepsilon$ and $\Pi \vdash_{\mathcal{V}} \Delta'$. By the DIP, for each $\delta \in \Delta'$, there exists $\Delta_\delta \subseteq \mathrm{Eq}(X)$ such that $\Pi \vdash_{\mathcal{V}} \Delta_\delta$ and $\Delta_\delta \vdash_{\mathcal{V}} \delta$ with $\mathrm{Var}(\Delta_\delta) \subseteq \mathrm{Var}(\Pi) \cap [\mathrm{Var}(\Sigma) \cup \mathrm{Var}(\varepsilon)] \subseteq \mathrm{Var}(\Pi) \cap \mathrm{Var}(\Sigma)$. Let $\Delta := \bigcup\{\Delta_\delta \mid \delta \in \Delta'\}$. Then $\Pi \vdash_{\mathcal{V}} \Delta$ and $\Delta \vdash_{\mathcal{V}} \Delta'$ with $\mathrm{Var}(\Delta) \subseteq \mathrm{Var}(\Pi) \cap \mathrm{Var}(\Sigma)$. Since $\Sigma \cup \Delta' \vdash_{\mathcal{V}} \varepsilon$ and $\Delta \vdash_{\mathcal{V}} \Delta'$, by transitivity, $\Sigma \cup \Delta \vdash_{\mathcal{V}} \varepsilon$. But also, because $\Pi \vdash_{\mathcal{V}} \Delta$ and $\mathrm{Var}(\Delta) \subseteq \mathrm{Var}(\Pi) \cap \mathrm{Var}(\Sigma)$, it follows from (ii) that $\Sigma \vdash_{\mathcal{V}} \Delta$. So, by transitivity, $\Sigma \vdash_{\mathcal{V}} \varepsilon$ as required. □

COROLLARY 8.3.2. *If a variety has the congruence extension property, then it has the amalgamation property if and only if it has the deductive interpolation property.*

The DIP and EP combine to form a stronger interpolation property. A variety $\mathcal{V}$ has the *Maehara interpolation property* (for short, MIP) if, for any set X, whenever

(i) $\Sigma \cup \Pi \cup \{\varepsilon\} \subseteq \mathrm{Eq}(X)$ and $\mathrm{Var}(\Sigma) \cap \mathrm{Var}(\Pi \cup \{\varepsilon\}) \neq \emptyset$;
(ii) $\Sigma \cup \Pi \vdash_{\mathcal{V}} \varepsilon$,

there exists $\Delta \subseteq \mathrm{Eq}(X)$ such that

(iii) $\Sigma \vdash_{\mathcal{V}} \Delta$;
(iv) $\Delta \cup \Pi \vdash_{\mathcal{V}} \varepsilon$;
(v) $\mathrm{Var}(\Delta) \subseteq \mathrm{Var}(\Sigma) \cap \mathrm{Var}(\Pi \cup \{\varepsilon\})$.

The MIP corresponds to an interesting algebraic property. A variety $\mathcal{V}$ has the *transferable injections property* (for short, TIP) if whenever $\mathbf{A}, \mathbf{B}, \mathbf{C} \in \mathcal{V}$, φ_B is an embedding of $\mathbf{A}$ into $\mathbf{B}$, and φ_C is a homomorphism from $\mathbf{A}$ to $\mathbf{C}$, there exist $\mathbf{D} \in \mathcal{V}$, a homomorphism ψ_B from $\mathbf{B}$ to $\mathbf{D}$, and an embedding ψ_C from $\mathbf{C}$ into $\mathbf{D}$ such that $\psi_B\varphi_B = \psi_C\varphi_C$.

THEOREM 8.3.3. *The following statements are equivalent for any variety $\mathcal{V}$:*

(1) *$\mathcal{V}$ has the Maehara interpolation property.*
(2) *$\mathcal{V}$ has the deductive interpolation property and the extension property.*
(3) *$\mathcal{V}$ has the amalgamation property and the congruence extension property.*
(4) *$\mathcal{V}$ has the transferable injections property.*

PROOF. The equivalence of (1) and (2) is left as a straightforward exercise.

The equivalence of (2) and (3) follows from Theorems 8.3.1 and 8.2.5, so it remains to prove the equivalence of (3) and (4). Suppose first that $\mathcal{V}$ has the TIP. Restricting to the case where φ_B and φ_C are embeddings, the TIP yields homomorphisms $\psi_{x,y}$ and $\psi_{x,y}$ as in Lemma 8.1.3, for all distinct x, y in B, C. So $\mathcal{V}$ has the AP. Moreover, restricting to the case where φ_C is surjective, yields the CEP. To see this, consider a subalgebra $\mathbf{A}$ of $\mathbf{B} \in \mathcal{V}$ and $\Theta \in \mathrm{Con}(\mathbf{A})$ and let $\varphi_C \colon \mathbf{A} \to \mathbf{A}/\Theta$ be the canonical map (usually denoted by π_Θ) and let $\varphi_B \colon \mathbf{A} \to \mathbf{B}$ be the inclusion map. Then, by the TIP, there exist $\mathbf{D} \in \mathcal{V}$, a homomorphism ψ_B from $\mathbf{B}$ to $\mathbf{D}$, and an embedding ψ_C from $\mathbf{A}/\Theta$ into $\mathbf{D}$ such that $\psi_B\varphi_B = \psi_C\varphi_C$. Let $\Phi := \ker(\psi_B)$. Then, for any $a, b \in A$, using the fact that ψ_C is injective,

$$\begin{aligned}\langle a, b\rangle \in \Phi &\iff \psi_B\varphi_B(a) = \psi_B\varphi_B(b)\\ &\iff \psi_C\varphi_C(a) = \psi_C\varphi_C(b)\\ &\iff \varphi_C(a) = \varphi_C(b)\\ &\iff \langle a, b\rangle \in \ker(\varphi_C) = \Theta.\end{aligned}$$

That is, $\Phi \cap A^2 = \Theta$. Hence $\mathcal{V}$ has the CEP.

For the converse, suppose that $\mathcal{V}$ has the AP and CEP. Let $\mathbf{A}$, $\mathbf{B}$, $\mathbf{C}$, φ_B, and φ_C be as in the definition of the TIP, assuming without loss of generality that $\mathbf{A}$ is a subalgebra of $\mathbf{B}$. Let $\mathbf{C}' := \varphi_C[\mathbf{A}]$. Then $\varphi_C \colon \mathbf{A} \to \mathbf{C}'$ is surjective and the Homomorphism theorem (Corollary B.3) yields $\mathbf{C}' \cong \mathbf{A}/\Theta$, where $\Theta := \ker(\varphi_C)$. By the CEP, there exists $\Phi \in \mathrm{Con}(\mathbf{B})$ such that $\Theta = \Phi \cap A^2$. Let $\mathbf{D}' := \mathbf{B}/\Phi$ and let $\pi_\Phi \colon \mathbf{B} \to \mathbf{D}'$ be the canonical map onto $\mathbf{D}'$. Then there exists also an embedding $\varphi_{C'} \colon \mathbf{C}' \to \mathbf{D}'$ such that $\pi_\Phi\varphi_B = \varphi_{C'}\varphi_C$. Since $\mathcal{V}$ has the AP, the V-formation $\langle \mathbf{C}', \mathbf{D}', \mathbf{C}, \varphi_{C'}, \chi\rangle$, where $\chi \colon \mathbf{C}' \to \mathbf{C}$ is the inclusion map, has an amalgam $\langle \mathbf{D}, \psi_{D'}, \psi_C\rangle$ as indicated by the commutative diagram

$$\begin{array}{ccccc}\mathbf{B} & \xrightarrow{\pi_\Phi} & \mathbf{D}' & \xhookrightarrow{\psi_{D'}} & \mathbf{D}\\ \uparrow{\scriptstyle \varphi_B} & & {\scriptstyle \varphi_{C'}}\uparrow & & {\scriptstyle \psi_C}\uparrow\\ \mathbf{A} & \xrightarrow{\varphi_C} & \mathbf{C}' & \xhookrightarrow{\chi} & \mathbf{C}.\end{array}$$

Hence $\psi_B := \psi_{D'}\pi_\Phi$ and ψ_C satisfy $\psi_B\varphi_B = \psi_C\varphi_C$ as required by the TIP. □

To provide an equivalent algebraic characterization of the DIP, we make a further assumption on a variety $\mathcal{V}$. By the *$\mathcal{V}$-free product* of a family of algebras $\{\mathbf{A}_i\}_{i\in I}$ in $\mathcal{V}$, we mean an algebra $\mathbf{A} \in \mathcal{V}$, denoted by $*_{i\in I}\mathbf{A}_i$, together with a family of embeddings $\{\varphi_i \colon \mathbf{A}_i \to \mathbf{A}\}_{i\in I}$ such that

(i) $\bigcup_{i\in I} \varphi_i[A_i]$ generates $\mathbf{A}$;

(ii) for any $\mathbf{B} \in \mathcal{V}$ and family of homomorphisms $\{\psi_i \colon \mathbf{A}_i \to \mathbf{B}\}_{i\in I}$, there exists a (unique) homomorphism $\psi \colon \mathbf{A} \to \mathbf{B}$ satisfying $\psi\varphi_i = \psi_i$, for each $i \in I$.

To simplify notation, we use the "internal" definition of a $\mathcal{V}$-free product: we identify each free factor $\mathbf{A}_i$ with its isomorphic image $\varphi_i[A_i]$ in $*_{i\in I}\mathbf{A}_i$, treating each $\mathbf{A}_i$ as a subalgebra of $*_{i\in I}\mathbf{A}_i$, and each φ_i as an inclusion map.

Note that $\mathcal{V}$-free products are $\mathcal{V}$-coproducts satisfying the additional assumption that the associated homomorphisms are injective. However, while $\mathcal{V}$-coproducts exist for any variety $\mathcal{V}$, this is not the case for $\mathcal{V}$-free products. Their existence is a consequence of the *embedding property* for $\mathcal{V}$: given any $\mathbf{A}, \mathbf{B} \in \mathcal{V}$, there exists $\mathbf{C} \in \mathcal{V}$ into which both $\mathbf{A}$ and $\mathbf{B}$ can be embedded. Many significant varieties admit free products, in particular, all varieties of residuated lattices. To see this, note that each residuated lattice has a one-element subalgebra $\{\mathrm{e}\}$, and so $\mathbf{C}$ can be taken to be the direct product $\mathbf{A} \times \mathbf{B}$. This property also holds for certain varieties of pointed residuated lattices, such as Boolean algebras and MV-algebras, although the verification is less obvious. In what follows, we write "$\mathcal{V}$ admits free products" to mean that the $\mathcal{V}$-free product exists for any family of algebras in $\mathcal{V}$.

$$
\begin{array}{ccccccc}
 & & \mathsf{FAP}+\mathsf{CEP} & & & & \\
 & & \Updownarrow & & & & \\
\mathsf{CEP} & \Longleftarrow & \mathsf{TIP} & \Longrightarrow & \mathsf{AP} & \Longrightarrow & \mathsf{FAP} \\
\Updownarrow & & \Updownarrow & & \Updownarrow & & \Updownarrow \\
\mathsf{EP} & \Longleftarrow & \mathsf{MIP} & \Longrightarrow & \mathsf{RP} & \Longrightarrow & \mathsf{DIP} \\
 & & \Updownarrow & & & & \\
 & & \mathsf{DIP}+\mathsf{EP} & & & &
\end{array}
$$

FIGURE 8.1. Bridges between algebraic and syntactic properties

A variety $\mathcal{V}$ admitting free products has the *flat amalgamation property* (for short, FAP) if, for any $\mathbf{B} \in \mathcal{V}$, subalgebra $\mathbf{A}$ of $\mathbf{B}$, and non-empty set Z, the induced homomorphism $\mathbf{A} * \mathbf{F}(Z) \to \mathbf{B} * \mathbf{F}(Z)$ is injective.

THEOREM 8.3.4. *A variety admitting free products has the deductive interpolation property if and only if it has the flat amalgamation property.*

PROOF. Let $\mathcal{V}$ be a variety admitting free products. It is straightforward to check first, using Proposition 3.2.7, that $\mathcal{V}$ has the DIP if and only if, for any disjoint sets X, Y, Z with $X \neq \emptyset$ and $\Theta \in \mathrm{Con}(\mathbf{F}(X \cup Y))$,

$$\mathrm{Cg}^{\mathbf{F}(X \cup Y \cup Z)}(\Theta) \cap \mathrm{F}(X \cup Z)^2 = \mathrm{Cg}^{\mathbf{F}(X \cup Z)}(\Theta \cap \mathrm{F}(X)^2).$$

($\Leftarrow$) Suppose that $\mathcal{V}$ has the FAP and consider any disjoint sets X, Y, Z with $X \neq \emptyset$ and $\Theta \in \mathrm{Con}(\mathbf{F}(X \cup Y))$. Let $\mathbf{A} := \mathbf{F}(X)/(\Theta \cap \mathrm{F}(X)^2)$ and $\mathbf{B} := \mathbf{F}(X \cup Y)/\Theta$. Let $\varphi\colon \mathbf{F}(X \cup Z) \to \mathbf{A} * \mathbf{F}(Z)$ be the surjective homomorphism extending the canonical map from $\mathbf{F}(X)$ onto $\mathbf{A}$, and let $\psi\colon \mathbf{F}(X \cup Y \cup Z) \to \mathbf{B} * \mathbf{F}(Z)$ be the surjective homomorphism extending the canonical map from $\mathbf{F}(X \cup Y)$ onto $\mathbf{B}$. By the FAP, the induced homomorphism $\chi\colon \mathbf{A} * \mathbf{F}(Z) \to \mathbf{B} * \mathbf{F}(Z)$ is injective. Hence the restriction ψ' of ψ to $\mathbf{F}(X \cup Z)$ is onto $\chi[\mathbf{A} * \mathbf{F}(Z)]$ with $\psi' = \chi \circ \varphi$ and

$$\mathrm{Cg}^{\mathbf{F}(X \cup Y \cup Z)}(\Theta) \cap \mathrm{F}(X \cup Z)^2 = \ker(\psi') = \ker(\varphi) = \mathrm{Cg}^{\mathbf{F}(X \cup Z)}(\Theta \cap \mathrm{F}(X)^2).$$

($\Rightarrow$) Suppose that $\mathcal{V}$ has the DIP. Let $\mathbf{A}$ be a subalgebra of some $\mathbf{B} \in \mathcal{V}$ and let Z be any non-empty set. We may assume that $B = A \cup Y$, where A, Y, Z are disjoint. Let $\varphi\colon \mathbf{F}(A \cup Y) \to \mathbf{B}$ be the homomorphism that extends the identity map on $A \cup Y$ with $\Theta := \ker(\varphi)$, and let $\psi\colon \mathbf{F}(A \cup Y \cup Z) \to \mathbf{B} * \mathbf{F}(Z)$ be the homomorphism that extends the identity map on $A \cup Y \cup Z$. Then $\ker(\psi) = \mathrm{Cg}^{\mathbf{F}(A \cup Y \cup Z)}(\Theta)$. Now let ψ' be the restriction of ψ to $\mathbf{F}(A \cup Z)$. Then, using the DIP,

$$\begin{aligned}
\ker(\psi') &= \ker(\psi) \cap \mathrm{F}(A \cup Z)^2 \\
&= \mathrm{Cg}^{\mathbf{F}(A \cup Y \cup Z)}(\Theta) \cap \mathrm{F}(A \cup Z)^2 \\
&= \mathrm{Cg}^{\mathbf{F}(A \cup Z)}(\Theta \cap \mathrm{F}(A)^2).
\end{aligned}$$

It follows that $\psi[\mathbf{F}(A \cup Z)] \cong \mathbf{A} * \mathbf{F}(Z)$. Hence $\mathcal{V}$ has the FAP. □

The bridges between algebraic and syntactic properties established in this and the two previous sections are displayed in Figure 8.1.

8.4. Amalgamation in varieties of residuated lattices

Having built bridges between various algebraic and logical properties, we put these now to good use in our study of varieties of residuated lattices. In particular, we show how to establish the DIP, and thereby the AP, for varieties with a suitable sequent calculus. We also employ algebraic methods to prove that certain varieties do not have the AP, and give an example of a non-commutative variety of residuated lattices that does have this property.

Recall first that, by Proposition 4.3.9, any variety $\mathcal{V}$ of commutative residuated lattices has the CEP and therefore, by Theorem 8.2.5, the EP. The latter property appears in more concrete form in Proposition 4.4.10 as a *local deduction theorem*, stating that, for any $T \cup \{\alpha, \beta\} \subseteq \mathrm{Fm}$,

$$T \cup \{\alpha\} \vdash_{\mathcal{V}} \beta \iff T \vdash_{\mathcal{V}} (\alpha \wedge \mathrm{e})^n \to \beta, \text{ for some } n \in \mathbb{N}.$$

Hence we can ignore the CEP in the statement of Corollary 8.3.2 for such varieties, and obtain the following useful equivalence.

COROLLARY 8.4.1. *Let $\mathcal{V}$ be any variety of commutative residuated lattices. Then $\mathcal{V}$ has the amalgamation property if and only it has the deductive interpolation property.*

Note, however, that the variety of ℓ-groups, and hence also the variety of residuated lattices, does not have the CEP or EP. Any simple ℓ-group (recalling from Chapter 4 that congruences are in bijective correspondence with normal convex subuniverses) that has a nonsimple subalgebra yields a counterexample.

In order to apply Corollary 8.4.1, it is helpful to consider a reformulation of the DIP tailored to consequence in residuated lattices. Let $\mathcal{V}$ be any variety of residuated lattices and note first that, for any $\alpha_1, \dots, \alpha_n, \beta \in \mathrm{Fm}$,

$$\{\alpha_1, \dots, \alpha_n\} \vdash_{\mathcal{V}} \beta \iff \{\alpha_1 \wedge \cdots \wedge \alpha_n\} \vdash_{\mathcal{V}} \beta.$$

It follows easily that $\mathcal{V}$ has the DIP if and only if, for any $\alpha, \beta \in \mathrm{Fm}$ satisfying $\{\alpha\} \vdash_{\mathcal{V}} \beta$, there exists $\gamma \in \mathrm{Fm}$ with $\mathrm{Var}(\gamma) \subseteq \mathrm{Var}(\alpha) \cap \mathrm{Var}(\beta)$ such that $\{\alpha\} \vdash_{\mathcal{V}} \gamma$ and $\{\gamma\} \vdash_{\mathcal{V}} \beta$. To prove that $\mathcal{V}$ has this version of the DIP, we often establish a form of interpolation based on the underlying lattice order.

A variety $\mathcal{V}$ of residuated lattices is said to have the *Craig interpolation property* (for short, CIP) if, for any $\alpha, \beta \in \mathrm{Fm}$ satisfying $\vdash_{\mathcal{V}} \alpha \leq \beta$, there exists $\gamma \in \mathrm{Fm}$ with $\mathrm{Var}(\gamma) \subseteq \mathrm{Var}(\alpha) \cap \mathrm{Var}(\beta)$ such that $\vdash_{\mathcal{V}} \alpha \leq \gamma$ and $\vdash_{\mathcal{V}} \gamma \leq \beta$.

LEMMA 8.4.2. *Let $\mathcal{V}$ be a variety of commutative residuated lattices. If $\mathcal{V}$ has the Craig interpolation property, then it has the deductive interpolation property and amalgamation property.*

PROOF. Suppose that $\mathcal{V}$ has the CIP. By Corollary 8.4.1, it suffices to show that $\mathcal{V}$ has the DIP, so consider any $\alpha, \beta \in \mathrm{Fm}$ satisfying $\{\alpha\} \vdash_{\mathcal{V}} \beta$. By the local deduction theorem, $\vdash_{\mathcal{V}} (\alpha \wedge \mathrm{e})^n \to \beta$, for some $n \in \mathbb{N}$. Hence, by the CIP, there exists $\gamma \in \mathrm{Fm}$ with $\mathrm{Var}(\gamma) \subseteq \mathrm{Var}(\alpha) \cap \mathrm{Var}(\beta)$ such that $\vdash_{\mathcal{V}} (\alpha \wedge \mathrm{e})^n \to \gamma$ and $\vdash_{\mathcal{V}} \gamma \to \beta$. Using the local deduction theorem again twice, $\{\alpha\} \vdash_{\mathcal{V}} \gamma$ and $\{\gamma\} \vdash_{\mathcal{V}} \beta$. So $\mathcal{V}$ has the DIP. $\square$

The CIP is often established using a suitable analytic sequent calculus. Below, we illustrate this method for the variety of commutative residuated lattices, using the multiset-sequent calculus CFL presented in Figure 2.5.

THEOREM 8.4.3. *The variety of commutative residuated lattices has the Craig interpolation, deductive interpolation, and amalgamation properties.*

PROOF. By Lemma 8.4.2, it suffices to show that $\mathcal{CRL}$ has the CIP. However, by Corollary 2.4.10, this is equivalent to proving the following claim for an arbitrary multiset-sequent $\Gamma, \Delta \Rightarrow \alpha$ in the signature of residuated lattices:

If $\vdash_{\mathsf{CFL}^\circ} \Gamma, \Delta \Rightarrow \alpha$, then there exists $\beta \in \mathrm{Fm}$ (in the signature of residuated lattices) with $\mathrm{Var}(\beta) \subseteq \mathrm{Var}(\Gamma) \cap \mathrm{Var}(\Delta, \alpha)$ such that $\vdash_{\mathsf{CFL}^\circ} \Gamma \Rightarrow \beta$ and $\vdash_{\mathsf{CFL}^\circ} \Delta, \beta \Rightarrow \alpha$.

We proceed by induction on the height of a CFL°-derivation of $\Gamma, \Delta \Rightarrow \alpha$. Suppose first for the base case that $\Gamma, \Delta \Rightarrow \alpha$ is an instance of (ID). If Γ is $[\alpha]$ and Δ is $[]$, let $\beta := \alpha$. If Γ is $[]$ and Δ is $[\alpha]$, let $\beta := \mathrm{e}$. If $\Gamma, \Delta \Rightarrow \alpha$ is an instance of $(\Rightarrow \mathrm{e})$, then Γ and Δ are $[]$ and α is e, so let $\beta := \mathrm{e}$.

For the inductive step, we consider the last rule applied in the derivation. Let us just treat the paradigmatic case of implication, other cases being very similar. Suppose first that α is $\alpha_1 \to \alpha_2$ and the derivation ends with

$$\dfrac{\dfrac{\vdots}{\Gamma, \Delta, \alpha_1 \Rightarrow \alpha_2}}{\Gamma, \Delta \Rightarrow \alpha_1 \to \alpha_2.}\ (\Rightarrow\to)$$

By the induction hypothesis, there exists $\beta \in \mathrm{Fm}$ with $\mathrm{Var}(\beta) \subseteq \mathrm{Var}(\Gamma) \cap \mathrm{Var}(\Delta, \alpha)$ such that $\vdash_{\mathsf{CFL}^\circ} \Gamma \Rightarrow \beta$ and $\vdash_{\mathsf{CFL}^\circ} \Delta, \beta, \alpha_1 \Rightarrow \alpha_2$, and hence, by an application of $(\Rightarrow\to)$, also $\vdash_{\mathsf{CFL}^\circ} \Delta, \beta \Rightarrow \alpha_1 \to \alpha_2$.

Suppose now that the derivation ends with

$$\dfrac{\dfrac{\vdots}{\Gamma_1, \Delta_1 \Rightarrow \gamma_1} \quad \dfrac{\vdots}{\Gamma_2, \Delta_2, \gamma_2 \Rightarrow \alpha}}{\Gamma_1, \Gamma_2, \gamma_1 \to \gamma_2, \Delta_1, \Delta_2 \Rightarrow \alpha.}\ (\to\Rightarrow)$$

There are two subcases. First, suppose that Γ is Γ_1, Γ_2 and Δ is $\gamma_1 \to \gamma_2, \Delta_1, \Delta_2$. By the induction hypothesis twice, there exist $\beta_1, \beta_2 \in \mathrm{Fm}$ with $\mathrm{Var}(\beta_1), \mathrm{Var}(\beta_2) \subseteq \mathrm{Var}(\Gamma) \cap \mathrm{Var}(\Delta, \alpha)$ such that the following sequents are CFL°-derivable:

$$\Gamma_1 \Rightarrow \beta_1, \quad \Gamma_2 \Rightarrow \beta_2, \quad \Delta_1, \beta_1 \Rightarrow \gamma_1, \quad \text{and} \quad \Delta_2, \gamma_2, \beta_2 \Rightarrow \alpha.$$

Defining $\beta := \beta_1 \cdot \beta_2$, we obtain the required CFL°-derivations

$$\dfrac{\dfrac{\vdots}{\Gamma_1 \Rightarrow \beta_1} \quad \dfrac{\vdots}{\Gamma_2 \Rightarrow \beta_2}}{\Gamma_1, \Gamma_2 \Rightarrow \beta_1 \cdot \beta_2}\ (\Rightarrow\cdot) \qquad \dfrac{\dfrac{\dfrac{\vdots}{\Delta_1, \beta_1 \Rightarrow \gamma_1} \quad \dfrac{\vdots}{\Delta_2, \gamma_2, \beta_2 \Rightarrow \alpha}}{\Delta_1, \Delta_2, \gamma_1 \to \gamma_2, \beta_1, \beta_2 \Rightarrow \alpha}\ (\to\Rightarrow)}{\Delta_1, \Delta_2, \gamma_1 \to \gamma_2, \beta_1 \cdot \beta_2 \Rightarrow \alpha.}\ (\cdot\Rightarrow)$$

Now suppose that Γ is $\Gamma_1, \Gamma_2, \gamma_1 \to \gamma_2$ and Δ is Δ_1, Δ_2. Here we must be a bit more careful. Considering the sequent $\Gamma_1, \Delta_1 \Rightarrow \gamma_1$, we associate Γ_1 with γ_1, and obtain, by the induction hypothesis, $\beta_1 \in \mathrm{Fm}$ with $\mathrm{Var}(\beta_1) \subseteq \mathrm{Var}(\Gamma) \cap \mathrm{Var}(\Delta, \alpha)$ such that the following sequents are CFL°-derivable:

$$\Delta_1 \Rightarrow \beta_1 \quad \text{and} \quad \Gamma_1, \beta_1 \Rightarrow \gamma_1.$$

For $\Gamma_2, \Delta_2, \gamma_2 \Rightarrow \alpha$, we apply the induction hypothesis to obtain $\beta_2 \in \mathrm{Fm}$ with $\mathrm{Var}(\beta_2) \subseteq \mathrm{Var}(\Gamma) \cap \mathrm{Var}(\Delta, \alpha)$ such that the following sequents are CFL°-derivable:

$$\Gamma_2, \gamma_2 \Rightarrow \beta_2 \quad \text{and} \quad \Delta_2, \beta_2 \Rightarrow \alpha.$$

Defining $\beta := \beta_1 \to \beta_2$, we obtain the required CFL°-derivations

$$\dfrac{\dfrac{\dfrac{\vdots}{\Gamma_1, \beta_1 \Rightarrow \gamma_1} \quad \dfrac{\vdots}{\Gamma_2, \gamma_2 \Rightarrow \beta_2}}{\Gamma_1, \Gamma_2, \gamma_1 \to \gamma_2, \beta_1 \Rightarrow \beta_2}\,(\to\Rightarrow)}{\Gamma_1, \Gamma_2, \gamma_1 \to \gamma_2 \Rightarrow \beta_1 \to \beta_2}\,(\Rightarrow\to) \qquad \dfrac{\dfrac{\vdots}{\Delta_1 \Rightarrow \beta_1} \quad \dfrac{\vdots}{\Delta_2, \beta_2 \Rightarrow \alpha}}{\Delta_1, \Delta_2, \beta_1 \to \beta_2 \Rightarrow \alpha.}\,(\to\Rightarrow) \qquad \square$$

The proof of Theorem 8.4.3 can be adapted to establish the CIP — and hence also, in the presence of the CEP, the DIP and AP — for many other significant varieties of residuated lattices, including $\mathcal{RL}$, $\mathcal{IRL}$, $\mathcal{CIRL}$, and $\mathcal{B}rw$. The availability of a suitable analytic sequent calculus is crucial here, however. Indeed, the lack of such a calculus is often a good indication that a variety does not have the CIP. This is the case for any variety $\mathcal{V}$ of semilinear residuated lattices that contains the additive ordered group $\mathbf{R}$ of real numbers or its negative cone $\mathbf{R}^-$. Just observe that $\mathcal{V}$ satisfies $x \wedge (x\backslash y) \leq z \vee (z\backslash y)$, but there is no formula γ with $\mathrm{Var}(\gamma) \subseteq \{y\}$ such that $x \wedge (x\backslash y) \leq \gamma$ and $\gamma \leq z \vee (z\backslash y)$ are satisfied by $\mathcal{V}$; otherwise, evaluating x and z as $-\frac{1}{2}$, and y as -1 in $\mathbf{R}$ or $\mathbf{R}^-$ would imply $\gamma^{\mathbf{R}}(-1) = \gamma^{\mathbf{R}^-}(-1) = -\frac{1}{2}$, contradicting the fact that $\mathbb{Z}$ and $\mathbb{Z}^-$ are subuniverses of $\mathbf{R}$ and $\mathbf{R}^-$, respectively.

Nevertheless, some varieties of residuated lattices that lack the CIP do have the DIP and AP; notably, this is the case for the variety $\mathcal{AbLG}$ of Abelian ℓ-groups.

THEOREM 8.4.4. *The variety of Abelian ℓ-groups has the amalgamation property.*

PROOF. By Corollary 8.4.1, it is enough to establish that $\mathcal{AbLG}$ has the DIP. Indeed, reasoning inductively and using Proposition 5.4.11, it suffices to prove the following:

For any $\alpha \in \mathrm{Fm}$ and $x \in \mathrm{Var}(\alpha)$, there exists $\gamma \in \mathrm{Fm}$ with $\mathrm{Var}(\gamma) \subseteq \mathrm{Var}(\alpha)\backslash\{x\}$ such that, for all $\beta \in \mathrm{Fm}$ with $x \notin \mathrm{Var}(\beta)$,

$$\{\alpha\} \vdash_{\mathbf{Z}} \beta \iff \{\gamma\} \vdash_{\mathbf{Z}} \beta.$$

Using distributivity properties of $\mathbf{Z}$, we may assume that $\alpha = \alpha_1 \vee \cdots \vee \alpha_m$, where each α_i is a meet of $\{\cdot, {}^{-1}, \mathrm{e}\}$-formulas. Suppose now that we have shown, for each $i \in \{1, \ldots, m\}$, that there exists $\gamma_i \in \mathrm{Fm}$ with $\mathrm{Var}(\gamma_i) \subseteq \mathrm{Var}(\alpha)\backslash\{x\}$ such that, for all $\beta \in \mathrm{Fm}$ with $x \notin \mathrm{Var}(\beta)$,

$$\{\alpha_i\} \vdash_{\mathbf{Z}} \beta \iff \{\gamma_i\} \vdash_{\mathbf{Z}} \beta.$$

Then $\gamma := \gamma_1 \vee \cdots \vee \gamma_m$ satisfies the requirements for α.

Hence we may assume that α is a meet of $\{\cdot, {}^{-1}, \mathrm{e}\}$-formulas. Reasoning then as in the proof of Proposition 5.4.11, we may assume that α is the meet of δ_0 and zero or more of $\delta_1 x^k$ and $\delta_2 x^{-k}$, for some $k \in \mathbb{N}^{>0}$ and $\delta_0, \delta_1, \delta_2 \in \mathrm{Fm}$ with $x \notin \mathrm{Var}(\delta_0, \delta_1, \delta_2)$. We consider just the case where $\alpha = \delta_0 \wedge \delta_1 x^k \wedge \delta_2 x^{-k}$, the other cases being very similar. Let $\gamma := \delta_0 \wedge \delta_1\delta_2$, noting that $\{\alpha\} \vdash_{\mathbf{Z}} \gamma$, by some simple calculations in $\mathbf{Z}$. Now consider any $\beta \in \mathrm{Fm}$ with $x \notin \mathrm{Var}(\beta)$. If $\{\gamma\} \vdash_{\mathbf{Z}} \beta$, then, since $\{\alpha\} \vdash_{\mathbf{Z}} \gamma$, also $\{\alpha\} \vdash_{\mathbf{Z}} \beta$. For the converse, suppose that $\{\gamma\} \nvdash_{\mathbf{Z}} \beta$. Then $\varphi(\gamma) \geq 0$ and $\varphi(\beta) < 0$, for some $\varphi \in \hom(\mathbf{Fm}, \mathbf{Z})$. Hence also $\varphi(\delta_0) \geq 0$ and $\varphi(\delta_1) + \varphi(\delta_2) \geq 0$. We define $\psi \in \hom(\mathbf{Fm}, \mathbf{Z})$ by setting

$\psi(x) := \max\{-\varphi(\delta_1), \varphi(\delta_2)\}$ and $\psi(y) := k\varphi(y)$, for each variable $y \neq x$. A simple induction yields $\psi(\delta) = k\varphi(\delta)$, for any $\delta \in \mathrm{Fm}$ with $x \notin \mathrm{Var}(\delta)$. Hence $\psi(\delta_0) = k\varphi(\delta_0) \geq 0$, $\psi(\delta_1 x^k) = k\varphi(\delta_1) + k\max\{-\varphi(\delta_1), \varphi(\delta_2)\} \geq k\varphi(\delta_1) - k\varphi(\delta_1) = 0$, and $\psi(\delta_2 x^{-k}) = -k\varphi(\delta_2) + k\max\{-\varphi(\delta_1), \varphi(\delta_2)\} \geq -k\varphi(\delta_2) + k\varphi(\delta_2) = 0$. So $\psi(\alpha) \geq 0$ and $\psi(\beta) = k\varphi(\beta) < 0$, yielding $\{\alpha\} \not\vdash_{\mathbf{Z}} \beta$ as required. □

A similar methodology can be used to show that the varieties of MV-algebras and commutative, integral GMV-algebras have the DIP and hence the AP (see the bibliographical remarks for references).

In our proof of Theorem 8.4.4, we made heavy use of the fact that the variety of Abelian ℓ-groups is generated as a quasivariety by $\mathbf{Z}$ (Proposition 5.4.11). In general, to investigate the AP for a variety of semilinear residuated lattices with the CEP, we can restrict our attention to its totally ordered members. The following criterion is a direct consequence of Proposition 8.2.6 and the fact that a semilinear residuated lattice is finitely subdirectly irreducible if and only if it is totally ordered.

COROLLARY 8.4.5. *Let $\mathcal{V}$ be a variety of semilinear residuated lattices that has the congruence extension property. If every V-formation of finitely generated totally ordered members of $\mathcal{V}$ has an amalgam in $\mathcal{V}$, then $\mathcal{V}$ has the amalgamation property.*

EXAMPLE 8.4.6. Recall that the variety $\mathcal{RSA}$ of relative Stone algebras consists of Brouwerian algebras satisfying the equation $(x \to y) \vee (y \to x) \approx \mathrm{e}$. It is not hard to show (e.g., by considering finitely generated free algebras) that $\mathcal{RSA}$ is locally finite. Hence, to show that $\mathcal{RSA}$ has the AP, it suffices, by Corollary 8.4.5, to show that any V-formation $\langle \mathbf{A}, \mathbf{B}, \mathbf{C}, \varphi_B, \varphi_C \rangle$ of finite, totally ordered relative Stone algebras has an amalgam in $\mathcal{RSA}$. Without loss of generality, we may assume that $B \cap C = A$, and define a total order $\leq$ on $D := B \cup C$ that extends the orders of $\mathbf{A}$, $\mathbf{B}$, and $\mathbf{C}$. Since the residual of any totally ordered relative Stone algebra $\mathbf{L}$ satisfies $a \to^{\mathbf{L}} b = \mathrm{e}$ if $a \leq b$, and $a \to^{\mathbf{L}} b = b$ otherwise, defining such an operation on $\langle D, \leq \rangle$ yields an algebra $\mathbf{D} \in \mathcal{RSA}$ that contains both $\mathbf{B}$ and $\mathbf{C}$ as subalgebras. Hence $\mathbf{D}$ with inclusion maps is an amalgam of $\langle \mathbf{A}, \mathbf{B}, \mathbf{C}, \varphi_B, \varphi_C \rangle$ as required.

Proposition 8.1.10 supplies a further useful criterion for investigating the AP — in this case, for joins of disjoint varieties of residuated lattices. Just observe that, since residuated lattices are congruence-permutable, two varieties of residuated lattices are independent if and only if they are disjoint, by Proposition 8.1.9.

COROLLARY 8.4.7. *Let $\mathcal{V}_1$ and $\mathcal{V}_2$ be disjoint varieties of residuated lattices. Then $\mathcal{V}_1 \vee \mathcal{V}_2$ has the amalgamation property if and only if both $\mathcal{V}_1$ and $\mathcal{V}_2$ have the amalgamation property.*

For example, since $\mathcal{CGMV}$ is the join of the disjoint varieties $\mathcal{AbLG}$ and $\mathcal{CIGMV}$, by Corollary 7.2.4, and both these varieties have the AP, also $\mathcal{CGMV}$ has the AP.

We can also use Corollary 8.4.7 to establish negative results for amalgamation, exploiting the following result.

THEOREM 8.4.8. *The variety of ℓ-groups does not have the amalgamation property.*

PROOF. Given any ℓ-group $\mathbf{L}$ and totally ordered group $\mathbf{M}$, the ℓ-group $\mathbf{L} \overleftarrow{\times} \mathbf{M}$ consists of the direct product of the groups $\mathbf{L}$ and $\mathbf{M}$ equipped with the dual order of the lexicographic product of $\langle L, \leq^{\mathbf{L}} \rangle$ and $\langle M, \leq^{\mathbf{M}} \rangle$.

Let $\mathbf{B}$ be the ℓ-group with universe $(\mathbb{Z} \times \mathbb{Z}) \times \mathbb{Z}$ and group operation

$$\langle a, b, c\rangle \circ \langle u, v, w\rangle := \begin{cases} \langle a+u, b+v, c+w\rangle & \text{if } c \text{ is even;} \\ \langle a+v, b+u, c+w\rangle & \text{if } c \text{ is odd,} \end{cases}$$

equipped with the dual order of the lexicographic product of three copies of $\langle \mathbb{Z}, \leq\rangle$. Now let $\mathbf{A}$ be the ℓ-subgroup of $\mathbf{B}$ generated by $\{x, y, z^2\}$, where

$$x := \langle 1, 0, 0\rangle, \quad y := \langle 0, 1, 0\rangle, \quad \text{and} \quad z := \langle 0, 0, 1\rangle.$$

It is easily checked that $\mathbf{A}$ is isomorphic to the Abelian ℓ-group $(\mathbf{Z} \times \mathbf{Z}) \overleftarrow{\times} \mathbf{Z}$. We also consider $\mathbf{C} := (\mathbf{Z} \overleftarrow{\times} \mathbf{Z}) \times (\mathbf{Z} \overleftarrow{\times} \mathbf{Z})$ containing the elements

$$x_1 := \langle 1, 0, 0, 0\rangle, \quad y_1 := \langle 0, 0, 1, 0\rangle, \quad z_1 := \langle 0, 1, 0, 0\rangle, \quad \text{and} \quad z_2 := \langle 0, 0, 0, 1\rangle.$$

Let $\varphi_B \colon \mathbf{A} \to \mathbf{B}$ be the inclusion map, and define the homomorphism $\varphi_C \colon \mathbf{A} \to \mathbf{C}$ by setting $\varphi_C(x) := x_1$, $\varphi_C(y) := y_1$, and $\varphi_C(z^2) := z_1 z_2$. Suppose towards a contradiction that the V-formation $\langle \mathbf{A}, \mathbf{B}, \mathbf{C}, \varphi_B, \varphi_C\rangle$ has an amalgam $\langle \mathbf{D}, \psi_B, \psi_C\rangle$ in $\mathcal{LG}$. By Corollary 5.3.4, we may assume that $\mathbf{D}$ is an ℓ-subgroup of $\mathbf{Aut}(\langle \Omega, \leq\rangle)$, for some chain $\langle \Omega, \leq\rangle$.

Observe now that $\mathrm{e} = x_1 \wedge z_2 = \varphi_C(x) \wedge z_2$, and so

$$\mathrm{e} = \psi_C(\varphi_C(x)) \wedge \psi_C(z_2) = \psi_B\varphi_B(x) \wedge \psi_C(z_2) = \psi_B(x) \wedge \psi_C(z_2).$$

Hence, for some $\lambda \in \Omega$,

$$\psi_B(x)(\lambda) > \lambda \quad \text{and} \quad \psi_C(z_2)(\lambda) = \lambda.$$

Similarly, $\mathrm{e} = y_1 \wedge z_1 = \varphi_C(y) \wedge z_1$, and so

$$\mathrm{e} = \psi_C(\varphi_C(y)) \wedge \psi_C(z_1) = \psi_B\varphi_B(y) \wedge \psi_C(z_1) = \psi_B(y) \wedge \psi_C(z_1).$$

It follows that

$$\psi_B(yz) \wedge \psi_C(z_1)\psi_B(z) = \psi_B(y)\psi_B(z) \wedge \psi_C(z_1)\psi_B(z) = \psi_B(z).$$

Note also that $z^{-1}yz = x$. Using the fact that $\psi_B(x)(\lambda) > \lambda$,

$$\psi_B(z^{-1}yz)(\lambda) > \lambda, \quad \text{and so} \quad \psi_B(yz)(\lambda) > \psi_B(z)(\lambda).$$

Hence, using the previous equality,

$$\psi_B(z)(\lambda) = \psi_C(z_1)\psi_B(z)(\lambda).$$

Since $z > x$ and ψ is injective, $\psi_B(z) > \psi_B(x)$ and $\psi_B(z)(\lambda) > \lambda$. It follows that

$$\begin{aligned} \psi_B(z)(\lambda) &> \psi_C(z_1)(\lambda) \\ &= \psi_C(\varphi_C(z^2)z_2^{-1})(\lambda) \\ &= (\psi_C\varphi_C(z^2)\psi_C(z_2^{-1}))(\lambda) \\ &= \psi_C\varphi_C(z^2)(\lambda) \\ &= \psi_B(z^2)(\lambda), \end{aligned}$$

so ψ_B is not a homomorphism, a contradiction. □

Corollaries 7.2.4 and 8.4.7 now yield the following result.

COROLLARY 8.4.9. *No variety of GBL-algebras containing the class of all ℓ-groups has the amalgamation property.*

The reader may be wondering if there exists any variety of residuated lattices that has the AP but does not satisfy the commutativity law. Using Corollary 8.4.5, we show that this is the case for the variety generated by the non-commutative, four-element, totally ordered residuated lattice $\mathbf{C}_4$ defined in Example 5.1.5.

PROPOSITION 8.4.10. $\mathbb{HSP}(\mathbf{C}_4)$ *has the amalgamation property.*

PROOF. Let $\mathcal{V} := \mathbb{HSP}(\mathbf{C}_4)$. We will prove first that $\mathcal{V} = \mathbb{ISP}(\mathbf{C}_4)$ and that $\mathcal{V}$ has the EP and hence the CEP. We show that, for any $\alpha, \beta, \gamma \in \mathrm{Fm}$, defining $P_\alpha := \mathrm{e} \wedge \alpha \wedge (\mathrm{e}/(\mathrm{e}/\alpha))$,

$$\{\gamma, \alpha\} \vdash_{\mathbf{C}_4} \beta \implies \{\gamma\} \vdash_{\mathbf{C}_4} P_\alpha \backslash \beta \quad \text{and} \quad \{\gamma\} \vdash_{\mathcal{V}} P_\alpha \backslash \beta \implies \{\gamma, \alpha\} \vdash_{\mathcal{V}} \beta.$$

It then follows, using Proposition 3.2.4 for the third step, that

$$\{\alpha\} \vdash_{\mathcal{V}} \beta \implies \{\alpha\} \vdash_{\mathbf{C}_4} \beta \implies \vdash_{\mathbf{C}_4} P_\alpha \backslash \beta \implies \vdash_{\mathcal{V}} P_\alpha \backslash \beta \implies \{\alpha\} \vdash_{\mathcal{V}} \beta,$$

and so, by Propositions 3.2.2 and 3.2.5, that $\vdash_{\mathcal{V}} = \vdash_{\mathbf{C}_4} = \vdash_{\mathbb{ISP}(\mathbf{C}_4)}$ and $\mathcal{V} = \mathbb{ISP}(\mathbf{C}_4)$. These implications also show that $\mathcal{V}$ has the EP and hence the CEP.

For the first implication, suppose that $\{\gamma, \alpha\} \vdash_{\mathbf{C}_4} \beta$ and $\mathrm{e} \leq \varphi(\gamma)$, for some $\varphi \in \hom(\mathbf{Fm}, \mathbf{C}_4)$. If $\mathrm{e} \leq \varphi(\alpha)$, then $\mathrm{e} \leq \varphi(\beta)$ and hence $\mathrm{e} \leq \varphi(P_\alpha \backslash \beta)$. If $\varphi(\alpha) = \bot$, then $\varphi(P_\alpha) = \bot \leq \varphi(\beta)$, so $\mathrm{e} \leq \varphi(P_\alpha \backslash \beta)$. If $\varphi(\alpha) = a$, then $\varphi(\mathrm{e}/(\mathrm{e}/\alpha)) = \bot$, so $\varphi(P_\alpha) = \bot \leq \varphi(\beta)$ and $\mathrm{e} \leq \varphi(P_\alpha \backslash \beta)$. Suppose now for the second implication that $\{\gamma\} \vdash_{\mathcal{V}} P_\alpha \backslash \beta$. It is easy to check that $\{\alpha\} \vdash_{\mathcal{V}} P_\alpha$, and hence $\{\gamma, \alpha\} \vdash_{\mathcal{V}} \beta$.

To show that $\mathcal{V}$ has the AP, it suffices by Corollary 8.4.5, to consider V-formations of totally ordered members of $\mathcal{V}$. Note, however, that, for any totally ordered algebra $\mathbf{A} \in \mathcal{V} = \mathbb{ISP}(\mathbf{C}_4)$, the congruence Δ_A is meet-irreducible in $\mathrm{Con}(\mathbf{A})$ and hence $\mathbf{A} \in \mathbb{IS}(\mathbf{C}_4)$, by the Relativized Jónsson lemma (Theorem B.15). But $\mathbf{C}_4$ has no non-trivial proper subalgebras, so all V-formations of totally ordered members of $\mathcal{V}$ are trivial and clearly have an amalgam. Hence $\mathcal{V}$ has the AP. □

Let us note finally that many questions regarding interpolation and amalgamation are still to be resolved for prominent varieties of residuated lattices. Most notably, although $\mathcal{RL}$ has the CIP, it is not known whether it has the DIP or AP. Table 8.1 summarizes the known landscape and open problems for the selection of subvarieties of $\mathcal{RL}$ and $\mathcal{PRL}$ explored in this chapter. For references and results for other varieties, we refer to the bibliographical remarks.

8.5. Bibliographical remarks

Amalgamation was first considered for groups by Schreier [**342**] in the form of amalgamated free products. The general form of the amalgamation property was first formulated by Fraïssé [**134**], and its significance for the study of algebraic structures was demonstrated in a series of papers by Jónsson [**207–211**]. The proof that lattices have this property given in Example 8.1.1 follows a general method for constructing amalgams of ordered structures described in [**207**].

The characterizations of the amalgamation property provided by Lemma 8.1.3 and Theorem 8.1.5 (by way of Lemma 8.1.4) are due to Grätzer [**168**] and Pigozzi [**308**], respectively, and have been formulated also in the settings of model theory [**15**] and abstract algebraic logic [**98**]. Our exposition here follows [**256**] in providing an

TABLE 8.1. Properties of varieties of residuated lattices

Variety	Label	CEP	CIP	DIP	AP
Residuated lattices	$\mathcal{RL}$	no	yes	?	?
Commutative $\mathcal{RL}$	$\mathcal{CRL}$	yes	yes	yes	yes
Semilinear $\mathcal{RL}$	$\mathcal{SemRL}$	no	no	?	no
Commutative $\mathcal{SemRL}$	$\mathcal{CSemRL}$	yes	no	?	?
GBL-algebras	$\mathcal{GBL}$	no	?	?	no
Commutative $\mathcal{GBL}$	$\mathcal{CGBL}$	yes	?	?	?
Semilinear $\mathcal{GBL}$	$\mathcal{SemGBL}$	no	no	?	no
Commutative $\mathcal{SemGBL}$	$\mathcal{CSemGBL}$	yes	no	yes	yes
GMV-algebras	$\mathcal{GMV}$	no	?	?	no
Commutative $\mathcal{GMV}$	$\mathcal{CGMV}$	yes	no	yes	yes
ℓ-groups	$\mathcal{LG}$	no	?	?	no
Abelian ℓ-groups	$\mathcal{AbLG}$	yes	no	yes	yes
Integral $\mathcal{RL}$	$\mathcal{IRL}$	no	yes	?	?
Commutative $\mathcal{IRL}$	$\mathcal{CIRL}$	yes	yes	yes	yes
Semilinear $\mathcal{IRL}$	$\mathcal{SemIRL}$	no	no	?	?
Commutative $\mathcal{SemIRL}$	$\mathcal{CSemIRL}$	yes	no	?	?
Integral $\mathcal{GBL}$	$\mathcal{IGBL}$	no	?	?	no
Commutative $\mathcal{IGBL}$	$\mathcal{CIGBL}$	yes	?	?	?
Semilinear $\mathcal{IGBL}$	$\mathcal{SemIGBL}$	no	no	?	no
Commutative $\mathcal{SemIGBL}$	$\mathcal{CSemIGBL}$	yes	no	yes	yes
Integral $\mathcal{GMV}$	$\mathcal{IGMV}$	no	?	?	?
Commutative $\mathcal{IGMV}$	$\mathcal{CIGMV}$	yes	no	yes	yes
Negative cones of ℓ-groups	$\mathcal{LG}^-$	no	?	?	no
Abelian $\mathcal{LG}^-$	$\mathcal{AbLG}^-$	yes	no	yes	yes
Brouwerian algebras	$\mathcal{Brw}$	yes	yes	yes	yes
Relative Stone algebras	$\mathcal{RSA}$	yes	yes	yes	yes

account of these results in the framework of universal algebra. An algebraic proof of Proposition 8.1.6 may be found in [**171**] (see also [**169**, V.4: Corollaries 2,3]), while a version of Proposition 8.1.7 (omitting the restriction to finitely generated algebras) first appeared in [**256**]. Proposition 8.2.6, which generalizes a result of [**256**] for varieties of semilinear residuated lattices, is taken from [**137**].

The general results for independence of varieties given in Proposition 8.1.9 can be found in [**170**], and the strengthened version for congruence-permutable varieties is taken from [**212**]. For Proposition 8.1.10, we refer to [**220**].

The extension property was studied in [**292**] under the name "limited GINT" and Theorem 8.2.5 here may be viewed as a refinement of Theorem 8 from this paper; it also appears in an abstract algebraic logic setting as the "extension interpolation property" in [**98**], and as one of the model-theoretic properties considered in [**15**].

The relationship between amalgamation and deductive interpolation described by Theorem 8.3.1 first appeared in [**211**], and is credited there to unpublished work of Keisler. As observed in [**308**], the essential ideas underlying the proof may be traced back to Magnus' work in group theory. It was also already noted in [**209**] that the (seemingly stronger) weak amalgamation property implies the amalgamation property in the presence of the congruence extension property.

Theorem 8.3.3 has a number of antecedents. The equivalence of the transferable injections property and Maehara interpolation property was established in [**373**]; their equivalence to the conjunction of the deductive interpolation and congruence extension properties and the conjunction of the Robinson and extension properties were shown in [**94**] and [**292**], respectively.

Amalgamation and interpolation properties have been investigated for a broad spectrum of varieties of (pointed) residuated lattices and substructural logics (see, e.g., [**148, 217, 251, 252, 256, 271, 296, 300, 373**]), mostly following the strategies described in this chapter. In particular, the proof method used for Theorem 8.4.3 originated in Maehara's proof of interpolation for classical logic [**244**] and was first extended to substructural logics in [**296, 300, 373**].

Theorems 8.4.4 and 8.4.8 are due to Pierce [**307**]. The proof of Theorem 8.4.4 given here, and a similar proof for MV-algebras (see also [**56, 280**]) may be found in [**256**]; also provided in this paper is a full description of the varieties of commutative GMV-algebras that have the amalgamation property. Corollary 8.4.5 is taken from [**137**] (generalizing a result of [**256**]), Corollaries 8.4.7 and 8.4.9 were first stated in [**256**], and Proposition 8.4.10 first appeared in [**161**].

It was shown in [**159**] that if a variety of residuated lattices contains all ℓ-groups, then the variety containing its semilinear members does not have the amalgamation property. Although it is not known if the variety of commutative GBL-algebras — equivalently, using Corollaries 7.2.4 and 8.4.7, the variety of commutative, integral GBL-algebras — has the amalgamation property, it was shown in [**256**] (adapting a proof in [**271**] for BL-algebras) that this is the case for the varieties of commutative, semilinear GBL-algebras and commutative, semilinear, integral GBL-algebras.

Let us mention finally that "uniform" interpolation properties also admit algebraic characterizations. It is easy to show (see [**364**]) that a variety $\mathcal{V}$ has the deductive interpolation property if and only if, for any non-empty set X, finite $\Sigma \subseteq \mathrm{Eq}(X)$, and $Y \subseteq \mathrm{Var}(\Sigma)$, there exists $\Delta \subseteq \mathrm{Eq}(Y)$ such that $\Sigma \vdash_{\mathcal{V}} \varepsilon \iff \Delta \vdash_{\mathcal{V}} \varepsilon$, for any $\varepsilon \in \mathrm{Eq}(X)$ with $\mathrm{Var}(\varepsilon) \cap \mathrm{Var}(\Sigma) \subseteq Y$. If Δ in the preceding condition is required to be finite, $\mathcal{V}$ is said to have the *right uniform deductive interpolation property*. In [**221**], it is shown that $\mathcal{V}$ has this property if and only if it has the deductive interpolation property (or, in the presence of the congruence extension property, the amalgamation property) and is *coherent*: that is, any finitely generated subalgebra of a finitely presented member of $\mathcal{V}$ is finitely presented. Every locally finite variety is coherent, as are the varieties of Abelian ℓ-groups, MV-algebras, and — by a famous result of Pitts [**310**] — Heyting algebras. However, as shown in [**221**], this property fails for a large family of varieties of (pointed) residuated lattices. For further details and references, an algebraic characterization of the left uniform deductive interpolation property, and the relationship of these properties to the existence of model completions, we refer to [**221, 261, 364**].

CHAPTER 9

Finite embeddings and finite models

In this chapter, we study various *finiteness* properties of a variety and how they relate to the decidability or otherwise of its word problem and equational theory.

In Section 9.1, we provide a general account of the finite embeddability property for a variety and its relationship to residual finiteness. We then show in Section 9.2 that, if the signature is finite, this property is equivalent to the strong finite model property, and, if the variety is finitely based, it yields a decidable word problem. We also provide an algebraic characterization of the finite model property and show that it implies that a finitely based variety has a decidable equational theory.

In Sections 9.3 and 9.4, we focus again on ordered structures. First, in Section 9.3, we study join-extensions and join-completions of posets and pomonoids, and provide a description of the join-completions of a pomonoid that form a residuated lattice. Finally, in Section 9.4, we explain how this description can be used to establish the finite embeddability property, strong finite model property, and decidability of the word problem for certain varieties of integral residuated lattices.

9.1. The finite embeddability property

The main result of this section (Theorem 9.1.14) shows that the finite embeddability property for a variety of finite signature is equivalent to a structural property of its finitely presentable members known as residual finiteness.

We begin by establishing some basic terminology. Let P be any non-empty set and let $n \in \mathbb{N}$. An *n-ary partial operation* on P is a map $f\colon R \to P$ from a subset R of P^n to P. If $\langle a_1, \dots, a_n\rangle \in R$, we say that $f(a_1, \dots, a_n)$ is *defined*, and if $R = P^n$, we say that f is *total*. Note that if $n = 0$, then f is either the empty map or can be identified with exactly one element of P.

Now let $\mathcal{L}$ be any signature, recalling that $\mathcal{L}_n$ denotes the set of operation symbols of $\mathcal{L}$ of arity $n \in \mathbb{N}$. A *partial $\mathcal{L}$-algebra* $\mathbf{P}$ consists of a non-empty set P and an n-ary partial operation $f^{\mathbf{P}}$ on P, for each $f \in \mathcal{L}_n$ ($n \in \mathbb{N}$). If every partial operation of $\mathbf{P}$ is total, then $\mathbf{P}$ is an $\mathcal{L}$-algebra in the usual sense employed in this book.

The standard notions of a homomorphism and subalgebra extend to this setting. Let $\mathbf{P}$ and $\mathbf{Q}$ be any partial $\mathcal{L}$-algebras. A map $\varphi\colon P \to Q$ is called a *homomorphism* from $\mathbf{P}$ to $\mathbf{Q}$, written $\varphi\colon \mathbf{P} \to \mathbf{Q}$, if, for any $f \in \mathcal{L}_n$ ($n \in \mathbb{N}$) and $\langle a_1, \dots, a_n\rangle \in P^n$ such that $f^{\mathbf{P}}(a_1, \dots, a_n)$ is defined, also $f^{\mathbf{Q}}(\varphi(a_1), \dots, \varphi(a_n))$ is defined and equal to $\varphi(f^{\mathbf{P}}(a_1, \dots, a_n))$. If, additionally, φ is bijective and φ^{-1} is a homomorphism, φ is called an *isomorphism*, and $\mathbf{P}$ and $\mathbf{Q}$ are said to be *isomorphic*, written $\mathbf{P} \cong \mathbf{Q}$.

An injective homomorphism $\varphi\colon \mathbf{P} \to \mathbf{Q}$ is called an *embedding* of $\mathbf{P}$ into $\mathbf{Q}$. If $P \subseteq Q$ and $\varphi\colon \mathbf{P} \to \mathbf{Q}$ is the inclusion homomorphism, $\mathbf{P}$ is called a *partial subalgebra* of $\mathbf{Q}$. A partial subalgebra $\mathbf{P}$ of $\mathbf{Q}$ called *full* if, for any $f \in \mathcal{L}_n$ $(n \in \mathbb{N})$ and $a_1, \dots, a_n \in P$, whenever $f^{\mathbf{Q}}(a_1, \dots, a_n)$ is defined and belongs to P, also $f^{\mathbf{P}}(a_1, \dots, a_n)$ is defined. In this case, $\mathbf{P}$ is determined by the set P, and denoted by $\mathbf{Q}{\restriction}P$.

Given a homomorphism $\varphi\colon \mathbf{P} \to \mathbf{Q}$, the partial subalgebra $\varphi[\mathbf{P}]$ of $\mathbf{Q}$ consists of the set $\varphi[P]$ together with partial operations $f^{\varphi[\mathbf{P}]}$ such that, for each $f \in \mathcal{L}_n$ $(n \in \mathbb{N})$ and $a_1, \dots, a_n \in P$, whenever $f^{\mathbf{P}}(a_1, \dots, a_n)$ is defined,

$$f^{\varphi[\mathbf{P}]}(\varphi(a_1), \dots, \varphi(a_n)) := \varphi(f^{\mathbf{P}}(a_1, \dots, a_n)).$$

The map $\varphi\colon \mathbf{P} \to \varphi[\mathbf{P}]$ is a surjective homomorphism, and an isomorphism if φ is an embedding. An embedding $\varphi\colon \mathbf{P} \to \mathbf{Q}$ is said to be *full* if $\varphi[\mathbf{P}]$ is a full partial subalgebra of $\mathbf{Q}$; this is the case if and only if $\mathbf{P} \cong \varphi[\mathbf{P}] = \mathbf{Q}{\restriction}\varphi[P]$.

Now consider any variety $\mathcal{V}$, letting $\mathcal{V}_F$ denote the class of finite members of $\mathcal{V}$. An algebra $\mathbf{A} \in \mathcal{V}$ is said to have the *finite embeddability property* (for short, FEP) in $\mathcal{V}$ if every finite partial subalgebra of $\mathbf{A}$ embeds into some member of $\mathcal{V}_F$. The variety $\mathcal{V}$ is said to have the FEP if every member of $\mathcal{V}$ has the FEP in $\mathcal{V}$.

EXAMPLE 9.1.1. Recall that an algebra is locally finite if all of its finitely generated subalgebras are finite, and a variety is locally finite if each of its members is locally finite. In any locally finite variety $\mathcal{V}$, every finite partial subalgebra of a member of $\mathcal{V}$ generates, and hence embeds into, a member of $\mathcal{V}_F$. Hence all locally finite varieties — for example, the varieties of distributive lattices, Boolean algebras, and relative Stone algebras — has the FEP.

EXAMPLE 9.1.2. The variety of Brouwerian algebras is not locally finite, but does have the FEP. Let $\mathbf{P}$ be a finite partial subalgebra of a Brouwerian algebra $\mathbf{B}$. Since the variety of distributive lattices is locally finite, the distributive lattice generated by $P \cup \{\mathrm{e}\}$ is finite, even if the Brouwerian algebra generated by this set is infinite. Moreover, since the meet operation of any finite distributive lattice is residuated, this distributive lattice forms a finite Brouwerian algebra $\mathbf{D}$. Finally, it is easily checked that the partial residual operation of $\mathbf{P}$ coincides, where defined, with the residual operation of $\mathbf{D}$, so $\mathbf{P}$ embeds into $\mathbf{D}$.

EXAMPLE 9.1.3. The variety of lattices is not locally finite, but has the FEP. Let $\mathbf{P}$ be a finite partial sublattice of a lattice $\mathbf{L}$, and let $\mathbf{M}$ be the Dedekind-MacNeille completion (see Section 9.3) of the poset $\langle P, \leq^{\mathbf{P}}\rangle$, where $\leq^{\mathbf{P}} := P^2 \cap \leq^{\mathbf{L}}$; that is, $\mathbf{M}$ consists of the finite set of subsets X of P satisfying $(X^u)^l = X$, ordered by set-inclusion, where Y^l and Y^u denote the sets of lower bounds and upper bounds of $Y \subseteq P$, respectively. Clearly, $a \mapsto \{a\}^l$ is an embedding of $\mathbf{P}$ into $\mathbf{M}$.

EXAMPLE 9.1.4. Since every non-trivial, cancellative residuated lattice is infinite (see Lemma 6.1.2.(d)), no non-trivial variety of cancellative residuated lattices can have the FEP. In particular, there is no non-trivial variety of ℓ-groups or negative cones of ℓ-groups that has the FEP.

EXAMPLE 9.1.5. By Lemma 7.1.3, every finite, semilinear GBL-algebra is integral and commutative. Hence no variety of semilinear GBL-algebras with a nonintegral or noncommutative member — notably, the varieties of semilinear GBL-algebras, semilinear, integral GBL-algebras, and integral GMV-algebras — have the FEP.

Let us now consider a further property that is sometimes used instead of the FEP. We say that an algebra $\mathbf{A}$ of a variety $\mathcal{V}$ has the *full finite embeddability property* (for short, FEP^+) in $\mathcal{V}$ if every finite full partial subalgebra of $\mathbf{A}$ fully embeds into some $\mathbf{B} \in \mathcal{V}_F$, and that $\mathcal{V}$ has the FEP^+ if every $\mathbf{A} \in \mathcal{V}$ has the FEP^+ in $\mathcal{V}$.

The next lemma clarifies the relationship between the FEP^+ and FEP.

LEMMA 9.1.6. *Let $\mathcal{V}$ be any variety of $\mathcal{L}$-algebras and let $\mathbf{A} \in \mathcal{V}$.*

(a) *If $\mathbf{A}$ has the FEP^+ in $\mathcal{V}$, then $\mathbf{A}$ has the FEP in $\mathcal{V}$.*
(b) *If $\mathcal{L}$ is finite and $\mathbf{A}$ has the FEP in $\mathcal{V}$, then $\mathbf{A}$ has the FEP^+ in $\mathcal{V}$.*

PROOF.

(a) Suppose that $\mathbf{A}$ has the FEP^+ in $\mathcal{V}$, and let $\mathbf{P}$ be any finite partial subalgebra of $\mathbf{A}$. The identity map is an embedding of $\mathbf{P}$ into the finite full partial subalgebra $\mathbf{A}{\restriction}P$ of $\mathbf{A}$. But by assumption, there exists a full embedding of $\mathbf{A}{\restriction}P$ into some $\mathbf{B} \in \mathcal{V}_F$, so also $\mathbf{P}$ embeds into $\mathbf{B}$.

(b) Suppose that $\mathcal{L}$ is finite and that $\mathbf{A}$ has the FEP in $\mathcal{V}$, and let $\mathbf{P}$ be any finite full partial subalgebra of $\mathbf{A}$. Define

$$Q := P \cup \{f^{\mathbf{A}}(a_1, \dots, a_n) \mid n \in \mathbb{N},\ f \in \mathcal{L}_n, \text{ and } a_1, \dots, a_n \in P\}.$$

Since $\mathcal{L}$ and P are finite, so is Q. Since $\mathbf{A}$ has the FEP in $\mathcal{V}$, there exists an embedding φ of $\mathbf{Q} := \mathbf{A}{\restriction}Q$ into some $\mathbf{B} \in \mathcal{V}_F$. Our goal is to prove that the map $\varphi{\restriction}_P \colon P \to B;\ a \mapsto \varphi(a)$ is a full embedding of $\mathbf{P}$ into $\mathbf{B}$. Clearly, $\varphi{\restriction}_P$ is an embedding, so it remains to show that $\varphi{\restriction}_P[\mathbf{P}]$ is a full partial subalgebra of $\mathbf{B}$. Consider any $n \in \mathbb{N}$, $f \in \mathcal{L}_n$, and $b_1, \dots, b_n \in \varphi[P]$ such that $f^{\mathbf{B}}(b_1, \dots, b_n) \in \varphi[P]$. Then $\varphi(a_1) = b_1, \dots,\ \varphi(a_n) = b_n$, and $\varphi(a) = f^{\mathbf{B}}(b_1, \dots, b_n)$, for some elements $a_1, \dots, a_n, a \in P$. Also, by definition, $f^{\mathbf{A}}(a_1, \dots, a_n) \in Q$. So $f^{\mathbf{Q}}(a_1, \dots, a_n) = f^{\mathbf{A}}(a_1, \dots, a_n)$, and, since φ is an embedding of $\mathbf{Q}$ into $\mathbf{B}$,

$$\varphi(f^{\mathbf{Q}}(a_1, \dots, a_n)) = f^{\mathbf{B}}(\varphi(a_1), \dots, \varphi(a_n)) = f^{\mathbf{B}}(b_1, \dots, b_n) = \varphi(a).$$

Hence $a = f^{\mathbf{Q}}(a_1, \dots, a_n) = f^{\mathbf{A}}(a_1, \dots, a_n)$, by the injectivity of φ. But $\mathbf{P}$ is a full partial subalgebra of $\mathbf{A}$, so $f^{\mathbf{P}}(a_1, \dots, a_n)$ is defined and equal to a. It follows therefore, as required, that $f^{\varphi{\restriction}_P[\mathbf{P}]}(b_1, \dots, b_n)$ is also defined. □

The finite embeddability property for a variety $\mathcal{V}$ of finite signature can be described equivalently via a property of its finitely presentable members. To justify this claim, we first introduce the notion of a presentation of an algebra in $\mathcal{V}$ and describe a useful relationship between different presentations of the same algebra.

Let $\langle X \mid \Sigma \rangle$ denote an ordered pair consisting of a set X such that $X \cup \mathcal{L}_0 \neq \emptyset$ and a set of equations $\Sigma \subseteq \mathrm{Eq}(X)$. Recall that $\overline{\Sigma} = \{\langle \overline{\alpha}, \overline{\beta} \rangle \mid \alpha \approx \beta \in \Sigma\}$, where $\overline{\gamma}$ denotes the image of $\gamma \in \mathrm{Fm}(X)$ under the canonical map from $\mathbf{Fm}(X)$ onto the $\mathcal{V}$-free algebra $\mathbf{F}(X)$ over $\overline{X}$. We call $\langle X \mid \Sigma \rangle$ a *presentation (in $\mathcal{V}$)* of $\mathbf{A} \in \mathcal{V}$ if $\mathbf{A}$ is isomorphic to $\mathbf{F}(X)/\mathrm{Cg}^{\mathbf{F}(X)}(\overline{\Sigma})$.

A presentation $\langle X \mid \Sigma \rangle$ in $\mathcal{V}$ of $\mathbf{A} \in \mathcal{V}$ is called *finite* if X and Σ are both finite, and $\mathbf{A} \in \mathcal{V}$ is called *finitely presentable (in $\mathcal{V}$)* if there is a finite presentation of $\mathbf{A}$ in $\mathcal{V}$.

LEMMA 9.1.7. *Let $\mathcal{V}$ be any variety and let $\langle X \mid \Sigma \rangle$ be a presentation in $\mathcal{V}$ of $\mathbf{A} \in \mathcal{V}$. Suppose that Z and Γ are sets obtained from X and Σ by applying the operations below or their inverses:*

(I) *If $\Pi \subseteq \mathrm{Eq}(X)$ satisfies $\Sigma \vdash_{\mathcal{V}} \Pi$, let $Z := X$ and $\Gamma := \Sigma \cup \Pi$.*

(II) *If Y is a set disjoint from X and $\psi\colon Y \to \mathrm{Fm}(X)$ is any map, let $Z := X \cup Y$ and $\Gamma := \Sigma \cup \{y \approx \psi(y) \mid y \in Y\}$.*

Then $\langle Z \mid \Gamma \rangle$ is a presentation in $\mathcal{V}$ of $\mathbf{A}$. Moreover, defining $\Theta := \mathrm{Cg}^{\mathbf{F}(X)}(\overline{\Sigma})$ and $\Psi := \mathrm{Cg}^{\mathbf{F}(Z)}(\overline{\Gamma})$, there exists an isomorphism $\varphi\colon \mathbf{F}(X)/\Theta \to \mathbf{F}(Z)/\Psi$; $[\overline{\alpha}]_\Theta \mapsto [\overline{\alpha}]_\Psi$.

PROOF. For (I), it follows from the assumptions $Z = X$, $\Sigma \vdash_{\mathcal{V}} \Gamma$, and $\Gamma \vdash_{\mathcal{V}} \Sigma$ that $\Theta = \Psi$. So the identity map is the required isomorphism.

For (II), let $\rho\colon \mathbf{Fm}(Z) \to \mathbf{Fm}(X)$ be the surjective homomorphism that maps each $x \in X$ to x and each $y \in Y$ to $\psi(y)$, and let $\widehat{\rho}\colon \mathbf{F}(Z) \to \mathbf{F}(X)$ be the homomorphism mapping each $z \in Z$ to $\overline{\rho(z)}$. Now let $\Gamma_\psi := \{y \approx \psi(y) \mid y \in Y\}$, so $\Gamma = \Sigma \cup \Gamma_\psi$. Let $\pi_\Theta\colon \mathbf{F}(X) \to \mathbf{F}(X)/\Theta$ and $\pi_\Psi\colon \mathbf{F}(Z) \to \mathbf{F}(Z)/\Psi$ be the canonical maps, and let $\pi := \pi_\Theta\widehat{\rho}$. We obtain the solid part of the following diagram:

$$\begin{array}{ccc}
\mathbf{Fm}(X) & \overset{\rho}{\twoheadleftarrow} & \mathbf{Fm}(Z) \\
\downarrow & & \downarrow \\
\mathbf{F}(X) & \overset{\widehat{\rho}}{\twoheadleftarrow} & \mathbf{F}(Z) \\
{\scriptstyle \pi_\Theta}\downarrow & \overset{\pi}{\swarrow} & \downarrow{\scriptstyle \pi_\Psi} \\
\mathbf{F}(X)/\Theta & \overset{\varphi}{\dashrightarrow} & \mathbf{F}(Z)/\Psi.
\end{array}$$

Clearly, $\overline{\Sigma} \subseteq \ker(\pi_\Theta)$ and $\widehat{\rho}(\overline{\alpha}) = \overline{\alpha}$, for all $\overline{\alpha} \in \mathrm{F}(X)$, so $\overline{\Sigma} \subseteq \ker(\pi)$. But also $\widehat{\rho}(y) = \overline{\psi(y)} = \widehat{\rho}(\overline{\psi(y)})$, for each $y \in Y$, so $\overline{\Gamma}_\psi \subseteq \ker(\pi)$. Hence $\Psi \subseteq \ker(\pi)$. Now let $\langle \overline{\alpha}, \overline{\beta} \rangle \in \ker(\pi)$. Since $\widehat{\rho}$ is a homomorphism satisfying $\langle z, \widehat{\rho}(z) \rangle \in \Psi$, for each $z \in Z$, also $\langle \overline{\alpha}, \widehat{\rho}(\overline{\alpha}) \rangle, \langle \overline{\beta}, \widehat{\rho}(\overline{\beta}) \rangle \in \Psi$. Moreover, $\langle \widehat{\rho}(\overline{\alpha}), \widehat{\rho}(\overline{\beta}) \rangle \in \Theta \subseteq \Psi$, so $\langle \overline{\alpha}, \overline{\beta} \rangle \in \Psi$. Hence $\ker(\pi) = \Psi$ and the Homomorphism theorem (Corollary B.3) yields an isomorphism $\varphi\colon \mathbf{F}(X)/\Theta \to \mathbf{F}(Z)/\Psi$; $[\overline{\alpha}]_\Theta \mapsto [\overline{\alpha}]_\Psi$. □

LEMMA 9.1.8. *Let $\mathcal{V}$ be any variety. Given two presentations $\langle X \mid \Sigma \rangle$ and $\langle Y \mid \Delta \rangle$ in $\mathcal{V}$ of an algebra $\mathbf{A} \in \mathcal{V}$, one can be obtained from the other by applying the operations of type (I) and (II) and their inverses described in Lemma 9.1.7.*

PROOF. Let $\Theta := \mathrm{Cg}^{\mathbf{F}(X)}(\overline{\Sigma})$, $\Phi := \mathrm{Cg}^{\mathbf{F}(Y)}(\overline{\Delta})$, and $Z := X \cup Y$, assuming without loss of generality that $X \cap Y = \emptyset$. Assuming that $\langle X \mid \Sigma \rangle$ and $\langle Y \mid \Delta \rangle$ are both presentations in $\mathcal{V}$ of $\mathbf{A} \in \mathcal{V}$, there exists an isomorphism $\psi\colon \mathbf{F}(Y)/\Phi \to \mathbf{F}(X)/\Theta$. Now, for each $y \in Y$, there exists $\alpha_y \in \mathrm{Fm}(X)$ such that $\psi([\overline{y}]_\Phi) = [\overline{\alpha}_y]_\Theta$. Let $\Gamma := \Sigma \cup \{y \approx \alpha_y \mid y \in Y\}$ and $\Psi := \mathrm{Cg}^{\mathbf{F}(Z)}(\overline{\Gamma})$. By Lemma 9.1.7, there exists an isomorphism $\varphi\colon \mathbf{F}(X)/\Theta \to \mathbf{F}(Z)/\Psi$; $[\overline{\alpha}]_\Theta \mapsto [\overline{\alpha}]_\Psi$. Moreover, $\varphi\psi([\overline{y}]_\Phi) = \varphi([\overline{\alpha}_y]_\Theta) = [\overline{\alpha}_y]_\Psi = [\overline{y}]_\Psi$, for each $y \in Y$, so $\varphi\psi([\overline{\alpha}]_\Phi) = [\overline{\alpha}]_\Psi$, for all $\alpha \in \mathrm{Fm}(Y)$. Hence, if $\langle \alpha, \beta \rangle \in \Delta$, then $[\overline{\alpha}]_\Psi = \varphi\psi([\overline{\alpha}]_\Phi) = \varphi\psi([\overline{\beta}]_\Phi) = [\overline{\beta}]_\Psi$, that is, $\overline{\Delta} \subseteq \Psi$. So $\Gamma \vdash_{\mathcal{V}} \Delta$ and it follows, by Lemma 9.1.7, that $\langle Z \mid \Gamma \cup \Delta \rangle$ is a presentation of $\mathbf{A}$ in $\mathcal{V}$.

Next, for each $x \in X$, there exists $\beta_x \in \mathrm{Fm}(Y)$ such that $[\overline{\beta}_x]_\Phi = \psi^{-1}([\overline{x}]_\Theta)$. Let $\Pi := \{x \approx \beta_x \mid x \in X\}$. Then $[\overline{x}]_\Psi = \varphi([\overline{x}]_\Theta) = \varphi\psi([\overline{\beta}_x]_\Phi) = [\overline{\beta}_x]_\Psi$, for each

$x \in X$, so $\overline{\Pi} \subseteq \Psi \subseteq \mathrm{Cg}^{\mathbf{F}(Z)}(\overline{\Gamma} \cup \overline{\Delta})$. Hence $\Gamma \cup \Delta \vdash_{\mathcal{V}} \Pi$ and, by Lemma 9.1.7 again, $\langle Z \,|\, \Gamma \cup \Delta \cup \Pi\rangle$ is a presentation of $\mathbf{A}$ in $\mathcal{V}$. Moreover, we have shown that $\langle Z \,|\, \Gamma \cup \Delta \cup \Pi\rangle$ is derived from $\langle X \,|\, \Sigma\rangle$ with the use of operations of type (I) and (II). Since $\langle Z \,|\, \Gamma \cup \Delta \cup \Pi\rangle$ is also derived, analogously, from $\langle Y \,|\, \Delta\rangle$ with the use of these operations, $\langle Y \,|\, \Delta\rangle$ is derived from $\langle X \,|\, \Sigma\rangle$ with the use of type (I) and (II) operations and their inverses, and vice versa. □

Sometimes it can be useful to consider presentations of a certain simple form. Let us call a presentation $\langle X \,|\, \Sigma\rangle$ in a variety $\mathcal{V}$ of an algebra $\mathbf{A} \in \mathcal{V}$ *flat* if the equations in Σ are of the form $f(x_1, \ldots, x_n) \approx x$, for some $f \in \mathcal{L}_n$ and $x_1, \ldots, x_n, x \in X$.

LEMMA 9.1.9. *Let $\mathcal{V}$ be any variety of finite signature and suppose that $\mathbf{A} \in \mathcal{V}$ is finitely presentable in $\mathcal{V}$. Then there exists a flat finite presentation of $\mathbf{A}$ in $\mathcal{V}$.*

PROOF. Let $\langle X \,|\, \Sigma\rangle$ be a finite presentation of $\mathbf{A}$ in $\mathcal{V}$, and let S be the set of all subformulas of formulas occurring in Σ. For each $\alpha \in S$, define $\hat{\alpha} := \alpha$ if $\alpha \in \mathcal{L}_0 \cup X$, and otherwise let $\hat{\alpha}$ be a new variable, so that $Y := \{\hat{\alpha} \mid \alpha \in S \backslash (\mathcal{L}_0 \cup X)\}$ is disjoint from X and $\hat{\alpha} \neq \hat{\beta}$ if $\alpha \neq \beta$. Let $\Gamma := \{\hat{\alpha} \approx \alpha \mid \hat{\alpha} \in Y\}$. Then $\langle X \cup Y \,|\, \Sigma \cup \Gamma\rangle$ is a finite presentation of $\mathbf{A}$ in $\mathcal{V}$, by Lemma 9.1.7. Let

$$\Sigma_1 := \{f(\hat{\alpha}_1, \ldots, \hat{\alpha}_n) \approx \hat{\alpha} \mid \alpha = f(\alpha_1, \ldots, \alpha_n) \in S\}$$
$$\Sigma_2 := \{\hat{\alpha} \approx \hat{\beta} \mid \alpha \approx \beta \in \Sigma\}.$$

It follows by an easy induction on formula complexity that $\Gamma \vdash_{\mathcal{V}} f(\hat{\alpha}_1, \ldots, \hat{\alpha}_n) \approx \hat{\alpha}$ and $\Sigma_1 \vdash_{\mathcal{V}} \hat{\alpha} \approx \alpha$, for all $\alpha = f(\alpha_1, \ldots, \alpha_n) \in S$. So $\Gamma \vdash_{\mathcal{V}} \Sigma_1$ and $\Sigma_1 \vdash_{\mathcal{V}} \Gamma$, and hence $\Sigma \cup \Gamma \vdash_{\mathcal{V}} \Sigma_1 \cup \Sigma_2$ and $\Sigma_1 \cup \Sigma_2 \vdash_{\mathcal{V}} \Sigma \cup \Gamma$. By Lemma 9.1.7 again, $\langle X \cup Y \,|\, \Sigma_1 \cup \Sigma_2\rangle$ is a finite presentation of $\mathbf{A}$ in $\mathcal{V}$. Now let $Z := X \cup Y$ and $\Delta := \Sigma_1 \cup \Sigma_2$. If Δ contains an equation $z_1 \approx z_2$ with $z_1, z_2 \in Z$, we remove z_2 from Z and $z_1 \approx z_2$ from Δ, and substitute uniformly all occurrences of z_2 in Δ by z_1. We repeat this process until we obtain a flat presentation $\langle Z \,|\, \Delta\rangle$ of $\mathbf{A}$ in $\mathcal{V}$. □

We now consider a generalization of the notion of a free algebra over a set to a free algebra over a partial algebra. Let $\mathcal{V}$ be any variety of $\mathcal{L}$-algebras. A *$\mathcal{V}$-free algebra over a partial $\mathcal{L}$-algebra* $\mathbf{P}$ consists of an algebra $\mathbf{A} \in \mathcal{V}$ and a homomorphism $\eta\colon \mathbf{P} \to \mathbf{A}$ such that, for any $\mathbf{B} \in \mathcal{V}$ and homomorphism $\varphi\colon \mathbf{P} \to \mathbf{B}$, there exists a unique homomorphism $\overline{\varphi}\colon \mathbf{A} \to \mathbf{B}$ satisfying $\varphi = \overline{\varphi}\eta$. Note that in the case where the operations of $\mathbf{P}$ are everywhere undefined, the algebra $\mathbf{A}$ is a $\mathcal{V}$-free algebra over the set P (see Appendix B).

It is straightforward to prove that any two $\mathcal{V}$-free algebras over a partial $\mathcal{L}$-algebra $\mathbf{P}$ are isomorphic, so let us just show now that there exists at least one such algebra. Let $\hat{}\colon P \to \hat{P}$ be a bijection between P and a set (of variables) $\hat{P}$, and define

$$\mathrm{Diag}\,\mathbf{P} := \{f(\hat{a}_1, \ldots, \hat{a}_n) \approx \hat{a} \mid n \in \mathbb{N},\ f \in \mathcal{L}_n,\ a_1, \ldots, a_n, a \in P, \text{ and } f^{\mathbf{P}}(a_1, \ldots, a_n) = a\}.$$

We define $\mathbf{F}(\mathbf{P}) := \mathbf{F}(\hat{P})/\mathrm{Cg}^{\mathbf{F}(\hat{P})}(\overline{\mathrm{Diag}\,\mathbf{P}})$, recalling that $\mathbf{F}(\hat{P})$ is the $\mathcal{V}$-free algebra over the set $\hat{P}$, and let $[\hat{a}]$ denote the element $[\hat{a}]_{\mathrm{Cg}^{\mathbf{F}(\hat{P})}(\overline{\mathrm{Diag}\,\mathbf{P}})}$ of $\mathbf{F}(\mathbf{P})$, for $a \in P$.

LEMMA 9.1.10. *Let $\mathcal{V}$ be any variety of $\mathcal{L}$-algebras and $\mathbf{P}$ any partial $\mathcal{L}$-algebra. Then $\mathbf{F}(\mathbf{P})$, together with the homomorphism $\eta\colon \mathbf{P} \to \mathbf{F}(\mathbf{P})$; $a \mapsto [\hat{a}]$, is the unique (up to isomorphism) $\mathcal{V}$-free algebra over $\mathbf{P}$.*

PROOF. Observe first that if $f^{\mathbf{P}}(a_1, \ldots, a_n) = a$ is defined for $n \in \mathbb{N}$, $f \in \mathcal{L}_n$, and $a_1, \ldots, a_n, a \in P$, then $f(\hat{a}_1, \ldots, \hat{a}_n) \approx \hat{a} \in \operatorname{Diag} \mathbf{P}$ and

$$\begin{aligned}\eta(f^{\mathbf{P}}(a_1, \ldots, a_n)) &= [\hat{a}] \\ &= [\overline{f(\hat{a}_1, \ldots, \hat{a}_n)}] \\ &= f^{\mathbf{F}(\mathbf{P})}([\hat{a}_1], \ldots, [\hat{a}_n]) \\ &= f^{\mathbf{F}(\mathbf{P})}(\eta(a_1), \ldots, \eta(a_n)).\end{aligned}$$

So η is a homomorphism. It remains to show that $\mathbf{F}(\mathbf{P})$ and η satisfy the desired universal property. Consider any $\mathbf{B} \in \mathcal{V}$, homomorphism $\varphi\colon \mathbf{P} \to \mathbf{B}$, and the maps $\psi\colon P \to \mathrm{F}(\hat{P})$; $a \mapsto \hat{a}$ and $\pi\colon \mathbf{F}(\hat{P}) \to \mathbf{F}(\mathbf{P})$; $\alpha \mapsto [\alpha]$. Note that $\pi\psi = \eta$. Since $\mathbf{F}(\hat{P})$ is the $\mathcal{V}$-free algebra over $\hat{P}$, there is a unique homomorphism $\widetilde{\varphi}\colon \mathbf{F}(\hat{P}) \to \mathbf{B}$ such that $\widetilde{\varphi}\psi = \varphi$, i.e., the outer part of the following diagram commutes:

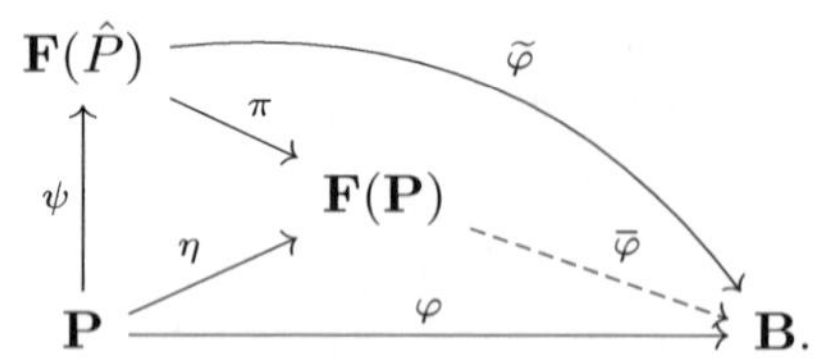

Observe next that if $f(\hat{a}_1, \ldots, \hat{a}_n) \approx \hat{a} \in \operatorname{Diag} \mathbf{P}$, then $f^{\mathbf{P}}(a_1, \ldots, a_n) = a$, and hence, since φ is a homomorphism,

$$\begin{aligned}\widetilde{\varphi}(\overline{f(\hat{a}_1, \ldots, \hat{a}_n)}) &= f^{\mathbf{B}}(\widetilde{\varphi}(\hat{a}_1), \ldots, \widetilde{\varphi}(\hat{a}_n)) \\ &= f^{\mathbf{B}}(\varphi(a_1), \ldots, \varphi(a_n)) \\ &= \varphi(f^{\mathbf{P}}(a_1, \ldots, a_n)) \\ &= \varphi(a) \\ &= \widetilde{\varphi}(\hat{a}),\end{aligned}$$

so $\langle \overline{f(\hat{a}_1, \ldots, \hat{a}_n)}, \hat{a} \rangle \in \ker(\widetilde{\varphi})$. That is, $\overline{\operatorname{Diag} \mathbf{P}} \subseteq \ker(\widetilde{\varphi})$. So $\ker(\pi) \subseteq \ker(\widetilde{\varphi})$ and, by the General homomorphism theorem (Theorem B.2), there exists a unique homomorphism $\overline{\varphi}\colon \mathbf{F}(\mathbf{P}) \to \mathbf{B}$; $[\overline{\alpha}] \mapsto \widetilde{\varphi}(\overline{\alpha})$ satisfying $\overline{\varphi}\pi = \widetilde{\varphi}$. But then also $\overline{\varphi}$ is the unique homomorphism satisfying $\varphi = \overline{\varphi}\eta$. $\square$

The following result provides a useful criterion for the embeddability of a partial algebra into an algebra belonging to some given variety.

PROPOSITION 9.1.11. *Let $\mathcal{V}$ be any variety of $\mathcal{L}$-algebras. A partial $\mathcal{L}$-algebra $\mathbf{P}$ embeds into an algebra in $\mathcal{V}$ if and only if the homomorphism $\eta\colon \mathbf{P} \to \mathbf{F}(\mathbf{P})$; $a \mapsto [\hat{a}]$ is an embedding.*

PROOF. The right-to-left direction is immediate. For the left-to-right direction, suppose that φ is an embedding of $\mathbf{P}$ into some $\mathbf{A} \in \mathcal{V}$. Then there exists a unique homomorphism $\overline{\varphi}\colon \mathbf{F}(\mathbf{P}) \to \mathbf{A}$ satisfying $\varphi = \overline{\varphi}\eta$. Since φ is injective, it follows that the same is true of η. $\square$

We now provide a further characterization of the finitely presentable algebras of a variety of finite signature.

PROPOSITION 9.1.12. *Let $\mathcal{V}$ be any variety of $\mathcal{L}$-algebras of finite signature. The following statements are equivalent for any algebra $\mathbf{A} \in \mathcal{V}$:*

(1) *There exists a finite subset $Q \subseteq A$ such that whenever $Q \subseteq P \subseteq A$, the algebra $\mathbf{A}$, together with the homomorphism $\psi\colon \mathbf{A}{\restriction}P \to \mathbf{A};\ a \mapsto a$, is a $\mathcal{V}$-free algebra over $\mathbf{A}{\restriction}P$.*

(2) *$\mathbf{A}$ is a $\mathcal{V}$-free algebra over some finite partial $\mathcal{L}$-algebra.*

(3) *$\mathbf{A}$ is finitely presentable in $\mathcal{V}$.*

PROOF.

(1) $\Rightarrow$ (2) Immediate.

(2) $\Rightarrow$ (3) If $\mathbf{A} = \mathbf{F}(\mathbf{P})$ is a $\mathcal{V}$-free algebra over some finite partial algebra $\mathbf{P}$, then $\langle \hat{P} \mid \mathrm{Diag}\,\mathbf{P}\rangle$ is a finite presentation of $\mathbf{A}$ in $\mathcal{V}$.

(3) $\Rightarrow$ (1) Suppose that $\mathbf{A}$ is finitely presentable in $\mathcal{V}$. By Corollary 9.1.9, there exists a finite flat presentation $\langle X \mid \Sigma\rangle$ of $\mathbf{A}$ in $\mathcal{V}$. Let $\pi_{\mathbf{A}}\colon \mathbf{F}(X) \to \mathbf{A}$ be a surjective homomorphism with kernel $\Theta := \mathrm{Cg}^{\mathbf{F}(X)}(\overline{\Sigma})$, and define $Q := \pi_{\mathbf{A}}[X]$. Suppose that $Q \subseteq P \subseteq A$ and let $\mathbf{P} := \mathbf{A}{\restriction}P$. We show that $\mathbf{A}$, together with the homomorphism $\psi\colon \mathbf{P} \to \mathbf{A};\ a \mapsto a$, is a $\mathcal{V}$-free algebra over $\mathbf{P}$. Consider any homomorphism φ from $\mathbf{P}$ to some $\mathbf{B} \in \mathcal{V}$, and let $\widetilde{\varphi}\colon \mathbf{F}(X) \to \mathbf{B}$ be the unique homomorphism satisfying $\widetilde{\varphi}(x) = \varphi\pi_{\mathbf{A}}(x)$, for each $x \in X$, as shown in the following diagram:

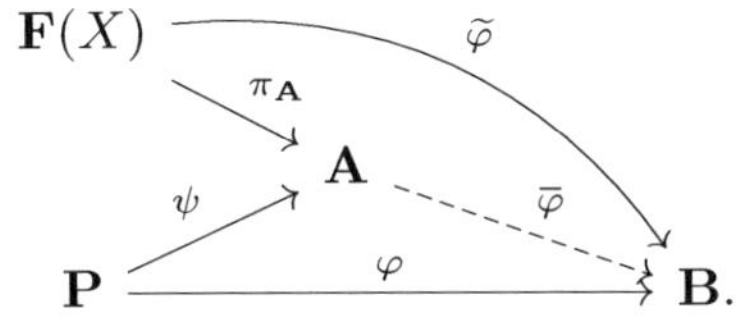

Now consider any $f(x_1, \dots x_n) \approx x \in \Sigma$ with $x_1, \dots, x_n, x \in X$. It follows that $f^{\mathbf{A}}(\pi_{\mathbf{A}}(x_1), \dots, \pi_{\mathbf{A}}(x_n)) = \pi_{\mathbf{A}}(x)$ and, since $\pi_{\mathbf{A}}(x_1), \dots, \pi_{\mathbf{A}}(x_n), \pi_{\mathbf{A}}(x) \in P$ and $\mathbf{P}$ is a full partial subalgebra of $\mathbf{A}$, also $f^{\mathbf{P}}(\pi_{\mathbf{A}}(x_1), \dots, \pi_{\mathbf{A}}(x_n)) = \pi_{\mathbf{A}}(x)$. Using the fact that φ and $\widetilde{\varphi}$ are homomorphisms,

$$\begin{aligned}
\widetilde{\varphi}(\overline{f(x_1, \dots, x_n)}) &= f^{\mathbf{B}}(\widetilde{\varphi}(x_1), \dots, \widetilde{\varphi}(x_n)) \\
&= f^{\mathbf{B}}(\varphi\pi_{\mathbf{A}}(x_1), \dots, \varphi\pi_{\mathbf{A}}(x_n)) \\
&= \varphi(f^{\mathbf{P}}(\pi_{\mathbf{A}}(x_1), \dots, \pi_{\mathbf{A}}(x_n))) \\
&= \varphi\pi_{\mathbf{A}}(x) \\
&= \widetilde{\varphi}(x),
\end{aligned}$$

that is, $\langle \overline{f(x_1, \dots, x_n)}, x\rangle \in \ker(\widetilde{\varphi})$. We have shown that $\overline{\Sigma} \subseteq \ker(\widetilde{\varphi})$, so $\Theta \subseteq \ker(\widetilde{\varphi})$. Hence, by the General homomorphism theorem (Theorem B.2), there exists a unique homomorphism $\overline{\varphi}\colon \mathbf{A} \to \mathbf{B}$ satisfying $\overline{\varphi}\pi_{\mathbf{A}} = \widetilde{\varphi}$. Since $\varphi\colon \mathbf{P} \to \mathbf{B}$ is a homomorphism, a simple inductive argument shows that $\widetilde{\varphi}(\overline{\alpha}) = \varphi\pi_{\mathbf{A}}(\overline{\alpha})$, for any $\alpha \in \mathrm{Fm}(X)$ with $\pi_{\mathbf{A}}(\overline{\alpha}) \in P$. It follows that $\varphi = \overline{\varphi}\psi$. Finally, the uniqueness of $\overline{\varphi}$ follows from the fact that Q generates $\mathbf{A}$. $\square$

Let $\mathcal{V}$ be any variety. An algebra $\mathbf{A} \in \mathcal{V}$ is said to be *residually finite (in $\mathcal{V}$)* if it is a subdirect product of a family of finite algebras in $\mathcal{V}$. The following equivalent characterization is left as an easy exercise for the reader.

LEMMA 9.1.13. *An algebra* $\mathbf{A}$ *is residually finite in a variety* $\mathcal{V}$ *if and only if, for every pair of distinct elements* $a, b \in A$, *there exist a finite algebra* $\mathbf{C} \in \mathcal{V}$ *and a homomorphism* $\varphi\colon \mathbf{A} \to \mathbf{C}$ *such that* $\varphi(a) \neq \varphi(b)$.

Finally, we prove the main result of this section.

THEOREM 9.1.14. *Let* $\mathcal{V}$ *be any variety of finite signature. The following statements are equivalent:*

(1) $\mathcal{V}$ *has the finite embeddability property.*
(2) $\mathcal{V}$ *has the full finite embeddability property.*
(3) *Every finite partial subalgebra of a finitely presentable algebra in* $\mathcal{V}$ *can be embedded into a finite member of* $\mathcal{V}$.
(4) *Every finitely presentable member of* $\mathcal{V}$ *is residually finite in* $\mathcal{V}$.

PROOF.

(1) $\Leftrightarrow$ (2) Follows directly from Lemma 9.1.6.

(2) $\Rightarrow$ (3) Follows by specialization.

(3) $\Rightarrow$ (4) Assume that (3) holds, and let $\mathbf{A}$ be any finitely presentable member of $\mathcal{V}$. Consider any $a, b \in A$ with $a \neq b$. By Proposition 9.1.12, there exists a finite partial subalgebra $\mathbf{P}$ of $\mathbf{A}$ with $a, b \in P$ such that $\mathbf{A}$, together with the homomorphism $\psi\colon \mathbf{P} \to \mathbf{A};\ c \mapsto c$, is a $\mathcal{V}$-free algebra over $\mathbf{P}$. By assumption, there exists an embedding φ of $\mathbf{P}$ into some $\mathbf{C} \in \mathcal{V}_F$. Since $\mathbf{A}$ is a $\mathcal{V}$-free algebra over $\mathbf{P}$, the homomorphism φ extends to a homomorphism $\overline{\varphi}\colon \mathbf{A} \to \mathbf{C}$ satisfying $\overline{\varphi}\psi = \varphi$. So $\overline{\varphi}(a) \neq \overline{\varphi}(b)$. Hence $\mathbf{A}$ is residually finite in $\mathcal{V}$, by Lemma 9.1.13.

(4) $\Rightarrow$ (1) Suppose that every finitely presentable member of $\mathcal{V}$ is residually finite in $\mathcal{V}$, and let $\mathbf{P}$ be any partial subalgebra of some $\mathbf{A} \in \mathcal{V}$. Then $\mathbf{P}$ is a partial subalgebra of $\mathbf{F}(\mathbf{P})$, by Proposition 9.1.11. Also, $\mathbf{F}(\mathbf{P})$ is finitely presentable in $\mathcal{V}$, by Proposition 9.1.12. Let $\psi\colon \mathbf{P} \to \mathbf{F}(\mathbf{P})$ be the inclusion homomorphism. By assumption and Lemma 9.1.13, there exist, for all $a, b \in P$ with $a \neq b$, some $\mathbf{C}_{a,b} \in \mathcal{V}_F$ and a homomorphism $\varphi_{a,b}\colon \mathbf{F}(\mathbf{P}) \to \mathbf{C}_{a,b}$ such that $\varphi_{a,b}(a) \neq \varphi_{a,b}(b)$. These homomorphisms induce a homomorphism $\varphi\colon \mathbf{F}(\mathbf{P}) \to \prod_{a \neq b} \mathbf{C}_{a,b}$. Clearly, $\varphi\psi\colon \mathbf{P} \to \prod_{a \neq b} \mathbf{C}_{a,b}$ is an embedding into a finite algebra. □

Note that it can be shown that the finiteness of the signature of $\mathcal{V}$ in Theorem 9.1.14 is necessary (see the bibliographical remarks).

9.2. Finite model properties

In this section, we show that the FEP for a variety of finite signature is equivalent to a property of equational consequence referred to as the strong finite model property (Theorem 9.2.8). We also provide an algebraic characterization of the (weaker) finite model property for such varieties and explore how these properties can be used to establish decidability results for finitely based varieties.

A variety $\mathcal{V}$ is said to have the *strong finite model property* (for short, SFMP) if $\Sigma \models_{\mathcal{V}} \varepsilon \iff \Sigma \models_{\mathcal{V}_F} \varepsilon$, for all finite $\Sigma \cup \{\varepsilon\} \subseteq \mathrm{Eq}$, and the *finite model property* (for short, FMP) if $\vdash_{\mathcal{V}} \varepsilon \iff \vdash_{\mathcal{V}_F} \varepsilon$, for all $\varepsilon \in \mathrm{Eq}$. Equivalently, it follows from Propositions 3.2.2, 3.2.4, and 3.2.5, and Corollary 3.2.8 that $\mathcal{V}$ has the SFMP if and only if $\mathcal{V} = \mathbb{ISPP}_U(\mathcal{V}_F)$, and the FMP if and only if $\mathcal{V} = \mathbb{HSP}(\mathcal{V}_F)$.

EXAMPLE 9.2.1. In Theorem 9.2.8, we will prove that a variety has the SFMP if it has the FEP, from which it follows, for example, that all locally finite varieties and the varieties of Brouwerian algebras and lattices have the SFMP. On the other hand, every non-trivial, cancellative residuated lattice is infinite, so no non-trivial variety of cancellative residuated lattices has the SFMP or even the FMP.

The following proposition provides a useful tool for proving that a variety does not have the SFMP.

PROPOSITION 9.2.2. *Let $\mathcal{V}$ be any variety such that each $\mathbf{A} \in \mathcal{V}$ has a semigroup reduct $\langle A, \cdot \rangle$ and semilattice reduct $\langle A, \vee \rangle$, and satisfies $x(y \vee z) \approx xy \vee xz$ and $(y \vee z)x \approx yx \vee zx$. Suppose also that $\langle \mathbb{Z}, \max, + \rangle$ is the $\{\vee, \cdot\}$-reduct of some $\mathbf{C} \in \mathcal{V}$. Then $\mathcal{V}$ does not have the strong finite model property.*

PROOF. Observe first that $1 \leq 1+1$ and $(1+1)+((-1)+(-1)) = 1+(-1)$ in $\mathbf{C}$, but $1+(-1) \neq (1+1)+(-1)$, and therefore

$$\{x \leq x^2,\ x^2y^2 \approx xy\} \nvdash_{\mathcal{V}} xy \approx x^2y.$$

Now consider any $\mathbf{A} \in \mathcal{V}_F$ and $a, b \in A$ such that $a \leq a^2$ and $a^2b^2 = ab$. Since $a \leq a^2$, it follows that $a \leq a^2 \leq a^3 \leq \ldots$ and hence, because $\mathbf{A}$ is finite, that $a^n = a^{n+1}$, for some $n \in \mathbb{N}$. Moreover, repeated applications of $a^2b^2 = ab$ yield $ab = a^nb^n = a^{n+1}b^n = a(a^nb^n) = aab = a^2b$ and hence

$$\{x \leq x^2,\ x^2y^2 \approx xy\} \vdash_{\mathcal{V}_F} xy \approx x^2y.$$

So $\mathcal{V}$ does not have the SFMP. □

EXAMPLE 9.2.3. It follows from Proposition 9.2.2 that no variety of residuated lattices containing the ordered group of integers — in particular, the varieties of commutative residuated lattices, semilinear residuated lattices, and GBL-algebras — can have the SFMP.

Clearly, if a variety has the SFMP, it must have the FMP, but the following example shows that the converse does not always hold.

EXAMPLE 9.2.4. The fact that the variety of distributive ℓ-monoids has the FMP was established in Corollary 5.4.4. However, since $\langle \mathbb{Z}, \min, \max, +, 0 \rangle$ is a member of this variety, it cannot have the SFMP, by Proposition 9.2.2.

The following result relates the FMP for a variety $\mathcal{V}$ to the residual finiteness of $\mathcal{V}$-free algebras in $\mathcal{V}$.

PROPOSITION 9.2.5. *The following statements are equivalent for any variety $\mathcal{V}$:*

(1) *$\mathcal{V}$ has the* FMP.
(2) *Every $\mathcal{V}$-free algebra is residually finite in $\mathcal{V}$.*
(3) *$\mathbf{F}(X)$ is residually finite in $\mathcal{V}$, for any countably infinite set X.*
(4) *Every finitely generated $\mathcal{V}$-free algebra is residually finite in $\mathcal{V}$.*

PROOF.

(1) $\Rightarrow$ (2) Suppose that $\mathcal{V}$ has the FMP. Let X be any set such that $X \cup \mathcal{L}_0 \neq \emptyset$. We show that $\mathbf{F}(X)$ is residually finite in $\mathcal{V}$ using Lemma 9.1.13. Consider any $\alpha, \beta \in \mathrm{Fm}(X)$ such that $\overline{\alpha} \neq \overline{\beta}$. Then $\not\models_{\mathcal{V}} \alpha \approx \beta$ and, by assumption, there exist $\mathbf{C} \in \mathcal{V}_F$ and a homomorphism $\psi\colon \mathbf{Fm}(X) \to \mathbf{C}$ such that $\psi(\alpha) \neq \psi(\beta)$. Now let $\varphi\colon \mathbf{F}(X) \to \mathbf{C}$ be the unique homomorphism extending $\psi{\restriction}X$. Then $\psi(\gamma) = \varphi(\overline{\gamma})$, for all $\gamma \in \mathrm{Fm}(X)$, so $\varphi(\overline{\alpha}) \neq \varphi(\overline{\beta})$ as required.

(2) $\Rightarrow$ (3) Immediate.

(3) $\Rightarrow$ (4) Follows from the fact that every finitely generated $\mathcal{V}$-free algebra embeds into $\mathbf{F}(X)$, for any countably infinite set X.

(4) $\Rightarrow$ (1) Assume (4) and consider any finite set X and $\alpha, \beta \in \mathrm{Fm}(X)$ such that $\not\models_{\mathcal{V}} \alpha \approx \beta$. Then $\overline{\alpha} \neq \overline{\beta}$ and, by (4) and Lemma 9.1.13, there exists a homomorphism φ from $\mathbf{F}(X)$ to some $\mathbf{C} \in \mathcal{V}_F$ such that $\varphi(\overline{\alpha}) \neq \varphi(\overline{\beta})$. So $\psi(\alpha) \neq \psi(\beta)$, where $\psi\colon \mathbf{Fm}(X) \to \mathbf{C}$ is the unique homomorphism extending $\varphi{\restriction}X$, and $\not\models_{\mathcal{V}_F} \alpha \approx \beta$. $\square$

A variety $\mathcal{V}$ is said to have a *decidable equational theory* if there exists an effective procedure for deciding $\models_{\mathcal{V}} \varepsilon$, for any $\varepsilon \in \mathrm{Eq}$.

PROPOSITION 9.2.6. *Let $\mathcal{V}$ be any finitely based variety. If $\mathcal{V}$ has the* FMP, *then $\mathcal{V}$ has a decidable equational theory.*

PROOF. Let $\mathcal{V}$ be any finitely based variety that has the FMP. Since $\mathcal{V}$ is finitely based, we may assume without loss of generality that the signature $\mathcal{L}$ is finite and that $\mathcal{V}$ is axiomatized by a finite set $T_{\mathcal{V}}$ of first-order $\mathcal{L}$-sentences. Let $\varepsilon \in \mathrm{Eq}(X)$, for some finite set X. To decide $\models_{\mathcal{V}} \varepsilon$, we run two processes in parallel.

Note first that $\models_{\mathcal{V}} \varepsilon$ if and only if ε is derivable from $T_{\mathcal{V}}$ in some calculus for first-order logic with equality. Process 1 constructs, for each $n \in \mathbb{N}$, the finitely many derivations from $T_{\mathcal{V}}$ of maximum length n built from first-order $\mathcal{L}$-formulas of maximum length n with object variables in $\{x_1, \dots, x_n\}$. If one of these derivations ends with ε, then Process 1 outputs "true". At the same time, Process 2 constructs, for each $n \in \mathbb{N}$, the finitely many (up to isomorphism) $\mathcal{L}$-algebras with n elements. If one of these $\mathcal{L}$-algebras satisfies $T_{\mathcal{V}}$ (and hence belongs to $\mathcal{V}$) but does not satisfy ε, then Process 2 outputs "false".

Process 1 outputs "true" if and only if $\models_{\mathcal{V}} \varepsilon$, and, since $\mathcal{V}$ has the FMP, Process 2 outputs "false" if and only if $\not\models_{\mathcal{V}} \varepsilon$. Hence running these two processes in parallel provides an effective procedure for deciding $\models_{\mathcal{V}} \varepsilon$. $\square$

We now turn our attention to the relationship between the SFMP and the FEP.

Let $\mathbf{P}$ be any partial $\mathcal{L}$-algebra and X any set such that $X \cup \mathcal{L}_0 \neq \emptyset$. For any map $\varphi\colon X \to P$, the *valuation* $\tilde{\varphi}\colon \mathrm{Fm}(X) \to P$ is a partial map defined recursively by

(a) $\tilde{\varphi}(x) := \varphi(x)$, for each $x \in X$.

(b) For each $f \in \mathcal{L}_n$ ($n \in \mathbb{N}$) and any $\alpha_1, \dots, \alpha_n \in \mathrm{Fm}(X)$, if $\tilde{\varphi}(\alpha_1), \dots, \tilde{\varphi}(\alpha_n)$ and $f^{\mathbf{P}}(\tilde{\varphi}(\alpha_1), \dots, \tilde{\varphi}(\alpha_n))$ are defined, then
$$\tilde{\varphi}(f(\alpha_1, \dots, \alpha_n)) := f^{\mathbf{P}}(\tilde{\varphi}(\alpha_1), \dots, \tilde{\varphi}(\alpha_n)),$$
otherwise $\tilde{\varphi}(f(\alpha_1, \dots, \alpha_n))$ is undefined.

Note that if $\mathbf{P}$ is an $\mathcal{L}$-algebra, then the map $\tilde{\varphi}$ is the unique homomorphism from $\mathbf{Fm}(X)$ to $\mathbf{P}$ extending φ.

LEMMA 9.2.7. *Let $\chi\colon \mathbf{P} \to \mathbf{Q}$ be any homomorphism between partial $\mathcal{L}$-algebras and X any set such that $X \cup \mathcal{L}_0 \neq \emptyset$. Then, for any map $\varphi\colon X \to P$ and $\psi := \chi\varphi$, whenever $\tilde{\varphi}(\alpha)$ is defined for some $\alpha \in \mathrm{Fm}(X)$, also $\tilde{\psi}(\alpha) = \chi(\tilde{\varphi}(\alpha))$ is defined.*

PROOF. We prove the claim by induction on the definition of α. If $\alpha = x \in X$, then $\tilde{\psi}(x) = \psi(x) = \chi(\varphi(x)) = \chi(\tilde{\varphi}(x))$ is defined. Now suppose that $\alpha = f(\alpha_1, \ldots, \alpha_n)$ and $\tilde{\varphi}(\alpha)$ is defined. Then $\tilde{\varphi}(\alpha_1), \ldots, \tilde{\varphi}(\alpha_n)$ and $f^{\mathbf{P}}(\tilde{\varphi}(\alpha_1), \ldots, \tilde{\varphi}(\alpha_n))$ are defined and also, by the induction hypothesis, $\tilde{\psi}(\alpha_i) = \chi(\tilde{\varphi}(\alpha_i))$, for each $i \in \{1, \ldots, n\}$. Since χ is a homomorphism, $f^{\mathbf{Q}}(\chi(\tilde{\varphi}(\alpha_1)), \ldots, \chi(\tilde{\varphi}(\alpha_n)))$ is defined, and hence also

$$\begin{aligned}\tilde{\psi}(\alpha) &= f^{\mathbf{Q}}(\tilde{\psi}(\alpha_1), \ldots, \tilde{\psi}(\alpha_n)) \\ &= f^{\mathbf{Q}}(\chi(\tilde{\varphi}(\alpha_1)), \ldots, \chi(\tilde{\varphi}(\alpha_n))) \\ &= \chi(f^{\mathbf{P}}(\tilde{\varphi}(\alpha_1), \ldots, \tilde{\varphi}(\alpha_n))) \\ &= \chi(\tilde{\varphi}(\alpha)).\end{aligned}$$ □

THEOREM 9.2.8. *Let $\mathcal{V}$ be any variety of $\mathcal{L}$-algebras.*

(a) *If $\mathcal{V}$ has the finite embeddability property, then $\mathcal{V}$ has the strong finite model property.*

(b) *If $\mathcal{L}$ is finite and $\mathcal{V}$ has the strong finite model property, then $\mathcal{V}$ has the full finite embeddability property.*

PROOF.

(a) Suppose that $\mathcal{V}$ has the FEP. To prove that $\mathcal{V}$ has the SFMP, it suffices to show that, for any finite set X and finite $\Sigma \cup \{\varepsilon\} \subseteq \mathrm{Eq}(X)$,

$$\Sigma \not\models_{\mathcal{V}} \varepsilon \implies \Sigma \not\models_{\mathcal{V}_F} \varepsilon.$$

Assume that $\Sigma \not\models_{\mathcal{V}} \varepsilon$; that is, there exist $\mathbf{A} \in \mathcal{V}$ and a map $\varphi\colon X \to A$ such that $\Sigma \subseteq \ker(\tilde{\varphi})$ and $\varepsilon \notin \ker(\tilde{\varphi})$. Consider the finite set

$$B := \{\tilde{\varphi}(\alpha) \mid \alpha \text{ is a subformula of a formula occurring in } X \cup \Sigma \cup \{\varepsilon\}\},$$

and let $\mathbf{P} := \mathbf{A}{\restriction}B$. By the FEP, there exists an embedding χ of $\mathbf{P}$ into some $\mathbf{C} \in \mathcal{V}_F$. Let $\psi := \chi\varphi$. Then $\tilde{\psi}$ is the unique homomorphism from $\mathbf{Fm}(X)$ to $\mathbf{C}$ extending ψ. Moreover, for any $\alpha, \beta \in \mathbf{Fm}(X)$, it follows from Lemma 9.2.7 that $\tilde{\psi}(\alpha) = \chi(\tilde{\varphi}(\alpha))$ and $\tilde{\psi}(\beta) = \chi(\tilde{\varphi}(\beta))$, and hence, by the injectivity of χ,

$$\tilde{\varphi}(\alpha) = \tilde{\varphi}(\beta) \iff \tilde{\psi}(\alpha) = \tilde{\psi}(\beta).$$

In particular, $\Sigma \subseteq \ker(\tilde{\psi})$ and $\varepsilon \notin \ker(\tilde{\psi})$, so $\Sigma \not\models_{\mathcal{V}_F} \varepsilon$ as required.

(b) Suppose now that $\mathcal{L}$ is finite and that $\mathcal{V}$ has the SFMP. Let $\mathbf{B}$ be a finite partial subalgebra of some $\mathbf{A} \in \mathcal{V}$. Since $\mathbf{B}$ and $\mathcal{L}$ are finite, also $\mathrm{Diag}\,\mathbf{B} \subseteq \mathrm{Eq}(\hat{B})$ is finite. Let $\varphi\colon \hat{B} \to B$ be the map sending each $\hat{b} \in \hat{B}$ to $b \in B$. Since $f^{\mathbf{A}}(b_1, \ldots, b_n) = f^{\mathbf{B}}(b_1, \ldots, b_n) = b$, for any $f(\hat{b}_1, \ldots, \hat{b}_n) \approx \hat{b} \in \mathrm{Diag}\,\mathbf{B}$,

$$\tilde{\varphi}(f(\hat{b}_1, \ldots, \hat{b}_n)) = f^{\mathbf{B}}(\varphi(\hat{b}_1), \ldots \varphi(\hat{b}_n)) = f^{\mathbf{B}}(b_1, \ldots, b_n) = b = \tilde{\varphi}(\hat{b}).$$

That is, $\mathrm{Diag}\,\mathbf{B} \subseteq \ker(\tilde{\varphi})$. Moreover, $\langle \hat{b}, \hat{b}' \rangle \notin \ker(\tilde{\varphi})$, for any distinct $b, b' \in B$, so

$$\mathrm{Diag}\,\mathbf{B} \not\models_{\mathcal{V}} \hat{b} \approx \hat{b}'.$$

Since $\mathcal{V}$ has the SFMP, it follows that there exists, for any distinct $b, b' \in B$, an algebra $\mathbf{C}_{b,b'} \in \mathcal{V}_F$ such that $\mathrm{Diag}\,\mathbf{B} \not\models_{\mathbf{C}_{b,b'}} \hat{b} \approx \hat{b}'$. Let $\mathbf{C}$ be the direct product of the algebras $\mathbf{C}_{b,b'}$ for distinct $b, b' \in B$. Clearly, $\mathbf{C} \in \mathcal{V}_F$.

We define a homomorphism $\chi\colon \mathbf{B} \to \mathbf{C}$ and prove that it is an embedding. Consider any distinct $b, b' \in B$. Since $\mathrm{Diag}\,\mathbf{B} \not\models_{\mathbf{C}_{b,b'}} \hat{b} \approx \hat{b}'$, there exists a map $\psi_{b,b'} : \hat{B} \to C_{b,b'}$ such that $\mathrm{Diag}\,\mathbf{B} \subseteq \ker(\tilde{\psi}_{b,b'})$ and $\langle \hat{b}, \hat{b}' \rangle \notin \ker(\tilde{\psi}_{b,b'})$. Let $\chi_{b,b'} : B \to C$ be the map that sends each $b \in B$ to $\psi_{b,b'}(\hat{b}) \in C$. Then $\chi_{b,b'}$ is a homomorphism from $\mathbf{B}$ to $\mathbf{C}_{b,b'}$, since, for any $f \in \mathcal{L}_n$ and $b_1, \ldots, b_n, b \in B$ such that $f^{\mathbf{B}}(b_1, \ldots, b_n) = b$, we have $f(\hat{b}_1, \ldots, \hat{b}_n) \approx \hat{b} \in \mathrm{Diag}\,\mathbf{B}$, and hence

$$\begin{aligned}
\chi_{b,b'}(f^{\mathbf{B}}(b_1, \ldots, b_n)) &= \chi_{b,b'}(b) \\
&= \tilde{\psi}_{b,b'}(\hat{b}) \\
&= \tilde{\psi}_{b,b'}(f(\hat{b}_1, \ldots, \hat{b}_n)) \\
&= f^{\mathbf{C}_{b,b'}}(\tilde{\psi}_{b,b'}(\hat{b}_1), \ldots, \tilde{\psi}_{b,b'}(\hat{b}_n)) \\
&= f^{\mathbf{C}_{b,b'}}(\chi_{b,b'}(b_1), \ldots, \chi_{b,b'}(b_n)).
\end{aligned}$$

Now let $\chi\colon \mathbf{B} \to \mathbf{C}$ be the unique homomorphism satisfying $\chi_{b,b'} = \pi_{b,b'}\chi$, for each projection map $\pi_{b,b'} : \mathbf{C} \to \mathbf{C}_{b,b'}$. Observe that if $b \neq b'$, then $\chi_{b,b'}(b) \neq \chi_{b,b'}(b')$, and hence $\chi(b) \neq \chi(b')$. So χ is an embedding. □

Combining Theorem 9.2.8 with Theorem 9.1.14 yields the following result.

COROLLARY 9.2.9. *The following statements are equivalent for any variety $\mathcal{V}$ of finite signature:*

(1) *$\mathcal{V}$ has the strong finite model property.*
(2) *$\mathcal{V}$ has the full finite embeddability property.*
(3) *$\mathcal{V}$ has the finite embeddability property.*
(4) *Every finitely presentable member of $\mathcal{V}$ is residually finite in $\mathcal{V}$.*

Let $\mathcal{V}$ be any variety. A finite presentation $\langle X \mid \Sigma \rangle$ of an algebra in $\mathcal{V}$ is said to have a *decidable word problem* if there exists an effective procedure for deciding $\bar{\varepsilon} \in \mathrm{Cg}^{\mathbf{F}(X)}(\overline{\Sigma})$, for any $\varepsilon \in \mathrm{Eq}(X)$. If this is the case for every finite presentation of an algebra in $\mathcal{V}$, then $\mathcal{V}$ is said to have a *decidable word problem.*

Recall also that $\bar{\varepsilon} \in \mathrm{Cg}^{\mathbf{F}(X)}(\overline{\Sigma})$ if and only if $\Sigma \vdash_{\mathcal{V}} \varepsilon$, for any $\Sigma \cup \{\varepsilon\} \subseteq \mathrm{Eq}(X)$, by Proposition 3.2.7. Hence $\mathcal{V}$ has a decidable word problem if and only if there exists an effective procedure for deciding $\Sigma \vdash_{\mathcal{V}} \varepsilon$, for any finite $\Sigma \cup \{\varepsilon\} \subseteq \mathrm{Eq}$, that is, if $\mathcal{V}$ has a decidable quasi-equational theory.

PROPOSITION 9.2.10. *Let $\mathcal{V}$ be any finitely based variety. If $\mathcal{V}$ has the strong finite model property, then $\mathcal{V}$ has a decidable word problem.*

PROOF. Let $\mathcal{V}$ be any finitely based variety that has the SFMP, and let $\langle X \mid \Sigma \rangle$ be a finite presentation of an algebra in $\mathcal{V}$. Since $\mathcal{V}$ is finitely based and Σ is finite, we may assume without loss of generality that the signature $\mathcal{L}$ is finite and that $\mathcal{V}$ is axiomatized by a finite set $T_{\mathcal{V}}$ of first-order $\mathcal{L}$-sentences. Let $\varepsilon \in \mathrm{Eq}(X)$, for some finite set X. To decide $\Sigma \vdash_{\mathcal{V}} \varepsilon$, we run two processes in parallel.

Note first that $\Sigma \vdash_{\mathcal{V}} \varepsilon$ if and only if the quasi-equation $\Sigma \implies \varepsilon$ is derivable from $T_{\mathcal{V}}$ in some calculus for first-order logic with equality. Process 1 constructs, for each $n \in \mathbb{N}$, the finitely many derivations from $T_{\mathcal{V}}$ of maximum length n built from first-order $\mathcal{L}$-formulas of maximum length n with object variables in $\{x_1, \ldots, x_n\}$. If one of these derivations ends with $\Sigma \implies \varepsilon$, then Process 1 outputs "true". At the same time, Process 2 constructs, for each $n \in \mathbb{N}$, the finitely many (up to isomorphism) $\mathcal{L}$-algebras with n elements. If one of these $\mathcal{L}$-algebras satisfies $T_{\mathcal{V}}$ (and hence belongs to $\mathcal{V}$) but does not satisfy ε, then Process 2 outputs "false".

Process 1 outputs "true" if and only if $\Sigma \vdash_{\mathcal{V}} \varepsilon$, and, since $\mathcal{V}$ has the SFMP, Process 2 outputs "false" if and only if $\Sigma \nvdash_{\mathcal{V}} \varepsilon$. Hence running these two processes in parallel provides an effective procedure for deciding $\Sigma \vdash_{\mathcal{V}} \varepsilon$. □

COROLLARY 9.2.11. *Every finitely based variety that has the finite embeddability property has a decidable word problem.*

Note that there are varieties such as the variety of Abelian ℓ-groups that have a decidable word problem (Corollary 5.4.12) but do not have the finite embeddability property or even the finite model property.

9.3. Join-extensions and join-completions

Whereas the two previous sections had a distinct universal algebraic flavor, in the rest of this chapter, we focus on the main objects of interest in this book: namely, residuated structures. In this section, we consider some order-theoretic properties of residuated pomonoids, related to *join-extensions* and *join-completions* of the underlying posets. The main result is Theorem 9.3.7, which provides a description of the join-completions of a pomonoid $\mathbf{P}$ that form a residuated lattice with respect to a (unique) multiplication extending the multiplication of $\mathbf{P}$. This description will then be used in the next section to establish the FEP for certain varieties of integral residuated lattices.

A poset $\mathbf{L}$ is called a *join-extension* of a poset $\mathbf{P}$ if $\mathbf{P}$ is a subposet of $\mathbf{L}$ and every element of $\mathbf{L}$ is a join of elements of $\mathbf{P}$; if $\mathbf{L}$ is also a complete lattice, it is called a *join-completion* of $\mathbf{P}$. *Meet-extensions* and *meet-completions* are defined dually.

LEMMA 9.3.1. *Let* $\mathbf{P}$ *be any poset.*

(a) *A join-extension* $\mathbf{L}$ *of* $\mathbf{P}$ *preserves all existing meets in* $\mathbf{P}$. *That is, if* $X \subseteq P$ *and* $\bigwedge^{\mathbf{P}} X$ *exists, then* $\bigwedge^{\mathbf{L}} X$ *exists and* $\bigwedge^{\mathbf{P}} X = \bigwedge^{\mathbf{L}} X$.
(b) *A meet-extension* $\mathbf{L}$ *of* $\mathbf{P}$ *preserves all existing joins in* $\mathbf{P}$. *That is, if* $X \subseteq P$ *and* $\bigvee^{\mathbf{P}} X$ *exists, then* $\bigvee^{\mathbf{L}} X$ *exists and* $\bigvee^{\mathbf{P}} X = \bigvee^{\mathbf{L}} X$.

PROOF. We only prove (a), since (b) then follows by duality. Suppose that $\bigwedge^{\mathbf{P}} X$ exists, for some $X \subseteq P$. We need to prove that $\bigwedge^{\mathbf{L}} X$ exists and $\bigwedge^{\mathbf{P}} X = \bigwedge^{\mathbf{L}} X$. Set $a := \bigwedge^{\mathbf{P}} X$ and let $b \in L$ be any lower bound of X. Since $\mathbf{L}$ is a join-extension of $\mathbf{P}$, there exists some $Y \subseteq P$ such that $b = \bigvee^{\mathbf{L}} Y$. Since b is both a lower bound of X and an upper bound of Y, each $y \in Y$ is also a lower bound of X in $\mathbf{P}$ and $y \leq \bigwedge^{\mathbf{P}} X = a$. So a is an upper bound of Y and $b = \bigvee^{\mathbf{L}} Y \leq a$. But b was an arbitrary lower bound of X in $\mathbf{L}$, so a is the greatest lower bound of X in $\mathbf{L}$. That is, $\bigwedge^{\mathbf{L}} X = a = \bigwedge^{\mathbf{P}} X$. □

One standard construction of a join-completion of a poset $\mathbf{P}$ involves taking the complete lattice $\mathcal{L}(\mathbf{P})$ of order-ideals of $\mathbf{P}$ (Example 1.3.1). We will show that, in fact, the closure systems of $\mathcal{L}(\mathbf{P})$ play a crucial role in providing a complete description of *all* join-completions of $\mathbf{P}$.

Recall from Section 1.3 that there exists a bijective correspondence between the closure operators on a poset $\mathbf{P}$ and the closure systems of $\mathbf{P}$ (Proposition 1.3.3), implemented by the map that sends any closure system $\mathbf{C}$ to the closure operator $\gamma_{\mathbf{C}} \colon P \to P$; $x \mapsto \min\{y \in C \mid x \leq y\}$, and the map sending any closure operator γ on $\mathbf{P}$ to its image $\mathbf{P}_\gamma$. Recall also that, for any poset $\mathbf{P}$, the map sending $X \subseteq P$ to the order-ideal $\downarrow X$ of $\mathbf{P}$ is a closure operator on $\langle \wp(P), \subseteq \rangle$, and the complete lattice $\mathcal{L}(\mathbf{P})$ of order-ideals of $\mathbf{P}$ is a closure system of $\langle \wp(P), \subseteq \rangle$ (Example 1.3.1).

Here we focus on the closure systems of $\mathcal{L}(\mathbf{P})$ that contain an isomorphic copy of $\mathbf{P}$, showing that these account for all the join-completions of $\mathbf{P}$ (up to isomorphism). Observe first that any join-extension $\mathbf{L}$ of $\mathbf{P}$ is order-isomorphic to its *principal ideal image* $\dot{\mathbf{L}} := \langle \{\downarrow x \cap P \mid x \in L\}, \subseteq \rangle$; in particular, $\mathbf{P}$ itself is order-isomorphic to $\dot{\mathbf{P}} = \langle \{\downarrow x \mid x \in P\}, \subseteq \rangle$. The set of subposets of $\mathcal{L}(\mathbf{P})$ that contain $\dot{P}$ must therefore contain an isomorphic copy of every join-extension of $\mathbf{P}$.

PROPOSITION 9.3.2. *Let $\mathbf{P}$ be any poset and let $\mathbf{L}$ be any subposet of $\mathcal{L}(\mathbf{P})$ such that $\dot{P} \subseteq L$. The poset $\mathbf{L}$ is a join-completion of $\dot{\mathbf{P}}$ if and only if it is a closure system of $\mathcal{L}(\mathbf{P})$. Moreover, $\mathcal{L}(\mathbf{P})$ is the largest join-completion of $\dot{\mathbf{P}}$.*

PROOF. Suppose first that the poset $\mathbf{L}$ is a join-completion of $\dot{\mathbf{P}}$. Then $\mathcal{L}(\mathbf{P})$ is a join-completion of $\mathbf{L}$ and, by Lemma 9.3.1, existing meets in $\mathbf{L}$ are preserved in $\mathcal{L}(\mathbf{P})$. Since $\mathbf{L}$ is a complete lattice, it is a closure system of $\mathcal{L}(\mathbf{P})$. Suppose now, conversely, that $\mathbf{L}$ is a closure system of $\mathcal{L}(\mathbf{P})$. An application of Proposition 1.3.6 yields $\bigvee^{\mathbf{L}} X = \gamma_{\mathbf{L}}(\bigvee^{\mathcal{L}(\mathbf{P})} X)$, for any $X \subseteq L$. Hence, for any $a \in L$, there exists $X \subseteq \dot{P}$ such that $a = \bigvee^{\mathcal{L}(\mathbf{P})} X$, and $a = \gamma_{\mathbf{L}}(a) = \gamma_{\mathbf{L}}(\bigvee^{\mathcal{L}(\mathbf{P})} X) = \bigvee^{\mathbf{L}} X$. So $\mathbf{L}$ is a join-completion of $\dot{\mathbf{P}}$. Finally, the fact that $\mathcal{L}(\mathbf{P})$ is the largest join-completion of $\dot{\mathbf{P}}$ follows from the remarks preceding this proposition. □

It is often the case in mathematics that the same entities can be regarded from two different perspectives: either concretely, as the results of a certain construction, or abstractly, as the unique objects (up to isomorphism) that fit a given description. In general, the latter is less cumbersome and more intuitive. In particular, we will often find the following abstract description of $\mathcal{L}(\mathbf{P})$ more convenient to use than the concrete construction in terms of order-ideals, noting that a lattice is said to be *dually algebraic* if its dual is algebraic.

PROPOSITION 9.3.3. *Let $\mathbf{P}$ be any poset. Then $\mathcal{L}(\mathbf{P})$ is, up to isomorphism, the unique algebraic and dually algebraic distributive lattice whose poset of completely join-prime elements is order-isomorphic to $\mathbf{P}$.*

PROOF. First, we show that $\mathcal{L}(\mathbf{P})$ is an algebraic and dually algebraic distributive lattice whose poset of completely join-prime elements is order-isomorphic to $\mathbf{P}$. The first part of this claim follows from Proposition 1.3.7 and the fact that the maps $X \mapsto {\downarrow}X$ and $X \mapsto {\uparrow}X$ are algebraic closure operators on $\langle \wp(P), \subseteq \rangle$.

Next, we show that $\dot{\mathbf{P}}$ is the poset of completely join-prime (equivalently, completely join-irreducible, since $\mathcal{L}(\mathbf{P})$ is a frame) members of $\mathcal{L}(\mathbf{P})$. Suppose that ${\downarrow}x = \bigcup \mathcal{F}$, for some $\mathcal{F} \subseteq \mathcal{L}(\mathbf{P})$. In particular, $x \in Y$, for some $Y \in \mathcal{F}$, so ${\downarrow}x \subseteq Y \subseteq \bigcup \mathcal{F} = {\downarrow}x$ and ${\downarrow}x = Y$. Conversely, let ${\downarrow}Y$ be any completely join-irreducible member of $\mathcal{L}(\mathbf{P})$. Since $\mathcal{L}(\mathbf{P})$ is algebraic, ${\downarrow}Y$ is a join of compact elements, and, since it is completely join-irreducible, there exists some $x \in P$ such that ${\downarrow}Y = {\downarrow}x$.

For uniqueness, let $\mathbf{L}$ be an algebraic and dually algebraic distributive lattice whose poset $\mathbf{J}$ of completely join-prime elements is order-isomorphic to $\mathbf{P}$. For each $a \in L$, the set $\varphi(a) := \{j \in J \mid j \leq a\}$ is an order-ideal of $\mathbf{J}$, and the map φ is clearly an order-homomorphism from $\mathbf{L}$ to $\mathcal{L}(\mathbf{J})$. If $\varphi(a) = \varphi(b)$, then a and b have the same completely join-irreducible elements below them, and hence $a = b$ by the dual of Proposition 1.3.8.(a). So φ is injective. Now let ${\downarrow}Y \in \mathcal{L}(\mathbf{J})$. Given any $j \in \varphi(\bigvee {\downarrow}Y)$, it follows, by Proposition 1.3.8.(b), that $j = j \wedge (\bigvee {\downarrow}Y) = \bigvee \{j \wedge k \mid k \in {\downarrow}Y\}$, and, since j is completely join-irreducible, $j \in {\downarrow}Y$. So ${\downarrow}Y = \varphi(\bigvee {\downarrow}Y)$ and φ is surjective. Hence $\mathbf{L}$ is order-isomorphic to $\mathcal{L}(\mathbf{J})$ and therefore also to $\mathcal{L}(\mathbf{P})$. □

We will switch freely between the concrete and abstract perspectives on $\mathcal{L}(\mathbf{P})$ in what follows. In particular, when adopting the latter perspective, we will identify join-completions of a poset $\mathbf{P}$ with the closure systems of $\mathcal{L}(\mathbf{P})$ that contain $\dot{P}$, viewing $\mathcal{L}(\mathbf{P})$ itself as the largest of these join-completions.

An equally satisfactory abstract description is available for another well-known completion of a poset $\mathbf{P}$. Given $X \subseteq P$, let $X^l := \{y \in P \mid y \leq x, \text{ for all } x \in X\}$ and $X^u := \{y \in P \mid y \geq x, \text{ for all } x \in X\}$. The map sending $X \subseteq P$ to $(X^u)^l$ is a closure operator on $\mathcal{L}(\mathbf{P})$, and the associated closure system $\mathcal{N}(\mathbf{P})$ is referred to as the *Dedekind-MacNeille completion* of $\mathbf{P}$.

PROPOSITION 9.3.4. *Let* $\mathbf{P}$ *be any poset.*

(a) $\mathcal{N}(\mathbf{P})$ *is, up to isomorphism, the only join- and meet-completion of* $\mathbf{P}$.

(b) *Any existing meets and joins in* $\mathbf{P}$ *are preserved in* $\mathcal{N}(\mathbf{P})$.

PROOF.

(a) Follows from the fact that $\mathcal{N}(\dot{\mathbf{P}})$ consists of all the intersections of principal order-ideals of $\mathbf{P}$.

(b) Follows from (a) and Lemma 9.3.1. □

From the perspective of this book, an especially interesting situation arises when pomonoids are considered instead of posets. Let us say that a pomonoid $\mathbf{L}$ is a *join-extension* of a pomonoid $\mathbf{P}$ if the poset reduct of $\mathbf{L}$ is a join-extension of the poset reduct of $\mathbf{P}$, and the monoid reduct of $\mathbf{P}$ is a submonoid of the monoid reduct of $\mathbf{L}$. Once again, $\mathbf{L}$ is said to be a *join-completion* of $\mathbf{P}$ if it is complete as a lattice.

The question then naturally arises as to whether we can obtain join-extensions of a given pomonoid from the join-extensions of the underlying poset. It turns out that, under certain conditions, the answer is yes. Indeed, these join-extensions, with suitable additional structure, form *residuated lattices.* The crucial observation here is that existing residuals are preserved by moving up the ladder of join-extensions.

Let $\mathbf{P}$ be any pomonoid. The residuated lattice $\wp(\mathbf{P}) = \langle \wp(P), \cap, \cup, \cdot, \backslash, /, \{e\}\rangle$ is built from the monoid reduct of $\mathbf{P}$ following the recipe given in Example 1.4.3. As explained in Example 1.5.7, the nucleus $\gamma_{\downarrow}\colon \wp(P) \to \mathcal{L}(\mathbf{P});\ X \mapsto {\downarrow}X$ on $\wp(\mathbf{P})$ yields a residuated lattice, whose lattice reduct is precisely $\mathcal{L}(\mathbf{P})$, where the product of two order-ideals $X, Y \in \mathcal{L}(\mathbf{P})$ is given by $X \circ_{\gamma_{\downarrow}} Y := {\downarrow}(X \cdot Y)$. When $\mathcal{L}(\mathbf{P})$ is viewed abstractly, this definition can be rephrased by setting, for any $a, b \in \mathcal{L}(\mathbf{P})$ and subsets X, Y of P with $a = \bigvee X$, $b = \bigvee Y$,

$$a \cdot^{\mathcal{L}(\mathbf{P})} b := \bigvee\{xy \mid x \in X,\ y \in Y\}.$$

More generally, we will show that a join-completion of a poset $\mathbf{P}$ can be endowed with the structure of a residuated lattice precisely when the associated closure operator is a *nucleus* on the pomonoid $\mathcal{L}(\mathbf{P})$. For convenience, we will henceforth commit a slight abuse of notation by using the same symbols for these enriched structures (pomonoids or residuated lattices) and their poset (or lattice) reducts.

LEMMA 9.3.5. *Let $\mathbf{P}$ be any pomonoid and let $\mathbf{L}$ be any join-extension of $\mathbf{P}$. For all $a, b \in P$, if $a\backslash^{\mathbf{P}}b$ exists (in $\mathbf{P}$), then $a\backslash^{\mathbf{L}}b$ exists (in $\mathbf{L}$) and*

$$a\backslash^{\mathbf{P}}b = a\backslash^{\mathbf{L}}b = a\backslash^{\mathcal{L}(\mathbf{P})}b.$$

The analogous statement holds for the other residual.

PROOF. Let $a, b \in P$ and assume that $a\backslash^{\mathbf{P}}b$ exists. Then $a\backslash^{\mathbf{P}}b \in \{x \in L \mid ax \le b\}$. We prove that $a\backslash^{\mathbf{P}}b = \max\{x \in L \mid ax \le b\}$ and hence $a\backslash^{\mathbf{P}}b = a\backslash^{\mathbf{L}}b$. To this end, consider any $x \in L$ such that $ax \le b$. Since $\mathbf{L}$ is a join-extension of $\mathbf{P}$, there exists $X \subseteq P$ such that $x = \bigvee^{\mathbf{L}} X$. Now, $ax \le b$ implies that $ac \le ax \le b$, for each $c \in X$, and hence $c \le a\backslash^{\mathbf{P}}b$. So $x = \bigvee^{\mathbf{L}} X \le a\backslash^{\mathbf{P}}b$ and $a\backslash^{\mathbf{P}}b = a\backslash^{\mathbf{L}}b$. Moreover, $a\backslash^{\mathbf{P}}b = a\backslash^{\mathcal{L}(\mathbf{P})}b$, since $\mathcal{L}(\mathbf{P})$ is a join-completion of $\mathbf{P}$ as a pomonoid. □

LEMMA 9.3.6. *Let $\mathbf{P} = \langle P, \cdot, e, \le\rangle$ be any pomonoid and $\mathbf{L}$ any join-extension of the poset $\langle P, \le\rangle$. Then there is at most one way to endow $\mathbf{L}$ with the structure of a residuated lattice such that its pomonoid reduct is a join-extension of $\mathbf{P}$.*

PROOF. Suppose that $\mathbf{L}$ is endowed with the structure of a residuated lattice in two ways with respect to multiplications $*_1$ and $*_2$ that extend the multiplication of $\mathbf{P}$ (indicated below by juxtaposition). Since $*_1$ and $*_2$ are residuated, they preserve arbitrary joins in $\mathbf{L}$. Given any $a, b \in L$, there exist non-empty subsets X, Y of P with $a = \bigvee^{\mathbf{L}} X$ and $b = \bigvee^{\mathbf{L}} Y$, and it follows that

$$\begin{aligned} a *_1 b &= (\textstyle\bigvee^{\mathbf{L}} X) *_1 (\textstyle\bigvee^{\mathbf{L}} Y) \\ &= \textstyle\bigvee^{\mathbf{L}}\{x *_1 y \mid x \in X,\ y \in Y\} \\ &= \textstyle\bigvee^{\mathbf{L}}\{xy \mid x \in X,\ y \in Y\} \\ &= \textstyle\bigvee^{\mathbf{L}}\{x *_2 y \mid x \in X,\ y \in Y\} \\ &= (\textstyle\bigvee^{\mathbf{L}} X) *_2 (\textstyle\bigvee^{\mathbf{L}} Y) = a *_2 b. \end{aligned}$$

That is, $*_1$ and $*_2$, and hence the corresponding residuated lattices, coincide. □

THEOREM 9.3.7. *Let $\mathbf{P} = \langle P, \cdot, \mathrm{e}, \leq \rangle$ be any pomonoid and $\mathbf{L}$ any join-extension of the poset $\langle P, \leq \rangle$. The following statements are equivalent:*

(1) *$\mathbf{L}$ can be endowed with the structure of a residuated lattice whose pomonoid reduct is a join-extension of $\mathbf{P}$.*
(2) *For all $a \in P$ and $b \in L$, the residuals $a \backslash^{\mathcal{L}(\mathbf{P})} b$ and $b /^{\mathcal{L}(\mathbf{P})} a$ are in L.*
(3) *$\mathbf{L}$ is a nucleus-system of $\mathcal{L}(\mathbf{P})$.*
(4) *The closure operator $\gamma_{\mathbf{L}}$ associated with $\mathbf{L}$ is a nucleus on $\mathcal{L}(\mathbf{P})$.*

Moreover, whenever the preceding conditions are satisfied, the residuated lattice expansion of $\mathbf{L}$ is uniquely determined and the inclusion map of $\mathbf{P}$ into $\mathbf{L}$ preserves, in addition to multiplication, all existing residuals and meets.

PROOF.

(1) $\Rightarrow$ (2) Suppose that $\mathbf{L}$ can be endowed with the structure of a residuated lattice whose pomonoid reduct is a join-extension of $\mathbf{P}$. By Lemma 9.3.6, this residuated lattice expansion of $\mathbf{L}$ is unique. Since $\mathcal{L}(\mathbf{P})$ is a join-extension of $\mathbf{L}$, also the pomonoid $\mathcal{L}(\mathbf{P})$ is a join-extension of the pomonoid $\mathbf{L}$. Let $a \in P$ and $b \in L$. Then $a \backslash^{\mathcal{L}(\mathbf{P})} b = a \backslash^{\mathbf{L}} b \in L$ and $b /^{\mathcal{L}(\mathbf{P})} a = b /^{\mathbf{L}} a \in L$, by Lemma 9.3.5.

(2) $\Rightarrow$ (3) Assume (2) and let $a \in \mathcal{L}(\mathbf{P})$ and $b \in L$. Since $\mathcal{L}(\mathbf{P})$ is a join-extension of $\mathbf{P}$, there exists $X \subseteq P$ such that $a = \bigvee^{\mathcal{L}(\mathbf{P})} X$. Hence, using Proposition 1.4.2 for the second equality and Lemma 9.3.1 with (2) for the third,

$$a \backslash^{\mathcal{L}(\mathbf{P})} b = (\bigvee\nolimits^{\mathcal{L}(\mathbf{P})} X) \backslash^{\mathcal{L}(\mathbf{P})} b = \bigwedge\nolimits^{\mathcal{L}(\mathbf{P})} \{c \backslash^{\mathcal{L}(\mathbf{P})} b \mid c \in X\} = \bigwedge\nolimits^{\mathbf{L}} \{c \backslash^{\mathcal{L}(\mathbf{P})} b \mid c \in X\}.$$

So $a \backslash^{\mathcal{L}(\mathbf{P})} b \in L$, and likewise, $b /^{\mathcal{L}(\mathbf{P})} a \in L$.

(3) $\Leftrightarrow$ (4) Follows from Lemma 1.5.2.

(3) $\Rightarrow$ (1) Suppose that $\mathbf{L}$ is a nucleus-system of $\mathcal{L}(\mathbf{P})$. Then $\mathbf{L}$ can be given the structure of a residuated lattice, following the recipe given in Lemma 1.5.4, and it remains to show that its pomonoid reduct is a join-extension of $\mathbf{P}$. We have already observed that $\mathbf{P}$ is a submonoid of $\mathcal{L}(\mathbf{P})$. Since $P \subseteq L$, it follows that $x \cdot^{\mathbf{L}} y = \gamma_{\mathbf{L}}(x \cdot^{\mathbf{P}} y) = x \cdot^{\mathbf{P}} y$, for all $x, y \in P$. Hence $\mathbf{P}$ is also a submonoid of $\mathbf{L}$.

Finally, suppose that conditions (1)–(4) are satisfied. The fact that the residuated lattice expansion of $\mathbf{L}$ is uniquely determined follows from Lemma 9.3.6, and the fact that the inclusion map of $\mathbf{P}$ into $\mathbf{L}$ preserves all existing residuals and meets follows from Lemmas 9.3.5 and 9.3.1, respectively. □

We note that Theorem 9.3.7 provides a simple proof of the fact that the Dedekind-MacNeille completion of a residuated pomonoid is a residuated lattice.

COROLLARY 9.3.8. *The Dedekind-MacNeille completion $\mathcal{N}(\mathbf{P})$ of any residuated pomonoid $\mathbf{P}$ can be uniquely endowed with the structure of a residuated lattice.*

PROOF. In view of Theorem 9.3.7, it suffices to show that $a \backslash^{\mathcal{L}(\mathbf{P})} b \in \mathcal{N}(\mathbf{P})$ and $b /^{\mathcal{L}(\mathbf{P})} a \in \mathcal{N}(\mathbf{P})$, for any $a \in P$ and $b \in \mathcal{N}(\mathbf{P})$. We just prove that $a \backslash^{\mathcal{L}(\mathbf{P})} b \in \mathcal{N}(\mathbf{P})$. Since $\mathcal{N}(\mathbf{P})$ is a join- and meet-completion of $\mathbf{P}$, by Proposition 9.3.4, there exists a subset $X \subseteq P$ such that $b = \bigwedge^{\mathcal{N}(\mathbf{P})} X$. Note that $b = \bigwedge^{\mathcal{N}(\mathbf{P})} X = \bigwedge^{\mathcal{L}(\mathbf{P})} X$, by Lemma 9.3.1. Hence, using Proposition 1.4.2 and Lemma 9.3.5,

$$\begin{aligned} a\backslash^{\mathcal{L}(\mathbf{P})} b &= a\backslash^{\mathcal{L}(\mathbf{P})} \bigwedge\nolimits^{\mathcal{L}(\mathbf{P})} X \\ &= \bigwedge\nolimits^{\mathcal{L}(\mathbf{P})} \{a\backslash^{\mathcal{L}(\mathbf{P})} x \mid x \in X\} \\ &= \bigwedge\nolimits^{\mathcal{N}(\mathbf{P})} \{a\backslash^{\mathcal{L}(\mathbf{P})} x \mid x \in X\} \\ &= \bigwedge\nolimits^{\mathcal{N}(\mathbf{P})} \{a\backslash^{\mathbf{P}} x \mid x \in X\} \in \mathcal{N}(\mathbf{P}). \end{aligned}$$

□

9.4. The FEP for varieties of residuated lattices

In this section, we use Theorem 9.3.7 to establish the FEP — and hence also the SFMP and decidability of the word problem — for the variety of integral residuated lattices (Theorem 9.4.6) and some of its subvarieties.

Consider any pomonoid $\mathbf{P}$ and non-empty subset $D \subseteq P$, viewed as a subset of $\dot{P}$. Let $\overline{D} := \{\bigwedge^{\mathcal{L}(\mathbf{P})} X \mid X \subseteq D\} \subseteq \mathcal{L}(\mathbf{P})$. Since $\overline{D}$ is closed under arbitrary meets in $\mathcal{L}(\mathbf{P})$, it forms a closure system $\overline{\mathbf{D}}$ of $\mathcal{L}(\mathbf{P})$ with corresponding closure operator $\gamma_{\overline{\mathbf{D}}} \colon \mathcal{L}(\mathbf{P}) \to \mathcal{L}(\mathbf{P});\ a \mapsto \bigwedge({\uparrow}a \cap D)$. Since any closure system of $\mathcal{L}(\mathbf{P})$ containing D also contains $\overline{D}$, we call $\overline{\mathbf{D}}$ the *closure system of* $\mathcal{L}(\mathbf{P})$ *generated by* D.

Using Theorem 9.3.7, we can identify conditions ensuring that $\overline{\mathbf{D}}$ forms a complete residuated lattice.

LEMMA 9.4.1. *Let* $\mathbf{P}$ *be any pomonoid and let* D *be any non-empty subset of* P *such that* $x\backslash^{\mathcal{L}(\mathbf{P})} y, y/^{\mathcal{L}(\mathbf{P})} x \in D$, *for all* $x \in P$ *and* $y \in D$. *Then the closure system* $\overline{\mathbf{D}}$ *of* $\mathcal{L}(\mathbf{P})$ *generated by* D *is a nucleus-system, and can be endowed with the structure of a complete residuated lattice whose pomonoid reduct is a join-completion of* $\mathbf{P}$.

PROOF. Since $\overline{\mathbf{D}}$ is a closure system of $\mathcal{L}(\mathbf{P})$, it is a join-completion of the poset $\mathbf{P}$, and hence, in view of Theorem 9.3.7, it suffices to prove that $a\backslash^{\mathcal{L}(\mathbf{P})} b, b/^{\mathcal{L}(\mathbf{P})} a \in \overline{D}$, for any $a \in P$ and $b \in \overline{D}$. Note first that $b = \bigwedge^{\mathcal{L}(\mathbf{P})} Y$, for some $Y \subseteq D$, and, by assumption, $a\backslash^{\mathcal{L}(\mathbf{P})} y \in D$, for each $y \in Y$. Hence

$$a\backslash^{\mathcal{L}(\mathbf{P})} b = a\backslash^{\mathcal{L}(\mathbf{P})} (\bigwedge\nolimits^{\mathcal{L}(\mathbf{P})} Y) = \bigwedge\nolimits^{\mathcal{L}(\mathbf{P})} \{a\backslash^{\mathcal{L}(\mathbf{P})} y \mid y \in Y\} \in \overline{D}.$$

Likewise, $b/^{\mathcal{L}(\mathbf{P})} a \in \overline{D}$. □

Our goal is to apply Lemma 9.4.1 to a (finite) partial subalgebra $\mathbf{B}$ of a residuated lattice $\mathbf{L}$ by extending B with "sufficiently many" elements from $\mathcal{L}(\mathbf{P})$ — where $\mathbf{P}$ is the pomonoid subreduct of $\mathbf{L}$ generated as a monoid by B — to obtain a subset $D \subseteq P$ satisfying $x\backslash^{\mathcal{L}(\mathbf{P})} y, y/^{\mathcal{L}(\mathbf{P})} x \in D$, for all $x \in P$ and $y \in D$.

To this end, let us call each variable y both a *residual formula* and *multiplicative formula of depth* 0 with *central variable* y, and define inductively, for a residual (multiplicative) formula α of depth $n \in \mathbb{N}$ and variable x not occurring in α, the residual formulas $x\backslash\alpha, \alpha/x$ (multiplicative formulas $x\alpha, \alpha x$) of depth $n+1$ with the same central variable as α. We denote by $\mathcal{R}$ and $\mathcal{M}$, the sets of all residual and multiplicative formulas, respectively, and write $\alpha(x_1, \ldots, x_n, y) \in \mathcal{R}$ or $\alpha(x_1, \ldots, x_n, y) \in \mathcal{M}$ to indicate that α has central variable y.

The following lemma is established via an easy induction on the depth of a residual or multiplicative formula.

LEMMA 9.4.2. *For any $\alpha(x_1,\ldots,x_n,z) \in \mathcal{R}$, there exists $\beta(x_1,\ldots,x_n,y) \in \mathcal{M}$, and for any $\beta(x_1,\ldots,x_n,y) \in \mathcal{M}$, there exists $\alpha(x_1,\ldots,x_n,z) \in \mathcal{R}$ such that, for any residuated pomonoid $\mathbf{P}$ and $a_1,\ldots,a_n,b,c \in P$,*

$$\beta^{\mathbf{P}}(a_1,\ldots,a_n,b) \leq c \iff b \leq \alpha^{\mathbf{P}}(a_1,\ldots,a_n,c).$$

The following result is a direct consequence of the previous result and Lemma 1.2.8.

COROLLARY 9.4.3. *Let $\mathbf{P}$ be any residuated pomonoid.*

(a) *For any $\alpha(x_1,\ldots,x_n,z) \in \mathcal{R}$ and $\{a_1,\ldots,a_n\} \cup C \subseteq L$ such that $\bigwedge^{\mathbf{P}} C$ exists,*
$$\alpha^{\mathbf{P}}(a_1,\ldots,a_n,\textstyle\bigwedge^{\mathbf{P}} C) = \textstyle\bigwedge^{\mathbf{P}}\{\alpha^{\mathbf{P}}(a_1,\ldots,a_n,c) \mid c \in C\}.$$

(b) *For any $\beta(x_1,\ldots,x_n,y) \in \mathcal{M}$ and $\{a_1,\ldots,a_n\} \cup B \subseteq L$ such that $\bigvee^{\mathbf{P}} B$ exists,*
$$\beta^{\mathbf{P}}(a_1,\ldots,a_n,\textstyle\bigvee^{\mathbf{P}} B) = \textstyle\bigvee^{\mathbf{P}}\{\beta^{\mathbf{P}}(a_1,\ldots,a_n,b) \mid b \in B\}.$$

Now let $\mathbf{L}$ be any residuated lattice and let $\mathbf{B}$ be any partial subalgebra of $\mathbf{L}$. Consider the pomonoid subreduct $\mathbf{P}$ of $\mathbf{L}$ generated as a monoid by B. Observe that, for any $u, v \in P$, if $u\backslash^{\mathbf{L}} v \in P$, then $u\backslash^{\mathbf{L}} v = \max\{w \in P \mid uw \leq v\}$, and, if $v/^{\mathbf{L}} u \in P$, then $v/^{\mathbf{L}} u = \max\{w \in P \mid wu \leq v\}$. We view the join-completion $\mathcal{L}(\mathbf{P})$ of $\mathbf{P}$ as a residuated lattice, recalling that the inclusion map $P \hookrightarrow \mathcal{L}(\mathbf{P})$ preserves multiplication and all existing residuals, and define

$$D_{\mathbf{L}}(\mathbf{B}) := \{\alpha^{\mathcal{L}(\mathbf{P})}(a_1,\ldots,a_n,b) \mid \alpha(x_1,\ldots,x_n,z) \in \mathcal{R},\ a_1,\ldots,a_n \in P, \text{ and } b \in B\}.$$

Clearly, $B \subseteq D_{\mathbf{L}}(\mathbf{B})$. Moreover, given any $\alpha^{\mathcal{L}(\mathbf{P})}(a_1,\ldots,a_n,b) \in D_{\mathbf{L}}(\mathbf{B})$, where $\alpha(x_1,\ldots,x_n,z) \in \mathcal{R}$, $a_1,\ldots,a_n \in P$, and $b \in B$, for any $c \in C$,

$$c\backslash^{\mathcal{L}(\mathbf{P})}\alpha^{\mathcal{L}(\mathbf{P})}(a_1,\ldots,a_n,b), \alpha^{\mathcal{L}(\mathbf{P})}(a_1,\ldots,a_n,b)/^{\mathcal{L}(\mathbf{P})}c \in D_{\mathbf{L}}(\mathbf{B}).$$

Hence, by Lemma 9.4.1, the closure system of $\mathcal{L}(\mathbf{P})$ generated by $D_{\mathbf{L}}(\mathbf{B})$ is a nucleus-system of $\mathcal{L}(\mathbf{P})$ that forms a complete residuated lattice $\overline{\mathbf{D}}_{\mathbf{L}}(\mathbf{B})$, where residuals and arbitrary meets in $\overline{\mathbf{D}}_{\mathbf{L}}(\mathbf{B})$ agree with those in $\mathcal{L}(\mathbf{P})$.

LEMMA 9.4.4. *For any partial subalgebra $\mathbf{B}$ of a residuated lattice $\mathbf{L}$, there exists an embedding of $\mathbf{B}$ into $\overline{\mathbf{D}}_{\mathbf{L}}(\mathbf{B})$ that preserves all existing meets and joins.*

PROOF. For convenience, let D and $\overline{\mathbf{D}}$ denote $D_{\mathbf{L}}(\mathbf{B})$ and $\overline{\mathbf{D}}_{\mathbf{L}}(\mathbf{B})$, respectively. Suppose first that $u, v \in B$ satisfy $u \cdot^{\mathbf{L}} v \in B$. Then $u \cdot^{\mathbf{P}} v = u \cdot^{\mathbf{L}} v \in B \subseteq \overline{D}$, and hence $u \cdot^{\overline{\mathbf{D}}} v = \gamma_{\overline{\mathbf{D}}}(u \cdot^{\mathbf{P}} v) = u \cdot^{\mathbf{P}} v = u \cdot^{\mathbf{L}} v \in B$. Suppose next that $u, v \in B$ satisfy $u\backslash^{\mathbf{L}} v \in B$. Then $u\backslash^{\mathbf{L}} v = u\backslash^{\mathbf{P}} v = u\backslash^{\mathcal{L}(\mathbf{P})} v = u\backslash^{\overline{\mathbf{D}}} v$, since $\mathcal{L}(\mathbf{P})$ is a join-completion of $\mathbf{P}$ and $\overline{\mathbf{D}}$ is a nucleus-system of $\mathcal{L}(\mathbf{P})$. The same argument works for $/$.

Now consider any $X \subseteq B$ such that $\bigwedge^{\mathbf{L}} X \in B$. Then $\bigwedge^{\mathbf{P}} X = \bigwedge^{\mathbf{L}} X$, since $\mathbf{P}$ is a subposet of $\mathbf{L}$. But also $\bigwedge^{\mathbf{P}} X = \bigwedge^{\mathcal{L}(\mathbf{P})} X = \bigwedge^{\overline{\mathbf{D}}} X$, since $\mathcal{L}(\mathbf{P})$ is a join-completion of $\mathbf{P}$ — and hence, by Lemma 9.3.1.(a), preserves existing meets in $\mathbf{P}$ — and $\overline{\mathbf{D}}$ is a nucleus-system of $\mathcal{L}(\mathbf{P})$.

Finally, consider any $X \subseteq B$ such that $\bigvee^{\mathbf{L}} X \in B$. Then $\bigvee^{\mathbf{P}} X = \bigvee^{\mathbf{L}} X$ and hence $\bigvee^{\overline{\mathbf{D}}} X = \gamma_{\overline{\mathbf{D}}}(\bigvee^{\mathcal{L}(\mathbf{P})} X) \leq \gamma_{\overline{\mathbf{D}}}(\bigvee^{\mathbf{L}} X) = \bigvee^{\mathbf{L}} X$. To show that in fact $\bigvee^{\overline{\mathbf{D}}} X = \bigvee^{\mathbf{L}} X$, let $\alpha^{\mathcal{L}(\mathbf{P})}(a_1,\dots,a_n,b)$ — with $\alpha(x_1,\dots,x_n,z)\in\mathcal{R}$, $a_1,\dots,a_n\in P$, and $b\in B$ — be an upper bound of the elements of X in $\overline{\mathbf{D}}$, and let $\beta(x_1,\dots,x_n,y)\in\mathcal{M}$ be the corresponding multiplicative formula given by Lemma 9.4.2. Then, for all $c\in B$,

$$\beta^{\mathcal{L}(\mathbf{P})}(a_1,\dots,a_n,c)\le b \iff c\le\alpha^{\mathcal{L}(\mathbf{P})}(a_1,\dots,a_n,b).$$

Hence $\beta^{\mathcal{L}(\mathbf{P})}(a_1,\dots,a_n,c)\le b$, for each $c\in X$. But then, using the fact that $\beta^{\mathcal{L}(\mathbf{P})}(a_1,\dots,a_n,\bigvee^{\mathbf{L}}X)\in P$, and Corollary 9.4.3,

$$\begin{aligned}\beta^{\mathcal{L}(\mathbf{P})}(a_1,\dots,a_n,\textstyle\bigvee^{\mathbf{L}}X)&=\beta^{\mathbf{P}}(a_1,\dots,a_n,\textstyle\bigvee^{\mathbf{L}}X)\\&=\beta^{\mathbf{L}}(a_1,\dots,a_n,\textstyle\bigvee^{\mathbf{L}}X)\\&=\textstyle\bigvee^{\mathbf{L}}\{\beta^{\mathbf{L}}(a_1,\dots,a_n,c)\mid c\in X\}\\&=\textstyle\bigvee^{\mathbf{L}}\{\beta^{\mathbf{P}}(a_1,\dots,a_n,c)\mid c\in X\}\\&=\textstyle\bigvee^{\mathbf{L}}\{\beta^{\mathcal{L}(\mathbf{P})}(a_1,\dots,a_n,c)\mid c\in X\}\\&\le b.\end{aligned}$$

So $\bigvee^{\mathbf{L}}X\le\alpha^{\mathcal{L}(\mathbf{P})}(a_1,\dots,a_n,b)$. Hence $\bigvee^{\overline{\mathbf{D}}}X=\bigvee^{\mathbf{L}}X$ as required. □

To deal with the case where $\mathbf{B}$ is a *finite* partial subalgebra of $\mathbf{L}$, we first consider a suitable partial order on the free monoid $\langle X^*,\circ,\varepsilon\rangle$ over a set X. Namely, for $s,t\in X^*$, define

$$s\preccurlyeq t :\iff s=u_1v_1u_2\cdots u_nv_nu_{n+1} \text{ and } t=v_1\cdots v_n$$
$$\text{for some } n\in\mathbb{N}^{>0} \text{ and } u_1,v_1,u_2,\dots,u_n,v_n,u_{n+1}\in X^*.$$

That is, $s\preccurlyeq t$ if the sequence t is the result of deleting some subsequences from s.

The following lemma collects some useful properties of this partial order.

LEMMA 9.4.5. *Let X be any set.*

(a) *If $su\preccurlyeq t$, for some $s,t,u\in X^*$, then there exist $t_1,t_2\in X^*$ such that $t=t_1t_2$, $s\preccurlyeq t_1$, and $u\preccurlyeq t_2$.*
(b) *$\langle X^*,\circ,\varepsilon,\preccurlyeq\rangle$ can be expanded with residual operations $\backslash$ and $/$ to obtain a residuated pomonoid $\mathbf{X}^*=\langle X^*,\circ,\backslash,/,\varepsilon,\preccurlyeq\rangle$ with greatest element ε.*
(c) *$\langle X^*,\preccurlyeq\rangle$ contains no infinite ascending chains.*
(d) *If X is finite, then $\langle X^*,\preccurlyeq\rangle$ contains no infinite antichains.*

PROOF.

(a) If t is the result of deleting some subsequences from su, then $t=t_1t_2$, where t_1 and t_2 are obtained by deleting subsequences from s and u, respectively.

(b) Clearly, ε is the greatest element of $\langle X^*,\preccurlyeq\rangle$ and $\circ$ is an order-preserving binary operation on $\langle X^*,\preccurlyeq\rangle$. Consider now any $s,t\in X^*$. The set $U:=\{t_2\in X^*\mid t_1t_2=t \text{ and } s\preccurlyeq t_1, \text{ for some } t_1\in X^*\}$ is finite and non-empty (since $t\in U$), and therefore contains a unique element u of shortest length. Clearly, $su\preccurlyeq t$. Moreover, if $su'\preccurlyeq t$, then, by (a), there exist $t_1,t_2\in X^*$ such that $t=t_1t_2$, $s\preccurlyeq t_1$, and $u'\preccurlyeq t_2$, yielding $u'\preccurlyeq u$. That is, $s\backslash t=u$. The case of the other residual is analogous.

(c) Follows immediately from the definition of $\preccurlyeq$.

(d) Suppose that $X = \{x_1, \dots, x_n\}$, for some $n \in \mathbb{N}$. We prove that every antichain in $\langle X^*, \preccurlyeq \rangle$ is finite, proceeding by induction on n. If $n = 0$, there are no antichains. For the inductive step, suppose that $n > 0$ and assume towards a contradiction that there exists an infinite antichain A in $\langle X^*, \preccurlyeq \rangle$. For each $k \in \{1, \dots, n\}$, the set of elements of A in which x_k does not occur is an antichain in $\langle (X \backslash \{x_k\})^*, \preccurlyeq \rangle$ and therefore, by the induction hypothesis, finite. Hence, since A is infinite, the set A' of elements of A where *all* the members of X occur is an infinite antichain in $\langle X^*, \preccurlyeq \rangle$. Consider any $a_1 \cdots a_m \in A'$ with $a_i \in X$, for each $i \in \{1, \dots, m\}$. Since there is no bound on the length of the sequences in A' and X is finite, there exists $l \in \mathbb{N}$ with $l > m$ and $b_1 \cdots b_l \in A'$ with $b_j \in X$, for each $j \in \{1, \dots, l\}$, such that $b_1 \cdots b_l = u_1 a_1 u_2 \cdots u_m a_m u_{m+1}$, for some $u_1, \dots, u_{m+1} \in A'$. So $b_1 \cdots b_l \preccurlyeq a_1 \cdots a_m$, contradicting the assumption that A is an antichain. □

We now have all the necessary ingredients to prove the main result of this section.

THEOREM 9.4.6. *The variety of integral residuated lattices has the finite embeddability property.*

PROOF. Let $\mathbf{B}$ be a finite partial subalgebra of an integral residuated lattice $\mathbf{L}$, and let $\mathbf{P}$ be the pomonoid subreduct of $\mathbf{L}$ generated as a monoid by B. Note that, since $\mathbf{L}$ is integral, e is the greatest element of $\mathbf{P}$. By Lemma 9.4.4, it suffices to show that the complete residuated lattice $\overline{\mathbf{D}}_{\mathbf{L}}(\mathbf{B})$, into which $\mathbf{B}$ embeds, is finite. To this end, let $B = \{b_1, \dots, b_k\}$ and $X := \{x_1, \dots, x_k\}$, and consider the surjective monoid homomorphism $\varphi \colon \langle X^*, \circ, \varepsilon \rangle \to \langle P, \cdot, \mathrm{e} \rangle$, defined by setting $\varphi(x_i) := b_i$, for each $i \in \{1, \dots, k\}$. Since e is the greatest element of $\mathbf{P}$, it follows that φ is also an order-homomorphism from $\langle X^*, \preccurlyeq \rangle$ to $\langle P, \leq \rangle$. Hence, by Lemma 1.1.8, given any order-ideal I of $\langle P, \leq \rangle$, the inverse image $\varphi^{-1}[I]$ is an order-ideal of $\langle X^*, \preccurlyeq \rangle$.

Clearly, $\overline{\mathbf{D}}_{\mathbf{L}}(\mathbf{B})$ is finite if and only if $D_{\mathbf{L}}(\mathbf{B})$ is finite, so it suffices to show that the set $D_b := \{\alpha^{\mathcal{L}(\mathbf{P})}(a_1, \dots, a_n, b) \mid \alpha(x_1, \dots, x_n, z) \in \mathcal{R} \text{ and } a_1, \dots, a_n \in P\}$ is finite, for any $b \in B$. Since $\langle X^*, \preccurlyeq \rangle$ has no infinite antichains, by Lemma 9.4.5.(d), there exists a finite antichain $U_b \subseteq X^*$ such that $\varphi^{-1}[{\downarrow} b] = {\downarrow} U_b$. Now consider any $\alpha(x_1, \dots, x_n, z) \in \mathcal{R}$ and, using Lemma 9.4.2, corresponding $\beta(x_1, \dots, x_n, y) \in \mathcal{M}$. Consider also any $a_1, \dots, a_n \in P$ and choose $t_1, \dots, t_n \in X^*$ such that $\varphi(t_i) = a_i$, for each $i \in \{1, \dots, n\}$. Then

$$
\begin{aligned}
s \in \varphi^{-1}[{\downarrow}\alpha^{\mathcal{L}(\mathbf{P})}(a_1, \dots, a_n, b)] &\iff \varphi(s) \leq \alpha^{\mathcal{L}(\mathbf{P})}(a_1, \dots, a_n, b) \\
&\iff \beta^{\mathcal{L}(\mathbf{P})}(a_1 \dots, a_n, \varphi(s)) \leq b \\
&\iff \varphi(\beta^{\mathbf{X}^*}(t_1 \dots, t_n, s)) \leq b \\
&\iff \beta^{\mathbf{X}^*}(t_1 \dots, t_n, s) \in \varphi^{-1}[{\downarrow} b] \\
&\iff \beta^{\mathbf{X}^*}(t_1 \dots, t_n, s) \preccurlyeq u, \text{ for some } u \in U_b \\
&\iff s \preccurlyeq \alpha^{\mathbf{X}^*}(t_1, \dots, t_n, u), \text{ for some } u \in U_b \\
&\iff s \in {\downarrow}\alpha^{\mathbf{X}^*}(t_1, \dots, t_n, u)), \text{ for some } u \in U_b.
\end{aligned}
$$

Hence, given any $a_1, \dots, a_n \in P$, there exist $t_1, \dots, t_n \in X^*$ such that

$$
\varphi^{-1}[{\downarrow}\alpha^{\mathcal{L}(\mathbf{P})}(a_1, \dots, a_n, b)] = \bigcup_{u \in U_b} {\downarrow}\alpha^{\mathbf{X}^*}(t_1, \dots, t_n, u).
$$

Consider now any $u \in U_b$. Since $\beta^{\mathbf{X}^*}(t_1, \ldots, t_n, u) \preccurlyeq \beta^{\mathbf{X}^*}(\varepsilon, \ldots, \varepsilon, u) = u$, it follows that $u \preccurlyeq \alpha^{\mathbf{X}^*}(t_1, \ldots, t_n, u)$. So $\alpha^{\mathbf{X}^*}(t_1, \ldots, t_n, u)$ belongs to $\uparrow U_b$, which is a finite set because U_b is finite and $\preccurlyeq$ has no infinite ascending chains (Lemma 9.4.5.(c)). Hence there are only finitely many sets of the form $\varphi^{-1}[\downarrow \alpha^{\mathcal{L}(\mathbf{P})}(a_1, \ldots, a_n, b)]$ with $\alpha(x_1, \ldots, x_n, z) \in \mathcal{R}$ and $a_1, \ldots, a_n \in P$. But also, since $\varphi[X^*] = P$, given any $\alpha(x_1, \ldots, x_n, z), \alpha(x_1, \ldots, x_m, z) \in \mathcal{R}$ and $a_1, \ldots, a_n, a'_1, \ldots, a'_m \in P$ such that $\alpha_1^{\mathcal{L}(\mathbf{P})}(a_1, \ldots, a_n, b) \neq \alpha_2^{\mathcal{L}(\mathbf{P})}(a'_1, \ldots, a'_m, b)$, the sets $\varphi^{-1}[\downarrow \alpha_1^{\mathcal{L}(\mathbf{P})}(a_1, \ldots, a_n, b)]$ and $\varphi^{-1}[\downarrow \alpha_2^{\mathcal{L}(\mathbf{P})}(a'_1, \ldots, a'_m, b)]$ are distinct. Hence there are only finitely many elements of the form $\alpha^{\mathcal{L}(\mathbf{P})}(a_1, \ldots, a_n, b)$ with $\alpha(x_1, \ldots, x_n, z) \in \mathcal{R}$ and $a_1, \ldots, a_n \in P$. That is, D_b is finite as required. □

Corollary 9.4.7. *The variety of integral residuated lattices has the strong finite model property and a decidable word problem.*

The approach described in this section serves, with very little embellishment, to establish the FEP, and hence also the SFMP and decidability of the word problem, for several other significant varieties of integral residuated lattices. We just observe that the algebra $\overline{\mathbf{D}}_{\mathbf{L}}(\mathbf{B})$ constructed for a partial subalgebra $\mathbf{B}$ of some integral residuated lattice $\mathbf{L}$ inherits certain properties of $\mathbf{L}$. Notably, if $\mathbf{L}$ is commutative or totally ordered, then so is $\overline{\mathbf{D}}_{\mathbf{L}}(\mathbf{B})$. Hence the variety of commutative, integral residuated lattices also has the FEP. Moreover, any finite partial subalgebra $\mathbf{B}$ of a semilinear, integral residuated lattice is a finite partial subalgebra of a product of finitely many totally ordered, integral residuated lattices, and hence embeds into a product of finitely many finite, totally ordered, integral residuated lattices. So the varieties of semilinear, integral residuated lattices and commutative, semilinear, integral residuated lattices also have the FEP.

The FEP, SFMP, and decidability of the word problem can be established for many other varieties of residuated lattices by implementing more significant changes to the construction described above. In Table 9.1, we provide an overview of the status of the FEP and decidability results for some of the varieties of residuated lattices considered in this chapter and elsewhere in the book. References for these results are given in the bibliographical remarks.

9.5. Bibliographical remarks

The general formulation of the FEP was introduced and studied systematically in a series of papers by Evans [**123**–**125**], following McKinsey and Tarski's proof for Heyting algebras [**249**] (see Example 9.1.2). Theorem 9.1.14 was proved in [**125**] and a finite signature was shown to be necessary for this result in [**19**]. Lemmas 9.1.7 and 9.1.8 are based on [**65**, Theorem III.8.4]. Lemma 9.1.10 and Proposition 9.1.11 appear to be folklore results, while Proposition 9.1.12 and an analysis of the subtle differences between the FEP and FEP$^+$ may be found in [**162**].

The proof of Theorem 9.2.8 is taken from [**40**], modulo a modification given in [**162**] to fill a small gap in the proof that the SFMP implies the FEP$^+$ (called the FEP in [**40**]), and Proposition 9.2.2 provides a slight generalization of another result from this paper (see also [**41**]). The straightforward proofs of Proposition 9.2.5 and Propositions 9.2.6 and 9.2.10 may be found in [**162**] and [**125**], respectively.

TABLE 9.1. Decidability and the FEP for varieties of residuated lattices

Variety	Label	FEP	Eq. Theory	Word Problem
Residuated lattices	$\mathcal{RL}$	no	decidable	undecidable
Commutative $\mathcal{RL}$	$\mathcal{CRL}$	no	decidable	undecidable
Semilinear $\mathcal{RL}$	$\mathcal{SemRL}$	no	?	?
Commutative $\mathcal{SemRL}$	$\mathcal{CSemRL}$	no	?	?
GBL-algebras	$\mathcal{GBL}$	no	?	undecidable
Commutative $\mathcal{GBL}$	$\mathcal{CGBL}$	no	decidable	decidable
Semilinear $\mathcal{GBL}$	$\mathcal{SemGBL}$	no	?	?
Commutative $\mathcal{SemGBL}$	$\mathcal{CSemGBL}$	no	decidable	decidable
GMV-algebras	$\mathcal{GMV}$	no	decidable	undecidable
Commutative $\mathcal{GMV}$	$\mathcal{CGMV}$	no	decidable	decidable
ℓ-groups	$\mathcal{LG}$	no	decidable	undecidable
Abelian ℓ-groups	$\mathcal{AbLG}$	no	decidable	decidable
Integral $\mathcal{RL}$	$\mathcal{IRL}$	yes	decidable	decidable
Commutative $\mathcal{IRL}$	$\mathcal{CIRL}$	yes	decidable	decidable
Semilinear $\mathcal{IRL}$	$\mathcal{SemIRL}$	yes	decidable	decidable
Commutative $\mathcal{SemIRL}$	$\mathcal{CSemIRL}$	yes	decidable	decidable
Integral $\mathcal{GBL}$	$\mathcal{IGBL}$	no	?	?
Commutative $\mathcal{IGBL}$	$\mathcal{CIGBL}$	yes	decidable	decidable
Semilinear $\mathcal{IGBL}$	$\mathcal{SemIGBL}$	no	?	?
Commutative $\mathcal{SemIGBL}$	$\mathcal{CSemIGBL}$	yes	decidable	decidable
Integral $\mathcal{GMV}$	$\mathcal{IGMV}$	no	decidable	?
Commutative $\mathcal{IGMV}$	$\mathcal{CIGMV}$	yes	decidable	decidable
Negative cones of ℓ-groups	$\mathcal{LG}^-$	no	decidable	?
Abelian $\mathcal{LG}^-$	$\mathcal{AbLG}^-$	no	decidable	decidable
Brouwerian algebras	$\mathcal{Brw}$	yes	decidable	decidable
Relative Stone algebras	$\mathcal{RSA}$	yes	decidable	decidable

Join-completions of posets, closely related to representations of complete lattices studied by Büchi [**52**], were introduced by Banaschewski in [**18**], who proved Propositions 9.3.2 and 9.3.4, and investigated extensively by Schmidt [**336–338**]. Dedekind-MacNeille completions for varieties of (pointed) residuated lattices have been studied in the context of algebraic proof theory in [**70, 71**].

Our account in Section 9.4 of the FEP for varieties of integral residuated lattices has its origins in the work of Blok and van Alten [**40, 41**]. Generalizations and variants of their approach that apply to different families of residuated structures have proliferated in recent years; see, e.g., [**54, 60, 126, 143, 144, 162, 178, 195, 362**]. In particular, our presentation follows [**162**], which provides a more general version of Theorem 9.3.7 and proves the FEP for the variety of algebras $\langle L, \wedge, \vee, \cdot, \backslash, /, \to, \mathrm{e}\rangle$ such that $\langle L, \wedge, \vee, \cdot, \backslash, /, \mathrm{e}\rangle$ is an integral residuated lattice and $\langle L, \wedge, \vee, \to, \mathrm{e}\rangle$ is a Brouwerian algebra. From this result it follows also that the variety of distributive, integral residuated lattices has the FEP (proved independently in [**144**]).

Proofs of the FEP for the varieties of commutative, integral GMV-algebras (term-equivalent to the variety of Wajsberg hoops), commutative, semilinear, integral

GBL-algebras, and commutative, integral GBL-algebras may be found in **[33]**, **[4]**, and **[204]**, respectively. The varieties of commutative GMV-algebras, commutative, semilinear GBL-algebras, and commutative GBL-algebras do not have the FEP, but the variety of Abelian ℓ-groups has a decidable word problem (Corollary 5.4.12), so, by Corollary 7.2.4, the same is true for these varieties. The fact that every finite GBL-algebra is integral and commutative, and hence no variety of GBL-algebras with a nonintegral or noncommutative member has the FEP, was first proved in **[203]**. Although the variety of commutative, cancellative, semilinear, integral residuated lattices has a decidable word problem **[193]**, decidability of even the equational theory is open for many other significant varieties of cancellative and/or semilinear residuated lattices.

Decidability of the equational theory for the varieties of residuated lattices and commutative residuated lattices (and many others) follows from the existence of a suitable analytic sequent calculus, as described in Chapter 2. Also, the algebraic proof of cut elimination can be adapted in some cases to establish the finite model property for these varieties **[143, 230, 265, 291, 295]**. Decidability of the equational theory of the variety of ℓ-groups **[189]** (Corollary 5.4.10) can be used to show that the same holds for the varieties of negative cones of ℓ-groups **[16]**, integral GMV-algebras **[152]**, and (by Corollary 7.2.4), GMV-algebras **[152]**.

Finally, the word problem — and, in some cases, the equational theory — is known to be undecidable for some varieties of residuated lattices. Indeed, any variety of residuated lattices containing all algebras defined on the power set of a finite monoid as in Example 1.4.3 has an undecidable word problem **[206]**: every semigroup is a subreduct of such an algebra and the word problem for the variety of semigroups is undecidable **[311]**. Undecidability of the word problem for the variety of ℓ-groups was proved in **[165]**, and implies that the same is true of any variety of GBL-algebras containing all ℓ-groups **[204]**. The fact that the variety of commutative residuated lattices has an undecidable word problem can be deduced from a related result for multiplicative additive linear logic **[236]**. Further negative results have been obtained for varieties of distributive residuated lattices **[138]**, involutive residuated lattices **[371]**, and residuated lattices satisfying "knotted" equations **[151, 195]**. Varieties that possess an undecidable equational theory include distributive, square-increasing, commutative involutive residuated lattices **[360]**, modular residuated lattices **[206]**, and square-increasing residuated lattices **[69]**.

APPENDIX A

Open problems

In this appendix we briefly describe some open problems for residuated structures in algebra and logic with the aim of stimulating further research in the area.

Structure theory

The equational property of prelinearity plays a key role in the structure theory of e-cyclic residuated lattices developed in Chapter 4. However, unlike semilinearity — described in Chapter 5 in terms of minimal, prime convex subuniverses and (principal) polars — a structural description of prelinearity is lacking. In particular, as mentioned in the bibliographical remarks of Chapter 4, the lattice of principal convex subuniverses of any prelinear, e-cyclic residuated lattice is relatively normal, but there exist other e-cyclic residuated lattices (e.g., those that are right prelinear) that also have this property.

PROBLEM 1. Study varieties that form subclasses of the class of e-cyclic residuated lattices whose lattice of principal convex subuniverses is relatively normal.

Many notable classes of ℓ-groups (e.g., divisible Abelian ℓ-groups and finite-valued ℓ-groups) fail to be varieties or even quasivarieties. A non-empty class $\mathcal{K}$ of ℓ-groups is called a *radical class* if it is closed under taking convex subuniverses, isomorphic images, and, for any ℓ-group $\mathbf{G}$, arbitrary joins of convex subuniverses of $\mathbf{G}$ that belong to $\mathcal{K}$. A radical class of ℓ-groups is said to be a *torsion class* if it is also closed under homomorphic images **[253]**. The same definitions can be adopted in the setting of residuated lattices but, as yet, nothing is known about these classes.

PROBLEM 2. Investigate torsion and radical classes of residuated lattices.

Free algebras play a significant role in the proof theory for varieties of residuated lattices developed in Chapter 2 and the account of consequence given in Chapter 3, but explicit constructions of these algebras are known only for specific cases such as MV-algebras **[250]**, Abelian ℓ-groups **[369]**, ℓ-groups **[88]** (see also **[283]**), and Heyting algebras **[359]**.

PROBLEM 3. Describe free commutative, cancellative residuated lattices.

PROBLEM 4. Describe free commutative GBL-algebras.

PROBLEM 5. Describe the free residuated lattice on one generator.

PROBLEM 6. Determine whether any non-trivial, free (commutative) residuated lattice is directly indecomposable.

Theorem 6.1.3 shows that every lattice is a subreduct of a cancellative, integral residuated lattice, and hence that the variety of cancellative residuated lattices does not satisfy any non-trivial lattice equations, but it is not known if these results transfer to the commutative setting.

PROBLEM 7. Determine whether every lattice is a subreduct of a commutative, cancellative residuated lattice.

PROBLEM 8. Determine whether the variety of commutative, cancellative residuated lattices satisfies any non-trivial lattice equations.

The construction presented in Section 6.2 of an ℓ-group of left quotients of the monoid reduct of a cancellative residuated lattice applies only when the latter has a right reversible monoid reduct, excluding, e.g., the cancellative residuated lattices constructed in Theorem 6.1.3.

PROBLEM 9. Describe the class of monoids that are monoid reducts of cancellative residuated lattices.

As shown in Chapters 6 and 7 of this book, many important classes of residuated structures arise by applying nuclei and co-nuclei to well-studied structures such as ℓ-groups. The question of which classes can be obtained in this way has been studied in **[313]**, but many challenges remain. In particular, it was proved in **[376]** that the unit intervals of integral, commutative, cancellative residuated lattices are precisely the conuclear images of MV-algebras, but an equational description of these algebras is lacking.

PROBLEM 10. Describe the conuclear images of MV-algebras.

Proof systems

Analytic sequent and hypersequent calculi have been obtained for broad families of varieties of pointed residuated lattices. Indeed, the ideas presented in Section 2.4 form part of a general methodology for generating analytic calculi and algebraic completeness proofs for varieties defined relative to pointed residuated lattices by equations of a certain simple syntactic form **[70, 71, 143, 144]**. Analytic calculi with "non-standard" operation rules have also been provided for MV-algebras, Abelian ℓ-groups, and related varieties **[257–259]**, but there remain several challenges.

PROBLEM 11. Provide analytic calculi for varieties of (commutative, semilinear, and/or integral) cancellative residuated lattices.

PROBLEM 12. Provide analytic calculi for varieties of (commutative, semilinear, and/or integral) GBL-algebras.

A related issue concerns the lack of an algebraic completeness proof for certain analytic calculi. Notably, an elegant analytic hypersequent calculus for ℓ-groups was presented in **[146]**, but completeness for this calculus was established only via a rather intricate cut-elimination proof.

PROBLEM 13. Obtain algebraic completeness proofs for analytic calculi for ℓ-groups and related varieties of residuated lattices.

Let us also mention two open problems that are algebraic in nature, but might be tackled using the proof-theoretic methods originating in [**72**, **255**] and described in Chapters 3 and 5.

PROBLEM 14. Determine whether the variety of commutative, semilinear involutive residuated lattices is densifiable.

PROBLEM 15. Provide an equational basis for the variety generated by dense, totally ordered residuated lattices.

As explained in Chapter 8, a variety of commutative residuated lattices has the amalgamation property if and only if it has the Craig interpolation property, which can sometimes be established using an analytic sequent calculus. Despite specialized results for such varieties (see, e.g. [**145**, **296**]), a general description is lacking.

PROBLEM 16. Describe the varieties of commutative residuated lattices axiomatized relative to $\mathcal{CRL}$ by simple equations that have the Craig interpolation property.

Amalgamation and interpolation

The bridge theorems presented in Chapter 8 provide — in combination with the proof systems and structure theory developed in previous chapters — useful tools for establishing various algebraic and logical properties of varieties of residuated lattices, notably amalgamation and interpolation. As indicated, however, by the question marks in Table 8.1, there remain many intriguing gaps in our knowledge. Perhaps most notably, although the variety of residuated lattices has the Craig interpolation property, it is not known if it has the amalgamation property.

PROBLEM 17. Determine whether the variety of residuated lattices has the amalgamation property.

The landscape of varieties of commutative, semilinear GBL-algebras that have the amalgamation property has been studied in some detail (see, e.g., [**2**, **91**, **256**, **271**]), using the structure theory for these algebras developed in Chapter 7, and it may be hoped that the more general structure theory developed in [**204**, **205**] can be used to obtain similar results for commutative GBL-algebras.

PROBLEM 18. Determine whether the variety of commutative GBL-algebras has the amalgamation property.

If a variety of residuated lattices contains all ℓ-groups, then, as proved in [**159**], the variety containing its semilinear members does not have the amalgamation property. This result leaves the status of amalgamation open, however, for many significant varieties of commutative or integral, semilinear residuated lattices. Such problems may plausibly be tackled either by modifying the approach of [**159**] or using structure theorems for totally ordered, commutative, involutive residuated lattices established in [**201**].

PROBLEM 19. Determine whether the variety of integral, semilinear residuated lattices has the amalgamation property.

Problem 20. Determine whether the variety of commutative, semilinear residuated lattices has the amalgamation property.

Proposition 8.4.10 gives an example of a variety of residuated lattices that has the amalgamation property, but does not satisfy the commutativity law. However, since this variety is generated by a finite algebra, it provides no indication of how to resolve the corresponding longstanding open problem for ℓ-groups.

Problem 21. Determine whether the variety of Abelian ℓ-groups is the only non-trivial variety of ℓ-groups that has the amalgamation property.

Let us also remark that for varieties of residuated lattices that lack the congruence extension property, the failure of the amalgamation property does not imply the failure of the deductive interpolation property.

Problem 22. Determine whether the variety of ℓ-groups has the deductive interpolation property.

For certain varieties of residuated lattices the amalgamation property has been established using an analytic sequent calculus and a syntactic proof of deductive interpolation, but an algebraic proof is still lacking.

Problem 23. Give an algebraic proof that the variety of commutative residuated lattices has the amalgamation property.

Decidability

The algebraic and algorithmic methods presented in this book have been used to address decidability issues for a wide range of varieties of residuated lattices, but many problems remain open, as indicated by the question marks in Table 9.1. In particular, although the variety of commutative, cancellative, semilinear, integral residuated lattices is known to have a decidable word problem [**193**], decidability of the equational theory is yet to be resolved for several varieties in its vicinity.

Problem 24. Prove that the equational theory of cancellative residuated lattices is (un)decidable.

Problem 25. Prove that the equational theory of commutative, semilinear, cancellative residuated lattices is (un)decidable.

Problem 26. Prove that the equational theory of commutative, integral, cancellative residuated lattices is (un)decidable.

In the case of divisibility, it is known that the word problem is decidable for the variety of commutative GBL-algebras, but undecidable for the variety of GBL-algebras [**204**]. This leaves open the question as to whether the latter — or, equivalently, as noted in Chapter 9, the variety of integral GBL-algebras — has a decidable equational theory.

Problem 27. Prove that the equational theory of GBL-algebras is (un)decidable.

Unlike the aforementioned varieties, analytic hypersequent calculi are available for important varieties of semilinear residuated lattices. Unfortunately, however, these calculi do not seem to be of much use in designing decision procedures.

PROBLEM 28. Prove that the equational theory of semilinear residuated lattices is (un)decidable.

PROBLEM 29. Prove that the equational theory of commutative, semilinear residuated lattices is (un)decidable.

Moreover, it is not known if the equational theories of even the residual-free reducts of these varieties — that is, the equational theories of totally ordered monoids and commutative, totally ordered monoids — are decidable.

PROBLEM 30. Prove that the equational theory of totally ordered monoids is (un)decidable.

PROBLEM 31. Prove that the equational theory of commutative, totally ordered monoids is (un)decidable.

First-order and modal substructural logics

The residuated structures studied in this book provide algebraic semantics for propositional substructural logics. There exists also, however, a long tradition of using these same structures to define semantics for corresponding first-order logics (see, e.g., [**75, 79, 218, 294, 299**]). In this setting, first-order formulas are defined for a predicate language $\mathcal{P}$ using the (substructural) propositional connectives and quantifiers $\forall$ and $\exists$, and a $\mathcal{P}$-structure $\mathcal{S}$ consists of a (complete) residuated lattice $\mathbf{L}$ together with a non-empty set S and functions from S^n to L for each n-ary relation symbol of $\mathcal{P}$. Formulas are assigned values in L with respect to $\mathcal{S}$, where $\forall$ and $\exists$ are interpreted by taking infima and suprema in $\mathbf{L}$, respectively. Analytic proof systems for these logics can be obtained in many cases by adding standard quantifier rules to a sequent or hypersequent calculus for the propositional logic (see, e.g., [**218, 259, 294**]), but an algebraic counterpart along the lines of Tarski's cylindric algebras for first-order classical logic, is lacking.

PROBLEM 32. Develop an algebraic semantics for first-order substructural logics.

Whereas first-order substructural logics typically interpret the quantifiers $\forall$ and $\exists$, respectively, as lattice meet and join operations, a more substructural approach, introducing a quantifier understood as a generalization of the product operation of a residuated lattice, was proposed in [**270**]. Related work that follows a more proof-theoretic approach includes [**128, 287, 324**]. More generally, an indexed algebraic approach to providing semantics for non-standard quantifiers for first-order logics has been developed in the setting of hyperdoctrines in [**42, 282**]. What is lacking, however, is an understanding of how substructural quantifiers fit into the algebraic and proof-theoretic approaches described in this book.

PROBLEM 33. Give algebraic and proof-theoretic characterizations of substructural quantifiers.

A further active but still quite under-developed area of research concerns modal substructural logics: extensions of propositional substructural logics with unary operations that have a wide range of interpretations and motivations (see, e.g., [**51**, **57**, **174**, **179**, **315**]). Completeness results have been obtained for certain cases — in particular, "S5" modal logics, corresponding to one-variable fragments of first-order logics [**58**, **77**, **174**, **333**] — but a general theory relating algebraic semantics (equivalently, axiomatizations) to a relational semantics is lacking.

PROBLEM 34. Provide relational semantics for modal substructural logics.

A particularly promising class of algebras for further study in this context are extensions of ordered groups with modal operators (see, e.g., [**58**, **100**, **263**]). Such operators include the nuclei and co-nuclei considered in Chapters 6 and 7, and may therefore provide some insight into properties of varieties of GMV-algebras and cancellative residuated lattices.

PROBLEM 35. Develop a structure theory for ordered groups with modal operators.

APPENDIX B

Basic notions of universal algebra

The aim of this appendix is to keep the present volume reasonably self-contained by providing a concise primer on the basic notions of universal algebra, referring to **[24, 53, 65]** for more detailed treatments and relevant citations. Elementary order-theoretic concepts introduced in the main text will not be repeated, however, and the reader is invited to consult the relevant parts of Chapter 1 as required.

Algebras and subalgebras

Let A and B be any sets, and let A^B denote the set of all maps from B to A. For $n \in \mathbb{N}$, we let A^n stand for $A^{\{1,\ldots,n\}}$ and call each $a \in A^n$ an *n-tuple*, denoted by $\langle a(1), \ldots, a(n) \rangle$. In particular, $A^0 = A^\emptyset = \{\emptyset\}$, and A^1 is identified with A.

A *signature* is a family $\mathcal{L} = \{\nu_f\}_{f \in I}$ of natural numbers indexed by a set I, where each $f \in I$ is referred to as an *operation symbol of* $\mathcal{L}$ *of arity* ν_f. For $n \in \mathbb{N}$, we let $\mathcal{L}_n$ denote the set of operation symbols of $\mathcal{L}$ of arity n. If $I = \{f_1, \ldots, f_m\}$, then the *finite signature* $\mathcal{L} = \{\nu_f\}_{f \in I}$ is often denoted by a sequence of natural numbers $\langle \nu_{f_1}, \ldots, \nu_{f_m} \rangle$. A signature $\mathcal{L}' = \{\nu'_f \mid f \in J\}$ is called a *subsignature* of $\mathcal{L}$ if $J \subseteq I$ and $\nu'_f = \nu_f$, for each $f \in J$.

An *algebra* of signature $\mathcal{L}$ (or $\mathcal{L}$*-algebra*) is an ordered pair $\mathbf{A} = \langle A, F \rangle$ consisting of a non-empty set A, called the *universe* of $\mathbf{A}$, and a function F assigning to each $f \in \mathcal{L}_n$ ($n \in \mathbb{N}$) a map $f^{\mathbf{A}} \colon A^n \to A$, called a *fundamental n-ary operation* of $\mathbf{A}$. We call $\mathbf{A}$ *finite* if A is finite, and *trivial* if $|A| = 1$. Given a subsignature $\mathcal{L}'$ of $\mathcal{L}$, the $\mathcal{L}'$*-reduct* of $\mathbf{A}$ is the $\mathcal{L}'$-algebra $\langle A, F' \rangle$, where F' is the restriction of F to $\mathcal{L}'$; in this case, $\mathbf{A}$ is said to be an *expansion* of $\langle A, F' \rangle$.

We call each $f \in \mathcal{L}_0$ a *constant symbol* and identify the 0-ary operation $f^{\mathbf{A}}$ of an $\mathcal{L}$-algebra $\mathbf{A}$, called a *constant* of $\mathbf{A}$, with $f^{\mathbf{A}}(\emptyset) \in A$. We also call 1-ary operation symbols and operations *unary* and 2-ary operation symbols and operations *binary*, often using infix notation for the latter, i.e., writing $af^{\mathbf{A}}b$ rather than $f^{\mathbf{A}}(a, b)$.

A subset $B \subseteq A$ is said to be a *subuniverse* of of an $\mathcal{L}$-algebra $\mathbf{A}$ if it is closed under the fundamental operations of $\mathbf{A}$, that is, if $f^{\mathbf{A}}(b_1, \ldots, b_n) \in B$, for all $f \in \mathcal{L}_n$ ($n \in \mathbb{N}$) and $b_1, \ldots, b_n \in B$. Since a subuniverse of $\mathbf{A}$ must contain all the constants of $\mathbf{A}$, the empty set is a subuniverse of $\mathbf{A}$ if and only if $\mathcal{L}_0 = \emptyset$.

Every non-empty subuniverse B of an $\mathcal{L}$-algebra $\mathbf{A}$ is the universe of an $\mathcal{L}$-algebra $\mathbf{B}$, called a *subalgebra* of $\mathbf{A}$, where $f^{\mathbf{B}}$ is the restriction of $f^{\mathbf{A}}$ to B^n, for each $f \in \mathcal{L}_n$ ($n \in \mathbb{N}$). Subalgebras of specific classes of algebras, such as groups, lattices etc., are referred to as subgroups, sublattices, etc. A subalgebra of an $\mathcal{L}'$-reduct of an $\mathcal{L}$-algebra $\mathbf{A}$, for some subsignature $\mathcal{L}'$ of $\mathcal{L}$, is called an $\mathcal{L}'$*-subreduct* of $\mathbf{A}$.

The set of all subuniverses of an $\mathcal{L}$-algebra $\mathbf{A}$ will be denoted by $\mathrm{Sub}(\mathbf{A})$. It is easily checked that $\langle \mathrm{Sub}(\mathbf{A}), \subseteq\rangle$ is a closure system of $\langle \wp(A), \subseteq\rangle$ and hence a complete lattice (see Proposition 1.1.2), satisfying, for any $D \subseteq \mathrm{Sub}(\mathbf{A})$,

$$\bigwedge D = \bigcap D \quad \text{and} \quad \bigvee D = \bigcap\{B \in \mathrm{Sub}(\mathbf{A}) \mid \bigcup D \subseteq B\}.$$

The associated closure operator on $\langle \wp(A), \subseteq\rangle$ is defined for $X \subseteq A$ by

$$\mathrm{Sg}^{\mathbf{A}}(X) := \bigcap\{B \in \mathrm{Sub}(\mathbf{A}) \mid X \subseteq B\}.$$

Given $X \subseteq A$ such that $\mathrm{Sg}^{\mathbf{A}}(X) \neq \emptyset$, the subalgebra of $\mathbf{A}$ with universe $\mathrm{Sg}^{\mathbf{A}}(X)$ is called the *subalgebra of* $\mathbf{A}$ *generated by* X. In particular, if $\mathrm{Sg}^{\mathbf{A}}(X) = A$, for some (finite) $X \subseteq A$, then $\mathbf{A}$ is said to be *(finitely) generated by* X. An $\mathcal{L}$-algebra such that any of its finitely generated subalgebras is finite is called *locally finite*, and this term is also applied to any class of locally finite $\mathcal{L}$-algebras.

Homomorphisms and congruences

Let us fix a signature $\mathcal{L}$ and let $\mathbf{A}$ and $\mathbf{B}$ be any $\mathcal{L}$-algebras. A map $\varphi\colon A \to B$ is called a *homomorphism* from $\mathbf{A}$ to $\mathbf{B}$, denoted by $\varphi\colon \mathbf{A} \to \mathbf{B}$, if, for all $f \in \mathcal{L}_n$ ($n \in \mathbb{N}$) and $\langle a_1, \ldots, a_n\rangle \in A^n$,

$$\varphi(f^{\mathbf{A}}(a_1, \ldots, a_n)) = f^{\mathbf{B}}(\varphi(a_1), \ldots, \varphi(a_n)).$$

It is said to be *trivial* if $\varphi(a_1) = \varphi(a_2)$, for all $a_1, a_2 \in A$.

An injective homomorphism from $\mathbf{A}$ to $\mathbf{B}$ is called an *embedding* of $\mathbf{A}$ into $\mathbf{B}$. If there exists a surjective homomorphism from $\mathbf{A}$ to $\mathbf{B}$, then $\mathbf{B}$ is said to be a *homomorphic image* of $\mathbf{A}$, and if there exists a bijective homomorphism, called an *isomorphism*, from $\mathbf{A}$ to $\mathbf{B}$, then $\mathbf{A}$ is said to be *isomorphic* to $\mathbf{B}$, denoted by $\mathbf{A} \cong \mathbf{B}$. A homomorphism (isomorphism) from $\mathbf{A}$ to $\mathbf{A}$ is called an *endomorphism* (*automorphism*) of $\mathbf{A}$.

Let us denote the set of homomorphisms from $\mathbf{A}$ to $\mathbf{B}$ by $\hom(\mathbf{A}, \mathbf{B})$. Given $\varphi \in \hom(\mathbf{A}, \mathbf{B})$ and $\psi \in \hom(\mathbf{B}, \mathbf{C})$, for some $\mathcal{L}$-algebra $\mathbf{C}$, the composition map $\psi \circ \varphi\colon A \to C;\ a \mapsto \psi(\varphi(a))$, often shortened to $\psi\varphi$, belongs to $\hom(\mathbf{A}, \mathbf{C})$.

Recall next that an *equivalence relation* on a set A is a binary relation Θ on A that is reflexive, symmetric, and transitive. The *equivalence class* of $a \in A$ with respect to Θ is the set $\{b \in A \mid \langle a, b\rangle \in \Theta\}$, denoted by $[a]_\Theta$ (or just $[a]$). The set of equivalence classes of Θ is denoted by A/Θ.

It is not hard to prove that the map sending $R \subseteq A^2$ to the smallest equivalence relation on A containing R is an algebraic closure operator on the poset $\langle A^2, \subseteq\rangle$ and hence (see Proposition 1.3.7) that the set of equivalence relations on A ordered by set-inclusion is an algebraic lattice with least element $\Delta_A := \{\langle a, a\rangle \mid a \in A\}$ and greatest element A^2. In particular, for any set E of equivalence relations on A, it can be shown that $\bigwedge E = \bigcap E$ and $\langle a, b\rangle \in \bigvee E$ if and only if there exist $n \in \mathbb{N}$, $c_1, \ldots, c_{n+1} \in A$, and $\Theta_1, \ldots, \Theta_n \in E$ such that $a = c_1$, $b = c_{n+1}$, and $\langle c_k, c_{k+1}\rangle \in \Theta_k$, for each $k \in \{1, \ldots, n\}$.

A *congruence* of an $\mathcal{L}$-algebra $\mathbf{A}$ is an equivalence relation Θ on A such that, for any $f \in \mathcal{L}_n$ ($n \in \mathbb{N}$),

$$\langle a_1, b_1\rangle, \ldots, \langle a_n, b_n\rangle \in \Theta \implies \langle f^{\mathbf{A}}(a_1, \ldots, a_n), f^{\mathbf{A}}(b_1, \ldots, b_n)\rangle \in \Theta.$$

Let $\mathrm{Con}(\mathbf{A})$ denote the set of all congruences of $\mathbf{A}$. Then $\langle \mathrm{Con}(\mathbf{A}), \subseteq \rangle$ is a complete sublattice of the lattice of equivalence relations on A, called the *congruence lattice of* $\mathbf{A}$ and also denoted by $\mathrm{Con}(\mathbf{A})$. In particular, the meet and join in $\mathrm{Con}(\mathbf{A})$ of any set of congruences of $\mathbf{A}$ coincide with the meet and join in the lattice of equivalence relations on A. An $\mathcal{L}$-algebra $\mathbf{A}$ is called *simple* if $\mathrm{Con}(\mathbf{A}) = \{\Delta_A, A^2\}$.

For each $R \subseteq A^2$, let $\mathrm{Cg}^{\mathbf{A}}(R) := \bigcap\{\Theta \in \mathrm{Con}(\mathbf{A}) \mid R \subseteq \Theta\}$. If $R = \{\langle a, b\rangle\}$, then $\mathrm{Cg}^{\mathbf{A}}(\{\langle a, b\rangle\})$ is denoted by $\mathrm{Cg}^{\mathbf{A}}(a, b)$. It is straightforward to show that $\mathrm{Cg}^{\mathbf{A}}$ is an algebraic closure operator on the poset $\langle A^2, \subseteq\rangle$ and hence (see Proposition 1.3.7) that $\mathrm{Con}(\mathbf{A})$ is an algebraic lattice with least element Δ_A and greatest element A^2.

The *kernel* of a map $\varphi\colon A \to B$ is $\ker(\varphi) := \{\langle a_1, a_2\rangle \in A^2 \mid \varphi(a_1) = \varphi(a_2)\}$. It is easily checked that $\ker(\varphi)$ is an equivalence relation on A, and that if φ is a homomorphism from an $\mathcal{L}$-algebra $\mathbf{A}$ to an $\mathcal{L}$-algebra $\mathbf{B}$, then $\ker(\varphi) \in \mathrm{Con}(\mathbf{A})$.

PROPOSITION B.1. *Let* $\mathbf{A}$ *be an* $\mathcal{L}$*-algebra,* B *a set, and* φ *a surjective map from* A *to* B*. Then there exists at most one* $\mathcal{L}$*-algebra* $\mathbf{B}$ *with universe* B *such that* φ *is a surjective homomorphism from* $\mathbf{A}$ *to* $\mathbf{B}$*. Moreover, such an* $\mathcal{L}$*-algebra exists if and only if* $\ker(\varphi) \in \mathrm{Con}(\mathbf{A})$.

PROOF. Suppose first that $\ker(\varphi) \in \mathrm{Con}(\mathbf{A})$. Then, for each $f \in \mathcal{L}_n$ ($n \in \mathbb{N}$), the following n-ary operation is well-defined:

$$f^{\mathbf{B}}\colon B^n \to B; \quad \langle \varphi(a_1), \dots, \varphi(a_n)\rangle \mapsto \varphi(f^{\mathbf{A}}(a_1, \dots, a_n)).$$

Let $\mathbf{B}$ be the $\mathcal{L}$-algebra with universe B and operations $f^{\mathbf{B}}$, for each $f \in \mathcal{L}$. It follows by definition that φ is a surjective homomorphism from $\mathbf{A}$ to $\mathbf{B}$.

Assume now that φ is a surjective homomorphism from $\mathbf{A}$ to an $\mathcal{L}$-algebra $\mathbf{B}$ with universe B. Then $\ker(\varphi) \in \mathrm{Con}(\mathbf{A})$ and it follows, because φ is a homomorphism, that, for all $f \in \mathcal{L}_n$ ($n \in \mathbb{N}$) and $a_1, \dots, a_n \in A$,

$$f^{\mathbf{B}}(\varphi(a_1), \dots, \varphi(a_n)) = \varphi(f^{\mathbf{A}}(a_1, \dots, a_n)).$$

Hence $\mathbf{B}$ is the only $\mathcal{L}$-algebra with this property. □

Let Θ be any congruence of an $\mathcal{L}$-algebra $\mathbf{A}$. The *quotient of* $\mathbf{A}$ *modulo* Θ is the $\mathcal{L}$-algebra $\mathbf{A}/\Theta$ with universe A/Θ and operations defined for $f \in \mathcal{L}_n$ ($n \in \mathbb{N}$) by

$$f^{\mathbf{A}/\Theta}\colon (A/\Theta)^n \to A/\Theta; \quad \langle [a_1]_\Theta, \dots, [a_n]_\Theta\rangle \mapsto [f^{\mathbf{A}}(a_1, \dots, a_n)]_\Theta.$$

The *canonical map* $\pi_\Theta\colon A \to A/\Theta$; $a \mapsto [a]_\Theta$ is a surjective homomorphism from $\mathbf{A}$ onto $\mathbf{A}/\Theta$. Indeed, it follows from Proposition B.1 that $\mathbf{A}/\Theta$ is the only $\mathcal{L}$-algebra with universe A/Θ such that π_Θ is a surjective homomorphism.

THEOREM B.2 (General homomorphism theorem). *Let* $\mathbf{A}$*,* $\mathbf{B}$*, and* $\mathbf{C}$ *be* $\mathcal{L}$*-algebras,* $\varphi\colon \mathbf{A} \to \mathbf{B}$ *a surjective homomorphism, and* $\chi\colon \mathbf{A} \to \mathbf{C}$ *a homomorphism. Then there exists at most one homomorphism* $\psi\colon \mathbf{B} \to \mathbf{C}$ *satisfying* $\psi\varphi = \chi$*, and exactly one such homomorphism if and only if* $\ker(\varphi) \subseteq \ker(\chi)$*. Moreover,* ψ *is surjective if and only if* χ *is surjective, and* ψ *is injective if and only if* $\ker(\varphi) = \ker(\chi)$.

PROOF. Since φ is surjective, there exists a map $\psi\colon B \to C$ sending each $b \in B$ to $\chi(a)$, for some $a \in A$ such that $\varphi(a) = b$. If $\ker(\varphi) \subseteq \ker(\chi)$, then $\psi(\varphi(a)) = \chi(a)$, for all $a \in A$, and, for any $n \in \mathbb{N}$, $f \in \mathcal{L}_n$, and $a_1, \ldots, a_n \in A$,

$$\begin{aligned}\psi(f^{\mathbf{B}}(\varphi(a_1), \ldots, \varphi(a_n))) &= \psi(\varphi(f^{\mathbf{A}}(a_1, \ldots, a_n))) \\ &= \chi(f^{\mathbf{A}}(a_1, \ldots, a_n)) \\ &= f^{\mathbf{C}}(\chi(a_1), \ldots, \chi(a_n)) \\ &= f^{\mathbf{C}}(\psi(\varphi(a_1)), \ldots, \psi(\varphi(a_n))),\end{aligned}$$

that is, ψ is a homomorphism satisfying $\psi\varphi = \chi$.

Now suppose that there exists a homomorphism $\psi\colon \mathbf{B} \to \mathbf{C}$ satisfying $\psi\varphi = \chi$. Then $\ker(\varphi) \subseteq \ker(\chi)$ and, since φ is surjective, ψ is the unique homomorphism satisfying $\psi\varphi = \chi$. We leave it to the reader to check that ψ is surjective if and only if χ is surjective, and ψ is injective if and only if $\ker(\varphi) = \ker(\chi)$. □

COROLLARY B.3 (Homomorphism theorem). *Let* $\mathbf{A}$ *and* $\mathbf{B}$ *be* $\mathcal{L}$*-algebras and let* $\varphi\colon \mathbf{A} \to \mathbf{B}$ *be a surjective homomorphism. Then there exists a unique isomorphism* $\psi\colon \mathbf{A}/\ker(\varphi) \to \mathbf{B}$ *such that* $\psi\pi_{\ker\varphi} = \varphi$.

For any sets A and B, map $\varphi\colon A \to B$, and binary relations $R \subseteq A^2$, $S \subseteq B^2$, let

$$\begin{aligned}\varphi[R] &:= \{\langle\varphi(a_1), \varphi(a_2)\rangle \mid \langle a_1, a_2\rangle \in R\} \\ \varphi^{-1}[S] &:= \{\langle a_1, a_2\rangle \in A^2 \mid \langle\varphi(a_1), \varphi(a_2)\rangle \in S\}.\end{aligned}$$

THEOREM B.4 (Correspondence theorem). *Let* $\mathbf{A}$ *and* $\mathbf{B}$ *be* $\mathcal{L}$*-algebras and let* $\varphi\colon \mathbf{A} \to \mathbf{B}$ *be a surjective homomorphism.*

(a) *If* $\Theta_A \in \mathrm{Con}(\mathbf{A})$ *and* $\ker(\varphi) \subseteq \Theta_A$*, then* $\varphi[\Theta_A] \in \mathrm{Con}(\mathbf{B})$.
(b) *If* $\Theta_B \in \mathrm{Con}(\mathbf{B})$*, then* $\varphi^{-1}[\Theta_B] \in \mathrm{Con}(\mathbf{A})$ *and* $\ker(\varphi) \subseteq \varphi^{-1}[\Theta_B]$.
(c) *The map* $\Theta_A \mapsto \varphi[\Theta_A]$ *is a lattice isomorphism from the interval* $[\ker(\varphi), A^2]$ *of* $\mathrm{Con}(\mathbf{A})$ *to* $\mathrm{Con}(\mathbf{B})$ *with inverse map* $\Theta_B \mapsto \varphi^{-1}[\Theta_B]$.
(d) *If* $R \subseteq A^2$*, then* $\varphi^{-1}[\mathrm{Cg}^{\mathbf{B}}(\varphi[R])] = \mathrm{Cg}^{\mathbf{A}}(R) \vee \ker(\varphi) \in \mathrm{Con}(\mathbf{A})$.

PROOF. We leave (a), (b), and (c) as routine exercises for the reader. For (d), let $\Theta := \varphi^{-1}[\mathrm{Cg}^{\mathbf{B}}(\varphi[R])]$. Clearly, $R \subseteq \Theta \in \mathrm{Con}(\mathbf{A})$, so $\mathrm{Cg}^{\mathbf{A}}(R) \subseteq \Theta$. Since also $\ker(\varphi) \subseteq \Theta$, it follows that $\mathrm{Cg}^{\mathbf{A}}(R) \vee \ker(\varphi) \subseteq \Theta$. For the converse inclusion, since $\mathrm{Cg}^{\mathbf{A}}(R) \vee \ker(\varphi) \in [\ker(\varphi), A^2]$, it follows using (c) that

$$\varphi^{-1}[\mathrm{Cg}^{\mathbf{B}}(\varphi[R])] \subseteq \varphi^{-1}[\mathrm{Cg}^{\mathbf{B}}(\varphi[\mathrm{Cg}^{\mathbf{A}}(R) \vee \ker(\varphi)])] = \mathrm{Cg}^{\mathbf{A}}(R) \vee \ker(\varphi). \qquad \square$$

For any set A and binary relations $R \subseteq A^2$, $S \subseteq A^2$, let

$$R \circ S := \{\langle a, b\rangle \in A^2 \mid \langle a, c\rangle \in R \text{ and } \langle c, b\rangle \in S, \text{ for some } c \in A\}.$$

An $\mathcal{L}$-algebra $\mathbf{A}$ is said to be *congruence-distributive* if $\mathrm{Con}(\mathbf{A})$ is distributive, and *congruence permutable* if $\Theta_1 \circ \Theta_2 = \Theta_2 \circ \Theta_1$ (and hence $\Theta_1 \circ \Theta_2 = \Theta_1 \vee \Theta_2$), for any $\Theta_1, \Theta_2 \in \mathrm{Con}(\mathbf{A})$. A class $\mathcal{K}$ of $\mathcal{L}$-algebras is said to be *congruence-distributive* (respectively, *congruence-permutable*) if every $\mathbf{A} \in \mathcal{K}$ has this property.

PROPOSITION B.5. *Any* $\mathcal{L}$*-algebra with a lattice reduct is congruence-distributive.*

PROOF. Let $\mathbf{L}$ be an $\mathcal{L}$-algebra with a lattice reduct $\langle L, \wedge, \vee\rangle$. It suffices to show that $\Theta_1 \wedge (\Theta_2 \vee \Theta_3) \subseteq (\Theta_1 \wedge \Theta_2) \vee (\Theta_1 \wedge \Theta_3)$, for any $\Theta_1, \Theta_2, \Theta_3 \in \mathrm{Con}(\mathbf{L})$, since the converse inclusion holds in any lattice. Suppose that $\langle a, b\rangle \in \Theta_1 \wedge (\Theta_2 \vee \Theta_3)$.

Then $a\ \Theta_1\ b$ and $a\ \Theta_2\ c_1\ \Theta_3\ c_2\ \Theta_2 \cdots \Theta_2\ c_n\ \Theta_3\ b$, for some $c_1, \ldots, c_n \in L$. Now, for any $x, y, z \in L$, let $M(x, y, z) := (x \wedge y) \vee (x \wedge z) \vee (y \wedge z)$, observing that $M(x, x, y) = M(x, y, x) = M(y, x, x) = x$. Then, since $a\ \Theta_1\ b$, also $M(a, c_i, b)\ \Theta_1\ M(a, c_i, a) = a$, for each $i \in \{1, \ldots, n\}$. But $x\ \Theta_j\ y$ implies $M(a, x, b)\ \Theta_j\ M(a, y, b)$, for each $j \in \{2, 3\}$, so

$$\begin{aligned} a = M(a, a, b)\ (\Theta_1 \wedge \Theta_2) M(a, c_1, b)\ (\Theta_1 \wedge \Theta_3)\ M(a, c_2, b) \cdots \\ \cdots M(a, c_n, b)\ (\Theta_1 \wedge \Theta_3)\ M(a, b, b) = b. \end{aligned}$$

Hence $\langle a, b\rangle \in (\Theta_1 \wedge \Theta_2) \vee (\Theta_1 \wedge \Theta_3)$. □

Direct and subdirect products

The *Cartesian product* $\prod_{i \in I} A_i$ of a family of sets $\{A_i\}_{i \in I}$ consists of all maps $\varphi\colon I \to \bigcup_{i \in I} A_i$ satisfying $\varphi(i) \in A_i$, for each $i \in I$. The *direct product* of a family of $\mathcal{L}$-algebras $\{\mathbf{A}_i\}_{i \in I}$ is an $\mathcal{L}$-algebra $\mathbf{A} = \prod_{i \in I} \mathbf{A}_i$ with universe $\prod_{i \in I} A_i$ such that, for all $f \in \mathcal{L}_n$ $(n \in \mathbb{N})$, $a_1, \ldots, a_n \in \prod_{i \in I} A_i$, and $i \in I$,

$$f^{\mathbf{A}}(a_1, \ldots, a_n)(i) = f^{\mathbf{A}_i}(a_1(i), \ldots, a_n(i)).$$

For each $i \in I$, the map $\pi_i\colon \prod_{i \in I} A_i \to A_i$; $a \mapsto a(i)$ is a surjective homomorphism from $\mathbf{A}$ onto $\mathbf{A}_i$, called the *ith projection* of $\mathbf{A}$. Note that for $I = \emptyset$, the direct product is a trivial $\mathcal{L}$-algebra that has no projections.

An $\mathcal{L}$-algebra $\mathbf{B}$ is called a *subdirect product* of a family of $\mathcal{L}$-algebras $\{\mathbf{A}_i\}_{i \in I}$ if it is a subalgebra of $\prod_{i \in I} \mathbf{A}_i$ and $\pi_i[B] = A_i$, for each $i \in I$. In this case, $\bigcap_{i \in I} \ker(\pi_i{\restriction}B) = \Delta_B$. Conversely, suppose that $\{\Theta_i\}_{i \in I}$ is a family of congruences of an $\mathcal{L}$-algebra $\mathbf{B}$ satisfying $\bigcap_{i \in I} \Theta_i = \Delta_B$. Then there exists an isomorphism φ from $\mathbf{B}$ onto a subdirect product of the family of $\mathcal{L}$-algebras $\{\mathbf{B}/\Theta_i\}_{i \in I}$ such that $\pi_i \varphi = \pi_{\Theta_i}$, for each $i \in I$.

An $\mathcal{L}$-algebra $\mathbf{B}$ is said to be *(finitely) subdirectly irreducible* if whenever φ is an isomorphism from $\mathbf{B}$ onto a subdirect product of a (non-empty finite) family of $\mathcal{L}$-algebras $\{\mathbf{A}_i\}_{i \in I}$, then $\pi_i \varphi$ is an isomorphism, for some $i \in I$. Note that a trivial $\mathcal{L}$-algebra is finitely subdirectly irreducible, but not subdirectly irreducible.

PROPOSITION B.6. *An $\mathcal{L}$-algebra $\mathbf{B}$ is* (a) *subdirectly irreducible if and only if Δ_B is completely meet-irreducible in* $\mathrm{Con}(\mathbf{B})$, *and* (b) *finitely subdirectly irreducible if and only if Δ_B is meet-irreducible in* $\mathrm{Con}(\mathbf{B})$.

PROOF. (a) Let $I := \mathrm{Con}(\mathbf{B}) \backslash \{\Delta_B\}$. Suppose first that Δ_B is not completely meet-irreducible in $\mathrm{Con}(\mathbf{B})$. Then $\bigcap I = \Delta_B$ and there is an isomorphism of $\mathbf{B}$ onto a subdirect product of $\{\mathbf{B}/\Theta \mid \Theta \in I\}$ such that $\pi_\Theta\colon \mathbf{B} \to \mathbf{B}/\Theta$; $b \mapsto [b]_\Theta$ is not an isomorphism for any $\Theta \in I$. So $\mathbf{B}$ is not subdirectly irreducible. Suppose now that Δ_B is completely meet-irreducible. Then $\bigcap I \neq \Delta_B$ and there exists $\langle a, b\rangle \in \bigcap I$ such that $a \neq b$. Let φ be any isomorphism from $\mathbf{B}$ to a subdirect product of a family of $\mathcal{L}$-algebras $\{\mathbf{A}_j\}_{j \in J}$. Since φ is an isomorphism, $\pi_j \varphi(a) = \varphi(a)(j) \neq \varphi(b)(j) = \pi_j \varphi(b)$, for some $j \in J$. Hence $\langle a, b\rangle \notin \ker(\pi_j \varphi)$ and $\bigcap I \not\subseteq \ker(\pi_j \varphi)$, yielding $\ker(\pi_j \varphi) = \Delta_B$, i.e., $\pi_j \varphi$ is an isomorphism. The proof of (b) is analogous. □

THEOREM B.7 (Subdirect representation theorem). *Every $\mathcal{L}$-algebra is isomorphic to a subdirect product of subdirectly irreducible $\mathcal{L}$-algebras.*

PROOF. Every trivial $\mathcal{L}$-algebra is isomorphic to the direct product of the empty set of $\mathcal{L}$-algebras, so let $\mathbf{B}$ be any non-trivial $\mathcal{L}$-algebra. Each element of an algebraic lattice is a meet of completely meet-irreducible elements (Proposition 1.3.8), so $\Delta_B = \bigcap I$, for some set I of completely meet-irreducible elements of $\mathrm{Con}(\mathbf{B})$ and $\mathbf{B}$ is isomorphic to a subdirect product of the family of $\mathcal{L}$-algebras $\{\mathbf{B}/\Theta \mid \Theta \in I\}$. Since each $\Theta \in I$ is completely meet-irreducible in $\mathrm{Con}(\mathbf{B})$, also $\Delta_{B/\Theta}$ is completely meet-irreducible in $\mathrm{Con}(\mathbf{B}/\Theta)$, by the Correspondence theorem (Theorem B.4.(c)). Hence $\mathbf{B}$ is isomorphic to a subdirect product of subdirectly irreducible $\mathcal{L}$-algebras, by Proposition B.6. □

Varieties and free algebras

Let $\mathcal{K}$ be a class of $\mathcal{L}$-algebras. We let $\mathbb{I}(\mathcal{K})$, $\mathbb{H}(\mathcal{K})$, $\mathbb{S}(\mathcal{K})$, and $\mathbb{P}(\mathcal{K})$ denote the classes of isomorphic copies, homomorphic images, subalgebras, and direct products of members of $\mathcal{K}$, respectively. The class $\mathcal{K}$ is called a *variety* if it is closed under $\mathbb{H}$, $\mathbb{S}$, and $\mathbb{P}$, that is, if $\mathbb{H}(\mathcal{K}) \subseteq \mathcal{K}$, $\mathbb{S}(\mathcal{K}) \subseteq \mathcal{K}$, and $\mathbb{P}(\mathcal{K}) \subseteq \mathcal{K}$. It is not hard to prove that $\mathbb{SH}(\mathcal{K}) \subseteq \mathbb{HS}(\mathcal{K})$, $\mathbb{PS}(\mathcal{K}) \subseteq \mathbb{SP}(\mathcal{K})$, and $\mathbb{PH}(\mathcal{K}) \subseteq \mathbb{HP}(\mathcal{K})$, and hence that $\mathbb{HSP}(\mathcal{K})$ is the smallest variety containing $\mathcal{K}$, called the variety *generated by* $\mathcal{K}$.

Let $\mathcal{V}$ be any variety of $\mathcal{L}$-algebras, and let X be any set. An $\mathcal{L}$-algebra $\mathbf{A} \in \mathcal{V}$ is said to be a *$\mathcal{V}$-free algebra over* X if it satisfies the following conditions:

(i) $\mathrm{Sg}^{\mathbf{A}}(X) = A$;

(ii) for any $\mathbf{B} \in \mathcal{V}$ and map $\varphi\colon X \to B$, there exists a unique homomorphism $\overline{\varphi}\colon \mathbf{A} \to \mathbf{B}$ extending φ.

The set X is then said to *freely generate* $\mathbf{A}$.

It is easy to see that any two $\mathcal{V}$-free algebras $\mathbf{A}_1$ and $\mathbf{A}_2$ over X are isomorphic. Observe first that the identity maps $\varphi_1\colon X \to A_1$ and $\varphi_2\colon X \to A_2$ extend to unique homomorphisms $\overline{\varphi}_1\colon \mathbf{A}_2 \to \mathbf{A}_1$ and $\overline{\varphi}_2\colon \mathbf{A}_1 \to \mathbf{A}_2$. But then, since $\varphi_1\varphi_2$ and $\varphi_2\varphi_1$ also extend the identity map on X, by uniqueness, $\overline{\varphi}_1\overline{\varphi}_2$ and $\overline{\varphi}_2\overline{\varphi}_1$ coincide with the identity maps on A_1 and A_2, respectively. So $\overline{\varphi}_1$ and $\overline{\varphi}_2$ are mutual isomorphisms.

We now establish the existence of *free semigroups* — free algebras in the variety of semigroups — which will both illustrate the general principles and provide the basis for proving that free algebras exist in every variety.

Let $\mathbf{A} = \langle A, \cdot\rangle$ be any semigroup and let $X \subseteq A$, denoting the product $a \cdot b$ of elements $a, b \in X$ in $\mathbf{A}$ by ab and omitting brackets for iterated products. We say that $\mathbf{A}$ is *uniquely representable* by X if each $a \in A$ is uniquely represented as $a = x_1 \cdots x_n$, for some $x_1, \ldots, x_n \in X$.

For any non-empty set X, the semigroup $\mathbf{S}_X = \langle \bigcup_{n\in\mathbb{N}^{>0}} X^n, \cdot\rangle$, with $\langle a_1, \ldots, a_n\rangle \cdot \langle b_1, \ldots, b_m\rangle := \langle a_1, \ldots, a_n, b_1, \ldots, b_m\rangle$, is clearly uniquely representable by X. In fact, $\mathbf{S}_X$ is the unique (up to isomorphism) free semigroup over X.

LEMMA B.8. *Let $\mathbf{A} = \langle A, \cdot\rangle$ be any semigroup and let $X \subseteq A$. Then $\mathbf{A}$ is uniquely representable by X if and only if it is a free semigroup over X.*

PROOF. Suppose first that $\mathbf{A}$ is uniquely representable by X. Clearly, $\mathrm{Sg}^{\mathbf{A}}(X) = A$ and given any semigroup $\mathbf{B}$, every map $\varphi\colon X \to B$ extends to a homomorphism $\overline{\varphi}\colon \mathbf{A} \to \mathbf{B}$; $x_1 \cdots x_n \mapsto \varphi(x_1) \cdots \varphi(x_n)$. So $\mathbf{A}$ is a free semigroup over X.

Now suppose conversely that $\mathbf{A}$ is a free semigroup over X. Then $\mathrm{Sg}^{\mathbf{A}}(X) = A$ and there exists a homomorphism $\overline{\varphi}\colon \mathbf{A} \to \mathbf{S}_X$ extending the identity map on X. Hence if $x_1 \cdots x_n = x'_1 \cdots x'_m$, for some $x_1, \ldots, x_n, x'_1, \ldots, x'_m \in X$, then $\varphi(x_1) \cdots \varphi(x_n) = \varphi(x'_1) \cdots \varphi(x'_m)$ and, since $\mathbf{S}_X$ is uniquely representable by X, we obtain $n = m$ and $x_i = x'_i$, for each $i \in \{1, \ldots, n\}$. So $\mathbf{A}$ is uniquely representable by X. □

An $\mathcal{L}$-algebra $\mathbf{A}$ is said to be a *formula algebra of signature $\mathcal{L}$ over a set X* (also often called a *term algebra*) provided it satisfies the following postulates, for all $f \in \mathcal{L}_n, g \in \mathcal{L}_m$ $(n, m \in \mathbb{N})$ and $a \in A^n, b \in A^m$:

(P1) $f(a) \notin X$;

(P2) $f(a) = g(b)$ implies $f = g$ and $a = b$;

(P3) $\mathrm{Sg}^{\mathbf{A}}(X) = A$.

Note that (P1) states that the images of operations are disjoint to the generating set, whereas (P2) states that the images of operations are pairwise disjoint and each operation is injective.

PROPOSITION B.9. *Let $\mathcal{L}$ be any signature.*

(a) *For any set $X \neq \emptyset$, there exists a formula algebra of signature $\mathcal{L}$ over X.*

(b) *There exists a formula algebra of signature $\mathcal{L}$ over $\emptyset$ if and only if $\mathcal{L}_0 \neq \emptyset$.*

PROOF. Suppose that $\mathcal{L} = \{\nu_i \mid i \in I\}$ and X is any set such that $X \cup \mathcal{L}_0 \neq \emptyset$, assuming without loss of generality that $I \cap X = \emptyset$. Let $\mathbf{S}$ be the free semigroup over $I \cup X$ as constructed above and define inductively $B_0 := X$ and $B_{k+1} := \{fb_1 \cdots b_n \mid n \in \mathbb{N},\ f \in \mathcal{L}_n, \text{ and } b_1, \ldots, b_n \in B_k\}$, for each $k \in \mathbb{N}$. We define an $\mathcal{L}$-algebra $\mathbf{B}$ with universe $B := \bigcup\{B_k \mid k \in \mathbb{N}\}$ and, for each $f \in \mathcal{L}_n$ $(n \in \mathbb{N})$, an operation $f^{\mathbf{B}}\colon B^n \to B;\ \langle b_1, \ldots, b_n\rangle \mapsto fb_1 \cdots b_n$. That $\mathbf{B}$ satisfies conditions (P1), (P2), and (P3) in the definition of a formula algebra follows from the fact that $\mathbf{S}$ is uniquely representable by $I \cup X$.

To conclude the proof of (b), just observe that if $X \cup \mathcal{L}_0 = \emptyset$, then $\mathrm{Sg}^{\mathbf{A}}(X) = \emptyset$, for any $\mathcal{L}$-algebra $\mathbf{A}$, so there is no formula algebra of signature $\mathcal{L}$ over $\emptyset$. □

It follows that any formula algebra of signature $\mathcal{L}$ over a set X is a free algebra of the class of all $\mathcal{L}$-algebras over X, and hence unique up to isomorphism. We denote such an $\mathcal{L}$-algebra by $\mathbf{Fm}_{\mathcal{L}}(X)$, or simply $\mathbf{Fm}(X)$ when $\mathcal{L}$ is understood, and call the members of $\mathrm{Fm}(X)$, denoted by lowercase Greek letters $\alpha, \beta, \gamma, \ldots$, *$\mathcal{L}$-formulas*. *Subformulas* are defined inductively by the conditions (i) α is a subformula of α; and (ii) if $\alpha = f(\beta_1, \ldots, \beta_n)$ and γ is a subformula of β_i, for some $i \in \{1, \ldots, n\}$, then γ is a subformula of α.

For any $\mathcal{L}$-algebra $\mathbf{A}$, $k \in \mathbb{N}$, and $\alpha \in \mathrm{Fm}(\{x_1, \ldots, x_k\})$, the function $\alpha^{\mathbf{A}}\colon A^k \to A$ is defined inductively as follows:

(i) $x_i^{\mathbf{A}}(a_1, \ldots, a_k) = a_i$, for each $i \in \{1, \ldots, k\}$;

(ii) for all $n \in \mathbb{N}$, $f \in \mathcal{L}_n$, and $\beta_1, \ldots, \beta_n \in \mathrm{Fm}(\{x_1, \ldots, x_k\})$,

$$f(\beta_1, \ldots, \beta_n)^{\mathbf{A}}(a_1, \ldots, a_k) = f^{\mathbf{A}}(\beta_1^{\mathbf{A}}(a_1, \ldots, a_k), \ldots, \beta_n^{\mathbf{A}}(a_1, \ldots, a_k)).$$

An easy induction then shows that, for any $\Theta \in \mathrm{Con}(\mathbf{A})$ and $\alpha \in \mathrm{Fm}(\{x_1, \ldots, x_k\})$,

$$\langle a_1, b_1\rangle, \ldots, \langle a_k, b_k\rangle \in \Theta \implies \langle \alpha^{\mathbf{A}}(a_1, \ldots, a_k), \alpha^{\mathbf{A}}(b_1, \ldots, b_k)\rangle \in \Theta.$$

Given any variety $\mathcal{V}$ of $\mathcal{L}$-algebras and a set X such that $X \cup \mathcal{L}_0 \neq \emptyset$, let

$$I_{\mathcal{V}}(X) := \{\Theta \in \mathrm{Con}(\mathbf{Fm}(X)) \mid \mathbf{Fm}(X)/\Theta \in \mathcal{V}\} \quad \text{and} \quad \Theta_{\mathcal{V}}(X) := \bigcap I_{\mathcal{V}}(X).$$

Set $\bar{\alpha} := [\alpha]_{\Theta_{\mathcal{V}}(X)}$, for each $\alpha \in \mathrm{Fm}(X)$ and $\overline{X} := \{\bar{x} \mid x \in X\}$, and let

$$\mathbf{F}_{\mathcal{V}}(X) := \mathbf{Fm}(X)/\Theta_{\mathcal{V}}(X).$$

The $\mathcal{L}$-algebra $\mathbf{F}_{\mathcal{V}}(X)$ is generated by $\overline{X}$ and determined up to isomorphism by $\mathcal{V}$ and $|X|$. Moreover, if $\mathcal{V}$ contains a non-trivial $\mathcal{L}$-algebra $\mathbf{B}$, then distinct elements of X can be distinguished by a homomorphism from $\mathbf{Fm}(X)$ to $\mathbf{B}$, and $|X| = |\overline{X}|$. If the variety $\mathcal{V}$ is clear from the context, we omit the subscript and write $\mathbf{F}(X)$.

Now let $\mathbf{A} := \prod\{\mathbf{Fm}(X)/\Theta \mid \Theta \in I_{\mathcal{V}}(X)\} \in \mathcal{V}$ and let $\varphi\colon \mathbf{Fm}(X) \to \mathbf{A}$ be the induced homomorphism satisfying $\varphi(\alpha)(\Theta) = \pi_\Theta(\alpha)$, for each $\alpha \in \mathrm{Fm}(X)$ and $\Theta \in I_{\mathcal{V}}(X)$. Then $\ker(\varphi) = \Theta_{\mathcal{V}}(X)$ and $\mathbf{F}(X)$ is isomorphic to a subalgebra of $\mathbf{A}$, by the Homomorphism theorem (Corollary B.3). So $\mathbf{F}(X) \in \mathcal{V}$.

To confirm that $\mathbf{F}(X)$ is indeed a $\mathcal{V}$-free algebra over $\overline{X}$, it remains to show that it is freely generated by $\overline{X}$. Consider any $\mathbf{B} \in \mathcal{V}$ and map $\varphi\colon \overline{X} \to B$. Let $\pi_{\Theta_{\mathcal{V}}}\colon \mathbf{Fm}(X) \to \mathbf{F}(X);\ \alpha \mapsto \bar{\alpha}$ be the canonical map. Then $\psi = \varphi(\pi_{\Theta_{\mathcal{V}}}\restriction X)$ is a map from X to B and extends to a unique homomorphism $\overline{\psi}\colon \mathbf{Fm}(X) \to \mathbf{B}$. Since $\ker(\pi_{\Theta_{\mathcal{V}}}) = \Theta_{\mathcal{V}}(X) \subseteq \ker(\overline{\psi})$, there exists, by the General homomorphism theorem (Theorem B.2), a homomorphism $\overline{\varphi}\colon \mathbf{F}(X) \to \mathbf{B}$ such that $\overline{\psi} = \overline{\varphi}\pi_{\Theta_{\mathcal{V}}}$. But then $\overline{\varphi}(\bar{x}) = \overline{\varphi}\pi_{\Theta_{\mathcal{V}}}(x) = \overline{\psi}(x) = \varphi\pi_{\Theta_{\mathcal{V}}}(x) = \varphi(\bar{x})$, for each $x \in X$.

Now consider any set Y such that $X \subseteq Y$. It is easily checked that $\Theta_{\mathcal{V}}(X) = \Theta_{\mathcal{V}}(Y) \cap \mathrm{Fm}(X)^2$ and hence that $\mathbf{F}(X)$ embeds into $\mathbf{F}(Y)$. We may therefore harmlessly assume that $\mathbf{F}(X)$ is a subalgebra of $\mathbf{F}(Y)$.

LEMMA B.10. *Let $\mathcal{V}$ be a variety of $\mathcal{L}$-algebras, X and Y sets such that $X \cup \mathcal{L}_0 \neq \emptyset$ and $X \subseteq Y$, and $\Theta \in \mathrm{Con}(\mathbf{F}(X))$. Then $\mathrm{Cg}^{\mathbf{F}(Y)}(\Theta) \cap \mathrm{F}(X)^2 = \Theta$.*

PROOF. Let $\pi_\Theta\colon \mathbf{F}(X) \to \mathbf{F}(X)/\Theta$ be the canonical map. Fix some $c \in X \cup \mathcal{L}_0$ and let $\psi\colon \mathbf{F}(Y) \to \mathbf{F}(X)/\Theta$ be the homomorphism satisfying $\psi(\bar{x}) = \pi_\Theta(\bar{x})$, for each $x \in X$, and $\psi(\bar{y}) = [\bar{c}]_\Theta$, for all $y \in Y \backslash X$. Then $\Theta = \ker(\pi_\Theta) = \ker(\psi) \cap \mathrm{F}(X)^2$ and hence $\Theta \subseteq \mathrm{Cg}^{\mathbf{F}(Y)}(\Theta) \cap \mathrm{F}(X)^2 \subseteq \ker(\psi) \cap \mathrm{F}(X)^2 = \Theta$. □

Equational classes and the HSP theorem

Let us fix a signature $\mathcal{L}$ and set X such that $X \cup \mathcal{L}_0 \neq \emptyset$. An *$\mathcal{L}$-equation over X* is an ordered pair of $\mathcal{L}$-formulas $\alpha, \beta \in \mathrm{Fm}(X)$, usually written $\alpha \approx \beta$. The set of all $\mathcal{L}$-equations over X is denoted by $\mathrm{Eq}_{\mathcal{L}}(X)$ or just $\mathrm{Eq}(X)$ when $\mathcal{L}$ is understood.

An $\mathcal{L}$-algebra $\mathbf{A}$ *satisfies* $\alpha \approx \beta \in \mathrm{Eq}(X)$, written $\mathbf{A} \models \alpha \approx \beta$, if $\varphi(\alpha) = \varphi(\beta)$ for every homomorphism $\varphi\colon \mathbf{Fm}(X) \to \mathbf{A}$. More generally, a class $\mathcal{K}$ of $\mathcal{L}$-algebras *satisfies* $\Sigma \subseteq \mathrm{Eq}(X)$, denoted by $\mathcal{K} \models \Sigma$ (omitting brackets when $\mathcal{K}$ or Σ is a singleton), if every member of $\mathcal{K}$ satisfies every equation in Σ.

It is straightforward to prove that, for any class of $\mathcal{L}$-algebras $\mathcal{K}$, the classes $\mathcal{K}$, $\mathbb{I}(\mathcal{K})$, $\mathbb{H}(\mathcal{K})$, $\mathbb{S}(\mathcal{K})$, and $\mathbb{P}(\mathcal{K})$ satisfy the same $\mathcal{L}$-equations over any set. In particular, $\mathcal{K}$ and the variety $\mathbb{HSP}(\mathcal{K})$ satisfy the same $\mathcal{L}$-equations.

LEMMA B.11. *Let $\mathcal{V}$ be a variety of $\mathcal{L}$-algebras. For any $\alpha \approx \beta \in \mathrm{Eq}(X)$,*

$$\mathcal{V} \models \alpha \approx \beta \iff \mathbf{F}(X) \models \alpha \approx \beta \iff \langle \alpha, \beta \rangle \in \Theta_{\mathcal{V}}(X).$$

PROOF. If $\mathcal{V} \models \alpha \approx \beta$, then $\mathbf{F}(X) \models \alpha \approx \beta$, since $\mathbf{F}(X) \in \mathcal{V}$. If $\mathbf{F}(X) \models \alpha \approx \beta$, then $\langle \alpha, \beta \rangle \in \ker(\pi_{\Theta_{\mathcal{V}}(X)}) = \Theta_{\mathcal{V}}(X)$. Finally, let $\langle \alpha, \beta \rangle \in \Theta_{\mathcal{V}}(X)$. For any $\mathbf{A} \in \mathcal{V}$ and homomorphism $\varphi\colon \mathbf{Fm}(X) \to \mathbf{A}$, the Homomorphism theorem (Corollary B.3) yields $\mathbf{Fm}(X)/\ker(\varphi) \in \mathcal{V}$. So $\langle \alpha, \beta \rangle \in \Theta_{\mathcal{V}}(X) \subseteq \ker(\varphi)$. □

Given any $\Sigma \subseteq \mathrm{Eq}(X)$, let $\mathrm{Mod}(\Sigma)$ denote the class of $\mathcal{L}$-algebras that satisfy Σ. A class $\mathcal{K}$ of $\mathcal{L}$-algebras is called an *equational class* if $\mathcal{K} = \mathrm{Mod}(\Sigma)$, for some set X and $\Sigma \subseteq \mathrm{Eq}(X)$, and Σ is then called an *equational basis* for $\mathcal{K}$. If $\mathcal{K} = \mathrm{Mod}(\Sigma)$, for some finite set $\Sigma \subseteq \mathrm{Eq}(X)$, then $\mathcal{K}$ is said to be *finitely based*. We also say that a class $\mathcal{K}'$ of $\mathcal{L}$-algebras is *defined relative to $\mathcal{K}$ by $\Sigma \subseteq \mathrm{Eq}(X)$* if $\mathcal{K}'$ consists of all the members of $\mathcal{K}$ that satisfy Σ.

THEOREM B.12 (HSP theorem). *A class of $\mathcal{L}$-algebras is an equational class if and only if it is a variety.*

PROOF. Suppose first that $\mathcal{K}$ is an equational class. Then $\mathcal{K} = \mathrm{Mod}(\Sigma)$, for some $\Sigma \subseteq \mathrm{Eq}(X)$. Since $\mathcal{K}$ and $\mathbb{HSP}(\mathcal{K})$ satisfy the same equations over X, it follows that $\mathbb{HSP}(\mathcal{K}) \subseteq \mathrm{Mod}(\Sigma) = \mathcal{K} \subseteq \mathbb{HSP}(\mathcal{K})$. Hence $\mathcal{K} = \mathbb{HSP}(\mathcal{K})$ is a variety.

Now let $\mathcal{V}$ be a variety and X any infinite set, and define

$$\Sigma := \{\alpha \approx \beta \in \mathrm{Eq}(X) \mid \mathcal{V} \models \alpha \approx \beta\} \quad \text{and} \quad \mathcal{V}' := \mathrm{Mod}(\Sigma).$$

Since $\mathcal{V}'$ is an equational class, it is a variety, by the first part of this proof. Also, $\mathcal{V}' \models \alpha \approx \beta$ if and only if $\mathcal{V} \models \alpha \approx \beta$, for all $\alpha \approx \beta \in \mathrm{Eq}(X)$. We prove that $\mathcal{V} = \mathcal{V}'$. Clearly, $\mathcal{V} \subseteq \mathcal{V}'$, so consider any $\mathbf{A} \in \mathcal{V}'$ and set Y satisfying $|Y| \geq |A|$. There exists a surjective map $\varphi\colon \overline{Y} \to A$ extending to a surjective homomorphism $\overline{\varphi}\colon \mathbf{F}_{\mathcal{V}'}(Y) \to \mathbf{A}$, so $\mathbf{A} \in \mathbb{H}(\mathbf{F}_{\mathcal{V}'}(Y))$. It remains to show that $\Theta_{\mathcal{V}}(Y) = \Theta_{\mathcal{V}'}(Y)$ and hence $\mathbf{A} \in \mathbb{H}(\mathbf{F}_{\mathcal{V}'}(Y)) = \mathbb{H}(\mathbf{F}_{\mathcal{V}}(Y)) \subseteq \mathcal{V}$. Clearly, $\Theta_{\mathcal{V}'}(Y) \subseteq \Theta_{\mathcal{V}}(Y)$, so let $\langle \alpha, \beta \rangle \in \Theta_{\mathcal{V}}(Y)$. Then $\mathcal{V} \models \alpha \approx \beta$, by Lemma B.11. But also $\alpha \approx \beta \in \mathrm{Eq}(Y')$, for some finite $Y' \subseteq Y$, and, since X is infinite, we may assume that $Y' \subseteq X$. Hence $\mathcal{V}' \models \alpha \approx \beta$ and Lemma B.11 yields $\langle \alpha, \beta \rangle \in \Theta_{\mathcal{V}'}(Y)$ as required. □

Congruence lattice properties for the algebras of a variety are in some cases reflected by the satisfaction of certain equations, as illustrated by the following result.

PROPOSITION B.13. *Let $\mathcal{V}$ be a variety of $\mathcal{L}$-algebras and suppose that there exists a formula $\alpha \in \mathrm{Fm}(\{x, y, z\})$ satisfying $\mathcal{V} \models \alpha(x, x, z) \approx z$ and $\mathcal{V} \models \alpha(x, z, z) \approx x$. Then $\mathcal{V}$ is congruence-permutable.*

PROOF. Let $\mathbf{A} \in \mathcal{V}$ and $\Theta_1, \Theta_2 \in \mathrm{Con}(\mathbf{A})$, and suppose that $\langle a, b \rangle \in \Theta_1 \circ \Theta_2$. Then $a\, \Theta_1\, c$ and $c\, \Theta_2\, b$, for some $c \in A$, and

$$a = \alpha^{\mathbf{A}}(a, b, b)\; \Theta_2\; \alpha^{\mathbf{A}}(a, c, b)\; \Theta_1\; \alpha^{\mathbf{A}}(c, c, b) = b.$$

Hence $\langle a,b\rangle \in \Theta_2 \circ \Theta_1$. So $\Theta_1 \circ \Theta_2 \subseteq \Theta_2 \circ \Theta_1$ and it follows by symmetry that $\Theta_1 \circ \Theta_2 = \Theta_2 \circ \Theta_1$. □

Similarly, the proof of Proposition B.5 can be adapted to show that a variety $\mathcal{V}$ of $\mathcal{L}$-algebras is congruence-distributive if there exists a $\beta \in \mathrm{Fm}(\{x,y,z\})$ satisfying $\mathcal{V} \models \beta(x,x,y) \approx x$, $\mathcal{V} \models \beta(x,y,x) \approx x$, and $\mathcal{V} \models \beta(y,x,x) \approx x$.

Ultraproducts

An *ultrafilter* on a set I is a lattice-filter F of $\langle \wp(I), \subseteq\rangle$ such that $\emptyset \notin F$ and, for any $J \subseteq I$, either $J \in F$ or $I \backslash J \in F$. The following principle is proved by a standard application of Zorn's lemma.

LEMMA B.14 (Ultrafilter lemma). *Let I be any set and suppose that $X \subseteq \wp(I)$ satisfies $Y_1 \cap \cdots \cap Y_n \neq \emptyset$, for any $Y_1, \ldots, Y_n \in X$. Then there exists an ultrafilter U on I such that $X \subseteq U$.*

Let us fix a signature $\mathcal{L}$. Given any family of $\mathcal{L}$-algebras $\{\mathbf{A}_i\}_{i\in I}$ and an ultrafilter U on I, let Θ_U be the binary relation on $\prod_{i\in I} A_i$ defined by

$$\langle a,b\rangle \in \Theta_U \;:\Longleftrightarrow\; \{i \in I \mid \pi_i(a) = \pi_i(b)\} \in U.$$

Then Θ_U is a congruence of $\prod_{i\in I} \mathbf{A}_i$ and the quotient $\prod_{i\in I} \mathbf{A}_i/\Theta_U$, also denoted by $\prod_{i\in I} \mathbf{A}_i/U$, is called the *ultraproduct* of $\{\mathbf{A}_i \mid i \in I\}$ modulo U.

Let $\mathcal{K}$ be a class of $\mathcal{L}$-algebras and let $\mathbb{P}_U(\mathcal{K})$ denote the class of ultraproducts of members of $\mathcal{K}$. The class $\mathcal{K}$ is said to be a *quasivariety* if it is closed under $\mathbb{I}$, $\mathbb{S}$, $\mathbb{P}$, and $\mathbb{P}_U$, that is, if $\mathbb{I}(\mathcal{K}) \subseteq \mathcal{K}$, $\mathbb{S}(\mathcal{K}) \subseteq \mathcal{K}$, $\mathbb{P}(\mathcal{K}) \subseteq \mathcal{K}$, and $\mathbb{P}_U(\mathcal{K}) \subseteq \mathcal{K}$. It is straightforward to prove that $\mathbb{P}_U\mathbb{S}(\mathcal{K}) \subseteq \mathbb{S}\mathbb{P}_U(\mathcal{K})$ and $\mathbb{P}_U\mathbb{P}(\mathcal{K}) \subseteq \mathbb{S}\mathbb{P}\mathbb{P}_U(\mathcal{K})$, and hence that $\mathbb{I}\mathbb{S}\mathbb{P}\mathbb{P}_U(\mathcal{K})$ is the smallest quasivariety containing $\mathcal{K}$, called the quasivariety *generated by* $\mathcal{K}$. In the case where $\mathcal{K}$ is a finite set of finite $\mathcal{L}$-algebras, it is closed under taking ultraproducts, and the quasivariety $\mathbb{I}\mathbb{S}\mathbb{P}\mathbb{P}_U(\mathcal{K}) = \mathbb{I}\mathbb{S}\mathbb{P}(\mathcal{K})$ is said to be *finitely generated.*

Let X be any set. An *$\mathcal{L}$-quasi-equation over X* is an ordered pair consisting of a finite set $\{\varepsilon_1, \ldots, \varepsilon_n\}$ of $\mathcal{L}$-equations over X and an $\mathcal{L}$-equation ε over X, written

$$\varepsilon_1 \,\&\, \cdots \,\&\, \varepsilon_n \implies \varepsilon.$$

An $\mathcal{L}$-algebra $\mathbf{A}$ *satisfies* an $\mathcal{L}$-quasi-equation $\varepsilon_1 \,\&\, \cdots \,\&\, \varepsilon_n \implies \varepsilon$ over X if for every homomorphism $\varphi\colon \mathbf{Fm}(X) \to \mathbf{A}$, whenever $\{\varepsilon_1, \ldots, \varepsilon_n\} \subseteq \ker(\varphi)$, also $\varepsilon \in \ker(\varphi)$. More generally, a class $\mathcal{K}$ of $\mathcal{L}$-algebras *satisfies* a set Σ of $\mathcal{L}$-quasi-equations over X, written $\mathcal{K} \models \Sigma$ (omitting brackets when $\mathcal{K}$ or Σ is a singleton), if every member of $\mathcal{K}$ satisfies every $\mathcal{L}$-quasi-equation in Σ.

It is straightforward to prove that, for any class $\mathcal{K}$ of $\mathcal{L}$-algebras, the classes $\mathcal{K}$, $\mathbb{I}(\mathcal{K})$, $\mathbb{S}(\mathcal{K})$, $\mathbb{P}(\mathcal{K})$, and $\mathbb{P}_U(\mathcal{K})$ satisfy the same $\mathcal{L}$-quasi-equations over any set. In particular, $\mathcal{K}$ and the quasivariety $\mathbb{I}\mathbb{S}\mathbb{P}\mathbb{P}_U(\mathcal{K})$ satisfy the same $\mathcal{L}$-quasi-equations.

We conclude with a useful result first proved in [**96**, Lemma 1.5].

THEOREM B.15 (Relativized Jónsson lemma). *Let $\mathcal{K}$ be a class of $\mathcal{L}$-algebras and let $\mathbf{A} \in \mathbb{I}\mathbb{S}\mathbb{P}(\mathcal{K})$. If Δ_A is meet-irreducible in* $\mathrm{Con}(\mathbf{A})$*, then $\mathbf{A} \in \mathbb{I}\mathbb{S}\mathbb{P}_U(\mathcal{K})$.*

Proof. Let $\mathbf{A}$ be a subalgebra of $\mathbf{B} = \prod_{i \in I} \mathbf{B}_i$, where $\mathbf{B}_i \in \mathcal{K}$, for each $i \in I$, and suppose that Δ_A is meet-irreducible in $\mathrm{Con}(\mathbf{A})$. For each lattice-filter F of $\langle \wp(I), \subseteq \rangle$, consider the congruence Θ_F of $\mathbf{B}$ defined by

$$a \,\Theta_F\, b \;:\Longleftrightarrow\; \{i \in I \mid a(i) = b(i)\} \in F.$$

Then $\Theta_F \cap A^2$ is a congruence of $\mathbf{A}$ and, for any lattice-filters F_1, F_2 of $\langle \wp(I), \subseteq \rangle$,

$$\Theta_{F_1} \cap \Theta_{F_2} = \Theta_{F_1 \cap F_2}.$$

We consider the set of lattice-filters of $\langle \wp(I), \subseteq \rangle$

$$\mathcal{F} := \{F \subseteq \wp(I) \mid F \text{ is a lattice-filter of } \langle \wp(I), \subseteq \rangle \text{ and } \Theta_F \cap A^2 = \Delta_A\}.$$

Observe that $\{I\} \in \mathcal{F} \neq \emptyset$. Moreover, if $\mathcal{C}$ is a non-empty chain in $\langle \mathcal{F}, \subseteq \rangle$, then $\Theta_{\bigcup \mathcal{C}} \cap A^2 = \Delta_A$, so $\mathcal{C}$ has an upper bound in $\langle \mathcal{F}, \subseteq \rangle$. Hence, by Zorn's lemma, $\langle \mathcal{F}, \subseteq \rangle$ has a maximal element U. We claim that U is an ultrafilter on I.

Let $\emptyset \neq J \subseteq I$. Let $F_1 := U \vee {\uparrow} J$ and $F_2 := U \vee {\uparrow}(I \backslash J)$. Then $(\Theta_{F_1} \cap A^2) \cap (\Theta_{F_2} \cap A^2) = (\Theta_{F_1} \cap \Theta_{F_2}) \cap A^2 = (\Theta_{F_1 \cap F_2}) \cap A^2 = \Theta_U \cap A^2 = \Delta_A$. Since Δ_A is meet-irreducible in $\mathrm{Con}(\mathbf{A})$, either $\Theta_{F_1} \cap A^2 = \Delta_A$ or $\Theta_{F_2} \cap A^2 = \Delta_A$. But then the maximality of U implies that $J \in U$ or $I \backslash J \in U$, establishing that U is an ultrafilter on I. Since $\Theta_U \cap A^2 = \Delta_A$, it follows by the Homomorphism theorem (Corollary B.3) that $\mathbf{L}$ is isomorphic to a subalgebra of $\mathbf{B}/\Theta_U$. □

Index

Bibliography

[1] W. Ackermann, *Begründung einer strengen implikation*, Journal of Symbolic Logic **21** (1956), 113–128.

[2] S. Aguzzoli and M. Bianchi, *Amalgamation property for varieties of BL-algebras generated by one chain with finitely many components*, Proceedings of RAMiCS 2021 (U. Fahrenberg, M. Gehrke, L. Santocanale, and M. Winter, eds.), Lecture Notes in Computer Science, vol. 13027, Springer, 2021, pp. 1–18.

[3] P. Aglianò and F. Montagna, *Varieties of BL-algebras I: general properties*, Journal of Pure and Applied Algebra **181** (2003), 105–129.

[4] P. Aglianò, F. Montagna, and I.M.A. Ferreirim, *Basic hoops: an algebraic study of continuous t-norms*, Studia Logica **87** (2007), no. 1, 73–98.

[5] A.R. Anderson and N.D. Belnap Jr., *Entailment: the logic of relevance and necessity*, Princeton University Press, 1975.

[6] A.R. Anderson, N.D. Belnap Jr., and J.M. Dunn, *Entailment: the logic of relevance and necessity II*, Princeton University Press, 1992.

[7] M. Anderson, P. Conrad, and J. Martinez, *The lattice of convex ℓ-subgroups of a lattice-ordered group*, Lattice-ordered groups (A.M.W. Glass and W.C. Holland, eds.), Reidel, 1989, pp. 105–127.

[8] M. Anderson and C.C. Edwards, *A representation theorem for distributive ℓ-monoids*, Canadian Mathematical Bulletin **27** (1984), 238–240.

[9] M. Anderson and T. Feil, *Lattice-ordered groups: an introduction*, Reidel, 1988.

[10] A. Avron, *A constructive analysis of RM*, Journal of Symbolic Logic **52** (1987), no. 4, 939–951.

[11] A. Avron, *The semantics and proof theory of linear logic*, Theoretical Computer Science **57** (1988), 161–184.

[12] A. Avron, *Hypersequents, logical consequence and intermediate logics for concurrency*, Annals of Mathematics and Artificial Intelligence **4** (1991), no. 3–4, 225–248.

[13] S. Awodey, *Category theory*, Oxford University Press, 2006.

[14] M. Baaz, A. Ciabattoni, and F. Montagna, *Analytic calculi for monoidal t-norm based logic*, Fundamenta Informaticae **59** (2004), no. 4, 315–332.

[15] P.D. Bacsich, *Amalgamation properties and interpolation theorems for equational theories*, Algebra Universalis **5** (1975), 45–55.

[16] P. Bahls, J. Cole, N. Galatos, P. Jipsen, and C. Tsinakis, *Cancellative residuated lattices*, Algebra Universalis **50** (2003), no. 1, 83–106.

[17] P. Baldi and K. Terui, *Densification of FL chains via residuated frames*, Algebra Universalis **75** (2016), no. 2, 169–195.

[18] B. Banaschewski, *Hüllensysteme und erweiterung von quasi-ordnungen*, Zeitschrift für Mathematische Logik und Grundlagen der Mathematik **2** (1956), 117–130.

[19] B. Banaschewski and E. Nelson, *On residual finiteness and finite embeddability*, Algebra Universalis **2** (1972), 361–364.

[20] E. Barrio, L. Rosenblatt, and D. Tajer, *The logics of strict-tolerant logic*, Journal of Philosophical Logic **44** (2015), 551–571.

[21] J.C. Beall and J. Murzi, *Two flavors of Curry's paradox*, Journal of Philosophy **110** (2013), 143–165.

[22] F. Belardinelli, P. Jipsen, and H. Ono, *Algebraic aspects of cut elimination*, Studia Logica **77** (2004), 209–240.

[23] N.D. Belnap Jr., *Display logic*, Journal of Philosophical Logic **11** (1982), 375–417.

[24] C. Bergman, *Universal algebra: fundamentals and selected topics*, CRC Press, 2012.

[25] L. Beran, *Orthomodular lattices: algebraic approach*, Reidel, 1985.

[26] J. Berman, *Homogeneous lattices and lattice-ordered groups*, Colloquium Mathematicae **32** (1974), no. 1, 13–24.

[27] A. Bigard, K. Keimel, and A. Wolfenstein, *Groupes et anneaux réticulés*, Springer, 1977.

[28] K. Bimbó, *Proof theory: sequent calculi and related formalisms*, CRC Press, 2014.

[29] G. Birkhoff, *On the structure of abstract algebras*, Proceedings of the Cambridge Philosophical Society **31** (1935), 433–454.

[30] G. Birkhoff, *Lattice-ordered groups*, Annals of Mathematics **43** (1942), 298–331.

[31] G. Birkhoff, *Subdirect unions in universal algebra*, Bulletin of the American Mathematical Society **50** (1944), 764–768.

[32] G. Birkhoff, *Lattice theory*, American Mathematical Society, 1940.

[33] W.J. Blok and I.M.A. Ferreirim, *On the structure of hoops*, Algebra Universalis **43** (2000), 233–257.

[34] W.J. Blok and B. Jónsson, *Equivalence of consequence operators*, Studia Logica **83** (2006), 91–110.

[35] W.J. Blok and D. Pigozzi, *A. Tarski's work on general metamathematics*, Journal of Symbolic Logic **53** (1988), 36–50.

[36] W.J. Blok and D. Pigozzi, *Algebraizable logics*, American Mathematical Society, 1989.

[37] W.J. Blok and J.G. Raftery, *Assertionally equivalent quasivarieties*, International Journal of Algebra and Computation **18** (2008), no. 4, 589–681.

[38] W.J. Blok and J.G. Raftery, *Constructing simple residuated lattices*, Algebra Universalis **50** (2003), 385–389.

[39] W.J. Blok and J. Rebagliato, *Algebraic semantics for deductive systems*, Studia Logica **73** (2003), 1–29.

[40] W.J. Blok and C.J. van Alten, *The finite embeddability property for residuated lattices, pocrims and BCK-algebras*, Algebra Universalis **48** (2002), 253–271.

[41] W.J. Blok and C.J. van Alten, *On the finite embeddability property for residuated ordered groupoids*, Transactions of the American Mathematical Society **357** (2005), no. 10, 4141–4157.

[42] C. Bloomfield and Y. Maruyama, *Fibered universal algebra for first-order logics.* Manuscript.

[43] K. Blount, *On the structure of residuated lattices*, Ph.D. thesis, Vanderbilt University, 1999.

[44] K. Blount and C. Tsinakis, *The structure of residuated lattices*, International Journal of Algebra and Computation **13** (2003), no. 4, 437–461.

[45] T.S. Blyth, *Lattices and ordered algebraic structures*, Springer, 2005.

[46] T.S. Blyth and M.F. Janowitz, *Residuation theory*, Pergamon Press, 1972.

[47] B. Bosbach, *Concerning cone algebras*, Algebra Universalis **15** (1982), 58–66.

[48] B. Bosbach, *Residuationsstrukturen*, Universität Kassel, 1997.

[49] M. Botur, J. Kühr, L. Liu, and C. Tsinakis, *The Conrad program: from ℓ-groups to algebras of logic*, Journal of Algebra **450** (2016), 173–203.

[50] M. Botur, J. Kühr, and C. Tsinakis, *Strong simplicity and states in ordered algebras: pushing the limits.* Journal of Pure and Applied Algebra, to appear.

[51] F. Bou, F. Esteva, L. Godo, and R. Rodríguez, *On the minimum many-valued logic over a finite residuated lattice*, Journal of Logic and Computation **21** (2011), no. 5, 739–790.

[52] J.R. Büchi, *Representation of complete lattices by sets*, Portugaliae Mathematica **11** (1952), 151–167.

[53] S. Burris and H.P. Sankappanavar, *A course in universal algebra*, Springer, 1981.

[54] W. Buszkowski, *Interpolation and FEP for logics of residuated algebras*, Logic Journal of the IGPL **19** (2011), no. 3, 437–454.

[55] M. Busaniche and F. Montagna, *Hájek's logic BL and BL-algebras*, Handbook of mathematical fuzzy logic (P. Cintula, P. Hájek, and C. Noguera, eds.), College Publications, 2011, pp. 355–447.

[56] M. Busaniche and D. Mundici, *Geometry of Robinson joint consistency in Łukasiewicz logic*, Annals of Pure and Applied Logic **147** (2007), 1–22.

[57] X. Caicedo, G. Metcalfe, R. Rodríguez, and R. Rogger, *Decidability in order-based modal logics*, Journal of Computer System Sciences **88** (2017), 53–74.

[58] X. Caicedo, G. Metcalfe, R. Rodríguez, and O. Tuyt, *One-variable fragments of intermediate logics over linear frames*, Information and Computation **287** (2022).

[59] A. Cantini, *I fondamenti della matematica*, Loescher, 1979.

[60] R. Cardona and N. Galatos, *The finite embeddability property for noncommutative knotted extensions of RL*, International Journal of Algebra and Computation **25** (2017), no. 3, 349–379.

[61] E. Casari, *Comparative logics and Abelian l-groups*, Logic Colloquium '88 (R. Ferro, C. Bonotto, S. Valentini, and A. Zanardo, eds.), North Holland, 1989, pp. 161–190.

[62] E. Casari, *Bolzano's logical system*, Oxford Science Publications, 2016.

[63] J. Certaine, *Lattice-ordered groupoids and some related problems*, Ph.D. thesis, Harvard University, 1945.

[64] I. Chajda and H. Länger, *Orthomodular lattices can be converted into left-residuated ℓ-groupoids*, Miskolc Mathematical Notes **18** (2017), no. 2, 685–689.

[65] P.M. Cohn, *Universal algebra*, 2nd ed., Reidel, 1981.

[66] C.C. Chang, *Algebraic analysis of many-valued logics*, Transactions of the American Mathematical Society **88** (1958), 467–490.

[67] C.C. Chang, *A new proof of the completeness of the Łukasiewicz axioms*, Transactions of the American Mathematical Society **93** (1959), 74–90.

[68] W. Chen and Y. Chen, *Variety generated by conical residuated lattice-ordered idempotent monoids*, Semigroup Forum **98** (2019), 431–455.

[69] K. Chvalovský and R. Horčík, *Full Lambek calculus with contraction is undecidable*, Journal of Symbolic Logic **81** (2016), no. 2, 524–540.

[70] A. Ciabattoni, N. Galatos, and K. Terui, *Algebraic proof theory for substructural logics: cut-elimination and completions*, Annals of Pure and Applied Logic **163** (2012), no. 3, 266–290.

[71] A. Ciabattoni, N. Galatos, and K. Terui, *Algebraic proof theory: hypersequents and hypercompletions*, Annals of Pure and Applied Logic **168** (2017), no. 3, 693–737.

[72] A. Ciabattoni and G. Metcalfe, *Density elimination*, Theoretical Computer Science **403** (2008), no. 1–2, 328–346.

[73] R. Cignoli, I.M.L. D'Ottaviano, and D. Mundici, *Algebraic foundations of many-valued reasoning*, Kluwer, 2000.

[74] R. Cignoli, F. Esteva, L. Godo, and A. Torrens, *Basic fuzzy logic is the logic of continuous t-norms and their residua*, Soft Computing **4** (2000), 106–112.

[75] P. Cintula and P. Hájek, *On theories and models in fuzzy predicate logics*, Journal of Symbolic Logic **71** (2006), no. 3, 863–880.

[76] P. Cintula, R. Horčík, and C. Noguera, *Non-associative substructural logics and their semilinear extensions: axiomatization and completeness property*, Review of Symbolic Logic **6** (2013), no. 3, 394–423.

[77] P. Cintula, G. Metcalfe, and N. Tokuda, *Algebraic semantics for one-variable lattice-valued logics*, Proceedings of AiML 2022 (D. Fernandez-Duque, A. Palmigiano, and S. Pinchinat, eds.), Advances in Modal Logic, vol. 14, College Publications, 2022, pp. 237–257.

[78] P. Cintula and C. Noguera, *A general framework for mathematical fuzzy logic*, Handbook of mathematical fuzzy logic (P. Cintula, P. Hájek, and C. Noguera, eds.), College Publications, 2011, pp. 103–207.

[79] P. Cintula and C. Noguera, *Logic and implication*, Springer, 2021.

[80] A. Citkin and A. Muravitsky, *Consequence relations: an introduction to the Lindenbaum-Tarski method*, Oxford Science Publications, 2022.

[81] A.H. Clifford and G.H. Preston, *The algebraic theory of semigroups*, American Mathematical Society, 1961.

[82] A. Colacito, N. Galatos, G. Metcalfe, and S. Santschi, *From distributive ℓ-monoids to ℓ-groups, and back again*, Journal of Algebra **601** (2022), 129–148.

[83] A. Colacito and C. Tsinakis, *Nilpotency and the Hamiltonian property for cancellative residuated lattices*, International Journal of Algebra and Computation **32** (2022), no. 4, 629–652.

[84] P. Conrad, *The structure of a lattice-ordered group with a finite number of disjoint elements*, Michigan Mathematical Journal **7** (1960), 171–180.

[85] P. Conrad, *Some structure theorems for lattice-ordered groups*, Transactions of the American Mathematical Society **99** (1961), 212–240.

[86] P. Conrad, *The lattice of all convex ℓ-subgroups of a lattice-ordered group*, Czechoslovak Mathematical Journal **15** (1965), 101–123.

[87] P. Conrad, *Lex-subgroups of lattice-ordered groups*, Czechoslovak Mathematical Journal **18** (1968), 86–103.

[88] P. Conrad, *Free lattice-ordered groups*, Journal of Algebra **16** (1970), 191–203.

[89] P. Conrad, *Lattice ordered groups, an introduction*, Tulane University Lecture Notes, 1970.
[90] C. Constantinescu, *Some properties of spaces of measures*, Atti del Seminario Matematico e Fisico dell'Università di Modena **35** (1989), 1–286.
[91] T. Cortonesi, E. Marchioni, and F. Montagna, *Quantifier elimination and other model theoretic properties of BL-algebras*, Notre Dame Journal of Formal Logic **52** (2011), no. 4, 339–380.
[92] H.B. Curry, *The inconsistency of certain formal logics*, Journal of Symbolic Logic **7** (1942), 115–117.
[93] H.B. Curry, *The inferential approach to logical calculus*, Logique et Analyse **3** (1960), 119–136.
[94] J. Czelakowski, *Algebraic aspects of deduction theorems*, Studia Logica **44** (1985), 369–387.
[95] J. Czelakowski, *Protoalgebraic logics*, Kluwer, 2001.
[96] J. Czelakowski and W. Dziobiak, *Congruence distributive quasivarieties whose finitely subdirectly irreducible members form a universal class*, Algebra Universalis **27** (1990), 128–149.
[97] J. Czelakowski and J. Malinowski, *Key notions of Tarski's methodology of deductive systems*, Studia Logica **44** (1985), 321–351.
[98] J. Czelakowski and D. Pigozzi, *Amalgamation and interpolation in abstract algebraic logic*, Models, algebras, and proofs (Bogota 1995) (X. Caicedo and C.H. Montenegro, ed.), Marcel Dekker, 1999, pp. 187–265.
[99] M.R. Darnel, *Theory of lattice-ordered groups*, Marcel Dekker, 1995.
[100] D. Diaconescu, G. Metcalfe, and L. Schnüriger, *A real-valued modal logic*, Logical Methods in Computer Science **14** (2018), no. 1.
[101] R.P. Dilworth, *Abstract residuation over lattices*, Bulletin of the American Mathematical Society **44** (1937), 262–268.
[102] R.P. Dilworth, *The structure and arithmetical theory of non-commutative residuated lattices*, Ph.D. thesis, California Institute of Technology, 1939.
[103] R.P. Dilworth, *Non-commutative residuated lattices*, Transactions of the American Mathematical Society **46** (1939), 426–444.
[104] A. Di Nola, G. Georgescu, and A. Iorgulescu, *Pseudo-BL algebras I and II*, Journal of Multiple-Valued Logic and Soft Computing **8** (2002), no. 5–6, 673–714, 715–750.
[105] K. Došen and P. Schroeder-Heister, *Substructural logics*, Oxford University Press, 1993.
[106] P. Dubreil, *Sur les problèmes d'immersion et la théorie des modules*, Comptes Rendus des Séances de l'Académie des Sciences **216** (1943), 625–627.
[107] M. Dummett, *A propositional calculus with denumerable matrix*, Journal of Symbolic Logic **24** (1959), 97–106.
[108] M. Dummett, *Elements of intuitionism*, Clarendon Press, 1977.
[109] J.M. Dunn, *The algebra of intensional logics*, College Publications, 2019 (appeared as a Ph.D. thesis in 1966, University of Pittsburgh).
[110] J.M. Dunn, *A 'Gentzen' system for positive relevant implication*, Journal of Symbolic Logic **38** (1974), 356–357. (Abstract).
[111] J.M. Dunn, *Gaggle theory: an abstraction of Galois connections and residuation, with applications to negation, implication, and various logical operators*, Logics in AI: European workshop JELIA '90, (J. van Eijck, ed.), Springer, 1991, pp. 31–51.
[112] J.M. Dunn, *Partial gaggles applied to logics with restricted structural rules*, Substructural logics (P. Schroeder-Heister and K. Došen, ed.), Oxford University Press, 1993, pp. 63–108.
[113] J.M. Dunn, *Gaggle theory applied to intuitionistic, modal and relevance logics*, Logik und mathematik. Frege-Kolloquium Jena 1993, (I. Max and W. Stelzner, ed.), W. De Gruyter, 1995, pp. 335–368.
[114] A. Dvurečenskij, *States on pseudo MV-algebras*, Studia Logica **68** (2001), 301–327.
[115] A. Dvurečenskij, *Pseudo MV-algebras are intervals in ℓ-groups*, Journal of the Australian Mathematical Society **72** (2002), no. 3, 427–445.
[116] A. Dvurečenskij, *Aglianò-Montagna type decomposition of pseudo hoops and its applications*, Journal of the Australian Mathematical Society **211** (2007), 851–861.
[117] A. Dvurečenskij, *Every linear pseudo BL-algebra admits a state*, Soft Computing **11** (2007), 495–501.
[118] A. Dvurečenskij, R. Giuntini, and T. Kowalski, *On the structure of pseudo BL-algebras and pseudo hoops in quantum logics*, Foundations of Physics **40** (2010), 1519–1542.

[119] H.M. Edwards, *The genesis of ideal theory*, Archive for History of Exact Sciences **23** (1980), 321–378.

[120] M. Erné, *Adjunctions and Galois connections: origins, history and development*, Galois connections and applications (K. Denecke, M. Erné, and S.L. Wismath, eds.), Kluwer, 2004, pp. 1–138.

[121] F. Esteva and L. Godo, *Monoidal t-norm based logic: towards a logic for left-continuous t-norms*, Fuzzy Sets and Systems **124** (2001), no. 3, 271–288.

[122] F. Esteva, J. Gispert, L. Godo, and F. Montagna, *On the standard and rational completeness of some axiomatic extensions of the monoidal t-norm logic*, Studia Logica **71** (2002), no. 2, 199–226.

[123] T. Evans, *The word problem for abstract algebras*, Journal of the London Mathematical Society **26** (1951), 64–71.

[124] T. Evans, *Embeddability and the word problem*, Journal of the London Mathematical Society **28** (1953), 76–80.

[125] T. Evans, *Some connections between residual finiteness, finite embeddability and the word problem*, Journal of the London Mathematical Society **2** (1969), no. 1, 399–403.

[126] M. Farulewski, *Finite embeddability property for residuated groupoids*, Reports on Mathematical Logic **43** (2008), 25–42.

[127] K. Fine, *Models for entailment*, Journal of Philosophical Logic **3** (1974), 347–372.

[128] A. Fjellstad and J.F. Olsen, *IKT^{ω} and Łukasiewicz models*, Notre Dame Journal of Formal Logic **62** (2021), no. 2, 247–256.

[129] J.M. Font, *On the contributions of Helena Rasiowa to mathematical logic*, Multiple-Valued Logic **4** (1999), 159–179.

[130] J.M. Font, *Beyond Rasiowa's algebraic approach to non-classical logics*, Studia Logica **82** (2006), 1–31.

[131] J.M. Font, *Abstract algebraic logic: an introductory textbook*, College Publications, 2016.

[132] J.M. Font and R. Jansana, *A general algebraic semantics for sentential logics*, Springer, 1996.

[133] J.M. Font and T. Moraschini, *M-sets and the representation problem*, Studia Logica **103** (2015), no. 1, 21–51.

[134] R. Fraïssé, *Sur l'extension aux relations de quelques propriétés des ordres*, Annales Scientifiques de l'École Normale Supérieure **71** (1954), 363-388.

[135] O. Frink, *Pseudo-complements in semi-lattices*, Duke Mathematical Journal **29** (1962), 505–514.

[136] L. Fuchs, *Partially ordered algebraic systems*, Pergamon Press, 1963.

[137] W. Fussner and G. Metcalfe, *Transfer theorems for finitely subdirectly irreducible algebras*, 2022. Submitted.

[138] N. Galatos, *The undecidability of the word problem for distributive residuated lattices*, Ordered algebraic structures (J. Martinez, ed.), Kluwer, 2002, pp. 231–243.

[139] N. Galatos, *Minimal varieties of residuated lattices*, Algebra Universalis **52** (2005), no. 2, 215–239.

[140] N. Galatos and J. Gil-Férez, *Modules over quantaloids: applications to the isomorphism problems in algebraic logic and π-institutions*, Journal of Pure and Applied Algebra **221** (2017), no. 1, 1–24.

[141] N. Galatos and R. Horčík, *Cayley's and Holland's theorems for idempotent semirings and their applications to residuated lattices*, Semigroup Forum **87** (2013), 569–589.

[142] N. Galatos and R. Horčík, *Densification via polynomials, languages, and frames*, Journal of Pure and Applied Algebra **226** (2022), no. 3, 106852.

[143] N. Galatos and P. Jipsen, *Residuated frames with applications to decidability*, Transactions of the American Mathematical Society **364** (2013), 1219–1249.

[144] N. Galatos and P. Jipsen, *Distributive residuated frames and generalized bunched implication algebras*, Algebra Universalis **78** (2017), no. 3, 303–336.

[145] N. Galatos, P. Jipsen, T. Kowalski, and H. Ono, *Residuated lattices: an algebraic glimpse at substructural logics*, Elsevier, 2007.

[146] N. Galatos and G. Metcalfe, *Proof theory for lattice-ordered groups*, Annals of Pure and Applied Logic **167** (2016), no. 8, 707–724.

[147] N. Galatos and H. Ono, *Cut elimination and strong separation for substructural logics: an algebraic approach*, Annals of Pure and Applied Logic **161** (2010), no. 9, 1097–1133.

[148] N. Galatos and H. Ono, *Algebraization, parametrized local deduction theorem and interpolation for substructural logics over* **FL**, Studia Logica **83** (2006), 279–308.

[149] N. Galatos and J.G. Raftery, *Adding involution to residuated structures*, Studia Logica **77** (2004), no. 2, 181–207.

[150] N. Galatos and J.G. Raftery, *Idempotent residuated structures: some category equivalences and their applications*, Transactions of the American Mathematical Society **367** (2014), 3189–3223.

[151] N. Galatos and G. St. John, *Most simple extensions of* FL_e *are undecidable*, Journal of Symbolic Logic **87** (2022), no. 3, 1156–1200.

[152] N. Galatos and C. Tsinakis, *Generalized MV-algebras*, Journal of Algebra **283** (2005), no. 1, 254–291.

[153] N. Galatos and C. Tsinakis, *Equivalence of closure operators: an order-theoretic and categorical perspective*, Journal of Symbolic Logic **74** (2009), no. 3, 780–810.

[154] A. Galli, R.A. Lewin, and M. Sagastume, *The logic of equilibrium and Abelian lattice ordered groups*, Archive for Mathematical Logic **43** (2004), 141–158.

[155] G. Gentzen, *Untersuchungen über das logische schließen I, II*, Mathematische Zeitschrift **39** (1935), no. 1, 176–210, 405–431.

[156] G. Georgescu and A. Iorgulescu, *Pseudo-MV algebras*, Multiple-Valued Logic **6** (2001), no. 1–2, 95–135.

[157] G. Gierz, K.H. Hofmann, K. Keimel, J. Lawson, M. Mislove, and D. Scott, *Continuous lattices and domains*, Cambridge University Press, 2003.

[158] J. Gil-Férez, *Representation of structural closure operators*, Archive for Mathematical Logic **50** (2011), no. 1, 45–73.

[159] J. Gil-Férez, A. Ledda, and C. Tsinakis, *The failure of the amalgamation property for semilinear varieties of residuated lattices*, Mathematica Slovaca **65** (2015), no. 4, 817–828.

[160] J. Gil-Férez, F.M. Lauridsen, and G. Metcalfe, *Integrally closed residuated lattices*, Studia Logica **108** (2020), no. 5, 1063–1086.

[161] J. Gil-Férez, P. Jipsen, and G. Metcalfe, *Structure theorems for idempotent residuated lattices*, Algebra Universalis **81** (2020), 1–25.

[162] J. Gil-Férez, L. Spada, C. Tsinakis, and H. Zhou, *Join-completions of partially ordered algebras*, Annals of Pure and Applied Logic **171** (2020), 102842.

[163] J-Y. Girard, *Linear logic*, Theoretical Computer Science **50** (1987), 1–102.

[164] J-Y. Girard, *Proof theory and logical complexity*, Bibliopolis, 1987.

[165] A.M.W. Glass and Y. Gurevich, *The word problem for lattice-ordered groups*, Transactions of the American Mathematical Society **280** (1983), no. 1, 127–138.

[166] V. Glivenko, *Sur quelques points de la logique de M. Brouwer*, Bulletin Academie des Sciences de Belgique **15** (1929), 183–188.

[167] K. Gödel, *Zum intuitionistischen aussagenkalkül*, Anzeiger der Akademie der Wissenschaften in Wien **32** (1932), 65–66.

[168] G. Grätzer, *A note on the amalgamation property. Abstract.*, Notices of the American Mathematical Society **22** (1975), 453.

[169] G. Grätzer, *General lattice theory*, 2nd ed., Birkhäuser, 1998. New appendices by the author with B.A. Davey, R. Freese, B. Ganter, M. Greferath, P. Jipsen, H.A. Priestley, H. Rose, E.T. Schmidt, S.E. Schmidt, F. Wehrung, and R. Wille.

[170] G. Grätzer, H. Lakser, and J. Płonka, *Joins and direct products of equational classes*, Canadian Mathematical Bulletin **12** (1969), 741–744.

[171] G. Grätzer and H. Lakser, *The structure of pseudocomplemented distributive lattices. II: congruence extension and amalgamation*, Transactions of the American Mathematical Society **156** (1971), 343–358.

[172] V.N. Grišin, *Predicate and set-theoretical calculi based on logic without the contraction rule*, Mathematical USSR Izvestiya **18** (1982), 41–59; English transl., Izvéstia Akademii Nauk SSSR **45** (1981), 47–68.

[173] A. Grothendieck, *Sur quelques points d'algèbre homologique*, Tôhoku Mathematical Journal **9** (1957), 119–221.

[174] P. Hájek, *Metamathematics of fuzzy logic*, Trends in Logic, vol. 4, Kluwer, 1998.

[175] P. Hájek, *Basic fuzzy logic and BL-algebras*, Soft Computing **2** (1998), 124–128.

[176] P. Hájek, L. Godo, and F. Esteva, *A complete many-valued logic with product-conjunction*, Archive for Mathematical Logic **35** (1996), 191–208.

[177] T. Halperin, *Boole's algebra isn't Boolean algebra*, A Boole anthology (J. Gasser, ed.), Kluwer, 2000.

[178] Z. Hanikova and R. Horčík, *The finite embeddability property for residuated groupoids*, Algebra Universalis **72** (2014), 1–13.

[179] G. Hansoul and B. Teheux, *Extending Łukasiewicz logics with a modality: algebraic approach to relational semantics*, Studia Logica **101** (2013), no. 3, 505–545.

[180] J.B. Hart, L. Rafter, and C. Tsinakis, *The structure of commutative residuated lattices*, International Journal of Algebra and Computation **12** (2002), no. 4, 509–524.

[181] J.B. Hart and C. Tsinakis, *Decompositions of relatively normal lattices*, Transactions of the American Mathematical Society **341** (1994), 519–548.

[182] L. Henkin, J.D. Monk, and A. Tarski, *Cylindric algebras. Part II*, North-Holland, 1985.

[183] A. Heyting, *Die formalen regeln der intuitionistischen logik*, Sitzungsberichte der Preussischen Akademie der Wissenschaften. Physikalisch-mathematische Klasse (1930), 42–56.

[184] R. Hinnion and T. Libert, *Positive abstraction and extensionality*, Journal of Symbolic Logic **68** (2003), no. 3, 828–836.

[185] U. Höhle, *Commutative residuated ℓ-monoids*, Non-classical logics and their applications to fuzzy subsets (U. Höhle and P. Klement, eds.), Kluwer, 1995, pp. 53–106.

[186] O. Hölder, *Die axiome der quantität und die lehre vom mass*, Leipziger Berichte **53, 1** (1901), 1–64.

[187] W.C. Holland, *The lattice-ordered group of automorphisms of an ordered set*, Michigan Mathematical Journal **10** (1963), 399–408.

[188] W.C. Holland, *The largest proper variety of lattice-ordered groups*, Proceedings of the American Mathematical Society **57** (1976), 25–28.

[189] W.C. Holland and S.H. McCleary, *Solvability of the word problem in free lattice-ordered groups*, Houston Journal of Mathematics **5** (1979), no. 1, 99–105.

[190] C. Hollings, *Mathematics across the iron curtain: a history of the algebraic theory of semigroups*, AMS Publications, 2014.

[191] R. Horčík, *Algebraic semantics: semilinear FL-algebras*, Handbook of mathematical fuzzy logic (P. Cintula, P. Hájek, and C. Noguera, eds.), College Publications, 2011, pp. 283–353.

[192] R. Horčík, *Standard completeness theorem for ΠMTL*, Archive for Mathematical Logic **44** (2005), no. 4, 413–424.

[193] R. Horčík, *Decidability of cancellative extension of monoidal t-norm based logic*, Logic Journal of the IGPL **14** (2006), no. 6, 827–843.

[194] R. Horčík, *Structure of commutative cancellative integral residuated lattices on* $(0, 1]$, Algebra Universalis **57** (2007), 303–332.

[195] R. Horčík, *Word problem for knotted residuated lattices*, Journal of Pure and Applied Algebra **219** (2015), 1548–1563.

[196] W.A. Howard, *The formulae-as-types notion of construction*, To H.B. Curry: essays on combinatory logic, lambda calculus, and formalism (J.R. Hindley and J. Seldin, eds.), Academic Press, 1980, pp. 479–490.

[197] A. Hsieh and J.G. Raftery, *Semiconic idempotent residuated structures*, Algebra Universalis **61** (2009), 413–430.

[198] E.V. Huntington, *Sets of independent postulates for the algebra of logic*, Transactions of the American Mathematical Society **5** (1904), 288–309.

[199] A. Indrzejczak, *Sequents and trees: an introduction to the theory and applications of propositional sequent calculi*, Barnes and Noble, 2021.

[200] A. Indrzejczak, *Stanisław Jaśkowski and natural deduction systems*, The Lvov-Warsaw school. past and present (A. Garrido and U. Wybraniec-Skardowska, eds.), Birkhäuser, 2018, pp. 465–483.

[201] S. Jenei, *Group representation for even and odd involutive commutative residuated chains*, Studia Logica **110** (2022), 881–922.

[202] S. Jenei and F. Montagna, *A proof of standard completeness for Esteva and Godo's logic MTL*, Studia Logica **70** (2002), no. 2, 183–192.

[203] P. Jipsen and F. Montagna, *On the structure of generalized BL-algebras*, Algebra Universalis **55** (2006), 226–237.

[204] P. Jipsen and F. Montagna, *The Blok-Ferreirim theorem for normal GBL-algebras and its application*, Algebra Universalis **60** (2009), 381–404.

[205] P. Jipsen and F. Montagna, *Embedding theorems for normal GBL-algebras*, Journal of Pure and Applied Algebra **214** (2010), 1559–1575.

[206] P. Jipsen and C. Tsinakis, *A survey of residuated lattices*, Ordered algebraic structures (J. Martinez, ed.), Kluwer, 2002, pp. 19–56.

[207] B. Jónsson, *Universal relational structures*, Mathematica Scandinavica **4** (1956), 193–208.

[208] B. Jónsson, *Homogeneous universal relational structures*, Mathematica Scandinavica **8** (1960), 137–142.

[209] B. Jónsson, *Sublattices of a free lattice*, Canadian Journal of Mathematics **13** (1961), 146–157.

[210] B. Jónsson, *Algebraic extensions of relational systems*, Mathematica Scandinavica **11** (1962), 179–205.

[211] B. Jónsson, *Extensions of relational strucures*, The theory of models: Proceedings of the 1963 symposium at Berkley, North-Holland, 1965, pp. 146–157.

[212] B. Jónsson and C. Tsinakis, *Products of classes of residuated structures*, Studia Logica **77** (2004), 267–292.

[213] D.M. Kan, *Adjoint functors*, Transactions of the American Mathematical Society **87** (1958), 294–329.

[214] I. Kaplanski, *Fields and rings*, Chicago Lectures in Mathematics, Chicago University Press, 1965.

[215] K. Keimel, *Some trends in lattice-ordered groups and rings*, Lattice theory and its applications (K.A. Baker and R. Wille, eds.), Heldermann Verlag, 1995, pp. 131–161.

[216] O. Ketonen, *Untersuchungen zum prädikatenkalkül*, Annales Academiae Scientiarum Fennicae **I, 23** (1944).

[217] H. Kihara and H. Ono, *Interpolation properties, Beth definability properties and amalgamation properties for substructural logics*, Journal of Logic and Computation **20** (2010), no. 4, 823–875.

[218] Y. Komori, *Predicate logics without the structural rules*, Studia Logica **45** (1986), 393–404.

[219] T. Kovář, *A general theory of dually residuated lattice-ordered monoids*, Ph.D. Thesis, Palacký University Olomouc, 1996.

[220] T. Kowalski, A. Ledda, and F. Paoli, *On independent varieties and some related notions*, Algebra Universalis **70** (2013), 107–136.

[221] T. Kowalski and G. Metcalfe, *Uniform interpolation and coherence*, Annals of Pure and Applied Logic **170** (2019), no. 7, 825–841.

[222] R. Krömer, *Tool and object: a history and philosophy of category theory*, Birkhäuser, 2007.

[223] J. Kühr, *Ideals of noncommutative DRL monoids*, Czechoslovak Mathematical Journal **55** (2005), 97–111.

[224] J. Kühr, *Prime ideals and polars in DRℓ-monoids and pseudo BL-algebras*, Mathematica Slovaca **53** (2003), 233–246.

[225] J. Kühr, *Representable dually residuated lattice-ordered monoids*, Discussiones Mathematicae, General Algebra and Applications **23** (2003), 115–123.

[226] J. Kühr, *Pseudo BL-algebras and DRl-monoids*, Mathematica Bohemica **128** (2003), 199–208.

[227] J. Kühr, *Ideals of noncommutative DRℓ-monoids*, Czechoslovak Mathematical Journal **55** (2005), 97–111.

[228] J. Kühr, *Dually residuated ℓ-monoids having no non-trivial convex subalgebras*, Acta Universitatis Palackianae Olomucensis, Facultas Rerum Naturalium, Mathematica **45** (2006), 103–108.

[229] J. Kühr, *Representable pseudo-BCK-algebras and integral residuated lattices*, Journal of Algebra **317** (2007), 354–364.

[230] Y. Lafont, *The finite model property for various fragments of linear logic*, Journal of Symbolic Logic **62** (1997), 1202–1208.

[231] J. Lambek, *The mathematics of sentence structure*, American Mathematical Monthly **65** (1958), 154–170.

[232] A. Ledda, F. Paoli, and C. Tsinakis, *Semilinear varieties of lattice-ordered algebras*, P. Hájek on mathematical fuzzy logic (F. Montagna, ed.), Springer, 2014, pp. 207–221.

[233] A. Ledda, F. Paoli, and C. Tsinakis, *The Archimedean property: new horizons and perspectives*, Algebra Universalis **79** (2018), no. 4.

[234] J. Legris, *Paul Hertz's systems of propositions as a proof-theoretical conception of logic*, Advances in natural deduction (L. Pereira, E. Haeusler, and V. de Paiva, eds.), Springer, 2014, pp. 93–101.

[235] C.I. Lewis and C.H. Langford, *Symbolic logic*, The Century Company, 1932.

[236] P. Lincoln, J. Mitchell, A. Scedrov, and N. Shankar, *Decision problems for propositional linear logic*, Annals of Pure and Applied Logic **56** (1992), no. 1–3, 239–311.

[237] E. Linés Escardó and R. Mallol Balmaña, *On ℓ-groups (Spanish)*, Revista Matemática Hispanoamericana **12** (1952), no. 4, 129–136.

[238] J. Łos and R. Suszko, *Remarks on sentential logic*, Indagationes Mathematicae **20** (1958), 177–183.

[239] J. Łukasiewicz, *O logice trójwartościowej*, Polish logic 1920–1939 (S. McCall, ed.), Clarendon Press, 1967, 1920.

[240] J. Łukasiewicz and A. Tarski, *Untersuchungen über den aussagenkalkül*, Comptes Rendus des Séances de la Societé des Sciences et des Lettres de Varsovie, Classe III **23** (1930).

[241] R.C. Lyndon, *The representation of relational algebras*, Annals of Mathematics **51** (1950), no. 3, 707–729.

[242] S. MacLane, *Categories for the working mathematician*, Springer, 1971.

[243] R.D. Maddux, *Relation algebras*, North Holland, 2006.

[244] S. Maehara, *On the interpolation theorem of Craig (Japanese)*, Sugaku **12** (1960), 235–237.

[245] L.L. Maksimova, *On models of calculus E*, Algebra i Logika **6** (1967), 5–20.

[246] A.I. Mal'cev, *On the embedding of associative systems in groups* (Russian), Matematicheskii Sbornik **6** (1939), 331–336.

[247] A.I. Mal'cev, *The metamathematics of algebraic systems. Collected papers: 1936–1967*, North-Holland, 1971. Translated, edited, and provided with supplementary notes by B.F. Wells, III; Studies in Logic and the Foundations of Mathematics, Vol. 66.

[248] J.C.C. McKinsey, *Solution of the decision problem for Lewis S2 and S4 systems with an application to topology*, Journal of Symbolic Logic **6** (1941), 117–134.

[249] J.C.C. McKinsey and A. Tarski, *On closed elements in closure algebras*, Annals of Mathematics **47** (1946), no. 1, 122–162.

[250] R. McNaughton, *A theorem about infinite-valued sentential logic*, Journal of Symbolic Logic **16** (1951), 1–13.

[251] E. Marchioni, *Amalgamation through quantifier elimination for varieties of commutative residuated lattices*, Archive for Mathematical Logic **51** (2012), no. 1–2, 15–34.

[252] E. Marchioni and G. Metcalfe, *Craig interpolation for semilinear substructural logics*, Mathematical Logic Quarterly **58** (2012), no. 6, 468–481.

[253] J. Martinez, *Torsion theory for lattice-ordered groups*, Czechoslovak Mathematical Journal **25** (1975), no. 2, 284–299.

[254] T. Merlier, *Sur les demi-groupes réticulée et les o-demi-groupes*, Semigroup Forum **2** (1971), 64–70.

[255] G. Metcalfe and F. Montagna, *Substructural fuzzy logics*, Journal of Symbolic Logic **72** (2007), no. 3, 834–864.

[256] G. Metcalfe, F. Montagna, and C. Tsinakis, *Amalgamation and interpolation in ordered algebras*, Journal of Algebra **402** (2014), 21–82.

[257] G. Metcalfe, N. Olivetti, and D. Gabbay, *Analytic proof calculi for product logics*, Archive for Mathematical Logic **43** (2004), no. 7, 859–889.

[258] G. Metcalfe, N. Olivetti, and D. Gabbay, *Sequent and hypersequent calculi for Abelian and Łukasiewicz logics*, ACM Transactions of Computational Logic **6** (2005), no. 3, 578–613.

[259] G. Metcalfe, N. Olivetti, and D. Gabbay, *Proof theory for fuzzy logics*, Springer, 2008.

[260] G. Metcalfe, F. Paoli, and C. Tsinakis, *Ordered algebras and logic*, Uncertainty and rationality (H. Hosni and F. Montagna, eds.), Publications of the Scuola Normale Superiore di Pisa, Vol. 10, 2010, pp. 1–85.

[261] G. Metcalfe and L. Reggio, *Model completions for universal classes of algebras: necessary and sufficient conditions*, Journal of Symbolic Logic **88** (2023), no. 1, 381–417.

[262] G. Metcalfe and C. Tsinakis, *Density revisited*, Soft Computing **21** (2017), no. 1, 175–189.

[263] G. Metcalfe and O. Tuyt, *A monadic logic of ordered abelian groups*, Proceedings of AiML 2020 (S. Negri, N. Olivetti, G. Sandu, and R. Verbrugge, eds.), Advances in Modal Logic, vol. 13, College Publications, 2020, pp. 441–457.

[264] R.K. Meyer, *Topics in modal and many-valued logic*, Ph.D. thesis, University of Pittsburgh, 1966.

[265] R.K. Meyer and H. Ono, *The finite model property for BCK and BCIW*, Studia Logica **53** (1994), 107–118.

[266] R.K. Meyer and J.K. Slaney, *Abelian logic (from A to Z)*, Paraconsistent logic: essays on the inconsistent (G. Priest, R. Routley, and J. Norman, eds.), Philosophia, 1989, pp. 245–288.

[267] G. Mints, *Some calculi of modal logic*, Logical and logical-mathematical calculus, part I, Trudy Mathematical Institute Steklov, vol. 98, 1968, pp. 88–111.

[268] G. Mints, *Cut-elimination theorem for relevant logics*, Journal of Soviet Mathematics **6** (1976), 422–428.

[269] J.D. Monk, *The contributions of A. Tarski to algebraic logic*, Journal of Symbolic Logic **51** (1986), 899–906.

[270] F. Montagna, *Storage operators and multiplicative quantifiers in many-valued logics*, Journal of Logic and Computation **14** (2004), no. 2, 299–322.

[271] F. Montagna, *Interpolation and Beth's property in propositional many-valued logics: a semantic investigation*, Annals of Pure and Applied Logic **141** (2006), 148–179.

[272] F. Montagna and C. Tsinakis, *Ordered groups with a conucleus*, Journal of Pure and Applied Algebra **214** (2010), no. 1, 71–88.

[273] F. Montagna and H. Ono, *Kripke semantics, undecidability and standard completeness for Esteva and Godo's logic MTL∀*, Studia Logica **71** (2002), no. 2, 227–245.

[274] A. Monteiro, *L'arithmetique des filtres et les espaces topologiques*, Segundo Symposium de Matematics-Villavicencio (Mendoza, Buenos Aires) (1954), 129–162.

[275] A. Monteiro, *L'arithmetique des filtres et les espaces topologiques. I, II*, Notas Logica Mat. (1974), 29–30.

[276] E.H. Moore, *Introduction to a form of general analysis*, New Haven, 1910.

[277] T. Moraschini, *A study of truth predicates in matrix semantics*, Review of Symbolic Logic **11** (2018), no. 4, 780–804.

[278] P.S. Mostert and A.L. Shields, *On the structure of semigroups on a compact manifold with boundary*, Annals of Mathematics **65** (1957), 117–143.

[279] D. Mundici, *Interpretation of AF C*-algebras in Łukasiewicz sentential calculus*, Journal of Functional Analysis **65** (1986), no. 1, 15–63.

[280] D. Mundici, *Free products in the category of Abelian ℓ-groups with strong unit*, Journal of Algebra **113** (1988), 89–109.

[281] D. Mundici, *Advanced Łukasiewicz calculus and MV-algebras*, Springer, 2011.

[282] Y. Maruyama, *Fibred algebraic semantics for a variety of non-classical first-order logics and topological logical translation*, Journal of Symbolic Logic **86** (2021), no. 3, 1189–1213.

[283] S. McCleary, *Free lattice-ordered groups*, Ordered algebraic structures (W. Powell and C. Tsinakis, eds.), Lecture Notes in Pure and Applied Mathematics, Marcel Dekker, 1985, pp. 139–154.

[284] T. Nakano, *Rings and partly ordered systems*, Mathematische Zeitschrift **99** (1967), 355–376.

[285] S. Negri and J. von Plato, *Structural proof theory*, Cambridge University Press, 2001.

[286] B.H. Neumann and T. Taylor, *Subsemigroups of nilpotent groups*, Proceedings of the Royal Society of London Series A **274** (1963), no. 1356, 1–4.

[287] C. Nicolai, M. Piazza, and M. Tesi, *Non-contractive logics, paradoxes, and multiplicative quantifiers* (2023). Manuscript.

[288] M. Okada, *Phase semantics for higher order completeness, cut-elimination and normalization proofs (extended abstract)*, Electronic Notes in Theoretical Computer Science **3** (1996), 22 pp. (electronic).

[289] M. Okada, *An introduction to linear logic: expressiveness and phase semantics*, Theories of types and proofs (M. Takahashi, M. Okada, and M. Dezani, eds.), MSJ-Memoir 2, Mathematical Society of Japan, 1998.

[290] M. Okada, *Phase semantic cut-elimination and normalization proofs of first- and higher-order linear logic*, Theoretical Computer Science **227** (1999), 333–396.

[291] M. Okada and K. Terui, *The finite model property for various fragments of intuitionistic linear logic*, Journal of Symbolic Logic **64** (1999), 790–802.

[292] H. Ono, *Interpolation and the Robinson property for logics not closed under the Boolean operations*, Algebra Universalis **23** (1986), 111–122.

[293] H. Ono, *Structural rules and a logical hierarchy*, Mathematical logic (P.P. Petkov, ed.), Plenum, 1990, pp. 95–104.

[294] H. Ono, *Semantics for substructural logics*, Substructural logics (K. Došen and P. Schröder-Heister, eds.), Oxford University Press, 1993, pp. 259–291.

[295] H. Ono, *Decidability and the finite model property of substructural logics*, Tbilisi symposium on logic, language and computation: selected papers (J. Ginzburg, Z. Khasidashvili, C. Vogel, J.-J. Lévy, and E. Vallduví, eds.), CSLI, Stanford, 1998, pp. 263–274.

[296] H. Ono, *Proof-theoretic methods for nonclassical logic: an introduction*, Theories of types and proofs (M. Takahashi, M. Okada, and M. Dezani-Ciancaglini, eds.), Mathematical Society of Japan, 1998, pp. 207–254.

[297] H. Ono, *Substructural logics and residuated lattices: an introduction*, 50 years of Studia Logica (V.F. Hendricks and J. Malinowski, eds.), Kluwer, 2003, pp. 177–212.

[298] H. Ono, *Closure operators and complete embeddings of residuated lattices*, Studia Logica **74** (2003), no. 3, 427–440.

[299] H. Ono, *Crawley completions of residuated lattices and algebraic completeness of substructural predicate logics*, Studia Logica **100** (2012), no. 1–2, 339–359.

[300] H. Ono and Y. Komori, *Logics without the contraction rule*, Journal of Symbolic Logic **50** (1985), 169–201.

[301] Ø. Ore, *Linear equations in non-commutative fields*, Annals of Mathematics **32, 2** (1931), 463–477.

[302] Ø. Ore, *Galois connexions*, Transactions of the American Mathematical Society **55** (1944), 493–513.

[303] F. Paoli, *Substructural logics: a primer*, Kluwer, 2002.

[304] F. Paoli, **-autonomous lattices and fuzzy sets*, Soft Computing **10** (2006), 607–617.

[305] F. Paoli, M. Spinks, and B. Veroff, *Abelian logic and the logic of pointed lattice-ordered varieties*, Logica Universalis **2** (2008), 209–233.

[306] F. Paoli and C. Tsinakis, *On Birkhoff's common abstraction problem*, Studia Logica **100** (2012), 1079–1105.

[307] K.R. Pierce, *Amalgamations of lattice ordered groups*, Transactions of the American Mathematical Society **172** (1972), 249–260.

[308] D. Pigozzi, *Amalgamations, congruence-extension, and interpolation properties in algebras*, Algebra Universalis **1** (1972), 269–349.

[309] G. Pottinger, *Uniform, cut-free formulations of T, S4 and S5 (abstract)*, Journal of Symbolic Logic **48** (1983), no. 3, 900.

[310] A.M. Pitts, *On an interpretation of second order quantification in first order intuitionistic propositional logic*, Journal of Symbolic Logic **57** (1992), 33–52.

[311] E.L. Post, *Recursive unsolvability of a problem of Thue*, Journal of Symbolic Logic **12** (1947), no. 1, 1–11.

[312] A. Přenosil, *Cut elimination, identity elimination, and interpolation in super-Belnap logics*, Studia Logica **105** (2017), 1255–1289.

[313] A. Přenosil, *From partially ordered monoids to partially ordered groups via free nuclear preimages*, Journal of Algebra **610** (2022), 119–166.

[314] G. Priest, *The logic of paradox*, Journal of Philosophical Logic **8** (1979), 219–241.

[315] G. Priest, *Many-valued modal logics: a simple approach*, Review of Symbolic Logic **1** (2008), 190–203.

[316] A.P. Pynko, *Definitional equivalence and algebraizability of generalized logical systems*, Annals of Pure and Applied Logic **98** (1999), 1–68.

[317] A.P. Pynko, *Gentzen's cut-free calculus versus the logic of paradox*, Bulletin of the Section of Logic **39** (2010), 35–42.

[318] J. Rachůnek, *A non-commutative generalization of MV-algebras*, Czechoslovak Mathematical Journal **52** (2002), 255–273.

[319] J.G. Raftery, *Correspondences between Gentzen and Hilbert systems*, Journal of Symbolic Logic **71** (2006), no. 3, 903–957.

[320] J.G. Raftery, *The equational definability of truth predicates*, Reports on Mathematical Logic **41** (2006), 95–149.

[321] J.G. Raftery, *Representable idempotent commutative residuated lattices*, Transactions of the American Mathematical Society **359** (2007), no. 9, 4405–4427.

[322] J.G. Raftery, *Order algebraizable logics*, Annals of Pure and Applied Logic **164** (2013), no. 3, 251–283.

[323] H. Rasiowa, *An algebraic approach to non-classical logics*, North-Holland, 1974.

[324] B. Da R'e and L. Rosenblatt, *Contraction, infinitary quantifiers, and omega paradoxes*, Journal of Philosophical Logic **47** (2018), no. 4, 611–629.

[325] J. Rebagliato and V. Verdú, *On the algebraization of some Gentzen systems*, Fundamenta Informaticae **18** (1993), 319–338.

[326] V.B. Repnitskiĭ, *Bases of identities of varieties of lattice-ordered semigroups*, Algebra i Logika **22** (1983), no. 6, 649–665.

[327] G. Restall, *An introduction to substructural logics*, Routledge, 2000.

[328] A. Rose and J.B. Rosser, *Fragments of many-valued statement calculi*, Transactions of the American Mathematical Society **87** (1958), 1–53.

[329] K.I. Rosenthal, *Quantales and their applications*, Pitman Research Notes in Mathematics, vol. 234, Longman, 1990.

[330] R. Routley, V. Plumwood, R.K. Meyer, and R.T. Brady, *Relevant logics and their rivals*, Ridgeview, 1982.

[331] W. Rump and Y. Yang, *On Bosbach's cone algebras*, Studia Logica **98** (2011), 375–386.

[332] W. Rump, *The structure group of a generalized orthomodular lattice*, Studia Logica **106** (2018), no. 1, 85–100.

[333] J.D. Rutledge, *A preliminary investigation of the infinitely many-valued predicate calculus*, Ph.D. thesis, Cornell University, 1959.

[334] W. Scharlau, *Richard Dedekind 1831-1981, festschrift 1981*, Vieweg, 1982.

[335] J. Schmidt, *Beiträge zur filtertheorie, II*, Mathematische Nachrichten **10** (1953), 197–232.

[336] J. Schmidt, *Universal and internal properties of some extensions of partially ordered sets*, Journal für die reine und angewandte Mathematik **253** (1972), 28–42.

[337] J. Schmidt, *Universal and internal properties of some completions of k-join-semilattices and k-join-distributive partially ordered sets*, Journal für die reine und angewandte Mathematik **255** (1972), 8–22.

[338] J. Schmidt, *Each join-completion of a partially ordered set is the solution of a universal problem*, Journal of the Australian Mathematical Society **17** (1974), 406–419.

[339] J. Schmidt and C. Tsinakis, *Relative pseudo-complements, join-extensions and meet-retractions*, Mathematische Zeitschrift **157** (1977), 271–284.

[340] K.D. Schmidt, *A common abstraction of Boolean rings and lattice ordered groups*, Compositio Mathematica **54** (1985), 51–62.

[341] K.D. Schmidt, *Minimal clans: a class of ordered partial semigroups including Boolean rings and partially ordered groups*, Semigroups: theory and applications, Springer, 1988, pp. 300–341.

[342] O. Schreier, *Die untergruppen der freien gruppen*, Abhandlungen aus dem Mathematischen Seminar der Universität Hamburg **5** (1927), no. 1, 161–183.

[343] J. Schulte Mönting, *Cut elimination and word problems for varieties of lattices*, Algebra Universalis **12** (1981), 290–321.

[344] P. Simons, *Philosophy and logic in central Europe from Bolzano to Tarski*, Kluwer, 1992.

[345] T. Skolem, *Studies on the axiom of comprehension*, Notre Dame Journal of Formal Logic **4, 3** (1963), 162–179.

[346] T. Skolem, *Logisch-kombinatorische untersuchungen über die erfüllbarkeit und beweisbarkeit mathematischer Sätze, nebst einem theoreme über dichte mengen*, Videnskapsselskapets skrifter I. Matematisk-naturvidenskabelig klasse, 4 (1920).

[347] J. Snodgrass and C. Tsinakis, *The finite basis theorem for relatively normal lattices*, Algebra Universalis **33** (1995), 40–67.

[348] M.H. Stone, *The theory of representations for Boolean algebras*, Transactions of the American Mathematical Society **40** (1936), no. 1, 37–111.

[349] S. Surma, *On the origin and subsequent applications of the concept of the Lindenbaum algebra*, Logic, methodology and philosophy of science VI, North Holland, 1982, pp. 719–734.

[350] K.L.N. Swamy, *Dually residuated lattice-ordered semigroups*, Mathematische Annalen **159** (1965), 105–114.

[351] K.L.N. Swamy, *Dually residuated lattice-ordered semigroups II*, Mathematische Annalen **160** (1965), 64–71.

[352] K.L.N. Swamy, *Dually residuated lattice-ordered semigroups III*, Mathematische Annalen **167** (1966), 71–74.

[353] G. Takeuti, *Proof theory*, 2nd ed., North-Holland, 1987.

[354] A. Tarski, *Contributions to the theory of models*, Indagationes Mathematicae **17** (1955), 56–64.

[355] A. Tarski, *Logic, semantics, metamathematics*, Clarendon Press, 1956.

[356] K. Terui, *Which structural rules admit cut elimination? — an algebraic criterion*, Journal of Symbolic Logic **72** (2007), no. 3, 738–754.

[357] A.S. Troelstra and H. Schwichtenberg, *Basic proof theory*, 2nd ed., Cambridge University Press, 2000.

[358] A. Urquhart, *Semantics for relevant logic*, Journal of Symbolic Logic **37** (1972), 1059–1073.

[359] A. Urquhart, *Free Heyting algebras*, Algebra Universalis **3** (1973), 94–97.

[360] A. Urquhart, *The undecidability of entailment and relevant implication*, Journal of Symbolic Logic **49** (1984), no. 4, 1059–1073.

[361] C.J. van Alten, *Representable biresiduated lattices*, Journal of Algebra **247** (2002), 672–691.

[362] C.J. van Alten, *The finite model property for knotted extensions of propositional linear logic*, Journal of Symbolic Logic **70** (2005), no. 1, 84–98.

[363] R. Vaught, *Model theory before 1945*, Proceedings of the Tarski symposium, American Mathematical Society, 1974.

[364] S. van Gool, G. Metcalfe, and C. Tsinakis, *Uniform interpolation and compact congruences*, Annals of Pure and Applied Logic **168** (2017), 1827–1948.

[365] M. Ward, *Structure residuation*, Annals of Mathematics **39** (1938), 558–568.

[366] M. Ward, *The closure operations of a lattice*, Annals of Mathematics **43** (1942), 191–196.

[367] M. Ward and R.P. Dilworth, *Residuated lattices*, Proceedings of the National Academy of Sciences **24** (1938), 162–164.

[368] M. Ward and R.P. Dilworth, *Residuated lattices*, Transactions of the American Mathematical Society **45** (1939), 335–354.

[369] E.C. Weinberg, *Free lattice-ordered Abelian groups*, Mathematische Annalen **151** (1963), 187–199.

[370] P. Whitman, *Free lattices*, Annals of Mathematics **42** (1941), 325–329.

[371] A.M. Wille, *Residuated structures with involution*, Ph.D. thesis, Technische Universität Darmstadt, 2006.

[372] R. Wójcicki, *Logical matrices strongly adequate for structural sentential calculi*, Bulletin de l'Académie Polonaise des Sciences, Classe III **17** (1969), 333–335.

[373] A. Wroński, *Interpolation and amalgamation properties of BCK-algebras*, Mathematica Japonica **29** (1984), 115–121.

[374] O. Wyler, *Clans*, Compositio Mathematica **17** (1966), 172–189.

[375] S. Wang and B. Zhao, *HpsUL is not the logic of pseudo-uninorms and their residua*, Logic Journal of the IGPL **17** (2009), no. 4, 413–419.

[376] W. Young, *Heyting algebras as intervals of commutative, cancellative residuated lattices* (2014). Manuscript.

[377] W. Young, *From interior algebras to unital ℓ-groups: a unifying treatment of modal residuated lattices*, Studia Logica **103** (2015), no. 2, 265–286.

[378] J. Zygmunt and R. Purdy, *Adolf Lindenbaum: notes on his life, with bibliography and selected references*, Logica Universalis **8** (2014), no. 3–4, 285–320.

Selected Published Titles in This Series

277 **George Metcalfe, Francesco Paoli, and Constantine Tsinakis,** Residuated Structures in Algebra and Logic, 2023

276 **Balázs Bárány, Károly Simon, and Boris Solomyak,** Self-similar and Self-affine Sets and Measures, 2023

275 **Alekos Vidras and Alain Yger,** Multidimensional Residue Theory and Applications, 2023

274 **Willi Freeden and M. Zuhair Nashed,** Recovery Methodologies: Regularization and Sampling, 2023

273 **Gennadiy Feldman,** Characterization of Probability Distributions on Locally Compact Abelian Groups, 2023

272 **Tadashi Ochiai,** Iwasawa Theory and Its Perspective, Volume 1, 2023

271 **Hideto Asashiba,** Categories and Representation Theory, 2022

270 **Leslie Hogben, Jephian C.-H. Lin, and Bryan L. Shader,** Inverse Problems and Zero Forcing for Graphs, 2022

269 **Ralph S. Freese, Ralph N. McKenzie, George F. McNulty, and Walter F. Taylor,** Algebras, Lattices, Varieties, 2022

268 **Ralph S. Freese, Ralph N. McKenzie, George F. McNulty, and Walter F. Taylor,** Algebras, Lattices, Varieties, 2022

267 **Dragana S. Cvetković Ilić,** Completion Problems on Operator Matrices, 2022

266 **Kate Juschenko,** Amenability of Discrete Groups by Examples, 2022

265 **Nabil H. Mustafa,** Sampling in Combinatorial and Geometric Set Systems, 2022

264 **J. Scott Carter and Seiichi Kamada,** Diagrammatic Algebra, 2021

263 **Vugar E. Ismailov,** Ridge Functions and Applications in Neural Networks, 2021

262 **Ragnar-Olaf Buchweitz,** Maximal Cohen–Macaulay Modules and Tate Cohomology, 2021

261 **Shiri Artstein-Avidan, Apostolos Giannopoulos, and Vitali D. Milman,** Asymptotic Geometric Analysis, Part II, 2021

260 **Lindsay N. Childs, Cornelius Greither, Kevin P. Keating, Alan Koch, Timothy Kohl, Paul J. Truman, and Robert G. Underwood,** Hopf Algebras and Galois Module Theory, 2021

259 **William Heinzer, Christel Rotthaus, and Sylvia Wiegand,** Integral Domains Inside Noetherian Power Series Rings, 2021

258 **Pramod N. Achar,** Perverse Sheaves and Applications to Representation Theory, 2021

257 **Juha Kinnunen, Juha Lehrbäck, and Antti Vähäkangas,** Maximal Function Methods for Sobolev Spaces, 2021

256 **Michio Jimbo, Tetsuji Miwa, and Fedor Smirnov,** Local Operators in Integrable Models I, 2021

255 **Alexandre Boritchev and Sergei Kuksin,** One-Dimensional Turbulence and the Stochastic Burgers Equation, 2021

254 **Karim Belabas and Henri Cohen,** Numerical Algorithms for Number Theory, 2021

253 **Robert R. Bruner and John Rognes,** The Adams Spectral Sequence for Topological Modular Forms, 2021

252 **Julie Déserti,** The Cremona Group and Its Subgroups, 2021

251 **David Hoff,** Linear and Quasilinear Parabolic Systems, 2020

250 **Bachir Bekka and Pierre de la Harpe,** Unitary Representations of Groups, Duals, and Characters, 2020

249 **Nikolai M. Adrianov, Fedor Pakovich, and Alexander K. Zvonkin,** Davenport–Zannier Polynomials and Dessins d'Enfants, 2020

For a complete list of titles in this series, visit the
AMS Bookstore at **www.ams.org/bookstore/survseries/**.